III A	IV A	V A	VI A	VII A	VIII A
								2 **He** Helium 4.00260
			5 **B** Boron 10.81	6 **C** Carbon 12.011	7 **N** Nitrogen 14.0067	8 **O** Oxygen 15.9994	9 **F** Fluorine 18.998403	10 **Ne** Neon 20.179
	I B	II B	13 **Al** Aluminum 26.98154	14 **Si** Silicon 28.0855	15 **P** Phosphorus 30.97376	16 **S** Sulfur 32.06	17 **Cl** Chlorine 35.453	18 **Ar** Argon 39.948
28 **Ni** Nickel 58.70	29 **Cu** Copper 63.546	30 **Zn** Zinc 65.38	31 **Ga** Gallium 69.72	32 **Ge** Germanium 72.59	33 **As** Arsenic 74.9216	34 **Se** Selenium 78.96	35 **Br** Bromine 79.904	36 **Kr** Krypton 83.80
46 **Pd** Palladium 106.4	47 **Ag** Silver 107.868	48 **Cd** Cadmium 112.41	49 **In** Indium 114.82	50 **Sn** Tin 118.69	51 **Sb** Antimony 121.75	52 **Te** Tellurium 127.60	53 **I** Iodine 126.9045	54 **Xe** Xenon 131.30
78 **Pt** Platinum 195.09	79 **Au** Gold 196.9665	80 **Hg** Mercury 200.59	81 **Tl** Thallium 204.37	82 **Pb** Lead 207.2	83 **Bi** Bismuth 208.9804	84 **Po** Polonium (209)	85 **At** Astatine (210)	86 **Rn** Radon (222)

63 **Eu** Europium 151.96	64 **Gd** Gadolinium 157.25	65 **Tb** Terbium 158.9254	66 **Dy** Dysprosium 162.50	67 **Ho** Holmium 164.9304	68 **Er** Erbium 167.26	69 **Tm** Thulium 168.9342	70 **Yb** Ytterbium 173.04	71 **Lu** Lutetium 174.967
95 **Am** Americium (243)	96 **Cm** Curium (247)	97 **Bk** Berkelium (247)	98 **Cf** Californium (251)	99 **Es** Einsteinium (252)	100 **Fm** Fermium (257)	101 **Md** Mendelevium (258)	102 **No** Nobelium (259)	103 **Lr** Lawrencium (260)

QUANTITATIVE ANALYSIS

QUANTITATIVE ANALYSIS
Theory and Practice

Lawrence W. Potts
Gustavus Adolphus College

HARPER & ROW, PUBLISHERS, New York
Cambridge, Philadelphia, San Francisco, Washington,
London, Mexico City, São Paulo, Singapore, Sydney

To Elisabeth, Charlie, Annie, and Jill

Sponsoring Editor: Lisa S. Berger
Project Editor: Steven Pisano
Cover Design: Lucy Zakarian
Cover Photo: Dagmar Hailer-Hamann
Text Art: RDL Artset, Ltd.
Production: Willie Lane
Compositor: TAPSCO, Inc.
Printer and Binder: R. R. Donnelley & Sons Company

QUANTITATIVE ANALYSIS: Theory and Practice
Copyright © 1987 by Lawrence W. Potts

All rights reserved. Printed in the United States of America. No part of this book may be used or reproduced in any manner whatsoever without written permission, except in the case of brief quotations embodied in critical articles and reviews. For information address Harper & Row, Publishers, Inc., 10 East 53 Street, New York, NY 10022-5299.

Library of Congress Cataloging in Publication Data

Potts, Lawrence W.
   Quantitative analysis.

   Includes bibliographies and index.
   1. Chemistry, Analytic—Quantitative.   I. Title.
QD101.2.P68   1987   545   86-25713
ISBN 0-06-045271-4

87 88 89 90 9 8 7 6 5 4 3 2 1

BRIEF CONTENTS

CHAPTER 1 Chemical Analysis 1
CHAPTER 2 Errors in Chemical Analysis and the Statistical Treatment of Data 39
CHAPTER 3 The Equilibrium Condition 105
CHAPTER 4 Concepts of Acids and Bases 141
CHAPTER 5 Acid-Base Equilibria 157
CHAPTER 6 Acid-Base Titrations 211
CHAPTER 7 Titrations in Nonaqueous Solvents 261
CHAPTER 8 Solubility Equilibria 283
CHAPTER 9 Precipitation 315
CHAPTER 10 Gravimetric Methods 335
CHAPTER 11 Precipitation Titrations 359
CHAPTER 12 Complexes and Complexometric Methods of Analysis 383
CHAPTER 13 UV-Visible Absorption Spectrometry and Colorimetric Methods of Analysis 425
CHAPTER 14 Oxidation-Reduction Equilibria and Electrochemical Cells 485
CHAPTER 15 Potentiometry 517
CHAPTER 16 Redox Titrimetric Methods 547
CHAPTER 17 Analytical Separations and an Introduction to Chromatography 587
CHAPTER 18 Chromatographic Equipment 619
Appendixes 645
Answers to Selected Problems 691
Index 699

DETAILED CONTENTS

Preface xvii

CHAPTER 1
Chemical Analysis 1

    1.1   **Introduction** 1
          1.1.1   Qualitative and Quantitative Analysis   1
          1.1.2   Scope of Quantitative Analysis   2
    1.2   **The Process of Analysis** 3
          1.2.1   Obtaining the Sample   4
          1.2.2   Preparing the Sample   5
          1.2.3   The Determination   5
          1.2.4   Working with the Results   6
    1.3   **Basic Chemical Facts: Expressing Concentrations** 7
          1.3.1   Molarity   8
          1.3.2   Formality   8
          1.3.3   Normality   9
          1.3.4   Percentage Composition and Other Ratios   10
    1.4   **Scale of Analytical Operations**   11
    1.5   **Calculating the Results: Significant Digits**   13
    1.6   **Basic Laboratory Operations**   14
          1.6.1   Weighing   15
          1.6.2   Use and Calibration of Volumetric Glassware   19
          1.6.3   Filtering   24
    1.7   **Notes on Laboratory Safety**   28
    1.8   **Disposing of Chemical Wastes**   29
         Recommended Reading   31
         Problems   32
         Experiment: Calibration of Volumetric Glassware   35

CHAPTER 2
Errors in Chemical Analysis and the Statistical Treatment of Data 39

    2.1   **Introduction**   39
    2.2   **Error and the Concepts of Accuracy and Precision**   40
          2.2.1   Simple Descriptive Statistics   41

2.3 Errors in Chemical Analysis 43
    2.3.1 Systematic Errors 43
    2.3.2 Random Errors 49
    2.3.3 Quality Control 82
    2.3.4 Propagation of Error 84
    2.3.5 Least-Squares Line Fitting 89
    2.3.6 Sampling 94
References 98
Recommended Reading 99
Problems 100

CHAPTER 3
The Equilibrium Condition 105

3.1 Introduction 105
3.2 The Thermodynamic Approach 106
    3.2.1 The First Law of Thermodynamics 106
    3.2.2 Spontaneous Reactions 109
    3.2.3 Concentration Changes, Free Energy, and the Equilibrium Constant 113
    3.2.4 Characteristics of Chemical Equilibria 118
    3.2.5 Summary of the Thermodynamic Approach 122
3.3 The Kinetic Approach 123
    3.3.1 Catalysts and Reaction Rates 126
3.4 Solution Equilibria, Activity, and Concentration 127
    3.4.1 The Debye–Hückel Model 128
    3.4.2 Activity Coefficients at Very High Ionic Strengths 133
    3.4.3 Activity Coefficients of Uncharged Species 134
    3.4.4 Assumptions about Activity Coefficients 135
References 136
Study Question 136
Problems 137

CHAPTER 4
Concepts of Acids and Bases 141

4.1 Introduction 141
4.2 Early Views of Acids and Bases 141
4.3 The Arrhenius Definition 142
4.4 The Brønsted–Lowry Definition 144
4.5 The Solvent System Definition 149
4.6 The Lewis Definition 151
Recommended Reading 153
Problems 154

DETAILED CONTENTS

CHAPTER 5
Acid-Base Equilibria 157

- 5.1 **Introduction** 157
- 5.2 **Review of Basic Concepts** 157
    - 5.2.1 Definition of pH 157
    - 5.2.2 Reactions with Water as Solvent 159
    - 5.2.3 Acid-Base Equilibrium Constant Expressions 161
    - 5.2.4 Weak Base Reactions 162
- 5.3 **Basic Relationships** 163
    - 5.3.1 Material Balance 164
    - 5.3.2 Charge Balance (Electroneutrality) 165
    - 5.3.3 The Reference Level 166
- 5.4 **Monoprotic Acids and Monobasic Bases** 167
    - 5.4.1 Solutions of Strong Acids or Strong Bases and Their Salts 167
    - 5.4.2 Solution of a Weak Acid 169
    - 5.4.3 Solution of a Monobasic Weak Base 174
    - 5.4.4 Distribution of Species with Changing pH 175
    - 5.4.5 Drawing a Logarithmic Concentration Diagram 179
    - 5.4.6 Mixture of a Weak Acid and a Strong Acid 180
    - 5.4.7 Mixture of a Weak Acid and a Weak Base 182
- 5.5 **Polyprotic Acids** 185
- 5.6 **Buffers** 193
    - 5.6.1 Composition and Applications of Buffers 193
    - 5.6.2 Preparing Buffer Solutions 196
    - 5.6.3 Measures of the Effectiveness of a Buffer 198
- **References** 204
- **Recommended Reading** 204
- **Problems** 205

CHAPTER 6
Acid-Base Titrations 211

- 6.1 **Introduction** 211
- 6.2 **Standard Solutions** 212
    - 6.2.1 Standard Base Solutions 212
    - 6.2.2 Standard Acid Solutions 215
- 6.3 **Equivalent Weight and Normality** 218
- 6.4 **Acid-Base Indicators** 221
    - 6.4.1 Single Indicators 221
    - 6.4.2 Mixed and Screened Indicators 224
- 6.5 **Acid-Base Titration Curves** 225
    - 6.5.1 Experimental Considerations 225
    - 6.5.2 Titration of a Strong Acid with a Strong Base 227
    - 6.5.3 Titration of a Weak Acid with a Strong Base 231
    - 6.5.4 Titration of a Weak Base with a Strong Acid 236

    6.5.5 Titrations of Polyprotic Weak Acids 236
    6.5.6 Titration of a Polybasic Weak Base with a Strong Acid 241
 6.6 Applications 245
    6.6.1 Determination of Nitrogen by the Kjeldahl Method 246
 References 249
 Problems 249
 Experiment 6.1: Titration of a Pure Weak Acid 256
 Experiment 6.2: The Titrimetric Determination of Sodium Carbonate (Soda Ash) 258

CHAPTER 7
Titrations in Nonaqueous Solvents 261

 7.1 Introduction 261
 7.2 The Case of a Very Weak Acid 262
 7.3 Classification Scheme for Solvents 263
 7.4 A Closer Look at Four Critical Factors 264
    7.4.1 Inherent Acidity 265
    7.4.2 Autoprotolysis Constant 268
    7.4.3 Dielectric Constant 269
    7.4.4 Solvating Ability 272
 7.5 Applications 272
    7.5.1 Neutral Solvents: Alcohols 272
    7.5.2 Acidic Solvents: Acetic Acid 276
    7.5.3 Basic Solvents: Ammonia and Amines 277
    7.5.4 Aprotic Solvents 278
 References 279
 Recommended Reading 280
 Problems 280

CHAPTER 8
Solubility Equilibria 283

 8.1 Introduction 283
 8.2 Simple Solubility Relationships 283
 8.3 Concentrations: K_{sp}^0 vs. K_{sp} 287
 8.4 Solubility Calculations for More Complex Species 289
 8.5 Using K_{sp} to Predict Precipitation 290
 8.6 The Common Ion Effect 293
 8.7 Complexation 296
 8.8 Graphical Representations 300
 8.9 Foreign Ligand Competition 304
 8.10 Acid-Base Competition 307
 Problems 310

DETAILED CONTENTS

CHAPTER 9
Precipitation 315

- 9.1 Introduction 315
- 9.2 Nucleation and Growth of Precipitate Crystals 315
- 9.3 Colloids 318
- 9.4 Adsorption 319
- 9.5 Impurities in Precipitates 322
  - 9.5.1 Coprecipitation by Adsorption 322
  - 9.5.2 Coprecipitation by Occlusion 322
  - 9.5.3 Coprecipitation by Isomorphous Replacement 322
  - 9.5.4 Postprecipitation 323
  - 9.5.5 Methods Used to Reduce Contamination 324
- 9.6 Precipitation from Homogeneous Solution 327
  - 9.6.1 Precipitations of Hydroxides with Urea 327
  - 9.6.2 Precipitation with Sulfate Ion 328
  - 9.6.3 Precipitation with Sulfide Ion 328
  - 9.6.4 Precipitation of Ni(II) with Homogeneously Generated Dimethylglyoxime 329
  - 9.6.5 Limitations of Homogeneous Precipitation Methods 330
- 9.7 Precipitation Methods of Separation 330
- 9.8 References 332
- Problems 332

CHAPTER 10
Gravimetric Methods 335

- 10.1 Introduction 335
- 10.2 Properties of Samples and Precipitates 335
- 10.3 Gravimetry as a Process 336
- 10.4 Gravimetric Methods 338
- 10.5 Calculations for Gravimetric Methods 339
- 10.6 Advantages and Disadvantages of Gravimetric Methods 344
- References 344
- Recommended Reading 345
- Problems 345
- Experiment 10.1: Gravimetric Determination of Calcium as Calcium Oxalate Monohydrate 349
- Experiment 10.2: Gravimetric Determination of Chloride 351
- Experiment 10.3: Gravimetric Determination of Lead in Brass 354

CHAPTER 11
Precipitation Titrations 359

- 11.1 Introduction 359
- 11.2 The Model System: $Ag^+ + Cl^- = AgCl$ 359

11.3 Effects of Dilution 363
11.4 Real versus Ideal: Some Actual Data 363
11.5 Titration of a Mixture of Halides 365
11.6 Indicator Methods 367
    11.6.1 The Mohr Method 367
    11.6.2 The Fajans Method 369
    11.6.3 The Volhard Method 371
11.7 Applications 372
11.8 Endnote: Normality and the Use of Equivalent Weights 372
References 374
Recommended Reading 374
Problems 374
Experiment 11.1: Titrimetric Determination of Chloride by the Adsorption Indicator Method 378
Experiment 11.2: Titrimetric Determination of Chloride by the Volhard Method 379
Experiment 11.3: Titrimetric Determination of Chloride by the Mohr Method 381

CHAPTER 12
Complexes and Complexometric Methods of Analysis 383

12.1 Lewis Acids and Bases 383
12.2 Metal Ions in Solution 384
    12.2.1 Metal Ion Coordination Number and Geometry of Complexes 385
12.3 Ligands (Lewis Bases) 386
12.4 Stability of Complexes 390
    12.4.1 Size and Electronegativity 392
    12.4.2 Effects of Substituents 393
    12.4.3 Nature of the Bond 393
    12.4.4 Formation of Rings 395
    12.4.5 Effects of Solvent 396
12.5 Complexation Equilibria 396
    12.5.1 A Practical Question: Is The Reaction Complete? 399
12.6 Complexometric Titrations 402
12.7 Metallochromic Indicators 404
12.8 Special Techniques Using EDTA 411
    12.8.1 Auxiliary Complexing Agents 411
    12.8.2 Back Titrations 411
    12.8.3 Indirect Titrations 413
    12.8.4 Displacement Titrations 413
    12.8.5 Masking and Demasking 413
12.9 Endnote: Why EDTA Is So Popular 415

References   416
Recommended Reading   416
Problems   417
Experiment 12.1:   Titrimetric Determination of Calcium (Samples Containing No Magnesium)   421
Experiment 12.2:   Titrimetric Determination of Calcium (Samples Containing Magnesium)   422

CHAPTER   13
UV-Visible Absorption Spectrometry and Colorimetric Methods of Analysis   425

13.1   Introduction   425
13.2   Electromagnetic Radiation—Light   426
   13.2.1   Electromagnetic Radiation as Waves   426
   13.2.2   Electromagnetic Radiation as Particles   429
   13.2.3   The Electromagnetic Spectrum   430
   13.2.4   Emission and Absorption of Electromagnetic Radiation; Spectra   431
13.3   Absorption Spectroscopy   433
   13.3.1   Lambert's Law   435
   13.3.2   Beer's Law   435
13.4   Applications of Beer's Law   438
   13.4.1   The Analysis of Mixtures   438
   13.4.2   Photometric Titrations   441
   13.4.3   Determining Ligand:Metal Combining Ratios   442
13.5   Limitations to the Use of Beer's Law   445
   13.5.1   Physical Limitations   445
   13.5.2   Chemical Effects   446
   13.5.3   A Diversion: Taking Advantage of a Limitation   448
   13.5.4   Deviations Due to Instrumentation   449
13.6   Instrumentation   452
   13.6.1   Instrument Components   452
   13.6.2   Spectrophotometric Instruments   464
13.7   Summary   466
References   467
Recommended Reading   468
Problems   468
Experiment 13.1:   Colorimetric Determination of Manganese in Steel   473
Experiment 13.2:   Colorimetric Determination of Iron(II) (1,10-Phenanthroline Method)   475
Experiment 13.3:   Determination of the pK_a^0 of Bromcresol Green   478
Experiment 13.4:   Spectrophotometric Titration of Cu(II) with EDTA   481

CHAPTER 14
Oxidation-Reduction Equilibria and Electrochemical Cells 485

    14.1   Introduction 485
    14.2   Oxidation-Reduction Reactions 485
    14.3   Electrochemical Cells 486
            14.3.1   Cell Diagrams 487
            14.3.2   Conventional Wisdom 488
            14.3.3   Combining Half-Cell Reactions 489
            14.3.4   Cell Spontaneity 491
            14.3.5   Cell Reversibility 493
            14.3.6   Effects of Species Activities on Cell EMF: The Nernst Equation 496
            14.3.7   Effects of Complexation on Cell EMF 499
            14.3.8   An Additional Problem: Liquid Junction Potential 503
            14.3.9   Practical Cells 505
        References 511
        Problems 511

CHAPTER 15
Potentiometry 517

    15.1   Introduction 517
    15.2   Indicator Electrodes 517
            15.2.1   A Quick Review 517
            15.2.2   Ion Selective Electrodes 518
    15.3   Reference Electrodes 528
    15.4   Evaluating Activities and Concentrations by Potentiometry 532
            15.4.1   Analytical Methods 532
            15.4.2   Measuring Devices 533
    15.5   Potentiometric Titration Curves: Data Treatment 536
            15.5.1   If All Else Fails 541
            15.5.2   A Final Point about Titrimetric Errors 541
        References 542
        Recommended Reading 543
        Problems 544

CHAPTER 16
Redox Titrimetric Methods 547

    16.1   Introduction 547
    16.2   Fundamentals of Redox Titrations 547
            16.2.1   A Model System: Fe(II) + Ce(IV) = Fe(III) + Ce(III) 548
            16.2.2   Potentiometric Monitoring 551
            16.2.3   Redox Endpoint Indicators 554
    16.3   Analytical Methods and Reagents 556
            16.3.1   Analytical Case Studies: Iodometric Methods 556
            16.3.2   Redox Reagents 558
            16.3.3   Iodometric Methods—A Closer Look 567

References 574
Recommended Reading 574
Problems 574
Experiment 16.1: Redox Titrimetric Determination of Iron (Volumetric or Gravimetric Procedure) 581
Experiment 16.2: Iodometric Determination of Copper in Ore 583

CHAPTER 17
Analytical Separations and an Introduction to Chromatography 587

17.1 Introduction 587
17.2 Solvent Extraction 587
17.3 Separating Two Solutes 592
    17.3.1 Crosscurrent Separations 593
    17.3.2 Countercurrent Separations 596
    17.3.3 The Craig Countercurrent Apparatus 602
17.4 Simplified Model for Column Chromatography 603
17.5 Chromatographic Data 605
17.6 Efficiency of Column Separations 607
    17.6.1 Resolution 607
    17.6.2 Effects of Column Capacity 608
    17.6.3 Band Broadening in Real Chromatograms 609
17.7 Qualitative and Quantitative Analysis 611
References 615
Recommended Reading 615
Problems 616

CHAPTER 18
Chromatographic Equipment 619

18.1 Introduction 619
18.2 Gas Chromatography 620
    18.2.1 Equipment 620
    18.2.2 Applications 628
18.3 High-Performance Liquid Chromatography 629
    18.3.1 Equipment 629
    18.3.2 Separation Mechanisms and Applications 634
18.4 Comparison of the Chromatographic Techniques 640
Recommended Reading 641
Problems 642

Appendix I: Weak Acid Dissociation Constants 645
Appendix II: Solubility Product Constants Listed by Anions 655

Appendix III: Formation Constants for Metal Ion Complexes 657
Appendix IV: Standard and Formal Potentials (25°C) 665
Appendix V: Experimental Evaluation of Close pK_a Values for Diprotic Acids 675
Appendix VI: Solutions to Higher-Order Equations 681
Appendix VII: Some Compounds of Analytical Importance 687

Answers to Selected Problems 691

Index 699

PREFACE

This text is intended to serve the needs of second- and third-year undergraduate chemistry, biology, pre-medicine, and engineering majors. It is written with the assumption that students using it will have taken a full year of general chemistry aimed at science students and chemistry majors. While some knowledge of organic chemistry would be helpful, it is not essential to an understanding of the concepts in this text. The laboratory experience and problem-solving skills learned in a year-long course in organic chemistry will, of course, be valuable to anyone studying quantitative analysis. The treatment assumes that students are proficient in high-school-level algebra and have studied enough calculus to take derivatives of simple functions. The ability to use a computer and even to program in BASIC or Pascal is becoming an academic survival skill, and will be an advantage in solving numerical problems. Some of the problems at the ends of chapters involve writing short programs.

The information in this text will provide a foundation for advanced-level courses in chemical instrumentation and in solution equilibria. The orientation of the text and the laboratory exercises is toward classical "wet" chemistry, rather than instrumentation. UV-visible spectrometers, glass electrodes, pH meters, and gas chromatographs are all discussed, but only in the second half of the text and in terms of applications, after the fundamentals of wet analysis and chemical equilibria are well established. The experiments are a basic set of "tried and true" exercises, which work well with commercially prepared unknowns. I will share quality assurance information about the performance of my own students on these experiments with any instructor who is interested.

The unifying theme of this text is chemical equilibria. Chapters 3, 5, 8, 12, 14, and 17 approach the theory of quantitative analysis from the point of view of the equilibrium condition. These chapters are interspersed with chapters that focus on applications. The result is a weaving of theory and practice which should develop a solid basic understanding of chemical analysis. The material spans the contents of most existing courses in quantitative analysis: statistical analysis of experimental data, Brønsted acid-base equilibria, solubility equilibria, gravimetry, complexation equilibria, spectrophotometry, redox equilibria and electrochemistry, and separations and chromatography. In addition, the traditional titrimetric methods involving acid-base, precipitation, complexation, and redox reactions are covered in considerable detail. Following the trend many see in the teaching of analytical chemistry, I give less emphasis to gravimetry and to precipitation titrations, and put more emphasis on the theory of separations and chromatography.

I have found the sequence of topics as presented in this text to be very useful and effective. For example, teaching Brønsted acid-base equilibria before solubility equilibria allows me to teach competing equilibria in a rigorous way through the conditional solubility product constant. The concept of pH-dependent fractions of acid-base species established in Chapter 5 is reinforced in the treatment of titration curves (Chapter 6). When it is used again in Chapters 8 (solubility) and 12 (complexation), students find it quite easy to accept and apply.

When presented in detail, the material in this text is more than can fit into a regular 14-week semester course. Certain chapters have been written to require little lecture support. For example, Chapter 3 (The Equilibrium Condition) should serve as a review of thermodynamic concepts presented (but perhaps not absorbed by many students) in general chemistry. Chapter 4 (Concepts of Acids and Bases) traces the historical development of acid-base theories, and can also be read by students without much lecture support. At some schools the material in Chapters 9, 10, and 11 (relating to precipitation phenomena and gravimetric methods) has been absorbed by general chemistry, and might be replaced by greater emphasis on separations and chromatography.

It is an even greater challenge to fit the material into a 10-week course. This might be accomplished by limiting the detail in the treatment of statistics (limit depth in hypothesis testing and quality control), and omitting nonaqueous solvent equilibria, solubility titrations, and perhaps saving potentiometry, redox titrimetry, and separations for a later course.

In this text I have approached certain topics in somewhat different ways than have other authors. For example, I have taken a systematic, graphical approach in the analysis of solution equilibria. In Brønsted-Lowry equilibria, complicated algebraic expressions are developed after students are shown how to use graphical methods to visualize the composition of a solution. The use of graphical devices makes it possible to see immediately which species are important and which can be ignored, and emphasizes chemical interactions. I have found the graphical approach very helpful in avoiding the tendency to reduce acid-base equilibria to lectures on algebra.

Basic thermodynamics and kinetics are used to develop an understanding of chemical equilibria very early in the book. I have found few students who come from a course in general chemistry with a firm grasp of equilibrium and its relationship to free energy changes. Without apologies, I develop the relationship in detail in Chapter 3. The concepts are reinforced and developed further in each chapter in which theory is discussed. I have tried to use free energy changes and chemical activities in an honest, meaningful, and consistent way throughout the book, without discouraging students with added complexity.

The treatment of statistical analysis of data in this text is both broad and rather detailed, and includes discussions of the meaning of uses of inferential statistics. Statistics is a difficult and subtle topic, and analysts are expected to have a good conceptual grasp of many statistical methods. I hope this introductory treatment shows how useful statistical analysis can be, while conveying some of the subtlety of the discipline.

I have used real systems whenever possible to illustrate concepts, avoiding the use of hypothetical systems except when generalizations are required. Titration curves developed from equations are compared side-by-side with actual experimental results obtained by students. The complications which make analytical chemistry challenging and interesting are discussed, rather than glossed over. I hope students will look for complications and contradictions, even at the level of changing variables in examples to see what happens to results. Where appropriate I have given computer programming hints to encourage students to try calculations on computers.

I am indebted to many people for their help in the creation and production of this text. Professors Donald Poe (University of Minnesota, Duluth), Ned Daugherty (Colorado State University, Fort Collins), Warren Reynolds (University of Minnesota, Minneapolis), and Kelsey Cook (University of Tennessee, Knoxville) provided careful, critical, and scholarly reviews of the work. The staff at Harper & Row has been extremely cooperative and helpful.

I am indebted to the many fine students who wrestled with this book in class-testing form as it grew over the past five years. Fifty students at Oberlin College suffered through an early, much abbreviated version of the text during the spring of 1982. Four classes of Gustavus Adolphus students have helped me work out bugs in content and pedagogy. Many have contributed experimental data, and are cited in figure captions in various chapters. My thanks go to all of them for having taught me so much.

Finally, I am indebted to my own teachers, who sparked and focused my interest in chemistry: Arthur Root of Clarence (NY) Central High School; Martin Ackermann, Terry Carlton, Norman Craig, and Richard Schoonmaker of Oberlin College; and Stanley Bruckenstein (who first showed me the systematic approach to equilibrium calculations), Harold S. Swofford, Peter James Lingane, and the late Ernest B. Sandell, of the University of Minnesota.

L.W.P.

Chapter 1

Chemical Analysis

1.1 INTRODUCTION

1.1.1 Qualitative and Quantitative Analysis

Chemical analysis involves finding answers to the basic questions, "What is in this sample?" and "How much of it is there?" The first question is *qualitative* and is related to the detection and identification of the components of a sample. The second question is *quantitative* and is the subject of this text. Once a component is identified, the analytical chemist can find how much of it is actually present in a sample.

A variety of tools are available for qualitative analysis. You are probably familiar with some of the simplest: spot tests, flame tests, and the formation and color of precipitates are all used in inorganic qualitative analysis schemes taught to most general chemistry students. Studying colligative properties such as melting point, boiling point, and vapor pressure may be useful in identification. Spectroscopic methods are quick and rather simple ways to detect metals and organic compounds. Microscopic examination, which is generally ignored in undergraduate instruction, is also a powerful qualitative tool. It may be necessary to do a complete qualitative analysis on a sample before discovering how to determine just one or two components quantitatively. This is because many elements are sufficiently similar in their behavior that they interfere in a quantitative determination. Removing or masking interfering elements can be a very difficult task.

Quantitative analysis requires not only that a sample component be detected, but also that it be determined accurately and precisely. The practice of quantitative analysis and the development of quantitative methods are therefore very challenging undertakings, requiring both considerable manual skill (the "art" of analysis) and considerable education in chemistry. Although analytical chemistry is one of the most practical areas of chemistry ("What is in this sample?" is a very practical question), a successful analytical chemist must have a firm knowledge of theory to design experiments. I. M. Kolthoff, one of the greatest analytical chemists, put this quite succinctly: "Theory guides, experiment decides."

1.1.2 Scope of Quantitative Analysis

Chemistry is an experimental, quantitative science. Careful quantitative work has always been the basis for uncovering the fundamental laws of chemistry. In fact, almost everything we know about chemical stoichiometry and the chemistry of the elements is based on careful quantitative analysis. For example, the French chemist J.-S. Stas (1813–1891) synthesized and analyzed silver iodide in an experiment to test the law of conservation of matter. Stas's data for the precipitation and recovery of his samples are shown in Table 1.1. The quality of the quantitative analytical work is even more impressive when it is realized that this work was done in 1860.

Analytical chemistry has become vitally important in commerce and health care as well as many other aspects of modern life. The principal role of analytical chemistry in industry is in quality control and quality assurance. As new products are created

TABLE 1.1 TEST OF THE LAW OF CONSERVATION OF MATTER[a]
(Data of J.-S. Stas, 1860)

Weight of iodine taken (g)	Weight of silver taken (g)	Weight of silver iodide (g)	Difference (g)
32.4665	27.6223	60.0860	−0.0028
46.8282	39.8405	86.6653	−0.0034
44.7599	38.0795	82.8375	−0.0019
160.2752	136.3547	296.6240	−0.0059

[a] In the chemical reaction iodine was reduced with ammonium sulfite to form iodide. The iodide was precipitated with silver sulfate, which had been formed by dissolving the accurately weighed silver metal in nitric acid, heating the silver nitrate until it was dry, then redissolving and heating in ammonium bisulfate. All weights were corrected for buoyancy. Stas said that for samples weighing more than 100 g the analysis time was more than 46 hr, without interruptions!
Source: The work was published as "Nouvelles Recherches sur les Lois de Proportions Chimiques" in 1865. An account of the work can be found in I. Freund. *The Study of Chemical Composition. An Account of Its Method and Historical Development,* Dover, New York, 1968; originally published in 1904 by Cambridge University Press.

new methods of analysis must be devised. As new production lines are created, analytical chemists help in monitoring and optimizing them.

Clinical chemists and analytical chemists have developed highly sensitive and reliable methods for assaying components of blood without which the diagnosis and treatment of disease probably would not have progressed beyond the levels of a half-century ago. For less than $150 a patient can now have an analysis of 30 blood components and a profile of blood lipids (fats) within a few hours. These are, of course, fully automated determinations, the result of efforts of analytical chemists using instrumental (physicochemical) methods of analysis. A modern hospital laboratory will run thousands of analyses a day for hundreds of patients, a production rate that Stas could not have imagined more than 100 years ago.

Analytical chemistry is becoming more and more important in legal and environmental matters. Chemical and physical analyses related to the investigation of crime (*forensic chemistry*) are in the news almost daily. Sensitive analytical methods such as gas chromatography and mass spectrometry are used in investigations of arson and homicide and often provide decisive evidence. The same instrumental methods are used to identify threats to the environment such as concealed hazardous waste sites. The research involved in assessing and solving problems such as "acid rain" is basically analytical chemistry research.

Table 1.2 is a list of some of the applications discussed in special reports in the journal *Analytical Chemistry* between 1982 and 1985 and indicates the kinds of problems of interest to analytical chemists.

Strong preparation in analytical chemistry is also important for students of chemistry, regardless of the area they believe is their favorite. The typical course in analysis offers the first experiences in extremely careful experimental work: the proper use of sensitive balances, pipets and burets, and volumetric flasks. It is also a course in which chemical equilibria can be studied in a systematic way. The study of chemical equilibria provides a very convenient framework for quantitative analysis. Analytical chemists are always pushing around equilibria: we precipitate an ion quantitatively by suppressing its solubility with an excess of precipitating reagent; we determine a metal ion by complexing it with a species that gives it color and drive the reaction to completion by adding an excess of reagent. This connection between chemical analysis and what many chemists regard as a traditional part of physical chemistry has been an extremely fruitful area for learning about the behavior of chemical systems.

1.2 THE PROCESS OF ANALYSIS

A chemical analysis should be viewed as a process involving several steps: obtaining a sample, preparing the sample, planning and performing the analysis in the laboratory, and finally interpreting the results. We will consider these steps individually.

TABLE 1.2 CURRENT TOPICS IN ANALYTICAL CHEMISTRY (1982–1985)[a]

Topic	Citation
Art and History	
"The Analysis of Napoleonic Arsenic"	**54**, 1477A
"The Bust of Nefertiti"	**54**, 619A
"Papyrus"	**55**, 1221A
Environmental	
"Trace Analysis of the Dioxins"	**54**, 309A
"Risk Assessment and Analytical Chemistry"	**55**, 1438A
"Environmental Monitoring at Love Canal"	**55**, 943A
"Ethylene Dibromide Monitoring"	**56**, 573A
"Snow Peas and Acephate (a pesticide)"	**57**, 572A
Forensic	
"Tylenol Analysis"	**54**, 1474A
"Arson Analysis by Mass Spectrometry"	**54**, 1399A
"Forensic Toxicology in the 1980s"	**54**, 433A
"Linking Criminals to the Scene of the Crime with Glass Analysis"	**56**, 844A
"Solving Crimes with 3-D Fluorescence Spectroscopy"	**57**, 934A
"Chemical Tests for Intoxication"	**57**, 876A
Clinical	
"Analytical Chemistry in the Conquest of Diabetes"	**56**, 664A
"Enzyme Immunoassay"	**56**, 920A
"Electrochemical Sensors in Clinical Chemistry"	**57**, 345A

[a] Citations are volume and page from *Analytical Chemistry*.

1.2.1 Obtaining the Sample

This first step is never discussed in detail in introductory analytical texts, at least in part because so much of the knowledge needed to sample correctly is gained in on-the-job training. At first glance, sampling might appear to be trivial. Actually, many analytical chemists are never asked to go out in the field or the production facility to collect samples, but rather rely on long-standing sampling procedures. The problems in sampling are far from trivial, however. Imagine designing a procedure for the random and representative sampling of uranium in a carload of uranium ore. Ore samples are by nature quite inhomogeneous; that is, their composition is nonuniform. As an analyst, you would be expected to draw a few kilograms of material from a carload of several tons of ore consisting of everything from 10-kg boulders to dust. The portion of the sample you actually analyze may weigh only 1 or 2 g, and yet it must accurately reflect the composition of millions of grams of material. The price your firm pays or receives for the ore depends on how much uranium is present in the ore, and uranium is extremely expensive. How do you establish a sampling routine for this kind of sample? Some ideas are presented at the end of Chapter 2.

The amount of sample and its physical state will strongly influence your choice of an analytical method. Some methods require very little sample, while others require a great deal. The importance of the scale of analytical operations is taken up in Section 1.4.

1.2.2 Preparing the Sample

Samples must be treated before they are analyzed for components. Solid samples are ordinarily ground to powders to make them easier to dry before weighing, to mix them thoroughly (make them homogeneous), and to make them easier to dissolve. The grinding process must not, of course, contaminate the sample.

Dissolving a sample is often a challenging task that requires some research. Many inorganic samples can be dissolved in mineral acids such as hydrochloric acid and nitric acid. More resistant samples may have to be heated in mixed mineral acids in sealed high-pressure chambers. Others may be dissolved in melted potassium hydroxide with the help of strong oxidizers such as sodium peroxide. Samples containing a polymer matrix may be so difficult to dissolve that alternative analytical methods (not requiring dissolution) must be found.

If a sample can be dissolved, it must then be purified before analysis is begun. Purification is necessary because very few methods of analysis are specific for a particular element or species; all methods suffer from interferences, at least to some extent. Many methods of purification are available and are discussed in this text. Precipitation of either the *analyte* (the material analyzed) or the interference is an easy way to purify a sample. You probably studied qualitative inorganic analysis in an introductory chemistry course; precipitation is the principal way to purify and identify analytes in inorganic qualitative analysis. Masking is another method: the sample is treated with a reagent which will tie up interfering species in complexes and thus free the analyte for accurate determination.

If a sample cannot be purified by precipitation or masking, it may be necessary to perform a physical separation using either solvent extraction or chromatography. These techniques are discussed in Chapter 17. These methods must, of course, be tailored to the particular analytical problem, and this requires research.

1.2.3 The Determination

The focus of the analytical process is the actual determination, the chemical reaction or physical process by which the amount of analyte in the sample is found. In selecting a suitable method for analysis the analytical chemist must consider at least five criteria:

accuracy, reproducibility, selectivity, speed, and convenience. As nature would have it, some of these criteria are mutually exclusive. For example, methods which rely on the use of instruments tend to be quick and convenient; dozens of samples might be processed in an hour of laboratory time. They also save expensive analyst time. Unfortunately, these methods are often not very reproducible. Some of the highly accurate and reproducible classical methods are quite inconvenient and very labor-intensive and costly. The requirements of the analysis must be weighed against the costs.

Analytical methods are generally divided into two major categories, the *classical methods* and the *physicochemical methods*. The classical methods are also called wet methods or chemical methods and include gravimetric, volumetric, and gasometric methods. *Gravimetric* methods (Chapter 10) rely on measuring weight; the amount of water in a sample can be determined by how much weight the sample loses when it is heated. The amount of sulfate in a dissolved sample can be determined from the amount of barium sulfate precipitated from the solution. *Volumetric* methods rely on the accurate measurement of volume. A reactant is added to a sample in measurable volumes. When the reaction is judged complete, the volume and concentration of the reactant are used to calculate the composition of the sample. You probably performed at least one acid-base titration in introductory chemistry and know how to measure volumes with a buret. Titrations are a special class of volumetric method called *titrimetric* methods. *Gasometric* methods are among the oldest chemical methods, dating back centuries. For example, in 1775 Lavoisier measured the volume of "respirable air" (oxygen) given off by mercuric oxide when it was heated in a closed system.

Physicochemical methods rely on the measurement of some physical property to indicate the amount of material present. Examples of physical properties are melting and boiling points, color (ability to absorb or emit light), refractive index, and electrical conductivity. The more unique and discernible a particular physical property of the analyte, the simpler will be its analysis. Physicochemical methods are often quick, easy to adapt to chemical instrumentation, and therefore convenient. Instrumentation is more sensitive than human senses, and thus physicochemical methods can be used with lower concentrations and smaller samples than classical methods. Chapter 13 is devoted to *colorimetric* methods, which rely on the ability of an analyte to absorb light. *Chromatography,* the subject of Chapter 17, can be used to analyze mixtures of compounds according to their ability to adsorb on active columns. In Chapter 16 we will see how the electrical potential of an oxidation-reduction (*redox*) system can be used in analysis. Physicochemical methods are discussed in much greater detail in advanced-level courses in chemical analysis.

1.2.4 Working with the Results

The final step of the analytical process is the calculation and interpretation of results. At the introductory level calculations of results of classical methods are simple exercises in stoichiometry, much like those you performed in general chemistry. You must also learn to apply the basic principles of the statistical analyis of data to your results to assess the precision of repeated determinations (*replicates*) and learn what you can infer about the accuracy of your work. Statistical analysis is treated in Chapter 2. When working with chemical instrumentation you must use somewhat more sophisticated statistical methods. In colorimetry (Chapter 13) it is often necessary to analyze *working curves* (plots of instrument signal intensity vs. concentration) by using linear regression methods. When taking data with instruments at low signal levels it is often necessary to filter out instrument noise (electrical interference rather than chemical), either with electronic circuits or with computer signal processing. The rapidly growing discipline called *chemometrics* focuses on the acquisition, treatment, and understanding of chemical data.

One last aspect of the analytical process involves judging the results to see if they make sense in light of what the analyst knows about a sample. This can often be quite simple. For example, if an analyst calculates that a sample of brass is 60% lead, it is more likely that a decimal place was accidentally shifted in calculating the result than that such an alloy was actually prepared. Judging the results is often much more challenging, requiring that the analyst know a great deal about inorganic, organic, and physical chemistry. In any event, it is the responsibility of the analyst to learn about the possible reasonable outcomes of the analysis before starting work.

1.3 BASIC CHEMICAL FACTS: EXPRESSING CONCENTRATIONS

This text assumes that students using it understand the principles of chemical theory. We will not review such basics as the methods used to balance chemical equations or the laws of constant composition or multiple proportions. If you do not recognize those principles or do not feel comfortable with them, you should review them with the help of a general chemistry text. What is needed at this point is a quick recapitulation of some important units of concentration. We will take up the relationship between concentration and chemical activity in Chapter 3.

Chemists express concentrations in basically four ways: molarity, formality, normality, and percentage. Remember in the following descriptions that a *solute* is the material being dissolved in a *solvent* to prepare a *solution*. The *mole* is the basic unit for the amount of a chemical species: it is defined as the amount of material that contains as many elementary entities (atoms, molecules, ions, electrons, etc.) as there are atoms in exactly 0.012 kg (12 g) of ^{12}C. This number is called the Avogadro number and is 6.022045×10^{23}.

The *molecular weight* of a compound is the number of grams of the compound that contain an Avogadro number of molecules. It is easily calculated from the sum of the atomic weights of the atoms in the molecular formula. An example calculation is given below.

1.3.1 Molarity

The *molarity* of a solution, M, is defined as the number of moles of solute contained in 1 liter (L) of solution. It is the same as the number of millimoles (mmole) in 1 milliliter (mL) of solution,

$$\text{moles/liter} = \text{mmole/mL}$$

Since you will be working with milliliter quantities of solutions in the laboratory, you will probably find it easier to express molarity as millimoles per milliliter than as moles per liter.

The term "mole" is often used rather loosely. Strictly speaking, we should not use molecular weight when referring to compounds which are not molecules. To be more general we should speak of the *formula weight* of a salt rather than the molecular weight. The formula weight is simply the weight of one formula unit of a compound. The *gram formula weight (GFW)* is the formula weight taken in grams.

EXAMPLE 1.1 Preparing a 0.1M Solution

How would you prepare a liter of 0.1M sucrose ($C_{12}H_{22}O_{11}$) in water?

Solution

The molecular formula of sucrose is $C_{12}H_{22}O_{11}$ and its molecular weight is found from the sum of the atomic weights: $12(12.011) + 22(1.0079) + 11(15.9994) = 342.299$. To prepare a 0.1$M$ solution of sucrose, we weigh out 0.1 mole, or 34.23 g, and add enough water to make 1 liter of solution. Notice that we do not add 1 liter of water; we add enough water to make 1 liter of solution.

1.3.2 Formality

It is important to distinguish between the *formal concentration, F,* and the molar concentration, M. The formal concentration refers to the amount of a solute dissolved in solution, regardless of the chemical form it might take once in solution. It is also called the *molar analytical concentration*. In this text we will use molarity to refer to the actual concentration of a species in its particular form in solution, that is, its *molar equilibrium concentration*. The distinction is important in studies of chemical equilibria. We can prepare a solution by weighing out pure solute (prepare it "formally" to be a particular concentration), but we cannot know what its actual concentration will be once in solution, where it might dissociate or combine with some ion in solution. For

1.3 BASIC CHEMICAL FACTS: EXPRESSING CONCENTRATIONS

example, we can weigh out 0.10 formula weight ("mole") of sodium acetate trihydrate and dissolve it in enough water to make a liter of solution. As you will see later, acetate ion is a weak base and will react with water to form a weak acid, acetic acid. Although we added 0.10 formula weight of acetate to begin with, the reaction with water leaves less than 0.10 mole of acetate in the solution at equilibrium. The most accurate statement we can make at this point is that the sum of the concentrations of acetate and acetic acid is $0.1F$,

$$[CH_3COO^-] + [CH_3COOH] = 0.1F$$
$$\text{acetate} \qquad\qquad \text{acetic acid}$$

Later on you will learn how to calculate the molar concentration of each species in the solution.

1.3.3 Normality

Normality is a highly specialized way of referring to concentration that takes into account chemical reactivity. For example, we know that the following equation describes the oxidation of iron(II) by dichromate ion, Cr(VI):

$$Cr_2O_7^{2-} + 6Fe^{2+} + 14H^+ = 2Cr^{3+} + 6Fe^{3+} + 7H_2O$$

The balanced equation tells us that 6 moles of iron will react with 1 mole of dichromate. Another way to look at it might be that 1 mole of dichromate is six times more reactive than 1 mole of Fe^{2+}. If we have 1 mole of Fe^{2+} we can react it completely with $\frac{1}{6}$ mole of dichromate. The weight of dichromate corresponding to $\frac{1}{6}$ mole is called its *equivalent weight*. In oxidation-reduction reactions we calculate the equivalent weight by dividing the gram formula weight by the number of electrons in what is called the half-reaction. In this case the half-reactions are:

$$Cr_2O_7^{2-} + 14H^+ + 6e = 2Cr^{3+} + 7H_2O$$
and
$$Fe^{2+} = Fe^{3+} + e$$

The normality of a solution (unfortunately) depends on the specific reaction for which it is to be used. In the case of a redox reaction, when the half-reaction changes, the equivalent weight changes. This can become quite confusing and is a serious limitation to the usefulness of the normality system. A more serious limitation from our point of view is that equivalent weights do not give any useful information about chemical equilibria. It is much simpler and less confusing to use molar concentration units and always to check reaction stoichiometry.

1.3.4 Percentage Composition and Other Ratios

When concentrations are expressed in terms of percentages they are most often percentage by weight:

$$\text{weight percent (wt\%)} = \left(\frac{\text{weight of solute}}{\text{weight of solution}}\right) * 100$$

Other common units are percentage on a weight-to-volume basis:

$$\text{weight percent (wt/vol\%)} = \left(\frac{\text{weight of solute}}{\text{volume of solution}}\right) * 100,$$

and percentage on a volume-to-volume basis:

$$\text{volume percent (vol/vol\%)} = \left(\frac{\text{volume of solute}}{\text{volume of solution}}\right) * 100$$

Conversions between percentage composition units and molarity involve the use of density.

EXAMPLE 1.2 Percent by Weight and Molarity

A bottle of concentrated nitric acid is labeled 70% HNO_3. The density of 70% HNO_3 is 1.413 g/mL at 20°C. What is the molarity of the acid?

Solution

If we take 100 mL of the acid at 20°C, it will weigh 141.3 g. Of this weight, 70%, or 98.9 g, will be HNO_3; the rest will be water. The gram formula weight of HNO_3 is 63.01. The 98.9 g is 98.9/63.01 = 1.57 mole. The concentration of nitric acid is therefore 1.57 mole/0.100 liter = 15.7M.

Several other ratios are commonly used by analytical chemists. The most important are parts per thousand (ppt), parts per million (ppm), and parts per billion (ppb). Figure 1.1 lists these and other ratios and gives their equivalences in weight per volume and weight per weight. The equivalences are easy to understand if you keep in mind that the metric system is a rational system. According to SI (International System) rules, the basic unit of mass is the kilogram. The prefix "kilo" means thousand. There are many other prefixes: milli (one-thousandth), micro (one-millionth), nano (one-billionth), and pico (one-trillionth), are the most important. A ratio will have many equivalences. For example, a unit which is quite often in the news is parts per million (ppm). A catfish is considered unsafe to eat if it contains 5 ppm mercury. Here 5 ppm is the equivalent of 5 mg of mercury in a 1-kg fish, or 0.5 mg of mercury in a 100-g fish, or 0.5 g of mercury in a 100-kg fish (a real trophy fish). Some river waters contain 5 parts per billion mercury. A kilogram of this water (exactly 1 liter at 4°C) would contain 5 μg of mercury.

$$5 \ \mu g/1 \ kg = 5 \times 10^{-9} \ g/g$$

1.4 SCALE OF ANALYTICAL OPERATIONS

It is possible to begin to understand the scope of analytical chemistry by looking at the sizes of samples and concentrations within samples that can be dealt with using current technology. Figure 1.1 helps to illustrate this. The horizontal axis is the weight of sample, and the vertical axis shows the fraction of a particular constituent for which an analysis is sought. Notice that both scales are logarithmic. The horizontal scale covers sample sizes from 1 g down to 1 μg. The vertical scale covers fractions from 1.0 down to 10^{-12}. The abbreviations pph (parts per hundred), ppt (parts per thousand), ppm (parts per million), etc. appear at their correct fractions along the vertical axis. Also along the vertical axis are the names "major," "minor," "trace," and "ultratrace," which are general labels given to certain fraction sizes. For example, brass is about 70% copper; hence copper is a major component of brass. Many brasses are about 5% lead; lead is called a minor component of brass. There is very little carbon in brass, so carbon would be a trace component. Many elements are present in very low concentrations in a sample of brass and would be called ultratrace components. The names "macro," "meso," "micro," and "submicro" apply to sample sizes. Macro samples weigh more than about 0.1 g, and meso samples are generally between 10 and 100 mg. Milligram-size samples are called micro samples, while smaller samples are called submicro samples.

Diagonal lines are drawn across Fig. 1.1 to show combinations of sample size and component fraction which contain the same amount of material being analyzed. For example there is as much zinc in a 1-g sample containing 1% zinc as there is in a 10-mg sample of pure zinc. There is as much mercury in a 1-g sample containing 10^{-2}% mercury (the fraction is 10^{-4}) as there is in a 1-mg sample containing 10% mercury. Quite different technologies are required for work in different regions of Fig. 1.1. Most of the laboratory exercises in this text use samples that would appear in the lower left corner of the figure. Macro analytical balances can be used to weigh the samples, and methods such as gravimetry and titrimetry can be used to assay components. Working in the "macro-trace" domain requires more sensitivity than is

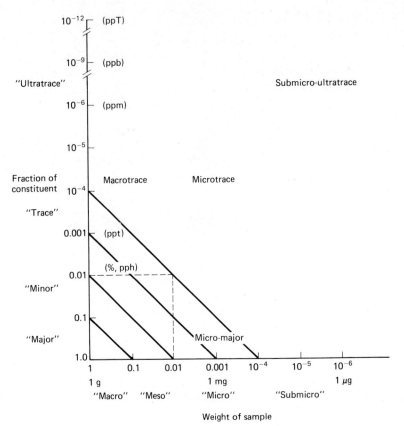

Figure 1.1 Scale of operations. The unit ppm means parts per million and represents a constituent fraction of 10^{-6}. If a sample weighs 1 g, a 1 ppm constituent has a weight of 1 μg in it. If a sample weighs 1 kg, a 1 ppm constituent has a weight of 1 mg in it. At room temperature 1 liter of water weighs about 1 kg. Fractions per unit weight sample and fractions per unit volume sample are about the same in water. Other equivalences: ppb (parts per billion) = 1 μg/kg = 10^{-3} mg/kg = 1 ng/g = 1 μg/liter water; ppT (parts per trillion) = 1 ng/kg = 10^{-3} μg/kg = 1 pg/g = 10^{-3} μg/liter water. This figure is based on ideas of E. B. Sandell and is similar to one which appears in his book, *Colorimetric Metal Analyses,* 3rd ed., Interscience, New York, 1959.

available in gravimetry and titrimetry, and chemical instrumentation is used in that region. For example, you may be asked to assay a sample for iron at the ppm level using a colorimetric method and an instrument called a spectrophotometer. Analytical work in the micro-major and micro-trace domains is extremely difficult and requires expensive and elaborate instrumentation and very pure chemical reagents. As you might imagine, these challenging domains are receiving considerable attention in research.

1.5 CALCULATING THE RESULTS: SIGNIFICANT DIGITS

People who try to communicate clearly and accurately with numbers face a simpler task than people who communicate only with words. While it is often hard to choose just the right words to express a thought accurately, a number expresses accuracy by itself. The number 5.23, for example, tells anyone who knows the basic rules of computation that the digit 3 is the first uncertain digit in the string of digits (the last *significant digit*), and that the number has an *absolute* uncertainty of ±0.01. The *relative* uncertainty of the number is expressed as (0.01/5.23) = 0.0019, or 1.9 parts per thousand (ppt).

The uncertainty of a number is expressed in an approximate way by the number of significant digits it has. All digits written to the right of the first nonzero digit are significant. An exception is when zeros lie to the right of the first nonzero digit and there is no decimal point. Zeros to the left of the first nonzero digit are not significant. Here are some examples:

- 1.0 has two significant digits and a relative uncertainty of 1 part in 10.
- 0.1050 has four significant digits and a relative uncertainty of 1 part in 1050.
- 0.0100 has three significant digits and a relative uncertainty of 1 part in 100.
- 100 has one significant digit. Its relative uncertainty is ambiguous. If it represents 100 pennies, there would be no uncertainty. If you were told in an experiment to weigh out 100 g, you would weigh to the nearest gram, and the uncertainty would be 1%.

There are certain defined quantities which can be known to as many significant digits as desired. Examples are the numbers pi and *e*, and integers. In determining the average weight of a penny, you might measure the weight of 100 pennies, then divide the result by 100. Since 100 is an integer (a counting number), it can used with as many significant digits as you wish.

Rounding Numbers   There are four rules for rounding numbers and retaining significant digits:

1. If the first nonsignificant digit (*residue*) from a calculation is greater than 5, increase the last retained digit by 1.
2. If the residue from a calculation is smaller than 5, retain the last digit.
3. If the residue is 5, increase the last digit by 1 if the last digit is odd.
4. If the residue is 5, leave the last digit unchanged if it is even.

Examples:

$$5.13465 \text{ rounds to } 5.1346$$
$$5.1346 \text{ rounds to } 5.135$$
$$5.135 \text{ rounds to } 5.14$$
$$5.14 \text{ rounds to } 5.1$$

Rules for Addition and Subtraction   In addition and subtraction retain only as many significant digits as appear in the least certain number being added or subtracted.
Examples:

$$1.05 + 2.613 + 4.1291 = 7.7921 \quad \text{or} \quad \underline{7.79}$$

Significant digits: 3   4   5   3

Notice that this preserves in the sum or difference the largest absolute uncertainty.

Rules for Multiplication and Division   In multiplication and division the result should have about the same relative uncertainty as that of the least certain number being worked on. More exactly, the result should be between one-fifth and twice as uncertain as the least certain number.
Example:

$$1.05 \times 2.513 = 2.744 \rightarrow 2.74 \text{ (rounded)}$$

The uncertainty in 1.05 is 1 part in 105. The uncertainty in 2.513 is 1 part in 2513. The product 2.744 must be rounded so that its relative uncertainty lies between 2 times 1 part in 105 (i.e., 1/63) and 1/5 times 1 part in 105 (i.e., 1/525). The properly rounded result is 2.74, which is uncertain to 1 part in 274; 2.744 expresses too much certainty (1/2744, much smaller than 1/525).
Example:

$$9.05 \times 2.723 = 24.643 \rightarrow 24.64$$

The less certain number is 9.05, which is certain to 1 part in 905. We must round the product so that its uncertainty falls between 1 part in 452 and 1 part in 4500. The correct way to report the product is 24.64. Notice that it is not necessary to round the product to three significant digits, even though one of the numbers carries only three significant digits.

Rules for Logarithms   When working with logarithms carry as many significant digits as there are in the *mantissa*. In the logarithm 2.3010, 2 is called the *characteristic* and .3010 is called the *mantissa*.
Example: the log of 540 is 2.732. The number 540 has three significant digits; the mantissa .732 must also have three significant digits.
Example: the antilog of 2.3010 is 2.000×10^2, not 2.0000×10^2. The mantissa has four significant digits.

1.6 BASIC LABORATORY OPERATIONS

While laboratory experiments are presented in many of the chapters, this text is in no sense a thorough laboratory manual. The only adequate way to teach laboratory tech-

1.6 BASIC LABORATORY OPERATIONS

nique is by demonstration; sketches and prose are simply not adequate substitutions for classroom instruction. What is presented here is the basic technical information students will need to work with balances and glassware provided in the usual quantitative analysis laboratory.

1.6.1 Weighing

The measurement of weight is the most common and perhaps most important operation performed by an analyst. When good equipment is used with care and thought, weighing can also be the most accurate measurement made. The single-pan analytical balances used in undergraduate quantitative analysis laboratories (called macro balances) can be used to measure weights up to about 160 g to the nearest 0.1 mg. The figure 160 g is the *capacity* of the balance, and the figure 0.1 mg is its *sensitivity*. The uncertainty of measuring such a weight is less than 1 part in a million (1.0×10^{-4} g/160 g), on a par with the accuracy of the most expensive quartz digital watches.

Whether the expense and care required to attain this accuracy is really necessary depends entirely on the requirements of the experiment and the nature of the analytical problem. For example, you might be told to weigh out enough potassium chloride to make 1 liter of $0.10F$ KCl solution (about 7.5 g). Since the concentration must be known with an uncertainty of 1 part in 10, it would make no sense to use an analytical balance and measure weight to five significant figures. Instead, you would weigh out the KCl on a triple-beam balance, which can be used to measure weight to the nearest 0.1 g. If, however, you were told to prepare a $0.1000F$ solution of KCl, a triple-beam balance would not be sufficiently sensitive, and a macro analytical balance would be needed. The rule is to *choose your weapons carefully;* it is a waste of time to use a balance with more sensitivity than the work requires, but all the effort of doing an analysis will be for naught if the balance is not sufficiently sensitive. Table 1.3 gives the capacities and sensitivities of a variety of balances.

Weighing Techniques There are two techniques used in the quantitative analysis laboratory: weighing by difference and weighing by addition.

Weighing by Difference All careful weighing of solids is done by difference. When weighing a sample, weigh a weighing bottle containing the sample, remove the bottle from the balance with clean dry fingers (or a folded paper strip, as your instructor directs), and pour some of the sample into a flask or beaker. You will then weigh the weighing bottle and its contents again. The weight of the sample transferred is the difference between the first and second weights measured.

It takes some practice to learn how to transfer the desired amount of sample when weighing by difference. It is quite harmless to weigh out less than the amount you need; you simply transfer a little more until you have transferred the right amount.

TABLE 1.3 CAPACITIES AND SENSITIVITIES OF SEVERAL BALANCES

Balance type	Capacity (g)	Sensitivity
Analytical		
Macro	160	0.1 mg
Semimicro	160	0.01 mg
Micro	30	0.001 mg
Ultramicro	3	0.0001 mg
Top-loading	160	1 mg
	300	10 mg
	1200	10 mg
	6000	100 mg
Triple-beam		
Hanging pan	300	10 mg
"Trip"	600	100 mg
Double pan	200	100 mg
	2000	1 g

If you vastly *overestimate* the amount, *do not* try to return some of the sample to the weighing bottle. Instead, throw out the transferred sample, wash the flask or beaker, and start over. Fortunately, it is almost never necessary to weigh out an exactly prescribed amount of sample. If directions call for a 0.4-g sample, samples weighing between 0.35 and 0.45 g will probably be acceptable. Of course, whatever a particular sample weighs, it must be weighed as accurately as the balance permits.

Never use spatulas or weighing paper when weighing by difference. It is crucial that every bit of sample that leaves the weighing bottle be used in the analysis.

Weighing by Addition This method is commonly used for weighing with triple-beam balances and less often used with analytical balances. The first weight measured will be that of an empty vessel or weighing paper. Sample is added to the vessel, and it is weighed again. The difference in weights is the amount of sample transferred. A spatula or scoop may be used to transfer sample when weighing by addition. The only important material is that which ends up in the vessel.

Weighing by addition is acceptable practice with an analytical balance if a specific weight of material must be measured. First the vessel is weighed, the desired weight is set on the balance dials, and the pan is set at half arrest. Sample is added to the vessel until the balance indicator jumps. Then small amounts of sample are added until the desired weight is obtained.

When working with volatile liquids there is really no choice but to weigh by addition into a sealable vessel. There is too great a risk of loss by evaporation when

1.6 BASIC LABORATORY OPERATIONS

weighing liquids by difference. Eyedroppers or Pasteur pipets can be used to transfer liquids when weighing by addition.

Practical Problems in Weighing   There are several potential sources of error that you must recognize and deal with when trying to make accurate weight measurements:

Adsorption of Moisture   Either the weighing bottle or the sample can adsorb moisture from the air. Adsorption by the sample is more serious than adsorption by the glassware, because finely divided samples have large surface areas and may be quite hygroscopic. Moisture can be removed from most samples by heating in an oven at 110°C for an hour. (Exact heating conditions vary from sample to sample.) After removing a sample from the oven, place the sample and weighing bottle in a desiccator and let them cool to room temperature. When the bottle and its contents are taken from the desiccator they may readsorb some moisture and slowly gain weight. Although this may be noticeable on hot, humid summer days, it is seldom very troublesome. If the sample is quite hygroscopic it should be transferred and weighed as rapidly as possible.

Static Electricity   While low humidity is desirable for avoiding adsorption of moisture, it promotes the electrostatic charging of glassware. If a weighing bottle becomes charged, when it is placed on a balance pan attractive forces between the bottle and the balance chamber walls may change the measured weight. The simplest cure for charging is to wipe the glassware with a damp (not wet!) chamois cloth. Alternatively, an ultraviolet light source or a low-intensity source of alpha-particle radiation (^{210}Po, for example) inside the balance chamber can help reduce static charging.

> CAUTION: Wear eye protection; UV light may damage your eyes.

Oscillations   Sometimes an object placed off-center on the balance pan can cause the pan to swing gently, resulting in a rocking motion of the optical scale. Try to center objects on the pan, and let swinging subside before weighing.

Temperature Differences   When a hot weighing bottle is placed on a balance pan, convection currents begin to rise around it and buoy it up. Also, warm air enclosed in a weighing bottle has a lower density than air in the balance chamber, and the surrounding cooler air buoys up the bottle. For example, in an actual experiment a 20-mL sealed bottle only 5°C warmer than the air in the balance chamber was found to weigh 0.5 mg less than it did when at the same temperature as the chamber air.

The relationship between weight loss (Δweight) due to buoyancy and temperature difference (ΔT) is

$$\Delta\text{weight} = \text{density} * \text{volume} * \Delta T/T$$
$$= (1.18 \times 10^{-3} \text{ g/cc}) * 20 \text{ cc} * 5 \text{ K}/298 \text{ K} = 0.4 \times 10^{-3} \text{ g}$$

Convection currents will increase the weight loss.

The way to avoid this kind of error is to let oven-dried objects cool to room temperature in a desiccator before weighing them. Cooling time depends on the mass and heat capacities of the sample, weighing bottle, and desiccator. In an actual experiment a 20-mL ground-glass-stoppered weighing bottle containing 5 g of a salt was heated to 110°C in an oven and then placed in a 500-cc glass jar desiccator to cool. After 30 minutes the sample was still 3°C above room temperature. After an additional 10 min the sample was at room temperature. More time is required if the object is large or its initial temperature is much higher. A 15-g crucible heated to dull redness will require at least an hour to cool to room temperature.

A more subtle source of temperature-related error is excessive handling of a weighing bottle during the weighing process. Transfer of body heat during handling can warm a weighing bottle by several degrees and result in slightly low weights. It is a good idea to make only brief contact with weighing bottles. Using a loop of paper to lift weighing bottles avoids heat transfer from the fingers.

Fingerprints Although a very greasy fingerprint may weigh as much as $\frac{1}{2}$ mg, if your fingers are clean and dry you may handle weighing bottles directly without worrying about increasing their weight. In an actual experiment a 29-g weighing bottle (with salt) was held between the thumb and two fingers for 30 seconds and appeared to *lose* 0.3 mg. Heat transfer (which lowers the weight) was more important than the weight of three fingerprints.

In work that requires the measurement of weight to be more accurate than 0.1 mg, finger cots or paper loops should be used to avoid both fingerprints and heat transfer.

Buoyancy Corrections Buoyancy corrections are described in Section 1.6.2 on glassware calibration. These corrections are generally unnecessary when weighing gram-size samples to the nearest 0.1 mg. When weighing by addition or difference, only the buoyancy of the material being transferred need be considered. The buoyancy of the weighing vessel influences the weight both before and after transfer, and subtracts out.

The Balance Room Analytical balances should be supported by heavy tables and be well isolated from building vibrations and rapid temperature changes. If the balances in your laboratory rest on ordinary tables you will find that measured weights will change when books, desiccators, arms, etc. are resting on the table. Pay attention to your surroundings as you make weight measurements.

If you spill sample on the balance pan or in the balance housing, clean it up before leaving the balance. Your instructor should supply a camel's hair brush for cleanup. Always reset the dials on the balance to zero and close the doors to the weighing chamber.

1.6.2 Use and Calibration of Volumetric Glassware

The basic set of volumetric glassware supplied to students of quantitative analysis consists of graduated cylinders, pipets, burets, and volumetric flasks.

Description and Routine Use

Graduated Cylinders These devices are used for approximate volume measurements and come in a variety of sizes. Use the smallest size consistent with your needs. For example, if you must measure 5 mL, use a 5- or 10-mL graduated cylinder rather than a 100-mL graduated cylinder.

Pipets Pipets are used to deliver specific volumes precisely and accurately. Both measuring pipets (graduated) and transfer pipets (marked for only one volume) are available. Measuring pipets provide less precision than transfer pipets but may be used when an error as large as a few percent is tolerable. Small transfer pipets are able to deliver volumes with errors smaller than 0.01 mL (see calibration below).

Some pipets are marked *to deliver* (TD) and others are marked *to contain* (TC). A TD pipet should be allowed to drain by itself; it has been calibrated taking into account solution left behind after transfer. The contents of a TC pipet, however, should be expelled forcibly into a delivery vessel; it has been calibrated without solution remaining.

Mechanical micropipets can be purchased, and they extend the range of precisely delivered volumes down to 10 μL or less. Volumetric errors of less than 0.1% are attainable with modern micropipets. Micropipets are all TC devices.

Your laboratory instructor will show you the correct procedure for using a pipet. Under no circumstances should you use mouth suction to draw solution into a pipet. It is far safer to use a rubber suction bulb, even though you will find it awkward at first.

Burets Burets are graduated pipets with a stopcock to control delivery of solution. In routine laboratory work 10-, 25-, or 50-mL burets are most often used. In instructional laboratories 50-mL burets graduated in tenths of a milliliter are the most common size used. With some practice you will be able to estimate volumes to the nearest 0.01 mL using a 50-mL buret.

An ordinary buret is filled by simply pouring solution into the top. When using smaller (10 mL) burets it is often easier to draw solution up into the buret with suction from a rubber bulb. Regardless of the filling method, you should rinse the buret with a few milliliters of solution before actually filling it for delivery. Add solution until the liquid level is well above the zero mark at the top of the buret. Then open the stopcock and let the solution level fall to the zero line. There should be no air bubbles in either the barrel or the tip of the buret. Wipe solution from the buret tip, read the initial volume, and begin delivering solution. When the desired amount has been delivered (say at the endpoint of a titration), read the final volume. Refill the buret to the zero line before delivering the next volume.

When you read a buret make sure your line of vision is directly along the solution meniscus (the curved solution surface) and perpendicular to the buret. If you look down on the meniscus or up at it, you may misread the level. It also helps to hold a dark object behind the buret, either above or below the meniscus; the image of the dark object will reflect off the meniscus and will make the solution level much easier to see. Try to read the buret to the nearest 0.01 mL.

Your laboratory instructor will show you the correct ways to use and read a buret. With patience and care you will find the buret very easy to use.

Volumetric Flasks A volumetric flask is designed to hold one certain volume. Volumetric flasks have a characteristic round shape with a flat bottom and a long narrow neck. Many are made with standard-taper ground glass tops for tight sealing. The volume in the neck is much smaller than the total volume of the flask to allow very accurate measurement of volume.

Solid samples are best transferred to a volumetric flask by first dissolving or suspending them in a minimum volume of solvent, and carefully pouring the solution into the volumetric flask with the help of a funnel. Add about half the total volume of solvent to the sample and completely dissolve it by swirling the solution. Add more solvent until the liquid level reaches the volumetric mark. Mix the contents of the flask by holding the stopper in place with your finger and inverting several times. The shape of the volumetric flask makes it rather difficult to mix solutions thoroughly, and you should invert the flask many times.

Your instructor will show you the proper method for filling and checking the volume of a volumetric flask. Calibration of a volumetric flask is discussed below.

Cleaning Glassware Glassware should be clean before calibration and use. An important sign of cleanness is that water drains from the glassware uniformly, leaving behind only a thin film of water without droplets. Routine cleaning of volumetric

1.6 BASIC LABORATORY OPERATIONS

flasks and burets can be done with dilute detergent solutions (1–2%) and brushes. Pipets can often be cleaned by rinsing with detergent solution. Heavier contamination can be removed with chromate/sulfuric acid cleaning solution. Your instructor may make this available.

> CAUTION: Under no circumstances should you work with chromate/sulfuric acid cleaning solution without supervision. Many students have been severely burned by spilling this material on their hands or clothing. You must wear eye protection and rubber gloves to protect yourself.

Oil and grease may cause water droplets to form in pipets and burets. Most light oils can be removed by rinsing glassware with a 50% (wt/vol) solution of sodium hydroxide in ethanol. Subsequent rinsing with water and dilute HCl and finally deionized water will neutralize the residual base, which tends to coat glass. Since base gradually dissolves (etches) glass, do not leave the ethanol/base solution in contact with the glassware for more than a few minutes.

> CAUTION: Ethanol/base solutions are quite caustic and will burn skin. Wash hands thoroughly after using this cleaning solution! Wear eye protection!

Grades of Volumetric Glassware Most chemical apparatus supply houses offer two grades of volumetric glassware. The first, class A glassware, is calibrated to very close tolerances, with relative uncertainties of about 1 part per thousand. This glassware is expensive but can be used without calibration in all but the most exacting work. The second grade, class B, is calibrated to tolerances generally twice as large as class A tolerances. Class B glassware may be adequate for many quantitative applications but is not recommended for exacting work unless it is first calibrated by the analyst. Tolerances for volumetric glassware are listed in Table 1.4.

Calibrating Volumetric Glassware The experiment at the end of this chapter gives directions for calibrating a 25-mL pipet, a 100-mL volumetric flask, and a 50-mL buret. Calibrations are done by weighing delivered quantities of deionized water at room temperature. The mass and volume of a liquid are related by density, which is temperature-dependent. Table 1.5 lists densities of water at a number of temperatures. In the most careful work corrections for thermal expansion of glass and buoyancy must be made. While the calculations involved in calibrations are not very difficult, they deserve explanation. The following example is given to introduce the correction factors and show how they are used.

TABLE 1.4 TOLERANCES OF CLASS A GLASSWARE[a]

Glassware and capacity (mL)	Limit of error (±mL)	Outflow time (sec)
Volumetric flasks		
5	0.02	
10	0.02	
25	0.03	
50	0.05	
100	0.08	
250	0.12	
500	0.15	
1000	0.30	
2000	0.50	
Transfer pipets		
2 (and smaller)	0.006	10
5 (and 4, 3)	0.01	10
10	0.02	15
20	0.03	25
25	0.03	25
50	0.05	25
100	0.08	30
Burets		
5	0.01[b]	
10	0.02	
25	0.03	
50	0.05	
100	0.08	

[a] Class A is precision grade glassware which conforms to these specifications. Class B glassware is also available and tolerances are generally twice as large as those for Class A.
[b] Limits of error for burets are for total or partial fill.
Source: Flasks: Fed. spec. NNN-F-289; American Society for Testing and Materials (ASTM) spec. E288, E542, E694. Burets: ASTM E694. Pipets: ASTM E969-83.

EXAMPLE 1.3 Glassware Calibration

A 100-mL volumetric flask weighing 55.100 g is filled to the mark with deionized water at 26°C. The weight when filled is 154.925 g. What is the volume of the flask at 20°C?

Solution

Step 1: The basic calculation. The density of water at 26°C is 0.99681 g/mL. The volume at 26°C is:

$$V_{26} = \text{weight/density} = (154.925 - 55.100)/0.99681 = 100.14 \text{ mL}$$

1.6 BASIC LABORATORY OPERATIONS

TABLE 1.5 DENSITY OF WATER BETWEEN 10° AND 40°C

Temperature (°C)	Density (g/mL)	Temperature (°C)	Density (g/mL)
10	0.99973	25	0.99707
11	0.99963	26	0.99681
12	0.99952	27	0.99654
13	0.99940	28	0.99626
14	0.99927	29	0.99597
15	0.99913	30	0.99567
16	0.99897	31	0.99537
17	0.99880	32	0.99505
18	0.99862	33	0.99473
19	0.99843	34	0.99440
20	0.99823	35	0.99406
21	0.99802	36	0.99371
22	0.99780	37	0.99336
23	0.99756	38	0.99299
24	0.99732	39	0.99262
		40	0.99224

Step 2: Correct for buoyancy. Archimedes' principle is that an object immersed in a fluid is buoyed up by a force equal to the weight of the fluid it displaces. When we use an analytical balance we are using a set of brass or stainless steel weights to weigh objects which generally have a much lower density. Since the lower-density object displaces more air than the weights, it will be measured to weigh less than it really does. We can correct for buoyancy by adding to the measured weight a density correction factor:

$$W_T = W_M + \rho_{air}[(W_M/\rho_{sample}) - (W_M/\rho_{bal.wt.})]$$

where W_T is the true (corrected) weight, W_M is the measured weight, ρ_{air} is the density of air (ca. 1.20×10^{-3} g/cc), ρ_{sample} is the density of the sample, and $\rho_{bal.wt.}$ is the density of the balance weights (8.4 g/cc for brass and 7.7 g/cc for steel).

The density of air may vary from 1.1×10^{-3} to 1.3×10^{-3} g/cc, depending on humidity and atmospheric pressure. We will use the average value 1.20×10^{-3} g/cc.

For our data, the buoyancy correction is

$$W_T = 99.825 \text{ g} + 0.00120[(99.825/0.99681) - (99.825/8.40)]$$
$$= 99.825 \text{ g} + 0.106 \text{ g} = 99.931 \text{ g}$$

The volume corrected for buoyancy is

$$V_{26} = W_T/\rho_{26} = 99.931/0.99681 = 100.25 \text{ mL}$$

The buoyancy correction is thus about 1 part per thousand, an error which might be significant in careful work.

We now know the volume of the volumetric flask at 26°C. Let us see if the volume at 20°C is much different.

Step 3: Correct for thermal expansion of the glass. The "cubical coefficient of expansion" of Pyrex glass is $1.0 \times 10^{-5}/°C$. The volume of a Pyrex flask as a function of temperature is given by

$$V_{20} = V_M[1 + 1.0 \times 10^{-5}(20 - T)]$$

where V_M is the measured volume at temperature T and V_{20} is the volume at 20°C.

For our volumetric flask the expansion effect is quite small, about 0.01 mL:

$$V_{20} = V_{26}[1 + 1.0 \times 10^{-5}(20 - 26)] = 100.25(1 - 6 \times 10^{-5})$$
$$= 100.24 \text{ mL}$$

In routine laboratory work the coefficient of expansion of glassware is usually small enough to be ignored safely.

1.6.3 Filtering

The filtering of solutions and the collection of precipitates are common parts of analytical laboratory work. In this section we discuss techniques of filtration, filtering media, and equipment.

Filtering with the Help of an Aspirator Before filtering a sample, you must assemble some equipment. Some filtrations, particularly those in which very small particles must be collected, need the help of a vacuum. Figure 1.2 shows a conventional filter flask and trap setup for use with a water aspirator. A trap is used between the filter flask and the aspirator to keep water from being drawn back from the aspirator to the filter flask. The filter flask is shown with a filter crucible. The rubber ring around the filter crucible ensures a tight seal. The tubing connecting the flask and trap to the aspirator must have an extra-thick wall to prevent it from collapsing when a vacuum is drawn in the system.

Filter Crucibles There are two types of filter crucibles. In one type, the fritted, or sintered, glass crucible, the filtering medium is a porous glass disk which is fused to the bottom of the glass crucible. The other type, called a Gooch crucible, is usually a glazed ceramic crucible whose bottom has rather large holes. The Gooch crucible is

1.6 BASIC LABORATORY OPERATIONS

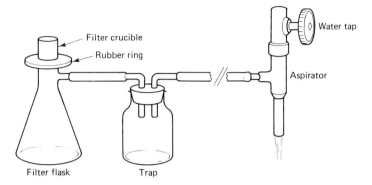

Figure 1.2 Filtering apparatus. The use of a rubber ring directly on the top of the filter flask is the simplest and least expensive way to use a filter crucible. Smaller rubber rings are made to fit inside the conical top of a short-stem funnel. A rubber stopper then holds the funnel in the filter flask.

made to support a fine glass wool or asbestos filtering mat that is not a part of the crucible.

Sintered glass crucibles are available in either Pyrex or fused silica and are available with several porosities, from extra coarse (EC) to very fine (VF), as shown in Table 1.6. Pyrex crucibles should not be heated much above 200°C. Fused silica crucibles may be heated to 500°C and can withstand more rapid cooling than Pyrex crucibles.

Gooch crucibles with asbestos mats can be heated to much higher temperatures. Unfortunately, asbestos is too dangerous a material to keep in undergraduate laboratories (it is a cancer-causing substance). Circular glass fabric mats are available for Gooch crucibles and provide excellent filtering capability. Glass mats, like glass crucibles, should not be heated above 500°C.

TABLE 1.6 SINTERED GLASS FRIT POROSITIES

Grade	Nominal pore size (μm)[a]	Use
Extra-coarse (EC)	170–220	Filtering very large crystals, supporting glass mats
Coarse (C)	40–60	Large crystals
Medium (M)	10–15	Crystalline precipitates
Fine (F)	4–5.5	Fine filtration
Very fine (VF)	2–2.5	Very fine crystals, some colloids

[a] The symbol μm means micrometer, that is, 10^{-6} m or 10^{-3} mm.

Unglazed porcelain crucibles with porous bottoms may be satisfactory for filtering fine precipitates. These crucibles are used in the same way as sintered glass crucibles, but have the advantage of being able to withstand much higher temperatures.

Sintered glass crucibles are recommended for the experiments in Chapter 10. Instructions are given there for cleaning and preparing them.

The proper technique for transferring solution and solid to a filter crucible is the same as that used with filter paper and is described below.

Filter Paper Paper is a very low cost and often very effective filtering medium. A number of porosities are available, as shown in Table 1.7. Coarse papers filter solutions quite quickly, while fine papers require a great deal of time and patience. Choose the largest porosity that will retain your precipitate.

Because paper absorbs water from the air quite readily, it cannot be used to collect precipitate for accurate weighing. Instead, it must be burned away from the precipitate before the precipitate is weighed. Special ashless papers are made which will leave hardly any residue after burning. A 9-cm disk (a very common size) of ashless paper will leave a residue weighing less than 0.1 mg after ignition, an amount which is negligible in macro analyses.

Some precipitates, notably iron and aluminum hydroxides, are gelatinous and tend to clog the pores of filter paper. This problem can be largely overcome by preparing a slurry of ashless filter paper pulp and mixing it with a gelatinous precipitate before filtering. The slurry may be made from pulp tablets (commercially available) or by breaking up a piece of ashless filter paper in concentrated hydrochloric acid (contact with solution only 2 or 3 minutes), and carefully washing away chloride before use.

TABLE 1.7 FILTER PAPER GRADES AND POROSITIES[a]

Property	Whatman number			
	41	40	44	42
Porosity	Coarse	Medium	Fine	Fine
Flow rate, sec[b]	12	75	175	240
Particle size retained, μm	20–25	8	3	2.5

[a] All grades 40–44 leave ash not exceeding 0.010% of weight of the paper. "Hardened" papers are also available: No. 54 is hardened No. 41, No. 50 is hardened No. 44. Hardened paper is made by treating ashless paper with nitric acid. It is smooth, withstands tearing when wet, and is preferred when filtering with vacuum.

[b] Flow rates are for 100 mL of prefiltered water passing through a 15 cm paper folded in quarters.

1.6 BASIC LABORATORY OPERATIONS

The pulp and ashless filter paper are burned away from the collected precipitate in a process called ignition (see below).

Filter paper may be used in conical (60-degree) funnels for filtering. A disk of filter paper is folded in half and firmly creased. It is then folded loosely and slightly off-center into quarters, and a small triangular corner is torn from one of the two single corners. The off-center fold will keep the bottom of the cone from clinging to the funnel, while the tear at the top will allow the wet paper to cling at the top of the funnel. The paper is then opened into a cone and fitted into the funnel. The cone is dampened with deionized water and gently pressed into place in the funnel. When the paper cone is in place there will be no air leaks between the funnel and the paper. When water is poured into the filter, the stem of the funnel should fill and the column of water should be held by surface tension. This column of water will speed filtration. Your instructor should demonstrate the proper way to fold and place filter paper in a funnel.

Unless you are told otherwise, you should heat solutions before filtering them through paper. Hot solutions have lower viscosity (resistance to flow) than cold solutions and can be filtered much more quickly. You should also let precipitates settle to the bottom of a solution before filtering; fine precipitates clog filter pores and dramatically slow down the filtering process.

When filter paper is in place in the funnel, put the funnel in a funnel stand and place a beaker under it to catch filtered solution. Touch the stem of the funnel to the wall of the beaker to avoid splashing, but keep the funnel stem out of the filtered solution in the beaker. Place a stirring rod across the beaker containing the sample to be filtered, grasp the beaker in your hand, holding the rod across the top with one finger. Tip the beaker toward the funnel and pour solution down the stirring rod into the funnel. Always use a stirring rod as a pathway for liquids in these operations. Do not fill the funnel to the top of the paper, but also avoid letting the liquid level fall into the stem. With practice you will learn how to achieve a "steady-state" solution transfer, with solution input keeping pace with solution output. Leave the solid material in the beaker until the very end of the filtering process. When precipitate finally enters the paper, the rate of filtering will decrease quite quickly.

When as much solution as possible has been passed through the filter and the solid and a little solution is left in the beaker, use a stream of wash solution (or water, as directed) from your wash bottle to rinse the remaining material into the filter paper. Any solid material left in the original beaker can be collected with the help of a rubber policeman if necessary. Rinse the original beaker and the solid in the filter paper as directed. Once again, your instructor should demonstrate these operations.

If the solid trapped in the filter paper is to be determined quantitatively, the filter paper must be burned away (ignited). The filter paper must be carefully loosened from the funnel, lifted out, and folded down around itself, something like an envelope. The folded paper is then placed carefully in a preweighed crucible. The crucible is placed upright in a ceramic triangle held on a ring stand, and the contents are dried under a heat lamp. When the paper appears to be dry, the crucible is covered and heated with a burner to slowly carbonize (char) the paper. Do not allow the paper to burst into flame, and avoid igniting gases that escape from the cover. When the charring is complete (when no more gases escape), turn the crucible at an angle of about 45 degrees from the vertical and partially remove the cover so that air can enter the crucible. Heat the bottom of the crucible until it is a dull red, but keep the burner flame from the mouth of the crucible; hot gases may sweep solid out of the crucible or may change the solid chemically.

After the carbon is completely gone, return the crucible to an upright position and place the cover directly on top. Adjust the burner flame for the particular method (specific instructions), and continue ignition for 15 to 30 minutes. It may be necessary to control the temperature carefully, in which case an electric muffle furnace is recommended for this last step. After the required time, remove the crucible from the heat with clean crucible tongs, and let it cool for several minutes in the air. Transfer the crucible to a desiccator and let it cool to room temperature (usually at least an hour).

1.7 NOTES ON LABORATORY SAFETY

Many of the experiments you will be asked to do in the quantitative analysis laboratory involve some element of personal risk. The experiments in this text and those used by most instructors are chosen in part because they are relatively safe and the risks that are present are quite well documented. Clearly, it is not possible to cover the topic of laboratory safety in only one section of a chapter. Nevertheless, there are a few simple rules which, if followed, will make your laboratory work much safer. Remember that you have an obligation not only to yourself but to those who work with you in the laboratory. Most of the rules reduce to common sense and courtesy, as you will see.

1. When you enter a new laboratory, locate the fire extinguishers, fire blankets, eye fountains, safety showers, and emergency exits. Your instructor will show you where everything is on the first day of laboratory.

2. Wear eye protection whenever you are working in the laboratory or whenever you are standing next to someone who is working. Many states require lab students to wear safety goggles at all times. Even if it is not a law in your state, it is a good idea. You must protect your eyes whenever you are working in or near a fume hood; anything that is foul enough to be kept in a hood will hurt your eyes, lungs, or skin.

3. Never work alone in the laboratory. Make sure someone is always within shouting distance.

4. Never mix things at eye level close to your face. Keep your work below the level of your head to avoid spattering on your face or neck and to avoid pouring solutions down your arms.

5. If you wear your hair long, tie it up when working in the laboratory to keep it away from flames and out of solutions.

6. Wear shoes in the laboratory. Shards of glass or spilled chemicals may be on the floor.

7. Never fill pipets by mouth suction; always use a pipet bulb.

8. Concentrate—do not lose sight of the work that you must do. Laboratory work involves social interactions. However, try to concentrate on every step. Horseplay in a laboratory can be extremely dangerous.

9. Never bring food or beverages into the laboratory. Never use a chemical refrigerator to store food.

10. Be absolutely sure of the chemistry you are doing. If you are following directions, follow them in sequence. If your instructor has you plan experiments, have them approved before coming to the laboratory. Do not mix together compounds without asking or thinking about possible reactions. Avoid surprises.

11. Look for warning labels on reagent bottles. Manufacturers now print warnings and precautions for storage on labels. If you are unsure about a compound, ask your instructor.

12. Do not remove reagents kept in the fume hood or bring them to a laboratory bench. These compounds need the extra ventilation provided by a fume hood.

13. If you are a chronic nail biter, wear gloves in the laboratory. Wash your hands frequently, especially just after cleaning up.

14. Report all injuries to your instructor, no matter how insignificant you think they are.

15. Clean up all spills immediately. If you leave a puddle of acid on your bench, you or your neighbor may put a hand down on it and get burned. In quantitative analysis laboratory it is a good idea to keep bench tops spotless to avoid contaminating your samples.

16. Be aware of what your neighbors are doing. This will help avoid, for example, accidentally mixing reactive compounds in the sink. Table 1.8 contains a list of incompatible compounds.

1.8 DISPOSING OF CHEMICAL WASTES

Academic chemical laboratories can be important sources of hazardous wastes. Although most academic laboratories are not regulated directly by the Environmental Protection Agency, it is important that we all recognize our responsibility for the careful disposal of chemical wastes. Here are some rules:

1. Before disposing of strong acids or bases, dilute them with water. Concentrated acids corrode plumbing, releasing lead solder from joints. They may also react at plumbing junctions with wastes from other laboratories to produce explosive or toxic gases.

TABLE 1.8 EXAMPLES OF INCOMPATIBLE CHEMICALS[a]

Chemical	Incompatible with:
Acetic acid	Nitric acid, perchloric acid, peroxides, permanganate
Acetone	Mixtures of concentrated nitric and sulfuric acids
Alkali, alkaline earth metals (Na, Ca)	Water, CCl_4, other chlorinated hydrocarbons, halogens
Ammonia (anhydrous)	Halogens, calcium hypochlorite, mercury
Ammonium nitrate	Acids, powdered metals, flammable liquids and combustibles
Arsenical compounds	Any reducing agent
Chromic acid (cleaning solution)	Acetic acid, alcohols, camphor, naphthalene
Copper	Acetylene, hydrogen peroxide
Cyanides	Acids
Fluorine	Everything (extremely reactive)
Halogens (others)	Ammonia, hydrogen, gaseous hydrocarbons, finely divided metals
Hydrogen peroxide	Most metals and their salts, alcohols, acetone, organic solvents in general
Nitrates	Sulfuric acid
Nitrites	Acids
Oxalic acid	Silver, mercury
Perchloric acid (concentrated)	Organics, paper, wood, grease (strong oxidizer)
Potassium chlorate	Sulfuric and other acids (strong oxidizer)
Permanganate	Glycerol, glycols (strong oxidizer)
Silver	Oxalic acid, tartaric acid, ammonium salts
Sodium peroxide	Ethanol, methanol, glacial acetic acid, glycerin, glycols, acetates
Sulfides	Acids
Sulfuric acid	Chlorates, permanganate, etc. (makes them stronger oxidizers)

[a] This list is by no means exhaustive. It is taken from a longer list compiled by the National Research Council in *Prudent Practices for Handling Hazardous Chemicals in Laboratories,* National Academy Press, Washington, D.C., 1981.

  2. In general, never dispose of transition metal compounds by flushing them down the sink. This is especially true for mercury, cadmium, chromium, lead, and arsenic compounds. Most other transition metals are sufficiently costly to interest people in recycling them. Sewage treatment systems can be harmed by chromium and other metals. If sewage sludge is used as fertilizer in your area, your laboratory wastes may end up in a farmer's field. Your instructor will provide a clearly labeled receptacle for metal wastes. The label should include a list of incompatible wastes. Cyanide wastes should have their own container and should be treated with care. Inadvertently adding strong acid to a solution containing cyanide will release HCN gas, a deadly poison. Your instructor can treat wastes to render them less harmful and can dispose of them properly. Check the compatibility list before mixing wastes.
  3. Recover organic solvents by distillation whenever possible. Never dispose of highly flammable solvents by pouring them down the sink. Solid still residues should be collected and destroyed chemically or by incineration at an approved facility.
  4. Today, very few undergraduate chemistry experiments involve cancer-causing

compounds. If your instructor gives you a special warning about a substance, be sure to use and dispose of the substance with great care.

RECOMMENDED READING

Analytical Chemistry as a Discipline and a Profession

The Analyst, December 1974 (vol. 99) issue, has a series of papers about analytical chemists and their various roles in science. Particularly interesting papers are: H. M. N. H. Irving, "One Hundred Years of Development in Analytical Chemistry"; R. Belcher, "Analytical Chemistry and Education"; J. Markland, "Analytical Chemistry in Public Service"; C. Whalley, "Analytical Chemistry in Industry"; H. A. Laitinen, "Analytical Chemistry in Interdisciplinary Environmental Science."

Isenhour, T. L. "The Future of Analytical Chemistry: Will There Be One?", *Analytical Chemistry,* 55, 824A (1983).

Laitinen, H. A., and G. W. Ewing. *A History of Analytical Chemistry.* American Chemical Society, Washington, D.C., 1977.

The Process and Techniques of Analytical Chemistry

Bogen, D. C. "Decomposing and Dissolving Samples: Inorganic," in *Treatise on Analytical Chemistry,* I. M. Kolthoff and P. Elving, eds., 2nd ed., part 1, vol. 5.

Dunlop, E. C., and C. R. Ginnard. "Decomposing and Dissolving Samples: Organic," *ibid.*

Kolthoff, I. M., E. B. Sandell, E. J. Meehan, and S. Bruckenstein. *Quantitative Chemical Analysis,* 4th ed., Macmillan, Toronto, 1969. This text is an encyclopedic treatment of the subject and is particularly strong in the traditional techniques. Chapters 17–21 contain a wealth of information about apparatus, balances and weighing, and volumetric glassware. The section on balances and weighing is quite outdated, and the following articles should be consulted.

Leonard, R. O. "Electronic Laboratory Balances," *Analytical Chemistry,* 48, 879A (1976).

Schoonover, R. M. "A Look at the Electronic Analytical Balance," *Analytical Chemistry,* 54, 973A (1982).

Laboratory Safety and Waste Disposal

Prudent Practices for Handling Hazardous Chemicals in Laboratories, National Research Council, National Academy Press, Washington, D.C., 1981.

Safety in Academic Chemistry Laboratories, 3rd ed., Committee on Chemical Safety, American Chemical Society, Washington, D.C., 1979.

PROBLEMS

1.1. Distinguish between the following:
   (a) molecular weight and gram formula weight;
   (b) normality and molarity;
   (c) a TC pipet and a TD pipet;
   (d) molarity and formality;
   (e) quantitative and qualitative;
   (f) class A and class B volumetric glassware.

1.2. What is the relative uncertainty of each of the following numbers: 0.2010, 1.90, 7.3183, 0.059, 5.90?

Use these uncertainties to express the answers to the following questions to the correct number of significant figures:
   (a) $0.2010 + 1.90 =$ ____.
   (b) $1.90 * 0.059 =$ ____.
   (c) $5.90 * 0.059 =$ ____.
   (d) antilog(7.3183) = ____.
   (e) log(100.2) = ____.
   (f) $0.2010/5.90 =$ ____.
   (g) $(7.3183 * 0.2010)/5.90 =$ ____.
   (h) $(7.3183 * 0.2010)/1.90 =$ ____.

1.3. The rule for carrying significant digits in multiplication and division is often simplified (incorrectly) to be "carry no more significant digits than appear in the least precise number being multiplied or divided." Show how this rule is incorrect for the following cases:
   (a) $4.82 * 0.72$.
   (b) $1.20/6.2$.

1.4. A modern macro analytical balance can measure weight with a precision of 0.0001 g up to its capacity, 160 g.
   (a) What is the relative precision of the measurement of 150 g, expressed in the ratio parts per million (ppm)?
   (b) If a relative precision of no greater than 1 part per thousand is satisfactory for the weighing step in an analysis, how small can a sample be? (In what weight of sample does 1 ppt correspond to a precision of 0.0001 g?)

1.5. Describe the technique you would use to weigh accurately 1-g samples of the following:
   (a) NaCl;
   (b) water;

(c) acetone;
(d) KOH pellets (absorb water and CO_2).

1.6. How do air convection currents change the observed weight of an object that is warmer than the air in the balance chamber? An object that is cooler?

1.7. What might cause the following phenomena?
(a) The measured weight of a finely divided salt gradually increases.
(b) The weight indicator on the balance rocks back and forth slowly.
(c) The measured weight of an empty weighing bottle gradually increases.

1.8. A sample is to be weighed in a weighing bottle. The weight of the empty bottle is already known.
(a) If a buoyancy correction must be made, is it necessary to correct for the buoyancy of the bottle as well as that of the sample?
(b) If the analyst can tolerate an error of 1 part in 10^4, is it necessary to correct for the buoyancy of a sample weighing 10 g? Assume that the density of the sample is 1.5 g/cc and that the weights are brass.

1.9. There is an old saying that improving the precision of a measurement by one digit increases the amount of work by at least a factor of 10. In the section on weighing, several potential problems were discussed, among them fingerprints, transfer of heat to weighing bottles, buoyancy corrections, and adsorption of moisture. Estimate the effects these factors might have on the accuracy and precision of weighing a 100-mg sample to either four significant digits (to the nearest 0.1 mg) or five significant digits (to the nearest 0.01 mg).

1.10. The precision of reading a buret is about 0.02 mL. What volume must be delivered to reduce the relative error arising from reading the buret to 1 part per thousand?

1.11. A chemist prepares a solution in a calibrated volumetric flask at 20°C (volume, 999.7 mL) at 8:00 a.m. By 3:00 p.m. the temperature of the laboratory, the flask, and the contents of the flask has risen to 30°C. Assuming that the solution has the same density at all temperatures as pure water, calculate the volume of the solution at 3:00 p.m.
(a) If the concentration was $0.1020M$ at 8:00 a.m., what would it be at 3:00 p.m.?
(b) Assume that the cubical coefficient of expansion of the glassware is 1.00×10^{-5} cc/°C. What is the volume of the flask itself (not the solution) at 30°C if its volume at 20°C is 1000.54 mL?

1.12. A chemist has a set of micropipets (less than 1 mL) which must be calibrated as TC pipets. Mercury is a very dense liquid (13.5939 g/mL at 20°C) which does not wet glassware (i.e., cling to the surface). What is the advantage of using mercury instead of water to calibrate micropipets? (Think of weighing.)

1.13. A 50-mL volumetric flask is filled to the mark with water at 24°C. The weight of the water is found to be 49.9657 g by using a balance with brass weights. Calculate the volume of the flask at 24°C and correct the value to 20°C.

1.14. A sample contains between 1 and 5% by weight insoluble material, which must be determined quite accurately. How much sample must be taken if at least 100 mg of insoluble material will be isolated?

1.15. A sample of a salt contains 0.10% (wt/wt) potassium iodide (GFW 166.0).
  (a) How many grams of KI are there in 56 g of the salt?
  (b) 56.0 g of the salt are dissolved in enough water to make 1 liter of solution. Assume that the density is 1.05 g/mL. What is the concentration of KI expressed as weight/weight and weight/volume ratios?
  (c) What is the formal concentration of potassium iodide in the solution in part (b)?

1.16. A 1.35×10^{-7} mole/liter solution of lead (gram atomic weight 207.2) must be prepared.
  (a) Express this concentration as a ratio of weight to volume.
  (b) Calculate the weight of lead you would have to dissolve and dilute in 1 liter of water to prepare the solution. Is it practical to weigh this amount of lead on an analytical balance? Explain.
  (c) Your laboratory is equipped with pipets and volumetric flasks. Suggest a way you could prepare the solution by starting with at least 100 mg of lead, dissolving it in nitric acid, diluting with water, and so on.

EXPERIMENT: Calibration of Volumetric Glassware

Introduction

This experiment will serve four useful purposes. First, you will learn how to use a pipet, volumetric flask, and buret for making very careful measurements of volume. Second, you will use an analytical balance to measure weights. Third, when the experiment is done you will have some well-calibrated glassware with which to perform experiments later on in your laboratory course. Finally, you will have perhaps your first set of replicate data on which to perform statistical calculations (discussed in Chapter 2).

Read the paragraphs in this chapter on cleaning and using volumetric glassware. Make sure your glassware is clean. Water droplets should not cling to the walls of the glassware. Your instructor should give you a demonstration of proper techniques for using volumetric glassware.

Water Supply

You will need a supply of water at room temperature for these calibrations. At the beginning of the laboratory period fill a 1-liter beaker with deionized water, insert a thermometer, and cover with a watch glass. Read the thermometer periodically to check for thermal equilibration with the room. A calibrated thermometer should be available to check your thermometer. You will measure volumes at whatever room temperature happens to be and then correct them to 20°C.

Calibrating the 25-mL Pipet

Begin by weighing a clean 125-mL Erlenmeyer flask to the nearest 0.1 mg on an analytical balance. Using your 25-mL pipet, carefully transfer its volume of deionized water to the Erlenmeyer flask. Weigh the flask and its contents, and convert the weight to volume by using the density of water at the experimental temperature and a buoyancy correction.

Repeat the transfer and weighing process several times. To save time, add several samples to the receiving flask without emptying the contents between additions. Use the final weight of the first sample and flask as the initial weight for the second determination. Note that the maximum capacity of most macro analytical balances is about 160 g.

Calculate the mean, standard deviation, and relative standard deviation of your volumes. See Chapter 2 for details on these calculations. Correct the volumes to 20°C, and use the mean as the volume of the pipet in future work. You should be able to obtain a precision of 1 part per thousand (relative standard deviation; see Chapter 2) for 25- and 10-mL pipets. It is more difficult to achieve this precision with smaller pipets.

Calibrating a 100-mL Volumetric Flask

The 100-mL volumetric flask must be clean and dry before it is calibrated. Use a 120°C drying oven to dry volumetric flasks; never heat a flask with a burner flame, lest you distort the flask and change its volume.

Weigh the clean, dry volumetric flask on an analytical balance. If the weight exceeds 60 g you will have to use another balance or another flask, since 160 g exceeds the capacity of most analytical macro balances. Record the weight of the flask to at least the nearest milligram (0.001 g), and return to your desk to fill the flask with room-temperature deionized water. Fill the flask to the mark. Remove any droplets of water clinging to the flask neck above the mark with a rolled up piece of filter paper. Weigh the filled volumetric flask again to the nearest milligram; record the weight and the water temperature. Calculate the weight of water and the volume corrected to 20°C.

To repeat the process, empty the volumetric flask, rinse it with some acetone, and dry it in the oven. Calculate the mean and standard deviation of your replicate data (see Chapter 2), and use the mean volume in future work. Be sure you label the flask to identify it.

Calibrating a Buret

The 50-mL buret in your laboratory desk should be calibrated at 10-mL intervals between 10 and 50 mL. The buret must be clean but need not be dry for calibration. Attach a buret clamp to a ring stand, and place your buret in the clamp. Pour deionized water at room temperature into the buret until the water level is a few centimeters above the top line (0.00 mL). Open the stopcock fully to force air bubbles from the buret tip. If necessary add more water to the buret until the top line is reached. Make sure that there are no bubbles in the buret tip and that the stopcock does not leak. If there is a leak, gently tighten the Teflon nut on the stopcock. Do not force the nut: Teflon threads strip easily. When the buret is finally ready, read the volume to ±0.01 mL. If the reading is more than 0.00 mL, do not try to add more water to reach 0.00 mL, but simply read the volume as it is. Record the volume in your notebook.

Weigh a clean but not necessarily dry 125-mL Erlenmeyer flask to at least the nearest 5 mg. Record this weight. Place the flask under the buret and open the stopcock slowly. Let out about 10 mL of water over a period of about 30 seconds. Wait another 30 seconds for water to drain down the walls, and read the volume on the buret. Touch the wall of the flask to the tip of the buret, remove the flask, and weigh to the same accuracy as before. Record the weight you measure.

Return to the buret and transfer the next 10-mL portion to the Erlenmeyer flask. Use the weight of the first 10-mL portion and flask as your initial weight for the second 10-mL portion. Dispense water at about the same rate (about 5 sec/mL), and wait 30 seconds to read the buret. Continue this process for 30, 40, and 50 mL.

Calculate the corrected volumes at 20°C for 10, 20, 30, 40, and 50 mL in the buret. You may repeat the process if time allows.

Chapter 2

Errors in Chemical Analysis and the Statistical Treatment of Data

2.1 INTRODUCTION

Few practicing chemists will deny the importance of basic statistical methods in collecting, describing, comparing, and judging the validity of experimental data. The field of statistics is quite broad, and every discipline takes methods from the field which it finds useful. In this chapter we will discuss statistical methods which are particularly important in analysis. While the coverage in one chapter cannot hope to be exhaustive, it is possible to describe a basic set of statistical tools and the logic behind their use.

Statistics are used in experimental science for description and inference. Descriptive statistics are the familiar kind of statistics we find in newspapers: a baseball player has a batting average of .427; the average automobile costs $7000. In chemistry we may describe analytical results by using a mean value and a measure of uncertainty, such as the standard deviation: "that cortisone pill contained 25.3 ± 0.5 milligrams of cortisone."

Statistical inference involves making inductive generalizations, that is, statements about a large system which are based on a small number of samples of it. For example, on the basis of the analysis of 20 vitamin C tablets drawn at random from a 5-pound lot of the tablets, we might conclude (with a certain probability) that the tablets in the lot contain 105 ± 10 mg of vitamin C per tablet. Notice that the conclusion involves a statement about probability. We try to estimate the vitamin C content of all the pills on the basis of only a small random sampling of the lot. Because there are variations from pill to pill and from one analysis to the next, we cannot be completely certain of the value and must instead speak in terms of a probability range around an estimate.

This range will allow us to have a certain amount of confidence that the true value for the tablets in the lot lies within the range.

Let us begin with some very basic descriptive statistics, including different measures of the "central value" of a set of data, simple quantitative measures of error, and the distinction between accuracy and precision. We will then examine in some detail the errors in chemical analysis which arise for definable reasons (the systematic errors), as well as those which arise simply by chance (random error). This will lead to a section on statistical inference, probability, and several tests of significance which are commonly used in experimental chemistry. The chapter concludes with sections on the propagation of errors, quality control, and a brief discussion of sampling.

2.2 ERROR AND THE CONCEPTS OF ACCURACY AND PRECISION

The terms accuracy and precision are used to describe two qualities of a set of replicate experimental data. The data are accurate if they agree well with the actual composition of the material being analyzed and precise if they are in close agreement with one another. While the best analyses produce data which are both precise and accurate, the presence of good precision by no means guarantees high accuracy, nor does a highly accurate method ensure extremely precise data.

To illustrate the difference between accuracy and precision, consider an analogy between running several replicate samples through an analytical procedure and shooting at a target with a pistol. Figure 2.1 shows four targets, each of which was shot at 10 times. The bull's-eye of each target corresponds to the *true value* (or *expected value*) in the chemical analysis. Target 1 shows the ideal outcome: all shots were in the bull's-eye and close to each other, representing highly accurate and precise shooting. In the laboratory an analogous outcome would mean that the analytical method was correct and that the analyst had used good technique. In target 2 all shots lie close to one another (precision is good) but far from the bull's-eye (accuracy is poor), a situation which is often called *bias* in the results. In target practice this outcome probably means that the gun is poorly sighted, while in laboratory work it may mean that volumetric glassware was incorrectly calibrated, a standard solution was improperly standardized, and so forth. Sources of bias in analyses will be taken up in detail later.

Target 3 shows widely scattered shots (poor precision) which nevertheless seem to center around the bull's-eye (good accuracy for the average of all shots). In the laboratory this situation might reflect a series of minor blunders in technique which cancel when large numbers of replicates are performed. The danger in relying on imprecise methods or techniques is that if only a few replicates are run (two or three, say), they may deviate in only one direction from the true value and give a mean result which is inaccurate. As you will see later, imprecise methods require large numbers of replicate analyses to ensure accuracy.

Target 4 shows the least desirable outcome: poor precision and poor accuracy.

2.2 ERROR AND THE CONCEPTS OF ACCURACY AND PRECISION

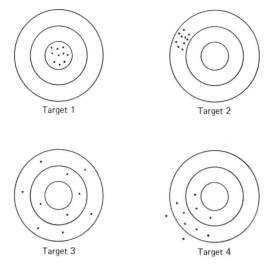

Figure 2.1 Accuracy and precision in target shooting.

The purpose of laboratory course work in analytical chemistry is to have chemists learn how to avoid this kind of outcome.

2.2.1 Simple Descriptive Statistics

Let us take target 1 and change it to a one-dimensional representation of a set of results for the analysis of the percentage of iodide in a sample known to contain 5.00% iodide (the true value, μ). In Fig. 2.2 the results are plotted in a line giving values for percent iodide. Replicate data such as these tend to cluster around a central value (the "central tendency" of the data), which can be expressed as a *mean* (arithmetic or geometric) or *median*. The central value along with a quantitative expression for precision and the number of replicate determinations constitutes the minimum amount of information needed for the proper communication of results.

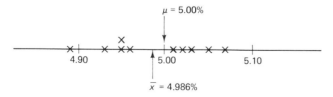

Figure 2.2 Line drawing of experimental results. Data: 4.89, 4.93, 4.95, 4.95, 4.96, 5.01, 5.02, 5.03, 5.05, 5.07%.

The mean of a set of data is given by the sum of the data divided by the number of data points, or in the general case for n results,

$$\bar{x} = \frac{x_1 + x_2 + x_3 + \cdots x_n}{n} = \frac{\Sigma_i^n x_i}{n}$$

In the case of the percent iodide data the mean value is 4.99%.

The median of a set of results is the middle result when the results are arranged in order of increasing size. When a sample contains an even number of results, such as the 10 iodide results shown in Fig. 2.2, the median is found by averaging the two central results, that is, $(4.96 + 5.01)/2 = 4.985\%$ I^-. For large samples (that is, a great number of replicate determinations) the median and mean are very nearly equal as long as the data follow what is called a *normal distribution,* which is discussed in Section 2.3.2.

The central value of a set of data is used in quantitative descriptions of accuracy. The simplest measure of accuracy is the absolute difference between the mean of the results and the true value,

$$E_{abs} = \mu - \bar{x}$$

This is called the *absolute error* and is expressed in the same units as the data. For the percent iodide data above,

$$E_{abs} = \mu - \bar{x} = 5.00 - 4.99 = 0.01\%$$

Whether an absolute error is considered serious depends on the size of the true value. For example, an absolute error of 0.1% in the determination of chloride at the 40% level may be quite acceptable. However, an absolute error of 0.1% in the determination of manganese in steel at the 0.4% level would be considered unacceptably large.

Many chemists prefer to express error relative to the true value. The *relative error* is given by

$$E_{rel} = \frac{\mu - \bar{x}}{\mu}$$

In the case of the percent iodide data,

$$E_{rel} = \frac{5.00 - 4.99}{5.00} = 0.002, \text{ or 2 ppt}$$

Relative error is dimensionless and is usually expressed in parts per thousand (ppt) or percent.

Errors are grouped as either random or systematic, as will be discussed shortly. The measures of random error called deviation and standard deviation are discussed in Section 2.3.2.

2.3 ERRORS IN CHEMICAL ANALYSIS

Errors which arise in experimental science can be conveniently grouped into two classes: *random error* and *systematic error*. These classes are generally distinguishable from one another but, as you will see, the distinction may at times be vague.

Random or *indeterminate* errors arise by chance from the minor uncertainties which are always a part of repetitive measurements, and these errors follow no regular pattern of behavior. Generally there will be about as many random errors larger than the true value (positive error) as there are random errors smaller than the true value (negative error). If random error is the only important error in a set of replicate experiments, the average of a large number of results will very closely approximate the true value.† The fact that random errors occur unpredictably makes a determination of their exact source, and therefore their elimination, impossible. Their influence can be minimized, however, by using very sensitive measuring devices when possible and by running as many replicate experiments as the costs in time and money will allow.

Systematic or *determinate* errors are errors of definite size and sign which can be traced (at least in theory) to specific sources. Systematic error is sometimes called bias, and leads to inaccuracy, even in large samples. There is an immense number of sources of systematic error in analytical chemistry, far too many to enumerate here in great detail. Many will be pointed out in the experiments in later chapters of this text. A discussion of other sources of error can be found at the end of this chapter.

2.3.1 Systematic Errors

Systematic errors can be categorized according to the ways in which they arise, either errors of the method, of the analyst, or of the equipment.

Types of Errors

Errors of the Method These are problems which exist to some extent in any analytical procedure and which arise mainly from the chemical properties of the substances. They are potentially the most serious errors in an analysis because they can be difficult to correct and control.

---

† This statement is valid if the results follow a normal distribution.

There are many examples. In gravimetric methods of analysis the objective is to isolate and purify by washing a slightly soluble compound. If the compound is a little too soluble, washing it may dissolve an appreciable quantity, producing a negative error in the analysis called *solubility loss*.

In titrimetric methods of analysis incomplete reactions (reactions with small equilibrium constants), unwanted side reactions, and improperly determined indicator blanks† may result in the addition of too much titrant and a determinate error in an analysis.

In clinical analysis of blood and urine samples one of the most important sources of systematic error is the unsuspected presence of drugs which can interfere with an analytical method. Clinical test results may be substantially raised or lowered when drugs are present. Summaries of drug effects are available in the literature (*1*). In addition, there are systematic factors such as age, sex, diurnal variations, and diet which can confound the interpretation of clinical results. Reference (*2*) should be consulted for a discussion of these factors.

There are certain methods that have been proved to involve negligible systematic error. The best of these methods are called *definitive methods* because they give "true values" with extremely high reliability. They require highly skilled analysts, expensive equipment, and are generally far too costly for routine work. These methods are preferred by the National Bureau of Standards for certification of *standard reference materials* (see below). As yet there are few definitive methods, and this is an active area of research.

Methods which have proven accuracy, as shown by comparison with definitive methods or standard reference materials, are called *reference methods*. These are generally less sophisticated and less costly than definitive method analyses, and they are most useful in periodic checks of quality control programs and in the evaluation of samples called *secondary reference materials,* that is, local references made up by individual laboratories for routine analyses. Reference methods may also be used to develop and evaluate what are called *field methods,* that is, methods routinely used to analyze large numbers of samples in industrial or clinical laboratories (often automated).

Definitive and reference methods are published by the National Bureau of Standards (NBS), the American Society of Testing and Materials (ASTM), and the Association of Official Analytical Chemists (AOAC). Standard methods for the analysis of water and wastewater are published jointly by the American Public Health Association (APHA), the American Water Works Association, and the Water Pollution Control Federation. Standard methods for air pollutants can be found in publications of the ASTM (e.g., see Recommended Reading section).

---

† When an indicator is used in a titration there are *two* reactions that take place; first, the desired chemical reaction, and second, the reaction of the titrant with the indicator to change its color. The second reaction requires a certain amount of titrant, which in some cases must be determined accurately. To do this a titration is performed containing everything except the material being analyzed. The volume measured is the *indicator blank*.

Errors of the Analyst These errors are the fault of the analyst rather than of the method and may result from carelessness in sampling, poor technique, fatigue, or lack of training and experience (the most important factors for novices). Errors in technique involve the introduction of contaminants, loss of material by spilling, allowing hot solutions to "bump" and spray sample, weighing hot weighing bottles, and even making mistakes in calculations. Practice and experience gradually eliminate most errors of technique.

A source of analyst error over which a person has much less control is what is called "personal error." Color blindness is an example. A color-blind analyst performing a titration with a visual indicator might fail to see a faint color change and add more titrant than required (a positive error). It is claimed that people may have subconscious preferences for certain numbers and may estimate an additional digit with bias when, for example, reading a buret. Beginners pressed for time in the laboratory will use the result of a first titration to calculate the approximate endpoints for successive titrations and thus risk introducing bias. It is difficult to control many of these forms of analyst error, especially when they are established behavior patterns or when they are shortcuts to save time and effort. A conscious effort must be made to avoid such errors, however, because they destroy the validity of experimental results.

Errors of Equipment and Reagents These are systematic errors which can be traced to the tools of the analyst. Instruments which are badly worn or uncalibrated, unleveled analytical balances, and uncalibrated volumetric glassware can all be sources of systematic error.

Improper use and storage of reagents can also be an important source of error. Standard solutions used in titrations should be checked routinely for changes in concentration. Bottles must be tightly stoppered to prevent evaporation of solvent or air oxidation of reducing agents. Light-sensitive materials (e.g., silver salts and solutions) should be kept out of the light, and temperature-sensitive compounds (certain biochemicals) should be refrigerated. Since many biochemicals are nutrients, it is important to keep them safe from attack by microorganisms.

A particularly important problem in the analysis of trace level contaminants involves storage of samples. Unless properly preserved by the addition of some potassium dichromate and nitric acid (to pH less than 2), metals such as mercury at parts-per-billion concentrations can be adsorbed onto the walls of the sample container in a matter of a few minutes. Preservation techniques are an important area of research in trace metal analysis at the present time.

Effects of Systematic Error on Results In addition to classifying errors by their origins, we may also classify them according to their effect on experimental results. Traditionally, errors have been placed in two categories: *constant errors* and *proportional errors*.

Constant Errors Absolute constant errors are errors that are not influenced by the size of the quantity being measured; they can be either positive or negative. An example of a negative constant error is solubility loss in gravimetric analysis. The solubility of a slightly soluble salt does not depend on how much of the salt is present (so long as there is any solid present). For example, silver chloride is soluble to the extent of about 0.2 mg in 100 mL of water, regardless of the amount of solid silver chloride present. If 0.2 mg dissolves during the washing of a silver chloride precipitate which weighs 500.0 mg, the absolute error is 0.2 mg and the relative error is 0.2/500.0 = 0.0004 or 0.04%. If, however, the precipitate weighs 50.00 mg, the absolute error is still 0.2 mg, but the relative error is 0.2/50.00 = 0.004 = 0.4%. This kind of result is perfectly general for constant errors: the absolute error remains constant when the size of the measured quantity is changed, but the relative error changes.

A positive constant error may be corrected by running a *blank,* that is, a sample containing everything except the sought substance and treated with all reagents in the same sequence as a regular sample. Indicator blanks are excellent examples. A titration is run without the sought substance and the volume of titrant required to change the indicator color is measured. The average of several replicates of this blank can then be subtracted from the volume required to titrate each sample.

Correcting negative constant errors presents a slightly different problem, since a blank determination would be negative and could not be detected. Instead, it may be possible to correct for such an error by analyzing replicate portions of *primary reference materials* or *standard reference materials* (SRM).

Proportional Errors Relative proportional errors are not influenced by the size of the quantity being measured. This kind of error is found when interfering substances are present in a sample. For example, Fe(III) interferes in the titration of Ca(II) with ethylenediaminetetraacetic acid (EDTA). If a sample contains, say, 0.1 mmole of Fe(III) for every 10 mmole of calcium, a 1% relative error can be expected regardless of the gross amount of sample taken. This can be seen from the data in Table 2.1. Notice that the absolute error increases with sample size, but the relative error remains

TABLE 2.1 RELATIVE PROPORTIONAL ERROR IN A TITRATION

Sample wt. (g)	Millimoles		0.1M EDTA (mL)	Error (mL)	Relative error (%)
	Ca^{+2}	Fe^{+3}			
1.000	5.00	0.050	50.5	.50	1.0
0.600	3.00	0.030	30.3	.30	1.0
0.200	1.00	0.010	10.1	.10	1.0

the same. Notice also that exactly the same effect would result from an improperly standardized solution of EDTA.

A great deal can be learned about the type of error present in a determination, perhaps even its source, if a series of analyses are run with different quantities of reagents, solvent, and sample. Such a process may be done in a series of carefully planned steps based on statistically sound experimental designs (*3*), but more often, especially with experienced analysts, intuition and educated guessing are used to uncover errors. If the source of an error can be found, steps may be taken to correct the problem. [In the example used here, cyanide can be added to the solution before titration to complex ("mask") the iron and keep it from interfering.]

It may be possible to correct for a proportional error by running samples of known composition called *controls*. A correction can be made if it is assumed that a proportionality exists between the quantity of material sought and the experimental result:

$$\frac{\text{quantity in unknown}}{\text{quantity in control}} = \frac{\text{result for unknown}}{\text{result for control}}$$

This method is most reliable if the ratio is close to unity, the sample and control are similar in nature (e.g., both soil or both seawater), and the conditions of analysis are identical.

It was mentioned in Chapter 1 that analysts who use instrumental methods of analysis often rely on "working curves." These are plots of an instrument signal (usually a voltage) recorded over a range of concentration of analyte, that is, the substance being analyzed. When the instrument signal and concentration are related linearly, the slope of a working curve is the proportionality constant between signal and concentration. Under ideal conditions the working curve will be a straight line with a zero intercept (a concentration of zero produces no signal). Constant and proportional errors will affect such a working curve in different ways. A positive constant error will simply raise the entire working curve without changing its slope. Figure 2.3 shows a simulated working curve for some substance over a weight range of 0 to 2.5 mg, in which there is a constant error of 0.1 mg. The intercept is the signal when the analyst has added no analyte, and is what is called a *blank signal*. The blank can be used to correct a working curve for positive proportional error.

Figure 2.4 shows the effect of a 10% proportional error on an ideal linear working curve. Positive proportional error increases the slope of the working curve, but does not change the intercept. Negative proportional error decreases the slope of the working curve.

EXAMPLE 2.1 Determinate Errors

A company is interested in purchasing a few hundred pounds of a special iron-containing alloy. One of the company's analysts is assigned the task of analyzing

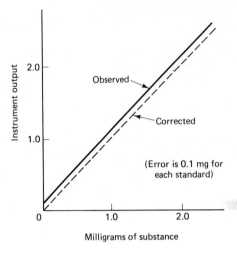

Figure 2.3 Effect of constant error on working curve.

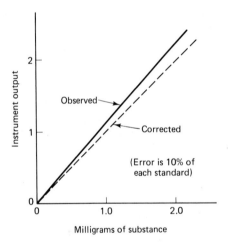

Figure 2.4 Effect of proportional error on working curve.

samples of the alloy for iron. The analyst chooses to titrate prepared samples with a standard solution of potassium dichromate and measure the endpoints with a colored indicator. The analyst standardizes the stock solution of dichromate with iron from a spool of wire from the stockroom shelf and obtains the following results:

Weight of iron wire (g)	Volume of dichromate solution (mL)
0.4350	12.21
0.4482	12.58
0.4162	11.67

The alloy samples titrated with this stock solution are all found to contain less iron than specified by the manufacturer, and the analyst's company rejects the shipment

2.3 ERRORS IN CHEMICAL ANALYSIS

of alloy. The manufacturer challenges the analysis, and the analyst begins a search for experimental errors.

To check for constant error, the analyst titrates more samples of the stockroom iron wire with the stock dichromate solution. The results are:

Iron wire (g)	Dichromate solution (mL)	Dichromate per gram of iron wire (mL)
0.2782	7.80	28.0
0.5080	14.26	28.07
0.7811	21.96	28.11
0.9108	25.52	28.02
1.2071	33.90	28.08
		mean 28.06

The nearly constant value of the ratio of milliliters of dichromate to grams of iron wire indicates that a constant error (such as an improper blank) is probably not present; a constant volume error added to each volume makes the ratio decrease as volume increases.

To check for proportional error, the analyst buys a sample of standard reference iron chips certified to be 99.99% iron and titrates several samples with the stock dichromate solution. The results are:

Iron chips (g)	Dichromate solution (mL)	Dichromate per gram of iron (mL)
0.4053	11.92	29.41
0.5133	15.11	29.44
0.4791	14.10	29.43
		mean 29.43

It is concluded that the ratio of milliliters of dichromate stock solution to grams of the stockroom spool of iron wire was too low by a factor of 0.049, or 4.9% (relative). The effect was to give results for the alloy samples which were 4.9% (relative error) too low. After renegotiation, the analyst's company is willing to accept the alloy shipment.

2.3.2 Random Errors

To understand the effect of random errors on experimental results we must first understand something about the nature of probability. We will begin by drawing a distinction between samples and the populations they represent, and by showing how titration results obtained by a quantitative analysis class distribute around a central (mean) value. We will then make the connection between probability and the distribution of data and discuss *normal distributions*. After describing in some detail several quantitative measures of the breadth of a distribution (the *dispersion* of the data), we will introduce the idea of the *confidence interval*, the foundation of statistical testing.

Samples and Populations As a first experiment in volumetric measurements a class was given a 20-liter vessel of hydrochloric acid solution. Each student was told to transfer by pipet five 20-ml portions to separate flasks and titrate each portion with a standard base solution (also provided) to the phenolphthalein endpoint. Seventeen students reported the results shown in the plot in Fig. 2.5. The 84 results make up a collection of data which can be called a *population*. A set of 5 results chosen at random from the 84 can be called a *sample* of the population.

The term *population* may have more than one meaning. It may refer to all measurements of a quantity that have been made. For example, the 84 molarity values determined for the HCl solution in the 20-liter vessel is a population which is fully known. Alternatively, the population might be taken to mean all the titration results which potentially could be obtained if all the acid in the vessel were titrated. In this case, knowledge of the population is limited by the fact that many titrations remain after 84 have been performed. A set of 5 titration results taken at random may be

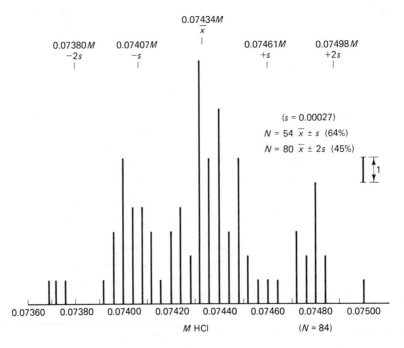

Figure 2.5 Distribution of 84 titration results.

thought of as a sample of the population of 84 titrations, or both sets of 5 and 84 titrations may be thought of as samples of a far larger population. While both kinds of populations may be dealt with in problems in chemical analysis, ordinarily population means the set of results that *might* be run if enough time and material were available, and conveys the idea of potential.

The term *sample* also has two meanings. To a statistician it is a collection of data which represents (ideally) the population. To a chemist a sample is also a material object, a quantity of material to be analyzed. In this discussion we will use the statistician's concept of sample. The question of how to take representative physical samples of material will be taken up in Section 2.3.6.

We use descriptive statistics to characterize samples and populations. Note that the 84 experimental values cluster around a middle value. We call this clustering the *central tendency* of the data, and use the term *sample mean* to describe it quantitatively. The closeness of the clustering is called the *dispersion* of the sample, and we use the term *standard deviation* to describe it quantitatively. Dispersion in a sample such as the 84 titration data is the result of random errors of different sizes. From the figure it should be clear that small random errors are much more likely to occur than are large random errors. Let us explore the connection between the sample distribution and probability.

Probability and the Distribution of Experimental Data The relationships between a sample and a population involve concepts of mathematical probability. The following ideas should help you recall some of these concepts.

In ordinary day-to-day situations we speak of probability in terms of the likelihood of an event of a certain kind occurring out of the set of all possible events. For example, if a fair coin is flipped there are *two* possible outcomes: it will land either "heads" up or "tails" up. The probability of the coin coming up heads is 1/2 or 50% (one outcome in the set of two possible outcomes). The probability of it coming up tails is also 1/2 or 50%. Since the coin must show heads or tails after being flipped, the probability of it coming up heads *or* tails is $\frac{1}{2} + \frac{1}{2} = 1$, or a certainty. Coin flipping is an example of classical probability, that is, a situation where all the outcomes are known.

In the statistical treatment of experimental data, the term probability is used in a slightly different sense. Since the entire set of outcomes (the population) is not known accurately, the probabilities of certain kinds of outcomes cannot be calculated accurately. What is done instead is to define the probability of one type of outcome as the ratio of the number of results of one type to the total number of results determined,

$$p = \frac{N_1}{N_{\text{total}}}$$

What is then needed is some assurance that for large values of N_{total} the statistical probability value approaches the classical probability value. In the statistical analysis of chemical data this assurance is found in the fact that most chemical experiments produce data which approximately conform to one of only a few model probability distributions.

Consider the possible sets of outcomes which can be expected when two, three, or four fair coins are flipped. When a single coin was flipped, the probability of heads (let us say a "positive" outcome) was exactly the same as that of tails (a "negative" outcome). When two coins are flipped there are three possible outcomes: heads-heads (positive-positive), heads-tails and tails-heads (both positive-negative), and tails-tails (negative-negative). There are four outcomes when three coins are flipped, and five outcomes when four coins are flipped. These possibilities are shown in Fig. 2.6 along with their corresponding probabilities. You may recognize this distribution as the *binomial distribution.* The number of outcomes of each type is given by the coefficients of the expression obtained when a function such as $(x + y)^4$ is expanded:

$$(x + y)^4 = \underline{1}x^4 + \underline{4}x^3y + \underline{6}x^2y^2 + \underline{4}xy^3 + \underline{1}y^4$$

The binomial distribution is symmetric, meaning that positive outcomes are as likely as negative outcomes. Furthermore, outcomes in which there are about as many heads as tails are far more common than outcomes in which either heads or tails predominate. Thus when four fair coins are flipped the probability of all four coins being tails (negative) is only 0.0625 (6.25%). A far more likely outcome is that two coins are heads and two tails (probability 37.5%). When 10 fair coins are flipped the probability of all coming up heads is only 0.000977.

The binomial distribution is important to experimentalists because it is typical of experiments in which positive and negative deviations are equally likely, that is,

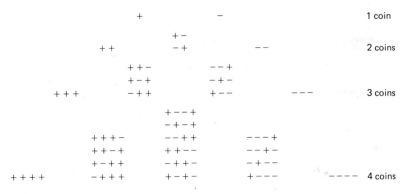

Figure 2.6 Coin flipping and the binomial distribution.

2.3 ERRORS IN CHEMICAL ANALYSIS

the kind of pattern expected from the effects of random error in measurements. For extremely large sets of data, in fact, the binomial distribution is quite similar to the *normal* or *Gaussian distribution,* shown in Fig. 2.7. This distribution is described by a formidable-looking equation,

$$f(x_i) = \frac{1}{\sigma\sqrt{2\pi}} \exp\left[-\frac{1}{2}\left(\frac{x_i - \mu}{\sigma}\right)^2\right]$$

In this expression the frequency of a particular outcome, $f(x_i)$, is related to the true value, μ, and the standard deviation of the population, σ. The notation "exp" means that the number which follows is raised to a power of e, the base of natural logarithms (e is approximately 2.7183). The number π is approximately 3.14159.

The normal distribution has some very important attributes. First, the most

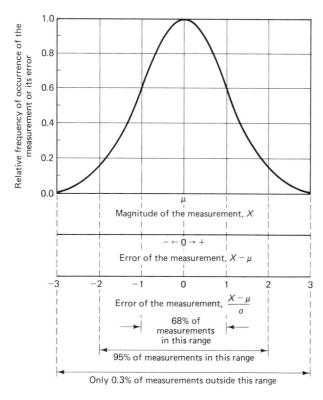

Figure 2.7 The normal distribution curve. (From W. J. Blaedel and V. W. Meloche, *Elementary Quantitative Analysis,* 2nd ed., Harper & Row, New York, 1963.)

frequently occurring element of the population has the same value as the mean. The normal curve equation predicts that the frequency of this element will be

$$f(x) = (1/\sigma\sqrt{2\pi}) \exp(-0) = 1/\sigma\sqrt{2\pi} \quad \text{(for } x_i = \mu\text{)}$$

and indicates that the frequency of the mean is inversely proportional to the standard deviation. The normal curve also has two symmetrically opposed inflection points (points where the slopes are a maximum), which are used as a measure of width and define the standard deviation of the population. The curve continues to fall off in both directions away from the mean, approaching (but in theory never reaching) the line $f(x) = 0$.

The area under the distribution curve contains the total number of elements in the population, and this invites questions about probabilities. What, for example, is the probability that an element of the population chosen at random (or an observation made when only random error affects a process) will lie within ±1 standard deviation of the mean value? Since 68.27% of a normally distributed population lies within this region, it follows that the probability is 0.6827. From a slightly different perspective, since 99.73% of the population lies within ±3 standard deviations of the mean value, the probability of selecting at random an element outside that range is less than 0.003 (3 parts per thousand). A most interesting question for students in laboratory is, given a set of three elements of a population, what is the probability that the mean of the set lies within ±2 standard deviations of the population mean? We will discuss a way to answer that question in a subsequent section.

It is necessary to somehow standardize normal distributions if we wish to compare two or more populations which involve different units of measurement and different standard deviations. Figure 2.8 shows two hypothetical normal distribution curves arising from populations with different standard deviations. While they are both normal distributions, one could hardly guess it from looking at them. However, by transforming the x-axis units from specific units (molarity, percent copper, test scores in percent, etc.) to units of multiples of the standard deviation, we reduce the two curves to a single *standard* normal curve. The factor z is used as a multiplying factor for the standard deviation and is defined as

$$z = (x_i - \mu)/\sigma$$

Including the z-factor changes the equation for the normal curve to

$$f(z) = (1/\sqrt{2\pi}) \exp(-z^2/2)$$

(see Ref. *4* for a detailed derivation). The horizontal axis is then expressed in units of z, and a standard curve is obtained, as shown in Fig. 2.7. The z-factor is quite useful in discussions of probability, where it is often called the "z-score," and it will be mentioned again shortly.

2.3 ERRORS IN CHEMICAL ANALYSIS

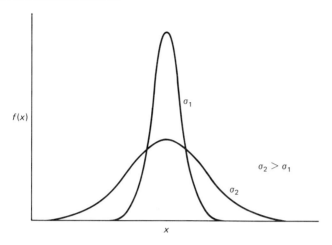

Figure 2.8 Two normal curves with different standard deviations.

Figure 2.9 shows the 84 experimental molarities mentioned previously plotted on a relative frequency scale† and superimposed on a normal distribution curve. The normal curve is drawn assuming that the mean and standard deviation of the population and the sample are the same. Clearly the normal curve is an idealized case and is approximated only roughly by a sample even as large as $n = 84$. The normal curve is asymptotic, a characteristic we would not expect to find in a real distribution. In real distributions we generally deal with data within $\pm 3\sigma$ of the mean value. The normal curve is also continuous, while all real experimental results will give only a limited number of discrete values for the variable. Note in Fig. 2.9 that the molarity values reported never differ by less than $0.00004M$.

Measures of Dispersion When treating experimental results in analytical chemistry we seldom deal with a fully known population. As a result, the population parameters μ and σ are never really known; they can only be approximated by parameters of the sample, the mean ($\bar{x}$), standard deviation (s), and the range (R). To show how these parameters are calculated, we will draw at random 5 molarity values from the sample of 84 titration results, $0.07454M$, $0.07432M$, $0.07424M$, $0.07428M$, and $0.07450M$. The mean of these five values is $0.07438M$. Recall that the mean of the set of 84 results was $0.07434M$.

Range The range of a sample is the difference between the largest and smallest data in the set. The range of the sample of 5 molarities is $(0.07454 - 0.07424) = 0.00030M$. Some very useful statistical tests make use of the range. The range can

† The number of data points at each molarity is divided by 84 and plotted as a percentage.

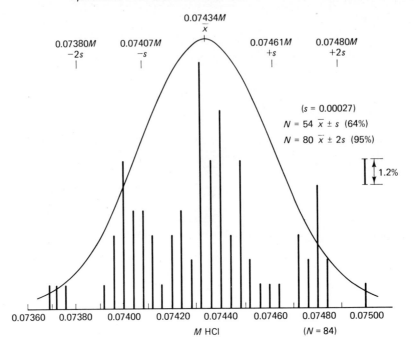

Figure 2.9 Eighty-four titration results and superimposed normal curve.

also be used to approximate the standard deviation of a set of data, as will be shown later in this section.

Deviation The difference between a single result and the mean of a sample is called the deviation of a result and is given by

$$d = \bar{x} - x_i$$

or its absolute value. Notice that the deviation has the same units as the mean. The deviation of a result from the true value was referred to above as the *absolute error* of the result,

$$E_{\text{abs}} = \mu - x_i$$

The *average deviation* of a sample is given by the sum of the absolute values of the deviations divided by the number of results,

$$\bar{d} = \frac{|\bar{x} - x_1| + |\bar{x} - x_2| + \cdots + |\bar{x} - x_n|}{N}$$

$$= \frac{\Sigma|\bar{x} - x_i|}{N}$$

Although the average deviation was used extensively in the past, it is almost never referred to in the current literature.

2.3 ERRORS IN CHEMICAL ANALYSIS

Variance and Standard Deviation of a Sample These measures of precision make use of the sum of the squares of the deviations of the results from the sample mean. For a sample containing N data, the variance is given by

$$\text{variance} = s^2 = \frac{\Sigma(\bar{x} - x_i)^2}{N - 1}$$

where $N - 1$ is the number of *degrees of freedom* (d.o.f.) of the sample data.† An equivalent expression which simplifies calculations is

$$s^2 = \frac{\Sigma x_i^2 - N(\bar{x})^2}{N - 1}$$

The variance of the population is given by

$$\sigma^2 = \frac{\Sigma(\mu - x_i)^2}{N}$$

This expression can be used when the true value is known, or when $\bar{x}$ is known to be a very close approximation of μ. Since the mean is known, it need not be calculated from the data, and a degree of freedom is not lost.

The standard deviation of a sample is given by the square root of the variance of the sample,

$$s = \left[\frac{\Sigma(\bar{x} - x_i)^2}{N - 1}\right]^{1/2}$$

The standard deviation is the most commonly used measure of dispersion (precision) for chemical data. The calculations of the variance and standard deviation of the 5 molarity data are summarized in the following table:

Result (M)	$(\bar{x} - x_i)$	$(\bar{x} - x_i)^2$
0.07424	+0.00014	1.96×10^{-8}
0.07428	+0.00010	1.0×10^{-8}
0.07432	+0.00006	0.36×10^{-8}
0.07450	−0.00012	1.44×10^{-8}
0.07454	−0.00016	2.56×10^{-8}
mean 0.07438	sum	7.32×10^{-8}

$$s^2 = \frac{7.32 \times 10^{-8}}{4} = 1.8 \times 10^{-8}$$

$$s = 0.00014\ M$$

---

† A set of 5 results is said to have 5 d.o.f. for computations. When the variance of a sample is calculated it is necessary to calculate the mean of the sample from the original 5 d.o.f. and then use the mean to calculate deviations. The data are thus used twice to calculate the variance, and one degree of freedom must be subtracted. This is sensible if you realize that the mean value and any 4 of the results will automatically determine what the fifth result must be. Thus calculating the mean costs 1 d.o.f.

A crude estimate of the standard deviation of a small sample can be obtained from the range, R, and the number of data points, N,

$$s \approx R/\sqrt{N} \quad \text{for} \quad 3 \leq N \leq 12$$

This approximation might be useful when you are working busily in a laboratory and need a quick answer. It should not be relied upon for final calculations.

The *relative standard deviation* of the sample is calculated from the standard deviation by dividing s by the sample mean. For the HCl molarity data,

$$s_{rel} = \frac{s}{\bar{x}} = \frac{0.00014M}{0.07438M} = 0.0019$$

or 1.9 parts per thousand.

It often appears difficult to compare the precision of two sets of data simply because they are expressed in different units. The use of relative standard deviations allows such comparisons to be made.

EXAMPLE 2.2 Comparing Precision

In the analysis of a portion of a sample for percent calcium oxide (CaO), a student standardizes a solution of EDTA in triplicate and then titrates six portions of the sample. The results are shown below:

Molarity of EDTA solution (standardization)	Percent of CaO by titration
0.1002	4.19
0.1014	4.14
0.1000	4.16
$\bar{x} = 0.1005M$	4.22
	4.17
	4.20
	$\bar{x} = 4.18\%$ CaO

Compare the precision of the standardization data and titration data.

Solution

Calculate the standard deviations for both sets of results:

$$\text{standardization:} \quad s = 7.6 \times 10^{-4}M$$
$$\text{titration \% CaO:} \quad s = 2.9 \times 10^{-2}\%$$

The standard deviation of the standardization data looks like a smaller number, but a comparison is impossible because the units are different.

Calculate the relative standard deviations of the data sets:

$$\text{standardization:} \quad \frac{s}{\bar{x}} = 7.6 \times 10^{-3} \text{ or } 7.6 \text{ ppt}$$

$$\text{titration \% CaO:} \quad \frac{s}{\bar{x}} = 6.9 \times 10^{-3} \text{ or } 6.9 \text{ ppt}$$

The relative standard deviations show the percent CaO data to be slightly more precise than the standardization data.

Precision of the Mean: Standard Deviation of the Mean A convenient and commonly used measure of the precision of the mean is the standard deviation of the mean, sometimes called the *standard error of the mean*,

$$s_{\bar{x}} = \frac{s}{\sqrt{N}}$$

The precision of the mean increases with the number of determinations, although the increase becomes more gradual when large samples are made even larger. For example, the precision of the mean doubles if 8 determinations are used instead of just 2. To double the precision of the mean again, it would be necessary to run 32 determinations $[(32)^{1/2} = 2(8)^{1/2} = 4(2)^{1/2}]$. The standard deviation of the mean will be used later in a discussion of confidence intervals and the comparison of experimental means.

EXAMPLE 2.3 Standard Deviation of the Mean

Calculate the standard deviation of the mean for the two sets of titration data examined in the previous example.

Solution

Taking the results of Example 2.2

$$\text{standardization:} \quad s = 7.6 \times 10^{-4} M$$
$$\text{titration \% CaO:} \quad s = 2.9 \times 10^{-2} \%$$

we calculate the standard deviations of the means to be

$$\text{standardization:} \quad s_{\bar{x}} = \frac{s}{\sqrt{N}} = \frac{7.6 \times 10^{-4} M}{\sqrt{3}} = 4.4 \times 10^{-4} M$$

$$\text{titration \% CaO:} \quad s_{\bar{x}} = 1.2 \times 10^{-2} \%$$

To compare the precision of the means, calculate the relative standard deviation of each mean,

$$\text{standardization:} \quad s_{\bar{x}\,\text{rel}} = \frac{4.4 \times 10^{-4}}{0.1005} = 4.4 \text{ ppt}$$

$$\text{titration \% CaO:} \quad s_{\bar{x}\,\text{rel}} = \frac{1.2 \times 10^{-2}\%}{4.18\%} = 2.9 \text{ ppt}$$

The mean of the percent CaO titration data is more precise than that of the standardization data.

Nonnormal Distributions Statistical tests based on a normal distribution of the population are applied to real samples *assuming* that the samples are normally distributed. While approximately normal distributions may be expected from many experiments, some experiments produce data whose distributions have peaks which may be either abnormally sharp or abnormally broad. Furthermore, certain experiments tend to produce data which lack symmetry about the mean, where, for example, there are more positive errors than negative errors. This may occur in titrations with visual indicators, where the tendency is to slightly overtitrate (to be certain of the color change) rather than to slightly undertitrate. The result will be that for large samples the median and mean values will be quite different. Asymmetric distributions are said to be *skewed*.

Lognormal Distribution A special skewed distribution which is found in trace level analyses is called the *lognormal* distribution. This kind of distribution may be found when results are taken near an instrument's limit of detection and when negative results are impossible. The arithmetic mean of a lognormal distribution is a poor estimator of the true value, μ. A better estimator is the *geometric mean,* the nth root of the product of n results:

$$\bar{x}_g = \sqrt[n]{x_1 \cdot x_2 \cdot x_3 \cdot \,\cdots\, \cdot x_n}$$

There are two different standard deviations for a lognormal distribution, one for positive deviations from $\bar{x}_g$, the other for negative deviations. The standard deviation can be determined from the expression

$$\log s = \pm \sqrt{\frac{\Sigma(\log x_i - \log \bar{x}_g)^2}{N-1}} = \pm \sqrt{\frac{\Sigma[\log(x_i/\bar{x}_g)]^2}{N-1}}$$

2.3 ERRORS IN CHEMICAL ANALYSIS

EXAMPLE 2.4 Lognormal Data

Ten results were obtained for the determination of mercury in a sample of industrial effluent at the parts-per-million level (milligrams of Hg per kilogram of water) by a method known to give lognormally distributed data. The results are tabulated with several steps of the determination of the standard deviation. Calculate the upper and lower standard deviations of the data.

Concentration Hg (ppm)	$x_i/\bar{x}_g$	$\log(x_i/\bar{x}_g)$	$[\log(x_i/\bar{x}_g)]^2$
1.4	0.73	−0.14	.0196
1.5	0.79	−0.10	.010
1.8	0.94	−0.03	.0009
2.3	1.21	0.08	.0064
1.7	0.89	−0.05	.0025
1.6	0.84	−0.08	.0064
2.9	1.52	0.18	.0324
2.3	1.21	0.08	.0064
1.6	0.84	−0.08	.0064
2.5	1.31	0.12	.0144
$\bar{x}_g = 1.9$ ppm		sum	0.1054

Solution

The log of the standard deviation is given by

$$\log s = \pm \sqrt{\frac{\Sigma[\log(x_i/\bar{x}_g)]^2}{N-1}}$$

$$\log s = \pm \sqrt{\frac{0.1054}{9}} = \pm 0.1082$$

Taking the antilog of both values gives the standard deviations:

$$s = 10^{+0.1082} = 1.28 \text{ ppm (upper } s\text{)}$$

$$s = 10^{-0.1082} = 0.78 \text{ ppm (lower } s\text{)}$$

Refer to Fig. 2.10 to interpret these results.

Several less important distributions are encountered in analytical chemistry. A distribution which is elongated toward low values may be inversely normal; that is, the inverse of a measured quantity may be normally distributed. "Compound" distributions may also be encountered when populations contain samples with different

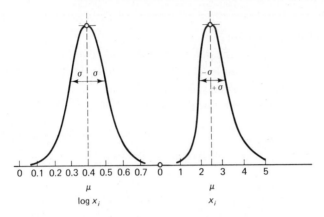

Figure 2.10 Normal and lognormal distributions. [From K. Eckschlager, in *Errors, Measurement and Results in Chemical Analysis* (R. A. Chalmers, ed.), Van Nostrand Reinhold, London, 1969, p. 104.]

characteristics. For example, grouping together the results of analyses performed on portions of a batch of iron ore by two different methods may (in an extreme case) generate a population distribution with two peaks. Obviously, it would be quite difficult to calculate reliable statistical parameters from a compound distribution. An interesting discussion of nonnormal distributions can be found in Reference (2).

Student t Distribution   The most important nonnormal distribution is called the Student t.† It is most useful to people who, like chemists, work with small numbers of samples. In such small sample cases, $\bar{x}$ and s are rather unreliable estimators of the population parameters μ and σ. In particular, statistics based on normal distributions tend to underestimate the standard deviation of a population represented by a small number of samples. To compensate for this underestimation, the Student t distribution allows for a greater frequency of results with large deviations. It also changes shape with sample size, becoming essentially identical to a normal distribution for sample sizes larger than about 30. Figure 2.11 compares Student t and normal distributions. In the following sections on confidence intervals and hypothesis testing we will make great use of the t distribution.

The Confidence Interval   The analyst who performs an analysis on triplicate portions of a sample is faced with a dilemma: in the absence of systematic error the true value of the sought quantity is only *approximated* by the mean of the experimental results.

---

† This distribution was the creation of W. S. Gosset, who published under the pen name "Student." Gosset was a chemist employed by the Guiness Brewery in Dublin around 1900. His article appeared in *Biometrika*, vol. 6, p. 1 (1908).

2.3 ERRORS IN CHEMICAL ANALYSIS

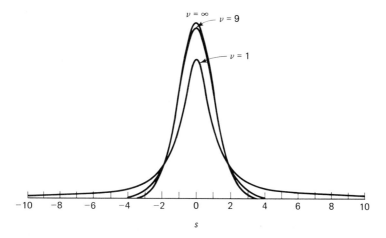

Figure 2.11 The t distribution for 1, 9, and ∞ degrees of freedom. (From G. E. P. Box, W. G. Hunter, and J. S. Hunter, *Statistics for Experimenters*, p. 50, © John Wiley & Sons, Inc., New York, 1978. Reprinted by permission of John Wiley & Sons, Inc.)

What kind of confidence can the analyst have in the mean of the results, particularly when the results are imprecise? The following questions have arisen from this kind of situation:

1. What are the chances (the probability) that μ lies within a certain interval (say $\pm 0.1\%$ or $\pm 0.5\%$) of the experimental mean? This question concerns the degree of confidence (the *confidence level*) one may have in the accuracy of results if only random error is present. If the interval around $\bar{x}$ is small compared to the standard deviation of the mean of the results, the probability is relatively low that μ will lie within it. However, if the interval is large compared to the standard deviation of the mean, the probability is high that μ will lie within it. The *risk* that μ lies outside the interval is designated by α. The probability that μ lies inside the range is therefore given by $1 - \alpha$.

2. How large is the interval over which the probability of finding μ is 0.95? This question concerns the size of the *confidence interval*. When the standard deviation of the mean of the results is small, this interval is small.

A confidence interval can be calculated by finding in a table the correct number of standard deviations of the mean, as represented by either of the coefficients z or t, depending on the situation. The coefficient z is used when the standard deviation of the mean of the population is known,† giving

---

† The population is assumed to be normally distributed. The coefficient z was seen earlier. Recall that in a normal distribution 95.4% of the data lie in the range $\mu \pm 2\sigma$. In that case $z = 2$.

$$\text{confidence interval} = \bar{x} \pm z(\sigma/\sqrt{N})$$

In most cases small samples are used and the population parameters are unknown, and σ must be estimated by s. It is then more accurate to assume that the data fit a Student t distribution and calculate the confidence interval using the t coefficient,

$$\text{confidence interval} = \bar{x} \pm t(s/\sqrt{N})$$

Table 2.2 gives z and t values at several confidence levels. To use the t tables it is necessary to choose a confidence level at which to work and then find the value of t which corresponds to the number of degrees of freedom in the sample. When the estimator s is calculated for N results, $N-1$ degrees of freedom remain, and $N-1$ must be used to find the correct t. Since the z values are independent of the size of the sample, only the confidence level must be decided to use the z table.

EXAMPLE 2.5 Confidence Interval

Sodium can be determined in human serum by an instrumental method called flame emission spectroscopy. An analyst found the following results for a sample of serum:

$$\text{millimoles Na}^+/\text{liter} = 140, 133, 144, 148, 132$$

What is the range of values in which the true value will lie with 95% probability, that is, the 95% confidence interval?

Solution

The mean and standard deviation of the mean of these results are $\bar{x} = 139$ mmole/liter and $s_{\bar{x}} = 3.09$ mmole/liter calculated as shown above. The confidence interval is calculated from

$$\text{c.i.} = \bar{x} \pm t\left(\frac{s}{\sqrt{N}}\right) = 139 \pm (2.78)(3.09 \text{ mmole/liter})$$

$$= 139 \pm 8.6 \text{ mmole/liter}$$

The true value (μ) should lie in the range 139 ± 8.6 mmole/liter with 95% probability. The risk that μ lies outside this range is 5%.

The confidence level of 95% is the conventional level at which decisions are made in statistical analysis. Most analysts find 5% risk levels acceptable. If greater

2.3 ERRORS IN CHEMICAL ANALYSIS

TABLE 2.2 LIMITS OF THE CONFIDENCE INTERVAL AROUND $\bar{x}$, WITHIN WHICH μ LIES[a]

$\mu = \bar{x} \pm z(\sigma/\sqrt{N})$ (σ known) $\mu = \bar{x} \pm t(s/\sqrt{N})$ (σ unknown)

Confidence level, or probability that μ lies in the stated interval

Degrees of freedom	80%		90%		95%		97.5%		99%	
	z	t	z	t	z	t	z	t	z	t
1	1.28	3.08	1.64	6.31	1.96	12.71	2.24	25.5	2.58	63.7
2		1.89		2.92		4.30		6.21		9.92
3		1.64		2.35		3.18		4.18		5.84
4		1.53		2.13		2.78		3.50		4.60
5		1.48		2.02		2.57		3.16		4.03
6		1.44		1.94		2.45		2.97		3.71
7		1.42		1.90		2.36		2.84		3.50
8		1.40		1.86		2.31		2.75		3.36
9		1.38		1.83		2.26		2.69		3.25
11		1.36		1.80		2.20		2.59		3.11
13		1.35		1.77		2.16		2.53		3.01
15		1.34		1.75		2.13		2.49		2.95
17		1.33		1.74		2.11		2.46		2.90
19		1.33		1.73		2.09		2.43		2.86
25		1.32		1.71		2.06		2.39		2.80
29		1.31		1.70		2.04		2.36		2.76
39		1.30		1.68		2.02		2.33		2.70
60		1.30		1.67		2.00		2.30		2.66
120		1.29		1.66		1.98		2.27		2.62
∞		1.28		1.64		1.96		2.24		2.58
	z	t	z	t	z	t	z	t	z	t
	20%		10%		5%		2.5%		1%	

Risk, or probability that μ lies outside the stated interval

[a] Values may also be used in two-tailed tests of significance.

Source: Based on a table in W. J. Blaedel and V. W. Meloche, *Elementary Quantitative Analysis,* 2nd ed., Harper & Row, New York, 1963, p. 622.

certainty is desired (say 99%) either the acceptable range must be increased, or the number of analyses increased.

EXAMPLE 2.6 Confidence Interval

The analyst who performed the serum sodium determinations needs to be more certain of the range in which the true value will fall. Calculate the 99% confidence interval for the data.

Solution

Calculate the 99% confidence interval with $t = 4.6$ (4 d.o.f.).

$$\text{c.i.} = \pm t\left(\frac{s}{\sqrt{N}}\right) = \pm(4.60)(3.09 \text{ mmole/liter})$$

$$= \pm 14.2 \text{ mmole/liter}$$

Notice that this interval is larger than the 95% confidence interval. This is the expected result: the more *confidence desired,* the wider the interval must be.

Another way to calculate the confidence interval, which is also appealing in its simplicity, is based only on the range of a set of results. The calculation is done with a single step:

$$\text{c.i.} = \bar{x} \pm KR$$

where R is the range of the results and K is a constant which, like the Student t, is a function of the number of results and the confidence level chosen. Table 2.3 gives K values at the .95 and .99 confidence levels.

EXAMPLE 2.7 Range-based Confidence Interval

Calculate the 95% confidence interval for the data in Example 2.5, using the range-based method.

Solution

The range is $148 - 132 = 16$ mmole/liter. Using $K = 0.51$ for $N = 5$ results, the confidence interval is calculated to be

2.3 ERRORS IN CHEMICAL ANALYSIS

$$\text{c.i.} = \pm KR = \pm 0.51(16 \text{ mmole/liter}) = \pm 8.2 \text{ mmole/liter}$$

Notice that this confidence interval is somewhat smaller than that calculated with the Student t value.

Statistical Testing Chemical analysts are often faced with the problem of making objective comparisons between bulk quantities of material (or large populations) on the basis of results obtained from the analysis of small samples. The kinds of situations that arise can be illustrated by the following hypothetical cases:

1. An analyst for the Food and Drug Administration must decide if a sample of a sulfa drug actually contains the promised amount of active ingredient. Does the mean of five analyses differ in a statistically significant way from the value printed on the bottle label (the expected value)?
2. A clinical laboratory manager wants to know if the values determined for sodium in a pooled serum sample are significantly different from the values reported by a different laboratory.
3. In perfecting a new method of analyzing samples of blood for ethanol, an analyst must decide if the mean values obtained for a single specimen by two different methods are statistically different.

TABLE 2.3 VALUES OF THE COEFFICIENT K FOR THE RANGE-BASED CONFIDENCE INTERVAL

	Confidence level	
(Results) N	0.95	0.99
2	6.4	31.8
3	1.3	3.01
4	0.92	1.32
5	0.51	0.84
6	0.40	0.63
7	0.33	0.51
8	0.29	0.43
9	0.26	0.37
10	0.23	0.33

Source: From R. B. Dean and W. J. Dixon, *Analytical Chemistry,* 23, 636 (1951).

In all three cases a judgment must be made between the two mean values. Is the difference between the means "small enough" to be explained by random errors in the analyses? Is the difference "large enough" to indicate that a batch of some product should be withdrawn from the market, or that a laboratory procedure should be revised? A goal of statistical analysis is to be able to substitute objective criteria for terms like "small enough" and "large enough."

To test two mean values we begin by making a hypothesis that there is really no difference between them (a *null hypothesis*). The hypothesis can be made in two ways:

1. "There is no statistically significant difference between the means." This kind of hypothesis allows us to deal with both extremes ("tails") of the distribution of the population in what is called a *two-tailed* test. Another way to state it is, "One mean is neither significantly larger nor smaller than the other." We will use this kind of hypothesis most often in comparing means.
2. "The mean of one sample is not significantly larger than the mean of the other sample." This hypothesis applies only to one tail of the distribution and is used in a *one-tailed* test. This kind of test will be used later in the F-test of precision.

In effect, the null hypothesis says that the means belong to the same population. Implicit is the assumption that the variances of the two samples are equal. This can be verified by applying the F-test to the variances, as shown in a later section about the use of the F-test in comparing the precision of two methods. If the variances are significantly different, a modified method of testing may be used, as will be discussed below.

In chemical analysis statistical samples are usually small, and so tests based on the Student t distribution are commonly used to test hypotheses. The mathematical expressions used are derived from the confidence interval expression for the mean of each sample. Using the numerical difference between the means we calculate a t value and then make a judgment about the t value which lets us decide to accept or reject the hypothesis.

If we reject a hypothesis we must recognize that there is a risk that we are making the *wrong* decision. This risk is called the *level of significance* of the test and is denoted by α. It is customary to use levels of significance of 0.05 or 0.01 in the analysis of experimental data, although other levels are also used. If a 0.05 level of significance is reached in testing a hypothesis, for example, there is only a 5% risk that we would reject the hypothesis when we should accept it. In other words, we are *95% confident* that our decision to reject the hypothesis is correct. A more conservative analyst might prefer to make decisions at the 99% confidence level (1% level of significance); here the risk of error in rejecting the hypothesis is only 1 in 100. Figure 2.12 shows 0.05 significance levels for one- and two-tailed tests based on the Student t distribution.

2.3 ERRORS IN CHEMICAL ANALYSIS

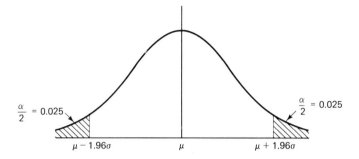

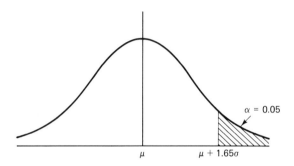

Figure 2.12 The 0.05 significance levels for one- and two-tailed tests.

Let us consider more carefully two important situations. The first involves comparing a sample mean with an expected value. The second involves comparing the means of two samples.

Comparing the Mean of a Sample with the Expected (True) Value In this application of the t-test the question is whether a statistically significant difference exists between a sample mean and the expected value. We start by postulating that there is no significant difference (the null hypothesis) and use the expression

$$t = \frac{(\mu - \bar{x})\sqrt{N}}{s}$$

where s is the standard deviation of the sample of N results.† We then compare this

† This expression comes from a mathematical statement of the null hypothesis

$$\mu = \bar{x} \pm ts/\sqrt{N}$$

which says that μ lies within the confidence interval of the sample mean.

calculated value of t with tabulated values at different confidence levels for $N - 1$ degrees of freedom† and make a judgment about the validity of the null hypothesis.

EXAMPLE 2.8 μ vs. $\bar{x}$

Twenty tablets of a commercial preparation of the sulfa drug sulfamethoxazole are crushed and ground to a powder. Five portions are weighed and titrated with standard sodium nitrite solution. The titration results are expressed as milligrams of sulfamethoxazole and percent sulfamethoxazole and are summarized below.

Weight of sample (mg)	Sulfamethoxazole (mg)	Sulfamethoxazole (%)
720.5	502	69.7
675.2	467	69.2
703.4	489	69.5
683.8	479	70.1
692.3	483	69.8

The manufacturer's specification for the preparation was 70.0% sulfamethoxazole. Does the sample of the 20 tablets meet the specification?

Solution

The question can be restated: "Is there a statistically significant difference between the mean value of 5 results, 69.7%, and the expected value, 70.0%?"

We begin by hypothesizing that there is no difference and test the hypothesis by the t-test. The standard deviation of the sample is calculated to be 0.34%. Using the expression for this kind of test,

$$t = \frac{(\mu - \bar{x})\sqrt{N}}{s} = \frac{(70.0 - 69.7)\sqrt{5}}{0.34} = 1.97$$

Now we consult the t table for various confidence levels at $N - 1 = 4$ degrees of freedom,

$$\begin{array}{lccccc} \alpha: & 0.20 & 0.10 & 0.05 & 0.025 & 0.01 \\ t\ (4\ \text{d.o.f.}): & 1.53 & 2.13 & 2.78 & 3.50 & 4.60 \end{array}$$

† A calculated mean value appears in the expression, and a degree of freedom must be lost.

2.3 ERRORS IN CHEMICAL ANALYSIS

The calculated t value (1.97) corresponds to a significance level of about 0.15. The probability that a difference this large or larger would arise from random error is thus about 15%. We can accept the null hypothesis at the 15% significance level. The manufacturer's specification is met.

Comparing Two Sample Means The Student t distribution is useful in deciding whether a difference between two mean values obtained from small data sets is statistically significant. Say there are two sets of data, set A with N_a results and set B with N_b results, mean values represented by $\bar{x}_a$ and $\bar{x}_b$, and standard deviations represented by s_a and s_b. Starting with the null hypothesis, we can write $\mu_a = \mu_b$

or

$$\bar{x}_a \pm \frac{ts_a}{\sqrt{N_a}} = \bar{x}_b \pm \frac{ts_b}{\sqrt{N_b}}$$

The null hypothesis means that $\bar{x}_a$ and $\bar{x}_b$ both arise from the same population and are both estimates of a single value, μ. If this is the case, it must also be true that s_a and s_b are both estimates of the population parameter, σ. When there is no reason to believe that s_a and s_b are significantly different from either the context of the problem or a test of precision (see the F-test, below), then we can pool the standard deviations of the sets by taking a weighted average:

$$s_{\text{pool}} = \left[\frac{(N_a - 1)s_a^2 + (N_b - 1)s_b^2}{N_a + N_b - 2} \right]^{1/2}$$

We can compare the mean values from the data sets by calculating a t value from the relationship

$$t_{\text{calc}} = \frac{|\bar{x}_a - \bar{x}_b|}{s_{\text{pool}}[(N_a + N_b)/N_a N_b]^{1/2}}$$

and then comparing the calculated value with tabulated values at particular significance levels for $(N_a + N_b - 2)$ degrees of freedom.

EXAMPLE 2.9 Comparing Two Means

An analyst is developing a method for the determination of calcium in serum by using atomic absorption spectroscopy and wishes to test whether the addition of ethanol to serum samples enhances the signal recorded from the instrument. Is there a statistically significant difference between the means of the experimental data sets?

Signal with method A (5% ethanol)	Signal with method B (no ethanol)
29.3	27.6
28.4	28.2
30.2	28.5
29.7	26.9
	27.5
$\bar{x}_a = 29.4$	$\bar{x}_b = 27.7$
$s_a = 0.76$	$s_b = 0.63$

Solution

From the context of the problem there is no reason to suspect that there should be a difference between the standard deviations of these sets of data. The sources of error are the same for both sets: the instrument, the analyst, the serum sample. Application of the F-test of precision, which we have not discussed yet, also shows that the standard deviations are not statistically different at the 0.05 significance level.

Using the expressions given above, we can calculate a value for the pooled standard deviation,

$$s_{\text{pool}} = \left[\frac{3(0.76)^2 + 4(0.63)^2}{4 + 5 - 2}\right]^{1/2} = 0.69$$

Using the pooled standard deviation, we can find a value for t_{calc}:

$$t_{\text{calc}} = \frac{|29.4 - 27.7|}{0.69(9/20)^{1/2}} = \frac{1.7}{0.46} = 3.7$$

We then look to Table 2.2 to find values for t,

significance levels:	0.20	0.10	0.05	0.025	0.01
t value, $(N_a + N_b - 2) = 7$ d.o.f.:	1.42	1.90	2.36	2.84	3.50

Since t_{calc} exceeds t_{tab} at the 0.01 significance level, we can reject the null hypothesis. More than 99% of the differences that would arise from random errors in measurement would be smaller than the observed difference, and we are justified in concluding that ethanol enhances the signal.

When we set out to compare means obtained by different methods of analysis it is no longer appropriate to assume that the standard deviations of the sets are estimators of the same value of σ. This results in an unfortunate complication of matters, since the statistic which can be used, represented by t^* and calculated from the equation

2.3 ERRORS IN CHEMICAL ANALYSIS

$$t^* = \frac{|\bar{x}_a - \bar{x}_b|}{[(s_a^2/N_a) + (s_b^2/N_b)]^{1/2}}$$

follows a distribution which differs from Gosset's Student t for small samples. Rather than introduce yet another table, we will use a method proposed by Welch (5) for approximating t^*. According to Welch, the Student t distribution may be used if the number of degrees of freedom is made a function of the standard deviation of the data sets. The adjusted d.o.f. can be calculated from the equation

$$\text{d.o.f.} = \frac{U^2}{[V^2/(N_a + 1)] + [W^2/(N_b + 1)]} - 2$$

where

$$V = \frac{s_a^2}{N_a} \qquad W = \frac{s_b^2}{N_b} \qquad U = V + W$$

In a manner analogous to that used in the previous example, we can calculate a value of t^* from the data sets (t^*_{calc}) and compare it to a tabulated value of t corresponding to the adjusted number of degrees of freedom. If it turns out that

$$t^*_{\text{calc}} > t_{\text{tab}}$$

at a particular level of significance, we can reject the null hypothesis at that level.

EXAMPLE 2.10 Comparing Two Means

Ethanol can be determined in blood by chemical or physical methods. The chemical method chosen (method A) involves separation of ethanol from the blood and its subsequent determination by oxidation with potassium dichromate solution. The physical method involves collection of ethanol vapor in the air space in a sealed vial with a gastight syringe and subsequent analysis of the vapor by gas chromatography (method B; see Chapter 18).

The following data were obtained from a sample of blood taken from an automobile driver thought to be intoxicated:

Method A (% ethanol)	Method B (% ethanol)
0.200	0.230
0.210	0.213
0.202	0.217
	0.230
	0.222
$\bar{x}_a = 0.204\%$	$\bar{x}_b = 0.222\%$
$s_a = 0.0053$	$s_b = 0.0076$

Is there a statistically significant difference between the mean values of the results obtained by the two methods?

Solution

We begin with the hypothesis that there is no difference between the means.
First determine t^*_{calc}:

$$t^*_{calc} = \frac{|\bar{x}_a - \bar{x}_b|}{\sqrt{s_a^2/N_a + s_b^2/N_b}}$$

$$= \frac{|0.204 - 0.222|}{\sqrt{\frac{2.81 \times 10^{-5}}{3} + \frac{5.78 \times 10^{-5}}{5}}} = 3.93$$

To use Welch's equation to calculate the proper number of degrees of freedom to use in finding t_{tab}, we must calculate the terms V, W, and U,

$$V = \frac{s_a^2}{N_a} = \frac{2.81 \times 10^{-5}}{3} = 9.37 \times 10^{-6}$$

$$W = \frac{s_b^2}{N_b} = \frac{5.78 \times 10^{-5}}{5} = 1.16 \times 10^{-5}$$

$$U = V + W = 2.09 \times 10^{-5}$$

Using Welch's equation, we find

$$\text{d.o.f.} = \frac{(2.09 \times 10^{-5})^2}{\frac{(9.37 \times 10^{-6})^2}{4} + \frac{(1.16 \times 10^{-5})^2}{6}} - 2$$

$$= 7.86$$

The nearest integer is 8.
Now we check the value calculated for t^* against values for t at several significance levels (from Table 2.2):

significance level: 0.20   0.10   0.05   0.025   0.01
t_{tab} (8 d.o.f.):   1.40   1.86   2.31   2.75   3.36

The calculated value of t^* is larger than 3.36 and we can therefore reject the hypothesis even at the 0.01 significance level. The risk is less than 1% that so large a difference between the means could arise simply by random errors in measurement.

Notice that in this example the number of degrees of freedom was coincidentally equal to the number of measurements. Had we assumed that s_a and s_b were not significantly different, we would have used 6 d.o.f. for the comparison. Would the conclusion have come out differently? Solve the problem by the method used in the previous example and see.

It is important to point out that one is not *forced* to accept or reject a null hypothesis on the basis of the outcome of a *t*-value calculation. No situation requires that a hypothesis be rejected at a 5% significance level but accepted at 1%. The probability range is, after all, continuous, and many people choose 5% significance as a convenient benchmark. Whether we reject a null hypothesis depends on nonstatistical factors: Is the decision a crucial decision? Does a life depend on it? Does the reputation of a company depend on it? In some cases rejecting a hypothesis that should have been accepted is a less serious error than accepting a hypothesis that should have been rejected. In circumstances where a decision is not clear, a doubtful outcome of a *t*-value calculation means that more determinations should be run. Remember that statistical tests do not tell us what decisions to make, only the probability that a certain decision will be correct. Information about statistical decision making can be found in Reference (6).

Comparing the Precision of Two Methods: The F-Test When the determinate error of a method is not significant the factor which limits the accuracy of the mean of a small number of analyses is the precision of the data. It is therefore important to select a method which is inherently highly precise as well as free of bias. Given a choice between two possible methods, what statistical test could be performed to decide objectively which method is more precise?

One test used by statisticians to judge precision is the *F*-test, a test based on the ratio of the variances of two sets of data. For the general case in which some method A (which gives N_a results with variance s_a^2) is compared with some method B (which gives N_b results with variance s_b^2), the ratio *F* may be calculated:

$$F = s_a^2/s_b^2$$

By convention *F* must always have a value greater than unity, and so the variance of the less precise data is always placed in the numerator. The hypothesis to be tested is then: "The precision of method B is not significantly better than that of method A." Notice that this is a one-tailed test.

The calculated *F* value may be compared with tabulated *F* values for $N_a - 1$ and $N_b - 1$ degrees of freedom at a particular level of significance. If the calculated value exceeds the tabulated value, the null hypothesis can be rejected at the particular level of significance. Table 2.4 has values for the *F* function for one-tailed tests at the 0.05 level of significance (0.95 confidence level).

TABLE 2.4 95th PERCENTILE VALUES (0.05 LEVELS), $F_{.95}$, FOR THE F DISTRIBUTION
(ν_1, Degrees of Freedom in Numerator; ν_2, Degrees of Freedom in Denominator)

ν_1 \ ν_2	1	2	3	4	5	6	7	8	9	10	12	15	20	24	30	40	60	120	∞
1	161	200	216	225	230	234	237	239	241	242	244	246	248	249	250	251	252	253	254
2	18.5	19.0	19.2	19.2	19.3	19.3	19.4	19.4	19.4	19.4	19.4	19.4	19.4	19.5	19.5	19.5	19.5	19.5	19.5
3	10.1	9.55	9.28	9.12	9.01	8.94	8.89	8.85	8.81	8.79	8.74	8.70	8.66	8.64	8.62	8.59	8.57	8.55	8.53
4	7.71	6.94	6.59	6.39	6.26	6.16	6.09	6.04	6.00	5.96	5.91	5.86	5.80	5.77	5.75	5.72	5.69	5.66	5.63
5	6.61	5.79	5.41	5.19	5.05	4.95	4.88	4.82	4.77	4.74	4.68	4.62	4.56	4.53	4.50	4.46	4.43	4.40	4.37
6	5.99	5.14	4.76	4.53	4.39	4.28	4.21	4.15	4.10	4.06	4.00	3.94	3.87	3.84	3.81	3.77	3.74	3.70	3.67
7	5.59	4.74	4.35	4.12	3.97	3.87	3.79	3.73	3.68	3.64	3.57	3.51	3.44	3.41	3.38	3.34	3.30	3.27	3.23
8	5.32	4.46	4.07	3.84	3.69	3.58	3.50	3.44	3.39	3.35	3.28	3.22	3.15	3.12	3.08	3.04	3.01	2.97	2.93
9	5.12	4.26	3.86	3.63	3.48	3.37	3.29	3.23	3.18	3.14	3.07	3.01	2.94	2.90	2.86	2.83	2.79	2.75	2.71
10	4.96	4.10	3.71	3.48	3.33	3.22	3.14	3.07	3.02	2.98	2.91	2.85	2.77	2.74	2.70	2.66	2.62	2.58	2.54
11	4.84	3.98	3.59	3.36	3.20	3.09	3.01	2.95	2.90	2.85	2.79	2.72	2.65	2.61	2.57	2.53	2.49	2.45	2.40
12	4.75	3.89	3.49	3.26	3.11	3.00	2.91	2.85	2.80	2.75	2.69	2.62	2.54	2.51	2.47	2.43	2.38	2.34	2.30
13	4.67	3.81	3.41	3.18	3.03	2.92	2.83	2.77	2.71	2.67	2.60	2.53	2.46	2.42	2.38	2.34	2.30	2.25	2.21
14	4.60	3.74	3.34	3.11	2.96	2.85	2.76	2.70	2.65	2.60	2.53	2.46	2.39	2.35	2.31	2.27	2.22	2.18	2.13
15	4.54	3.68	3.29	3.06	2.90	2.79	2.71	2.64	2.59	2.54	2.48	2.40	2.33	2.29	2.25	2.20	2.16	2.11	2.07
16	4.49	3.63	3.24	3.01	2.85	2.74	2.66	2.59	2.54	2.49	2.42	2.35	2.28	2.24	2.19	2.15	2.11	2.06	2.01
17	4.45	3.59	3.20	2.96	2.81	2.70	2.61	2.55	2.49	2.45	2.38	2.31	2.23	2.19	2.15	2.10	2.06	2.01	1.96
18	4.41	3.55	3.16	2.93	2.77	2.66	2.58	2.51	2.46	2.41	2.34	2.27	2.19	2.15	2.11	2.06	2.02	1.97	1.92
19	4.38	3.52	3.13	2.90	2.74	2.63	2.54	2.48	2.42	2.38	2.31	2.23	2.16	2.11	2.07	2.03	1.98	1.93	1.88
20	4.35	3.49	3.10	2.87	2.71	2.60	2.51	2.45	2.39	2.35	2.28	2.20	2.12	2.08	2.04	1.99	1.95	1.90	1.84
21	4.32	3.47	3.07	2.84	2.68	2.57	2.49	2.42	2.37	2.32	2.25	2.18	2.10	2.05	2.01	1.96	1.92	1.87	1.81
22	4.30	3.44	3.05	2.82	2.66	2.55	2.46	2.40	2.34	2.30	2.23	2.15	2.07	2.03	1.98	1.94	1.89	1.84	1.78
23	4.28	3.42	3.03	2.80	2.64	2.53	2.44	2.37	2.32	2.27	2.20	2.13	2.05	2.01	1.96	1.91	1.86	1.81	1.76
24	4.26	3.40	3.01	2.78	2.62	2.51	2.42	2.36	2.30	2.25	2.18	2.11	2.03	1.98	1.94	1.89	1.84	1.79	1.73
25	4.24	3.39	2.99	2.76	2.60	2.49	2.40	2.34	2.28	2.24	2.16	2.09	2.01	1.96	1.92	1.87	1.82	1.77	1.71
26	4.23	3.37	2.98	2.74	2.59	2.47	2.39	2.32	2.27	2.22	2.15	2.07	1.99	1.95	1.90	1.85	1.80	1.75	1.69
27	4.21	3.35	2.96	2.73	2.57	2.46	2.37	2.31	2.25	2.20	2.13	2.06	1.97	1.93	1.88	1.84	1.79	1.73	1.67
28	4.20	3.34	2.95	2.71	2.56	2.45	2.36	2.29	2.24	2.19	2.12	2.04	1.96	1.91	1.87	1.82	1.77	1.71	1.65
29	4.18	3.33	2.93	2.70	2.55	2.43	2.35	2.28	2.22	2.18	2.10	2.03	1.94	1.90	1.85	1.81	1.75	1.70	1.64
30	4.17	3.32	2.92	2.69	2.53	2.42	2.33	2.27	2.21	2.16	2.09	2.01	1.93	1.89	1.84	1.79	1.74	1.68	1.62
40	4.08	3.23	2.84	2.61	2.45	2.34	2.25	2.18	2.12	2.08	2.00	1.92	1.84	1.79	1.74	1.69	1.64	1.58	1.51
60	4.00	3.15	2.76	2.53	2.37	2.25	2.17	2.10	2.04	1.99	1.92	1.84	1.75	1.70	1.65	1.59	1.53	1.47	1.39
120	3.92	3.07	2.68	2.45	2.29	2.18	2.09	2.02	1.96	1.91	1.83	1.75	1.66	1.61	1.55	1.50	1.43	1.35	1.25
∞	3.84	3.00	2.60	2.37	2.21	2.10	2.01	1.94	1.88	1.83	1.75	1.67	1.57	1.52	1.46	1.39	1.32	1.22	1.00

Source: E. S. Pearson and H. O. Hartley, *Biometrika Tables for Statisticians*, vol. 2 (1972), table 5, p. 178. Cambridge University Press, London. Reproduced with the permission of the Biometrika Trustees.

2.3 ERRORS IN CHEMICAL ANALYSIS

EXAMPLE 2.11 *F*-test

The iron in a sample may be determined as Fe^{2+} by titration with a standard solution of potassium dichromate. Alternatively, iron may be complexed with a reagent such as sulfosalicylic acid to form a colored species, which can be determined quantitatively by the intensity of its color (using a spectrophotometer).

A student set out to compare the precision of the methods by analyzing replicate portions of two iron samples. The results were:

Titrimetric (method A)	Spectrophotometric (method B)
5.250	3.05
5.244	3.10
5.262	3.04
5.256	
5.238	
mean: 5.250	mean: 3.06

Is one method more precise than the other?

Solution

Calculate s_a^2 and s_b^2.

$$s_a^2 = 9.00 \times 10^{-5}$$
$$s_b^2 = 1.03 \times 10^{-3}$$

Calculate a value of *F*:

$$F_{calc} = \frac{s_b^2}{s_a^2} = \frac{1.03 \times 10^{-3}}{9.00 \times 10^{-5}} = 11.4$$

We hypothesize that the precision of the titration method is not better than that of the spectrophotometric method. Consult the table of *F* values and find $F = 6.94$ for *N* (numerator) = 2 d.o.f. and *N* (denominator) = 4 d.o.f. Since the calculated *F* is greater than the tabulated *F* at the 0.05 significance level, we can reject the one-tailed null hypothesis. We can conclude that the titrimetric method is more precise than the spectrophotometric method, with 95% confidence that our conclusion is correct. Notice that the samples contain different amounts of iron. If the *F*-test is to be reliable, precision ought not to be a function of concentration. It is probably safe to assume that this is the case here, since the concentrations are within a factor of 2 of each other.

TABLE 2.5 CRITICAL Q VALUES

N	0.90	0.95	0.99
3	0.886	0.941	0.988
4	0.679	0.765	0.889
5	0.557	0.642	0.760
6	0.482	0.560	0.698
7	0.434	0.507	0.637

Source: W. J. Dixon, *Annals of Mathematical Statistics,* 22, 69 (1951).

Rejecting Outlying Results In all the experiments described in this text and in most of those performed by analytical chemists, it is found that a set of replicate data points clusters rather closely around the mean. However, occasionally one point in a set is so different from the others that the analyst doubts its validity. Is it different enough from the others so that it can be rejected as simply a blunder? Or is that particular determination really as legitimate as the other values? The best solution to a dilemma such as this is to perform more analyses. If the outlying result really does represent a blunder, a few more close-lying points will lessen its effect on the mean. However, if it is not practical in terms of time or money to run any more analyses, a decision must be made to keep or discard the outlying result.

Several methods have been devised by which the decision to reject an outlying result can be made objectively. Most of them involve either prior knowledge of the standard deviation of the population, or an estimate of it based on a large set of parallel determinations made on similar samples. By far the most convenient test for small samples does not involve σ, but rather is based on the range of the results. This test uses quotients of factors involving the range, and is called the *Q-test*. If we arrange a set of N results in order of increasing size,

$$x_1 < x_2 < x_3 < x_4 < \cdots < x_{N-1} < x_N$$

then the ratio Q for the largest result (the Nth result) is given by

$$Q_N = \frac{x_N - x_{N-1}}{R}$$

where R is the range, $x_N - x_1$. The ratio Q for the smallest result is

$$Q_1 = \frac{x_2 - x_1}{R}$$

2.3 ERRORS IN CHEMICAL ANALYSIS

Either Q_N or Q_1 is tested, depending on whether the analyst suspects the result of being too high or too low. Q_N or Q_1 is calculated and compared with critical values of Q corresponding to the number of results (N) and the level of significance. If Q_N or Q_1 exceeds the critical Q, either x_N or x_1 may be rejected with a certain level of confidence. Values of Q are shown in Table 2.5.

EXAMPLE 2.12 Rejecting Outliers

The following results for percent chloride were obtained in the gravimetric determination of chloride in a sample: 29.21, 28.99, and 29.25%. The 28.99% value looks suspiciously low to the analyst. Use the Q-test to determine objectively whether the result may be rejected.

Solution

The range of the data is 29.25% − 28.99% = 0.26%. The difference between the smallest result and the middle result is 29.21% − 28.99% = 0.22%.

$$Q = \frac{x_2 - x_1}{x_3 - x_1} = \frac{0.22}{0.26} = 0.85$$

For three results the critical Q is 0.94 at the 95% confidence level. Since the calculated Q is smaller than the critical Q, the result 28.99% ought not to be rejected at the 95% confidence level.

EXAMPLE 2.13 Rejecting Outliers

If a fourth gravimetric analysis is performed after the calculation in the last example and it gives 29.23% chloride, could the 28.99% value be rejected at the 95% confidence level?

Solution

The calculated value of Q does not change, but the critical value becomes 0.76 (there are now four results). Therefore 28.99% can be rejected with 95% confidence.

If the 28.99% result is rejected, how different will the reported result be? The average of all four results is 29.17%, while that of the best three is 29.23%. The absolute difference is 0.06%, or about 2 parts per thousand relative difference.

It is best to use the Q-test conservatively, that is, to reject results as a last resort. It is usually good advice to run one or more additional analyses (if time and material permit) and then reevaluate the situation.

Students will often ask if the Q-test should be applied to the data again after one result has been rejected. The answer depends on the number of data in the sample and the analyst's knowledge of the method and his or her ability, but generally the test is applied only once to a set of data.

Determining the Correct Number of Replicates In most quantitative analysis laboratory courses, students are told to weigh out three or four portions of an unknown sample and run replicate analyses. The average of the results of these three or four determinations is reported as the analysis result. The number of parallel analyses run is determined by several factors, among the most important of which are the available time, the cost of equipment and reagents, and the number of replicates an inexperienced analyst can handle at one time. In industrial laboratories the importance of a particular analysis to the company or customer is weighed against the cost of running extra replicates.

From the point of view of statistics, the number of replicates is determined by the standard deviation *expected* for the results and the size of the confidence interval about the mean which will satisfy the requirements of the analyst. If the 95% confidence interval is represented by Δ, the relationship

$$\Delta = \frac{ts}{\sqrt{N}}$$

can be rearranged to solve for N:

$$N = \left(\frac{ts}{\Delta}\right)^2$$

With this relationship a series of values for N and t pairs can be calculated for specified values of s and Δ and compared with the t values which correspond to various N values in the Student t table (Table 2.2). When a calculated t value exceeds the tabulated t value, N is sufficiently large. This method requires a knowledge of the 95% confidence interval desired and an approximate standard deviation for the results. An experienced analyst may know his or her own abilities well enough to be able to anticipate an approximate standard deviation. An inexperienced analyst may be able to approximate s from the results of a standardization procedure. The following examples should help illustrate these ideas.

2.3 ERRORS IN CHEMICAL ANALYSIS

EXAMPLE 2.14 Number of Replicates

An experienced analyst is able to obtain a standard deviation of 0.10% in the titrimetric determination of iron with standard dichromate solution, at iron levels of about 12% of the sample weight. How many replicates of each sample should the analyst run if a 95% confidence interval of ±0.25% is desired?

Solution

$$N = \left(\frac{ts}{\Delta}\right)^2 = \left[\frac{t(0.10\%)}{0.25\%}\right]^2$$

Calculate values of t for several values of N and compare them with tabulated values of t:

N	t_{calc}	t_{tab}
2	3.5	4.3
3	4.3	3.2
4	5.0	2.8
5	5.6	2.6

The calculated t value exceeds the tabulated t at $N = 3$. At least three samples should be run.

EXAMPLE 2.15 Number of Replicates

An inexperienced analyst is told to determine iron in the same kind of sample as in the previous example, using the same analytical method. A 95% confidence interval of ±0.25% is still required. The results of standardizing the dichromate solution indicate that the novice can expect a standard deviation of 0.3%. How many replicates should the novice run to achieve the required confidence interval?

Solution

Using the expression

$$N = \left(\frac{ts}{\Delta}\right)^2 = \left[\frac{t(0.3\%)}{0.25\%}\right]^2$$

calculate values of t for a series of N values and compare them with tabulated t values. When the calculated value exceeds the tabulated value, N is sufficiently large.

N	t_{calc}	t_{tab}
2	1.18	4.30
⋮	⋮	⋮
5	1.86	2.57
⋮	⋮	⋮
7	2.20	2.37
8	2.36	2.31
9	2.50	2.20

The calculated t exceeds the tabulated t for $N = 8$. Eight replicates should be sufficient to achieve the required confidence interval.

From these examples it should be apparent that when the 95% confidence interval must be less than the standard deviation, the analyst must be prepared to do a tremendous amount of work. Such a requirement may be justified only when extremely important samples are being analyzed, or when the effort and cost per replicate are extremely low. This may be the case with modern automated instrumentation.

2.3.3 Quality Control

Among the most important services performed by analytical chemists in industry is the routine monitoring of the quality of production, an area of science called quality control.

Quality control operates on two levels. The first involves the routine analysis of samples of the manufactured product to see if specifications are being met. The second level involves regular checks on the performance of the analytical instrumentation and personnel to make sure the laboratory is turning out meaningful results.

At the heart of a quality control program is the routine analysis of standard reference materials (both intralab and interlab, or "round-robin" samples) and extremely careful keeping of records. The object of such a program is to keep determinate error from entering analytical procedures. A procedure is said to be "stable" or "in statistical control" when the daily or weekly variations in results can be ascribed to random error. Limits of variation can be set around the expected value for a standard which, if exceeded, can be taken as a signal that the procedure is out of control. Once

2.3 ERRORS IN CHEMICAL ANALYSIS

this occurs, sources of determinate error must be investigated and modifications made until the process returns to stability.

A graphical device called a *control chart* can be used to discover the loss of statistical control. A control chart is made by plotting a parameter of the experimental data as a function of time. The parameter may be the average of a sample of four or more replicates, the standard deviation of the sample, or its range, and plotting may be done each day or once a week. The target value is the expected value for the standard and is drawn in as a horizontal line. Upper (+) and lower (−) control limits† are drawn parallel to the target value line at

$$\mu \pm \frac{2\sigma}{\sqrt{N}}$$

In this expression N is the number of replicates run for each sample (not less than four, generally), and σ is the standard deviation of the population, a value which can be estimated from a large number of analyses.‡ Recall from the discussion of the confidence interval that about 5% of the means are expected to fall outside this interval if only random error influences the results.

As mentioned above, when a mean value falls outside the two-sigma limits, we would begin a search for a source of determinate error. A virtue of the graphical method is that we might even anticipate such an occurrence. For example, Duncan (6) recommends that a run of seven or more points above or below the target line or four or five points outside the one-sigma limits (upper and lower "warning limits" which some laboratories use) should be taken as a sign of loss of control. Similarly, the appearance of cycles or other nonrandom patterns may not be evident in a tabulation of results, but may be quite obvious on a control chart.

Figure 2.13 shows a control chart for the spectrophotometric determination of nitrate ion in water with brucine. The published relative standard deviation for the method is 5.5% at the level of 1 part per million (σ = 0.055 ppm; see Ref. 7). The mean of four replicate portions of a 1.000 ppm standard solution of nitrate is plotted daily. Between September 10 and September 23 the procedure appears to be in statistical control. However, on September 24 the upper control limit is crossed. In the week preceding September 24 there appears to be a run of five points above the target line. This might have been taken as a sign of impending loss of control.

Interlaboratory tests (round-robins) are designed either to test the performance of a group of cooperating laboratories, evaluate new methods of analysis, or evaluate

---

† These are the "two-sigma limits". Some laboratories use "three-sigma limits", $\pm \frac{3\sigma}{\sqrt{N}}$.

‡ μ and σ can be established initially by making at least 25 sets of 4 replicate determinations, and finding the average mean and average standard deviation. The average standard deviation may also be calculated from the average range, as shown in a previous section.

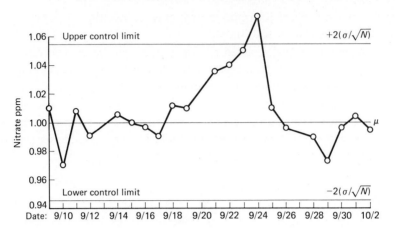

Figure 2.13 Control chart.

an established method on a new material. A round-robin program sponsored by the Minnesota Section of the American Chemical Society, the Minnesota Pollution Control Agency (MPCA), and the Minnesota Department of Health (MDH) serves as an interesting model. Part of the program was aimed at assessing the credibility of water and wastewater analyses performed by industrial, municipal, and commercial laboratories. Each laboratory that volunteered to take part in the study was sent a set of synthetic samples prepared by the MDH which were to be analyzed for 12 contaminants. The MDH ran a quality assurance test program during the test period to account for changes in sample composition during storage. Table 2.6 shows the results reported by the 16 participating commercial laboratories. The first column contains the expected values for the 12 contaminants. Results marked # were outside the EPA-approved limits of ±1 standard deviation of the mean of the data.† Notice that 46 of the 146 results (more than 30%) reported by the commercial laboratories failed to meet EPA criteria. One laboratory failed in 8 out of 12 procedures.

2.3.4 Propagation of Error

An analytical result is ordinarily obtained after the results of several different measurements are combined. The error of the computed result will be influenced by the errors in each of the measurements, the largest measurement error having the greatest effect. If errors in the measurements are determinate, then the error of the analytical result is well defined. However, if the errors in the measurements are random, the

---

† (14) These are the data which remain after rejecting data which fall outside ±3 standard deviations. The Environmental Protection Agency will not certify a laboratory for a test if results in a study such as this fall outside the ±1s limits.

TABLE 2.6 RESULTS REPORTED BY COMMERCIAL LABORATORIES

Determination	1	2	3	4	5	6	7	8	9	10	11	12	13	14	15	16
BOD[a], mg/liter	204	220	18.0#[b]	270#	14.5#	220		288	253#		202	110.6#	200	82.6#		125#
COD[c], mg/liter	300	301	180#	260		294	231.3#		192#[d]	324#	286	286	296	314.7		
pH	4.1	4.4	4.35	4.4		4.4	4.5	4.8#	4.8#	4.4	4.6[d]	4.4	4.5	4.21#	4.4	4.5
Ammonia nitrogen, mg/liter as N	2.5	2.36	6.4#	2.3	4.35	2.5	2.4	2.5	3.5#	2.3	2.6[d]	38.0#	2.6	2.5		
Total phosphate, mg/liter as P	3.0	2.88	4.26#	2.9		2.1#	2.5	3.0	2.63	0.11[d]	2.7	1.05#	2.3		0.05#[d]	3.47
Turbidity, NTU	40	42	30#			45	40	28#	40		41[d]	32#	48			43
Iron, mg/liter as Fe	1.0	0.98	1.0	1.25#			0.9	1.0	0.85	1.1	1.06	0.47#	1.0	1.4#	0.6#[d]	
Manganese, mg/liter as Mn	1.0	0.97	1.0	1.18#			0.75#	1.0	0.95		0.99	0.62#	1.0	0.91		
Chloride, mg/liter as Cl	36	35.2	40	36		33	11#	36	80#	76#	37[d]	36.2	33		35.0	
Nitrate, mg/liter as N	5.0	4.84	3.40#	4.8		4.7	4.6	4.80	4.0	3.8	5	3.0#	5.0	46#	8.0#	
Specific conductance, μmho	147	152	210#	115		142	150	125	160	140[d]	84#[d]	11,364#	137		157	
Total suspended solids, mg/liter	18	23	22.4	37	19.3	24	41	24	44#	61#[d]	14#	20.0	50#	31		23

[a] Biochemical oxygen demand (wastes which require O_2 for biodegradation).
[b] #, Outside Environmental Protection Agency acceptable ranges.
[c] Chemical oxygen demand (wastes which react with O_2 and require O_2 for biodegradation).
[d] Data exceeded time frame which MDH used to verify stability of samples.

TABLE 2.7 ERROR OF A CALCULATED RESULT[a]

Error of R	Sum or difference $R = X + Y - Z$	Product or quotient $R = \dfrac{X}{YZ}$
Determinate	$e_r = e_x + e_y - e_z$	$\dfrac{e_r}{R} = \dfrac{e_x}{X} - \dfrac{e_y}{Y} - \dfrac{e_z}{Z}$
Random	$s_r = \sqrt{s_x^2 + s_y^2 + s_z^2}$	$\dfrac{s_r}{R} = \sqrt{\left(\dfrac{s_x}{X}\right)^2 + \left(\dfrac{s_y}{Y}\right)^2 + \left(\dfrac{s_z}{Z}\right)^2}$

[a] Key: R, calculated result; X, Y, Z, experimentally measured quantities; e_r, e_x, e_y, e_z, absolute errors; s_r, s_x, s_y, s_z, absolute standard deviations; e_r/R, e_x/X, e_y/Y, e_z/Z, relative errors of R, X, Y, Z; s_r/R, s_x/X, s_y/Y, s_z/Z, relative standard deviation of R, X, Y, Z.

Source: Based on a table by A. A. Benedetti-Pichler, *Industrial and Engineering Chemistry, Analytical Edition*, 8, 373 (1936).

errors in each step are distributed over ranges and best described in terms of standard deviations.

Benedetti-Pichler has described the ways to estimate the error of a calculated result from the errors of the factors which are brought together to calculate it. These are summarized in Table 2.7.

Most of the calculations for error analysis expected in a first course in analytical chemistry can be performed with the equations in Table 2.7. In more complicated situations, say when both a subtraction and a multiplication are involved in a calculation, the equations can be used in a series of steps.

EXAMPLE 2.16 Propagation of Error

An analyst sets out to determine the equivalent weight of a very pure weak acid. To do this 0.8218 g of weak acid is titrated with 32.12 mL of 0.1253N base. If the standard deviation of the weight is 0.0002 g, that of the buret volume is 0.01 mL, and that of the base concentration is 0.0003N, what is the standard deviation of the equivalent weight? The expression used to solve for the equivalent weight is

$$\text{equivalent weight} = \frac{\text{weight of weak acid}}{(\text{volume of base})(\text{concentration of base})}$$

2.3 ERRORS IN CHEMICAL ANALYSIS

Solution

The expression has the form

$$R = \frac{X}{YZ}$$

Since the standard deviations are given, the expression

$$\frac{s_R}{R} = \sqrt{\left(\frac{s_x}{X}\right)^2 + \left(\frac{s_y}{Y}\right)^2 + \left(\frac{s_z}{Z}\right)^2}$$

can be used to calculate the standard deviation of the equivalent weight. Substituting in the values,

$$s_R = 204.2 \sqrt{\left(\frac{0.0002}{0.8218}\right)^2 + \left(\frac{0.01}{32.12}\right)^2 + \left(\frac{0.0003}{0.1253}\right)^2}$$

we find

$$s_R = 2.43 \times 10^{-3}(204.2) = 0.5 \text{ g (per equivalent)}$$

When the functional relationship between variables is more complicated, it is necessary to derive expressions from the fundamental equation for random error. Let a result, R, be calculated from the independent measured variables $X, Y, Z, \ldots$, that is,

$$R = f(X, Y, Z, \cdots)$$

If we represent the absolute error of the variables by $e_x, e_y, e_z, \ldots$, then the absolute determinate error of the result, e_r, is given by

$$e_r = \left(\frac{\partial R}{\partial X}\right)e_x + \left(\frac{\partial R}{\partial Y}\right)e_y + \left(\frac{\partial R}{\partial Z}\right)e_z + \cdots$$

that is, the sum of the partial derivatives of R with respect to the variables, each multiplied by the absolute error.

In the treatment of random error we make use of the additive properties of the variance. If we represent the variance of the variables as $s_x^2, s_y^2, s_z^2, \ldots$, then the variance of the result is given by

TABLE 2.8 DERIVATIVES OF SOME COMMON FUNCTIONS

R	$\partial R/\partial x$
k (constant)	0
x	1
x^n	nx^{n-1}
$x^{1/n}$	$(1/n)x^{(1/n-1)}$
e^x	e^x
10^x	$10^x (\ln 10)$
$\ln x$	$1/x$
$\log_{10} x$	$(\log_{10} e)(1/x)$
$\sin x$	$\cos x$
$\cos x$	$-\sin x$

$$s_r^2 = \left(\frac{\partial R}{\partial s_x}\right)^2 s_x^2 + \left(\frac{\partial R}{\partial s_y}\right)^2 s_y^2 + \left(\frac{\partial R}{\partial s_z}\right)^2 s_z^2 + \cdots$$

These general expressions are derived in most elementary calculus texts. Derivatives of some simple functions are included in Table 2.8.

EXAMPLE 2.17 Propagation of Error

The hydrogen ion activity of a solution is related to its pH by

$$a_{H^+} = 10^{-pH}$$

If an absolute determinate error of 1% is made in measuring the pH of a solution which has a pH of 5.00, what is the resulting error in a_{H^+}?

Solution

The derivative of a_{H^+} with respect to pH is

$$\frac{\partial a_{H^+}}{\partial pH} = 10^{-pH}(\ln 10) = 2.30 \times 10^{-pH}$$

The error in a_{H^+} is given by

2.3 ERRORS IN CHEMICAL ANALYSIS

$$e_r = \left(\frac{\partial a_{H^+}}{\partial pH}\right) 0.01 = 0.02303 \times 10^{-pH} = 2.30 \times 10^{-7}$$

This represents a 2.3% error in the activity.

2.3.5 Least-Squares Line Fitting

Many analytical methods involving instrumentation rely on the use of what is called a calibration curve, or *working curve*. A working curve is a plot of an instrument's response to the concentrations of analyte in a very carefully prepared set of standards. An example is shown in Fig. 2.14, a plot of optical absorbance versus the concentration of the iron complex tris(orthophenanthroline)-iron(II). The absorbance, A, is plotted as a dependent variable. It seems to depend linearly on concentration of the complex, that is,

$$A = k*C$$

Working curves for many methods are approximately straight lines. As a result of

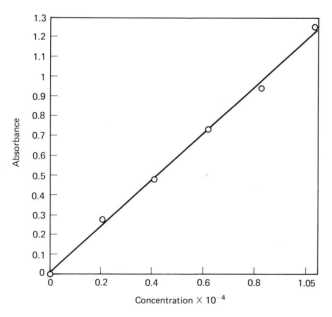

Figure 2.14 Beer's law plot for the determination of Fe(II) with 1,10-phenanthroline. (Data of T. Weber.)

random errors in the measuring process, however, individual data points may not fall exactly on a single line. The experimentalist must try objectively to determine the line which best "fits" the data. Statisticians recommend a technique for fitting the line called *regression analysis*. The simplest method of regression analysis is called the method of least squares and is the subject of this section.

We will make two assumptions to greatly simplify the treatment of the least-squares method. First, we assume that a linear relationship does indeed exist between the measured (dependent) variable and the analyte concentration. Some causes of nonlinearity in plots like Fig. 2.14 are discussed in Chapter 13. We will assume that they are unimportant in this application, however. The second assumption is that there is no significant error in the analyte concentrations used in the study to create the plot. The only error is assumed to be in the measurement of the dependent variable. We might assume that the error in the optical measurement is insignificant compared to that in the preparation of the standards and perform the regression analysis somewhat differently. We might even assume that random error is important in both variables, but the regression analysis would be considerably more difficult. See Reference (4) for a discussion of these alternatives.

Suppose we have a set of N data pairs, (x, y), which are plotted and conform to an approximately straight line. The average y value will be described by $\bar{y}$, and the average x value by $\bar{x}$. Statisticians call the point $(\bar{x}, \bar{y})$ the *centroidal point*. The line ultimately calculated by the least-squares method will pass through this point. Let us write the linear relationship we assume to exist between the variables in the familiar form

$$y = mx + b$$

where m is the slope and b the intercept, that is, the value of y when $x = 0$. For any measured value of y, say y_i, we can expect there to be a deviation from the value of y predicted by the assumed line,

$$y_i - (mx_i + b) = e_i$$

This deviation is called a *residual* and is denoted here by e_i. For N pairs of x, y data there will be N residuals. The method of least squares involves minimizing the squares of the residuals to find the straight line which best fits the data.

We can obtain three very useful terms from the variances of x and y and from a parameter called the *covariance* of x and y:

$$\Sigma X^2 = \Sigma(x_i - \bar{x})^2 \quad \text{or} \quad \Sigma(x_i)^2 - (\Sigma x_i)^2/N$$

$$\Sigma Y^2 = \Sigma(y_i - \bar{y})^2 \quad \text{or} \quad \Sigma(y_i)^2 - (\Sigma y_i)^2/N$$

$$\Sigma XY = \Sigma(x_i - \bar{x})(y_i - \bar{y}) \quad \text{or} \quad \Sigma x_i y_i - (\Sigma x_i \Sigma y_i)/N$$

2.3 ERRORS IN CHEMICAL ANALYSIS

We get these terms by multiplying the variances and the covariance by $N - 1$ degrees of freedom and taking square roots. Think of the covariance as being akin to the cross terms in the expansion of a quantity such as $(a + b)^2$. Although the resulting terms resemble standard deviations, they are different. We will use them to calculate the slope and intercept of the least-squares line describing the experimental data.

The following relationships are simply presented, not derived. Derivations appear in many of the references cited at the end of the chapter. The slope of the regression line (the best fit), m, can be found from $\Sigma\, XY$ and $\Sigma\, X^2$,

$$m = \Sigma\, XY / \Sigma\, X^2$$

The intercept of the regression line, b, is found from the linear equation by using the regression slope and the centroidal x and y values,

$$b = \bar{y} - m(\bar{x})$$

We can also use the three useful terms given above to calculate confidence intervals for points on the regression line and for the slope. Having confidence intervals then makes it possible to perform tests of significance on collections of plotted data. The *standard deviation about the regression* is given by

$$s_r = [(\Sigma\, Y^2 - m^2\, \Sigma\, X^2)/(N - 2)]^{1/2}$$

where $N - 2$ is the number of degrees of freedom. It can also be calculated from the sum of the squares of the residuals:

$$s_r = \left\{ \frac{[\Sigma\, y_i - (b + mx_i)]^2}{N - 2} \right\}^{1/2}$$

Remember that a residual is determined from a point y_i to the regression line ($b + mx_i$ here), rather than to the mean, $\bar{y}$, as we do when calculating $\Sigma\, X^2$ and $\Sigma\, Y^2$.

Knowing the standard deviation about the regression, we can calculate the *confidence interval* around the regression line at any particular point,

$$\text{c.i.} = y_i \pm t(s_{yi})/\sqrt{N}$$

The new term, s_{yi}, is the standard deviation of a point, y_i, and is determined from the standard deviation about the regression:

$$s_{yi} = \left\{ s_r^2 \left[1 + \frac{1}{N} + \frac{(x_i - \bar{x})^2}{\Sigma\, X^2} \right] \right\}^{1/2}$$

The confidence interval around the regression line has a peculiar shape, as predicted by the last expression. At the centroidal point $(\bar{x}, \bar{y})$, $(x_i - \bar{x})$ equals zero, and the value of s_{yi} (and therefore the confidence interval) has a minimum value. Confidence intervals for points lying far from the centroidal point are larger, as shown in the next example.

The standard deviation of the slope, m, is given by

$$s_m = (s_r^2 / \Sigma X^2)^{1/2}$$

The confidence interval of the slope is then

$$\text{c.i.} = m \pm t(s_m)$$

EXAMPLE 2.18 Regression Analysis

The following data were plotted in Figure 2.14:

Concentration (M)	Absorbance
0.00×10^{-5}	0.00
2.06×10^{-5}	0.28
4.12×10^{-5}	0.48
6.18×10^{-5}	0.73
8.24×10^{-5}	0.94
10.3×10^{-5}	1.25

Using the method of least squares, calculate the slope and intercept of the line which best fits the data. Calculate the 95% confidence interval about the line at the experimental concentrations. Calculate the 95% confidence interval for the slope.

Solution

The mean x value is found to be 5.150×10^{-5}. The mean y value is found to be 0.613. We will not round numbers until the final result. The following values are calculated for ΣX^2, ΣY^2, and ΣXY:

$$\Sigma X^2 = 7.426 \times 10^{-9}$$
$$\Sigma Y^2 = 1.0313$$
$$\Sigma XY = 8.735 \times 10^{-5}$$

2.3 ERRORS IN CHEMICAL ANALYSIS

Intermediate calculations are shown in the following table:

x	y	X²	Y²	XY
0.00×10^{-5}	0.00	2.652×10^{-9}	0.3763	3.157×10^{-5}
2.06×10^{-5}	0.28	0.9548×10^{-9}	0.1109	1.029×10^{-5}
4.12×10^{-5}	0.48	0.1061×10^{-9}	0.0177	0.137×10^{-5}
6.18×10^{-5}	0.73	0.1060×10^{-9}	0.0137	0.121×10^{-5}
8.24×10^{-5}	0.94	0.9548×10^{-9}	0.1069	1.010×10^{-5}
10.3×10^{-5}	1.25	2.652×10^{-9}	0.4058	3.281×10^{-5}
$\bar{x}\ 5.150 \times 10^{-5}$	$\bar{y}\ 0.613$	$\Sigma X^2\ 7.426 \times 10^{-9}$	$\Sigma Y^2\ 1.0313$	$\Sigma XY\ 8.735 \times 10^{-5}$

The slope is calculated as follows:

$$m = \Sigma XY / \Sigma X^2 = 1.176 \times 10^4$$

The intercept is calculated as

$$b = \bar{y} - m\bar{x} = 0.613 - (1.176 \times 10^4)(5.150 \times 10^{-5}) = 0.00736$$

In order to calculate the confidence intervals at the experimental concentrations we must calculate the standard deviation about the regression,

$$s_r = \left(\frac{\Sigma Y^2 - m^2 \Sigma X^2}{N - 2}\right)^{1/2} = 3.28 \times 10^{-2}$$

Then we calculate the standard deviation at each y value, s_{yi}, using the equation in the text. The results are tabulated below:

x	y	s_y^2	s_y	y (fit)	95% c.i.
0.00×10^{-5}	0.00	1.639×10^{-3}	0.0405	0.007	±0.046
2.06×10^{-5}	0.28	1.393×10^{-3}	0.0373	0.250	±0.042
4.12×10^{-5}	0.48	1.271×10^{-3}	0.0356	0.492	±0.040
6.18×10^{-5}	0.73	1.271×10^{-3}	0.0356	0.734	±0.040
8.24×10^{-5}	0.94	1.393×10^{-3}	0.0373	0.977	±0.042
10.3×10^{-5}	1.25	1.639×10^{-3}	0.0405	1.22	±0.046

The confidence intervals apply to the least-squares fitted line, which is used to calculate the numbers under "y (fit)" in the table. Notice that all the experimental y values fall within the 95% confidence intervals. Pencil in the confidence limits around the points in Fig. 2.14.

Finally, the standard deviation of the slope is calculated to be

$$S_m = (s_r^2 / \Sigma\, X^2)^{1/2} = 367$$

and its 95% confidence interval (4 d.o.f.) is

$$1.176 \times 10^4 \pm ts_m = 1.176 \times 10^4 \pm 2.78(367) = 1.176 \times 10^4 \pm 1.02 \times 10^3$$

2.3.6 Sampling

In the preceding sections of this chapter we have been dealing with errors in analysis. The emphasis has been on errors in operations and measurements and ways to uncover and treat those errors. We have assumed all along that the result we finally calculate will really answer questions about the composition of the material from which samples were taken. If the samples were representative of the material, then the answers will have meaning. If the samples were taken incorrectly, all the effort that went into the analysis was wasted.

The goal of sampling is to take a portion of a large quantity of material which is small enough for convenient laboratory work but which also has the same properties and composition as the bulk of the material. In order to accomplish this, several important questions must be considered:

1. What information is required from the samples? Are they supposed to represent the entire bulk of the material being tested, or just one part? For example, is a trace metal analysis of river water to include water at all depths? Should samples be taken from quiet pools as well as from quickly running water? Should samples include sediment or not? Is an average value to be reported, or a maximum level? Is the result to be an average of samples over a period of time? What is the acceptable precision ("sampling error")?

2. In what physical form is the material: gas, liquid, solid, suspension, solution? The answer will determine the tools used to draw samples and the strategy for using them.

3. What is the history of the material? Has it been exposed to weather, high temperatures, or intense light? How long and under what conditions has it been stored? Is the surface apt to have a different composition from the interior?

4. How many samples will have to be taken and what is the cost of analyzing each sample? If the cost is high, a sampling scheme which involves few analyses will be preferred to one which involves many replicates.

The answers to these questions allow the analyst to put together a sampling plan.

There are three basic strategies used in sampling: **random sampling**, **stratified**

2.3 ERRORS IN CHEMICAL ANALYSIS

sampling, and composite sampling. They are illustrated by the diagram shown in Fig. 2.15.

The method chosen for drawing samples depends on the physical nature of the material and where it is located. Materials such as coal, oil, or sugar beets can be found where they occur naturally or where they are stored (stockpiles, rail cars, holds of boats, bins, hoppers, drums, or bags). They may also be found in transit on conveyer belts or in pipelines, a situation which simplifies sampling considerably. The following is a description of some devices used to collect samples of various kinds of material.

1. *Gases.* The basic sampling device for gases is a glass cylinder with stopcock connections on both ends. Samples can be collected either by flushing the cylinder with 10 to 50 times its volume of gas to be sampled, by displacing a liquid in the cylinder, or by letting a sample expand into an evacuated cylinder. Taking representative samples of atmospheric gases in pollution studies presents many difficult problems. The ASTM has published recommended practices for atmospheric gas sampling (Ref. *8*).

2. *Liquids.* Homogeneous liquids are sampled quite easily by siphoning or pouring. Very few liquid samples should be assumed to be homogeneous, however, and sampling programs must be worked out for particular cases. Streams of liquid can be monitored continuously for species such as hydronium ion, dissolved oxygen, and chloride by using ion-selective electrodes. Individual or "grab" samples can be taken by lowering a weighted and stoppered bottle into the stream, pulling the stopper out by an attached line, letting the bottle fill, and removing it. The bottle can be lowered to predetermined depths when samples from stratified streams or bodies of water are being taken (e.g., sampling of freshwater lakes). A liquid sampling bottle is shown in Fig. 2.16.

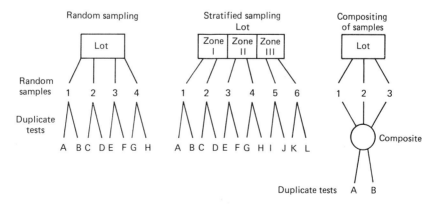

Figure 2.15 Diagrams of sampling strategies. [From C. A. Bicking, "Principles and Methods of Sampling," in *Treatise on Analytical Chemistry,* 2nd ed. (I. M. Kolthoff and P. J. Elving, eds.), © Copyright John Wiley & Sons, Inc., New York, 1978. Reproduced by permission of John Wiley & Sons, Inc.]

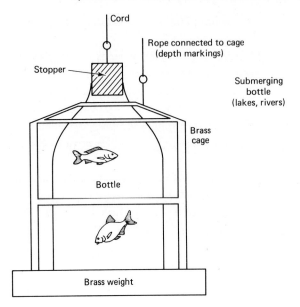

Figure 2.16 Sampling bottle for water. Cage and bottle are lowered into water using the marked rope. When they are at the desired depth, the cord is pulled, opening the stopper. The bottle fills with water, and the cage and bottle are brought to the surface.

3. *Solids.* Solids force an analyst to deal with an additional variable not present in liquids, that is, particle size. For example a sample of copper ore may consist of both large and small rocks, the large rocks perhaps containing a greater percentage of copper. An analyst who goes out and picks up a few lumps of ore from the top of a railroad car will not be taking a representative sample. In fact, the carload is probably quite stratified, with large rocks on top and "fines" near the bottom. To get a representative sample the analyst might divide the load into 8 or 16 imaginary cubes, dig down into each cube with a shovel, and remove a few kilograms from each. Alternatively, the rail car can be emptied onto a conveyer and ore removed periodically from the belt as it moves to be processed. Sampling 50 kg from each 20 tons may be adequate, depending on the size of the rocks and the desired certainty in sampling error. Samples from each load can then be ground for mixing and run by the stratified or composite strategies diagrammed in Fig. 2.15. A great deal of work has been done in the sampling of coal, as discussed in Reference (8).

While the most common sampling tools for materials like coal and ore are scoops and shovels, a device called a "sample thief" is better suited for powdered materials like fertilizer or portland cement. One thief design involves two close-fitting steel tubes with matching slots. The slots are opened and closed by rotating the inner tube. The thief is inserted with closed slots full length into the powdered material. The slots are

Figure 2.17 A sample thief.

then opened to let powder into the inner tube, then closed again as the thief is removed. A thief is shown in Fig. 2.17.

Solid samples are combined and reduced in size by two popular methods, "coning and quartering" and "riffling." In coning and quartering several samples are mixed and piled up in a cone on a board. The cone is flattened and divided into quarters. Two diagonally opposed quarters are discarded and the remaining quarters are piled into a new cone, flattened and quartered again. The process is continued until a sample of appropriate size is obtained.

Riffles are also used to reduce sample size in an unbiased manner. A riffle is a V-shaped trough divided into an even number of compartments. Every other compartment empties on one side; the alternate compartments empty on the other. Samples are thus divided in half on every pass through the riffle (see Fig. 2.18).

There are many more devices used in sampling, but these cannot be dealt with here in any detail. Specific sampling information can be obtained from many manufacturing associations and the American Society for Testing and Materials. A list appears in Reference (9).

A concern in sampling which we can only touch upon here, but which is of great importance in analysis, is proper preservation and storage of samples. Particularly

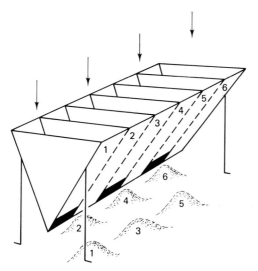

Figure 2.18 A riffle.

serious preservation problems arise with samples for trace level analysis. Biological degradation can cause the loss of carbon-, nitrogen-, and oxygen-containing materials. Chemical reactions can, for example, oxidize nitrite to nitrate, or reduce Cr(VI) to Cr(III). Precipitates which form slowly can adsorb metal ions from solution. Gases such as CO_2 and O_2 can be lost to the atmosphere or adsorbed by the walls of glass or plastic containers and be effectively removed from solutions containing them at the parts-per-billion level.

Proper care and correct handling of containers can minimize many of these problems. A great deal of time and effort is being spent in research to find ways to preserve samples for periods as short as a few days. Much of this work is sponsored by the U.S. Environmental Protection Agency and appears in their reports (Ref. *10*).

REFERENCES

1. D. S. Young, D. W. Thomas, R. B. Friedman, and L. C. Pestener, *Clinical Chemistry,* 18, 1041 (1972).
2. H. F. Martin, B. J. Gudzinowicz, and H. Fanger, *Normal Values in Clinical Chemistry,* Dekker, New York, 1975.
3. L. A. Currie, "Sources of Error and the Approach to Accuracy in Analytical Chemistry," in *Treatise on Analytical Chemistry,* 2nd ed., part 1, vol. 1, I. M. Kolthoff and P. J. Elving, eds., Wiley, New York, 1978, chapter 4.
4. J. B. Kennedy and A. M. Neville, *Basic Statistical Methods for Engineers and Scientists,* 2nd ed., Harper & Row, New York, 1976.
5. B. L. Welch, *Biometrika,* 34, 28–35 (1947).
6. A. J. Duncan, *Quality Control and Industrial Statistics,* 4th ed., Irwin, Homewood, IL, 1974.
7. *Standard Methods for the Examination of Water and Wastewater,* 14th ed., American Public Health Association, Washington, D.C., 1976.
8. *1979 Annual Book of ASTM Standards,* parts 1–48, ASTM, Philadelphia, 1979.
9. C. A. Bicking, "Principles and Methods of Sampling," in *Treatise on Analytical Chemistry,* 2nd ed., part 1, vol. 1, I. M. Kolthoff and P. J. Elving, eds., Wiley, New York, 1978, chapter 6.
10. *Manual for the Interim Certification of Laboratories Involved in Analyzing Public Drinking Water Supplies,* U.S. Environmental Protection Agency, Washington, D.C., 1976.
11. F. J. Langmyhr and I. Kjuus, "Direct Atomic Absorption Spectrometric Determination of Cadmium, Lead, and Manganese in Bone and of Lead in Ivory," *Analytica Chimica Acta,* 100, 139–144 (1978).
12. C. Y. Chan and P. N. Vijan, "Semi-automated Determination of Antimony in Rocks," *Analytica Chimica Acta,* 101, 33–43 (1978).
13. P. Goethals, C. Vandecasteele, J. Hoste, "Helium-4 and Neutron Activation

Analysis for Phosphorus in Aluminum–Silicon Alloys," *Analytica Chimica Acta,* 101, 63–70 (1978).

RECOMMENDED READING

Anders, O. U. "Representative Sampling and Proper Use of Reference Materials," *Analytical Chemistry,* 49, 33A (1977).

An Assessment of the Credibility of Data from Water and Wastewater Laboratories, Minnesota Section of the American Chemical Society, Minneapolis, 1979.

Barnard, A. J., R. M. Mitchell, and G. E. Wolf. "Good Analytical Practices in Quality Control," *Analytical Chemistry,* 50, 1079A (1978).

Box, G. E. P., W. G. Hunter, and J. S. Hunter, *Statistics for Experimenters,* Wiley, New York, 1978.

Cali, J. P., and W. P. Reed. "The Role of National Bureau of Standards SRM's in Accurate Trace Analysis," in *Proceedings, Symposium on Accuracy in Trace Analysis,* NBS Spec. Publ. No. 422, Government Printing Office, Washington, D.C., 1976.

Catalog of NBS Standard Reference Materials, NBS Spec. Publ. No. 260, National Bureau of Standards, Washington, D.C., 1975.

Eckslager, K. In *Errors, Measurement and Results in Chemical Analysis,* R. A. Chalmers, ed., Van Nostrand-Reinhold, London, 1969.

Grant, C. L., and P. A. Pelton. "Role of Homogeneity in Powder Sampling," in *Sampling, Standards and Homogeneity,* ASTM Symp. Ser. STP No. 540, ASTM, Philadelphia, 1972, pp. 16–29.

Jones, J. B., Jr. "Analytical Challenges for the Environmental Chemist," in *Environmental Chemistry and Cycling Processes,* D. C. Adriano and I. L. Brisbin, eds., Department of Energy, Washington, D.C., 1978, pp. 196–206.

Mandel, J. "Accuracy and Precision: Evaluation and Interpretation of Analytical Results," in *Treatise on Analytical Chemistry,* part 1, vol. 1, I. M. Kolthoff and P. J. Elving, eds., Wiley, New York, 1978, chapter 5.

Moore, D. S., *Statistics—Concepts and Controversies,* Freeman, San Francisco, 1979.

Natrella, M. G., *Experimental Statistics,* NBS Handbook No. 91, Government Printing Office, Washington, D.C., 1963.

Official Methods of Analysis, Association of Official Analytical Chemists, Washington, D.C., 1970.

Rhodes, R. C. "Components of Variation in Chemical Analysis," in *Validation of the Measurement Process,* J. R. DeVoe, ed., ASTM Symp. Ser. No. 63, ASTM, Philadelphia, 1977, chapter 6.

Uriano, G. A., and C. C. Gravott. "The Role of Reference Materials and Reference

Methods in Chemical Analysis," *Critical Reviews in Analytical Chemistry,* 6, 361–411 (1977).

Youdon, W. J., *Statistical Methods for Chemists,* Wiley, New York, 1951.

PROBLEMS

2.1. The following results were obtained for replicate determinations of the molecular weight of a weak acid: 119.3, 118.6, 123.4, 120.2, 121.4. Calculate the (a) range, (b) arithmetic mean, (c) median, (d) standard deviation, and (e) relative standard deviation of the results. If the weak acid is benzoic acid (molecular weight 122.1), what are the absolute and relative errors of the mean?

2.2. The following results were obtained for percent CaO in a sample: 4.10, 3.95, 4.15, 4.19, 3.90. Calculate the standard deviation, the relative standard deviation, and the standard deviation of the mean of the results.

2.3. An analyst set out to determine the equivalent weight of an acid by titrating with a standard base solution. Instead of recording the buret reading at the endpoint as 37.25 mL, however, he recorded 37.52 mL. If the equivalent weight of the acid is 204, what is the absolute error in equivalent weight resulting from the blunder?

2.4. A working curve for the determination of calcium in serum is shown below. Two serum calcium reference standards were run. From the working curve, reference A was found to be 10% too low and reference B was found to be 5% too low. Comment.

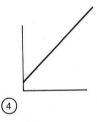

2.5. A working curve for the determination of cholesterol in serum is shown below. Two serum cholesterol reference samples whose concentrations differed by a factor of 2 were run using the working curve. Both were found to be 15% too low. Comment.

2.6. Ten results have a mean of 30.12% and standard deviation of 0.12%. Over what range will the true value lie with 95% confidence?

2.7. A student reported a mean value for three determinations of percent chloride in an unknown as 30.25% with a standard deviation of 0.30%. What are the absolute error, relative error, standard deviation of the mean, and relative standard deviation of the sample if the true value is 30.20%.

2.8. Langmyhr and Kjuus (*11*) determined lead in ivory tusks by atomic absorption. Five replicate determinations of one sample gave 0.45 ± 0.14 µg/g (standard deviation). Calculate the 90 and 95% confidence intervals. What can be said about the true value?

2.9. A student obtained the following results for percent manganese in a steel sample: 0.88, 0.68, 0.92, 0.95, 0.86, 0.83, 0.88.
  (a) Calculate the standard deviation by using the sum of the squares of the deviations and the range.
  (b) Should any of the data be rejected on statistical grounds (Q test)?
  (c) Calculate the 95% confidence interval. Do any of the data fall outside the interval?

2.10. Triplicate determinations of the molarity of a solution were 0.1117, 0.1115, and 0.1120M. Suppose a fourth titration was being contemplated. How low would the result have to be to be discarded according to the Q test ($Q_{.90}$)?

2.11. Lead is to be determined at the parts-per-billion level by a method known to give lognormally distributed results. Using the following replicate data, calculate and compare the arithmetic and geometric means and the "upper and lower" standard deviations: 8.4, 9.3, 8.0, 12.8, 8.8, 13.2, 14.5, 6.8, 17.0, 8.2, 12.7, 10.8, 9.9. The 95% confidence interval in a logarithmic distribution will not be symmetric around the geometric mean. Why? Calculate the 95% confidence interval from the data given.

2.12. Chan and Vijan (*12*) reported an atomic absorption method for determining antimony in rocks. They compared their results with those obtained by other investigators on the same standard reference material by the method of neutron activation.

	Atomic absorption	Neutron activation
ppm Sb:	4.37 ± 5.1 ($n = 9$)	4.17 ± 3.4 ($n = 6$)

The uncertainty reported is the relative standard deviation as a percentage.
  (a) Compare the mean values by a t-test.
  (b) Compare the precision of the methods. Is one more precise than the other? (Convert relative standard deviation to standard deviation.)

2.13. Goethals et al. (*13*) reported an analysis of the surface of an Al/Si alloy for phosphorus both before and after etching the alloy with strong mineral acids. The results were:

	Surface etched	Surface unetched
% P ± s:	4.22 ± 0.48 (n = 5)	5.03 ± 0.40 (n = 6)

Assuming that there is no determinate error in either set of data, is it safe to conclude that the composition of the surface of the alloy accurately represents the bulk of the material?

2.14. The following results were determined for glucose (milligrams per 100 mL) in serum: 94, 83, 86, 85. The serum was a standard certified to contain 92 mg/100 mL. Is there reason to suspect a determinate error in the analysis?

2.15. A standard sample of iron is analyzed both titrimetrically (method A) and colorimetrically (method B). The results (percent Fe) are:
Method A: 12.59, 12.37, 12.48, 12.62, 12.65
Method B: 12.29, 12.37, 12.34, 12.38, 12.24
(a) Is there a statistically significant difference between the means of these sets of data?
(b) The standard sample is certified to contain 12.45% Fe. Is there reason to suspect determinate error in either method?

2.16. The following results were obtained by two students for percent Cu determined by iodometric titration:
Student A: %Cu = 6.23, 6.54, 6.45, 6.09, 6.32, 6.30
Student B: %Cu = 5.92, 6.18, 5.70, 6.44
Is the precision of student A's work better than that of student B?

2.17. Arsenic can be determined by electrochemically generated iodine. The amount of electricity which passes (in coulombs) can be used to calculate the amount of As in the sample. The following replicate data were obtained (coulombs): 103.9, 104.2, 105.6, 103.8, 104.7, 104.4, 104.0, 103.8.
(a) Determine the mean, median, range, standard deviation, relative standard deviation, and 95% confidence interval (using a t value as well as a range coefficient).
(b) Each replicate contained exactly $\frac{1}{10}$ the number of moles in the original sample. If the original sample was a standard reference material which contained 0.4038 g of arsenic, is there statistical evidence to indicate the existence of determinate error in the method? (The number of moles of As in a replicate can be calculated from: moles As = coulombs/2(96486.6 coulombs)/mole.)

PROBLEMS

2.18. Suppose it is possible to read a buret with a standard deviation of 0.015 mL. What is the minimum volume one could measure and have an uncertainty of 1 part per thousand (relative standard deviation)? (Remember that two readings are needed to deliver the volume.)

2.19. Trace level methods are often not very precise. If an electrochemical method for cadmium is known to have a relative standard deviation of 10% at the level of 100 parts per billion, how many replicate determinations of a sample must be run to ensure a 95% confidence interval of ±12 ppb? How many should be run to ensure a 95% confidence interval of ±6 ppb?

2.20. Three measurements are performed in order to calculate a result: $X = 9.00$, $Y = 0.300$, $Z = 5.25$. If the relative determinate errors in X, Y, and Z are +0.1, +1.3, and +0.5%, what is the relative error in the result, R, if R is given by $R = XYZ$?

2.21. The standard deviation of the weight of a 1.0-g object is 0.0015 g. What is the absolute standard deviation of the weight of five of these objects?

2.22. The relationship between the potential of a glass electrode and the activity of hydrogen ion is given by the Nernst equation:

$$E_{\text{electrode}} = \text{constant} + 0.0257 \ln a_{H^+}$$

Calculate the relative error in a_{H^+} if an absolute error of 0.005 volt is made in measuring $E_{\text{electrode}}$.

2.23. The following data were obtained in the preparation of an analytical calibration curve for a luminescence analysis:

Concentration (M)	Intensity reading
1.20×10^{-3}	0.205
3.20×10^{-3}	0.402
3.80×10^{-3}	0.540
5.00×10^{-3}	0.621
5.60×10^{-3}	0.714
6.00×10^{-3}	0.840
8.15×10^{-3}	1.02
9.00×10^{-3}	1.16

(a) Calculate the mean concentration ($\bar{x}$) and mean intensity readings ($\bar{y}$) for the data.
(b) Calculate ΣX^2, ΣY^2, and ΣXY as defined in the text.
(c) Find the equation of the line that best fits the data, using the least-squares method. Plot the line you obtain along with the original data points.

(d) Calculate y values from the fitted equation corresponding to $x = 8.15 \times 10^{-3}$ and $5.00 \times 10^{-3} M$. Calculate the 95% confidence intervals around these calculated y values. Do the experimental points fall within the confidence intervals?

(e) Calculate the standard deviation and the 95% confidence interval for the slope. Plot the slopes at the 95% limits.

(f) The same set of eight standards was used to run a calibration curve for the instrument on a prior day. The slope of that curve was found to be 106.0. Is there a statistically significant difference in the slopes obtained on the two days? Assume that s_m values are the same on both days.

Chapter 3

The Equilibrium Condition

3.1 INTRODUCTION

This text puts great emphasis on applications of equilibria in chemical analysis. This chapter presents the theory of chemical equilibria, and it is written with the assumption that you have learned some basic thermodynamics in a course in general chemistry. If you discover that the treatment is too difficult, please find your general chemistry text and review its presentation. The concepts are vital to an understanding of analytical chemistry.

A state of equilibrium is a state of thermodynamic stability. It can be understood really only in terms of *thermodynamic* functions such as free energy, enthalpy, and entropy. While it is not a direct concern of thermodynamics, as a practical matter the *rate* at which a system reaches equilibrium is also quite important; for example, will your reaction proceed quickly enough for you to finish the laboratory in 3 hours? This is a problem of reaction mechanism or *kinetics* and is also important in understanding chemical equilibrium processes. It is characteristic of chemical systems that reactants are still being converted to products at equilibrium, but products are being converted back to reactants at the same rate. A chemical system at equilibrium thus appears to be unchanging, but is really *dynamic*. There is, however, no *net change* in the composition of a system at equilibrium.

We will examine equilibria from both thermodynamic and kinetic perspectives and then discuss the relationship between chemical activities and concentrations.

3.2 THE THERMODYNAMIC APPROACH

Thermodynamics is the science of heat and temperature and the laws which control the change of heat into mechanical or electrical energy. A system at equilibrium is said to be in a state of maximum stability. A chemical system can approach such a state spontaneously by reacting and changing its composition and either giving off or absorbing heat. Once equilibrium is reached there is no need for further change in composition because the system is as stable as it can become under the conditions. Let us begin with a description of the energy changes which are involved when a system equilibrates.

3.2.1 The First Law of Thermodynamics

Imagine an apparatus in which a small sealed reaction vessel is placed inside a very large water bath (see Fig. 3.1). We will insulate the water bath so well that energy can neither enter it nor leave it. We call such a closed system a *thermodynamic universe.* We will seal the reaction vessel tightly to keep it from exchanging material with the water bath, but will allow it to exchange thermal energy (heat).

The energy of a system is described in terms of *states* determined by the variables pressure, volume, temperature, and concentration. Consider a chemical system in the reaction vessel of the model universe. In any particular state the chemical species will have a certain *internal energy*, E_s, the energy of the molecules and their bonds. Similarly, in any state the water bath will have a certain internal energy, E_b. The first law of thermodynamics says that the energy of the model universe will stay constant during a change in the state of the vessel and bath. In other words, if the vessel loses energy when the chemical system reacts, the water bath gains exactly the same amount of energy. An equation that describes this law is

$$E(\text{universe}) = E_s + E_b = \text{constant}$$

Now if a change occurs between an initial state and a final state, it must be true that

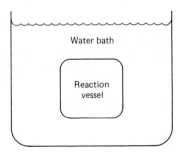

Figure 3.1 Model thermodynamic universe.

3.2 THE THERMODYNAMIC APPROACH

$$E(\text{universe})_{\text{initial}} = E(\text{universe})_{\text{final}}$$
or
$$(E_s + E_b)_{\text{initial}} = (E_s + E_b)_{\text{final}}$$

If the last expression is rearranged to show changes in the components of the universe, then

$$[(E_s)_{\text{final}} - (E_s)_{\text{initial}}] = -[(E_b)_{\text{final}} - (E_b)_{\text{initial}}]$$

If the Δ notation (meaning "change in ...") is introduced, then the mathematical statement of the first law is simply

$$\Delta E_s = -\Delta E_b$$

This says that energy is conserved in the model universe.

Enthalpy Consider the situation in which the reaction vessel of the model universe contains a half-mole of hydrogen gas (H_2) and a half-mole of chlorine gas (Cl_2) at a temperature of 25°C. In the chemical reaction that ensues, the pressure of the vessel will be maintained at a constant value. The reaction between hydrogen and chlorine will generate 92.31 kilojoules of energy for every mole of hydrogen chloride (HCl) produced:

$$\tfrac{1}{2}H_2(g) + \tfrac{1}{2}Cl_2(g) = HCl(g) + 92.31 \text{ kJ/mole} \qquad (25°C)$$

There are only two sources or sinks of energy in the model universe: the chemical reaction and the water bath. Any energy released by the chemical reaction must go to the water bath. Since in this example the reacting system releases energy, its final state must be of lower energy than its initial state.

Under conditions of constant pressure and temperature the energy change just described is called an *enthalpy of reaction*.† Enthalpies are symbolized by H. When a

---

† In the example volume was not fixed. Enthalpies must account for any volume changes during a reaction. In the case of the dissociation reaction for hydrogen,

$$H_2(g) = 2H(g) \qquad \Delta H = 436.0 \text{ kJ/mole}$$

there are twice as many fundamental particles in the products as there are in the reactants. At constant pressure the volume of the product is twice the volume of the reactant, and some of the energy supplied to the system would have to go into the work of expanding the volume. This work is given by the product of the pressure and the change in volume, $P\,\Delta V$. Thus the energy exchanged in the course of the reaction has two components:

$$\Delta E_s = \Delta H_{\text{reaction}} - P(\Delta V_{\text{reaction}})$$

Reactions that occur in solutions are accompanied by extremely small volume changes since liquids are difficult to compress, and the energy change in the system is very nearly equal to the enthalpy change.

reaction is accompanied by a positive change in enthalpy ($\Delta H > 0$), the reaction is called *endothermic*. When a reaction has a negative change in enthalpy ($\Delta H < 0$), it is called *exothermic*. In the reaction which forms HCl from H_2 and Cl_2, energy is released by the system, and the reaction is *exothermic:* $\Delta H(HCl) = -92.31$ kJ/mole.

Standard States and Reference States By general agreement among scientists, thermodynamic data are tabulated in standard states and reference states. A substance is in its standard state when it is at a pressure of 1 atm and a temperature of 298 K (25°C). For dissolved ionic or molecular substances a concentration (or, more accurately, an "activity"; see below) of 1 mole per liter is implied. Quantities which refer to standard states carry a superscript zero. For example, a standard enthalpy is signified by the symbol H^0.

Chemists find it useful to deal with thermodynamic quantities such as enthalpy relative to reference states rather than on an absolute scale.† Reference states for thermodynamic data are the chemical elements in their most stable form at standard conditions. For example, all compounds are assigned *standard molar enthalpies of formation*, ΔH_f^0 that correspond to the enthalpy change when 1 mole of a compound is formed from elements in their most stable form at 25°C. The standard molar enthalpies of the elements in this form are taken as zero.

EXAMPLE 3.1 Standard Molar Enthalpy Change

The standard molar enthalpies of formation of CO, O_2, and CO_2 are given below the balanced equation for the formation of CO_2:

$$\text{CO(g)} + \tfrac{1}{2}O_2(g) = CO_2(g)$$
$$H_f^0 \quad -110.53 \quad\quad 0.0 \quad\quad -393.51 \text{ (kJ/mole)}$$

What is ΔH^0 for this reaction?

Solution

The change in standard enthalpies is

$$\Delta H^0 \text{ (reaction)} = H_f^0 \text{ (products)} - H_f^0 \text{ (reactants)}$$

$$\Delta H^0 = -393.51 - (-110.53) = -282.98 \text{ kJ/mole}$$

---

† On an absolute scale the energies in chemical reactions are minuscule. For example, the absolute energy of 1 mole (18.02 g) of water is calculated from the Einstein relationship ($E = mc^2$) to be about 1.6×10^{12} kJ/mole. To heat this amount of water by 1°C would require only 0.075 kJ. To electrolyze it to hydrogen and oxygen at 25°C would require 286 kJ. On an absolute energy scale such trifling quantities would be lost in the uncertainty in measuring the speed of light, c.

3.2 THE THERMODYNAMIC APPROACH

The reaction in which CO_2 is formed from CO is thus *exothermic*. Notice that the reference state for oxygen is molecular oxygen at 25°C, the most stable form at that temperature, and ΔH_f^0 is set equal to zero.

Values of ΔH_f^0 for many compounds are available in reference books such as the *CRC Handbook of Chemistry and Physics* and *Lange's Handbook of Chemistry*.

State Functions Enthalpy is but one of a set of thermodynamic properties called *state functions*. These functions depend only on the initial and final states of a system and not on the mechanisms by which changes in the system take place. This is a very important property from the point of view of the chemist. When equations for chemical reactions are added or subtracted to describe a new reaction, the corresponding state function quantities can be added or subtracted to give state function quantities for the new reaction. You may recall from general chemistry that enthalpies can be added or subtracted (Hess's law). In the chapter on oxidation-reduction equilibria we will be adding the *free energies* of reactions, a state function to be described shortly.

3.2.2 Spontaneous Reactions

At some time every student of chemistry asks what makes a system react. Knowing a little about enthalpies and the spontaneous reactions that, for example, give off heat in a wood-burning stove, a novice might be tempted to say that enthalpy changes cause systems to react and that exothermic reactions ought to be spontaneous. However, there are many spontaneous *endothermic* reactions. For example, 1 mole (58.45 g) of NaCl is quite soluble in pure water (it dissolves spontaneously), but the standard enthalpy of the reaction†

$$NaCl(s) \stackrel{H_2O}{=} Na^+(aq) + Cl^-(aq)$$

is +3.89 kJ/mole. This spontaneous process is endothermic, that is, the solution becomes cool. Apparently some factor other than enthalpy is also involved.

Entropy The *second law of thermodynamics* allows us to define spontaneity in terms of the function *entropy (S)*. The second law says that if an isolated system (e.g., the model universe used above) is not in a state of equilibrium, the only changes that will occur spontaneously will be those which increase the entropy of the isolated system. Equilibrium will exist in the isolated system when entropy is as large as it can possibly be, and no further change in entropy is possible ($\Delta S = 0$).

---

† It is common in writing a reaction of a solid with a solvent to exclude the solvent from the balanced equation. The notation (aq) means that the ions are surrounded by water molecules, that is, are in solution.

At the microscopic level entropy is pictured as the amount of disorder a system has. For example, other things being equal, the entropy of a liquid substance is greater than that of a crystalline substance. The reason is that the atoms in the liquid are in considerable disorder, whereas the atoms in the crystal are in a very regular arrangement.

There is a simple relationship between the thermal energy that is transferred to a body (ΔE) and the change in the entropy of the body (ΔS) that results from the transfer,

$$\Delta S = \frac{\Delta E}{T} \quad \text{(at constant volume)}$$

In this relationship T is the absolute (Kelvin) temperature. Entropies are expressed in joules per degree Kelvin. Notice that the colder a body is, the greater is its entropy change per joule of energy added to it.

To investigate the relationship of entropy to spontaneity, consider a universe made up of two systems, a stone at temperature T_1 and a water bath at temperature T_2. The question is, if T_1 is greater than T_2, in which direction will heat flow, from stone to water bath, or from water bath to stone? Experience tells us that heat should flow from the warmer stone to the cooler water bath. The change in entropy of the stone (which *loses* heat) is given by

$$\Delta S_{stone} = \frac{-\Delta E_{stone}}{T_1}$$

while that of the water bath is

$$\Delta S_{bath} = \frac{\Delta E_{bath}}{T_2}$$

Since there are only two systems in this universe, the first law requires that

$$-\Delta E_{stone} = \Delta E_{bath}$$

The total change in entropy is given by

$$\Delta S_{universe} = \Delta S_{bath} + \Delta S_{stone}$$

or

$$\Delta S_{universe} = \Delta E \left(\frac{1}{T_2} - \frac{1}{T_1} \right)$$
$$\phantom{\Delta S_{universe} = \Delta E \ \ }\text{(bath)}\ \ \text{(stone)}$$

3.2 THE THERMODYNAMIC APPROACH

The spontaneous transfer of heat from the stone at T_1 to the water bath at T_2 gives a positive value of $\Delta S_{universe}$. The second law says that $\Delta S_{universe}$ must always be positive.

You should take note of the fact that eventually the stone and the water bath will be at the same temperature ($T_1 = T_2$). Under these conditions

$$\Delta S = \Delta E \left(\frac{1}{T_2} - \frac{1}{T_1} \right) = 0$$

and no further change occurs. This is a condition of equilibrium.

It is important to point out that the second law states that the entropy change of a *universe* must be positive for a change to be spontaneous. In the model universe consisting of a reaction vessel and a water bath, ΔS for the vessel *and* water bath must be positive for the change to be spontaneous. For the chemical reaction inside the vessel, ΔS could be negative if enough thermal energy is transferred between the vessel and the bath. This brings us to the discussion of free energy, a function of both entropy and thermal energy (enthalpy).

Free Energy Entropy and enthalpy considerations taken together allow a chemist to make accurate predictions about the direction and extent of chemical reactions. In the case of the dissociation of hydrogen,

$$H_2 = 2H(g)$$

the large positive standard enthalpy (436 kJ/mole) is partially offset by a positive standard entropy change (125.6 J/mole).[†] As the temperature of the system is raised, the entropy factor becomes more and more important, until at a temperature of about 5000 K the reaction proceeds readily to the right.

The thermodynamic function that drives a reaction is *free energy (G)*, a function of enthalpy, entropy, and temperature,[‡]

$$G = H - TS$$

A change in free energy of a system between two states at constant temperature is given by

$$\Delta G = \Delta H - T \Delta S$$

[†] Dissociation creates disorder, making two atoms for each molecule. The entropy change is therefore positive.

[‡] This is the *Gibbs free energy*, the free energy of a system at constant temperature and pressure. The Helmholtz free energy (*A*) is the free energy of a system at constant volume and temperature. Chemists are usually interested in the Gibbs free energy since constant-pressure conditions are far more common than constant-volume conditions in the laboratory.

For a process in an isolated system (a "universe") to occur spontaneously, the free energy change which accompanies it must be negative. A process for which ΔG is positive will not occur spontaneously. At equilibrium the free energy of a system is at its lowest possible value, and no further change in free energy will occur ($\Delta G = 0$).

EXAMPLE 3.2 Free Energy and Spontaneity

An isolated reaction vessel kept at constant pressure contains exactly 1 mole of nitrosyl chloride (ClNO), 1 mole of molecular chlorine (Cl$_2$), and 1 mole of nitric oxide (NO). Will the reaction

$$\text{ClNO(g)} \rightarrow \tfrac{1}{2}\text{Cl}_2(g) + \text{NO(g)}$$

be spontaneous at 25°C? At 400°C?
Use the data:

$S^0(\text{Cl}_2) = 222.97$ J/mole K $H^0(\text{Cl}_2) = 0$ kJ/mole
$S^0(\text{NO}) = 210.60$ J/mole K $H^0(\text{NO}) = 90.25$ kJ/mole
$S^0(\text{ClNO}) = 263.6$ J/mole K $H^0(\text{ClNO}) = 52.59$ kJ/mole

Assume that these values apply at both 25 and 400°C.

Solution

First we calculate ΔH^0 and ΔS^0 for the reaction:

$$\Delta H^0(\text{reaction}) = H^0(\text{products}) - H^0(\text{reactants})$$
$$= [H^0(\text{NO}) + \tfrac{1}{2}H^0(\text{Cl}_2)] - H^0(\text{ClNO})$$
$$= (90.25 + 0.00) - 52.59 = 37.66 \text{ kJ/mole}$$
$$\Delta S^0(\text{reaction}) = S^0(\text{products}) - S^0\text{reactants}$$
$$= [S^0(\text{NO}) + \tfrac{1}{2}S^0(\text{Cl}_2)] - S^0(\text{ClNO})$$
$$= 210.60 + 111.49 - 263.6 = 58.5 \text{ J/mole}$$

We then use the free energy relationship to calculate ΔG^0 at 25 and 400°C:

25°C (298 K): $\Delta G^0 = \Delta H^0 - T\Delta S^0$
$\qquad\qquad\qquad = 37.66$ kJ/mole $- 298$ K $(58.5 \times 10^{-3}$ kJ/mole$)$
$\qquad\qquad\qquad = 20.2$ kJ/mole
400°C (673 K): $\Delta G^0 = \Delta H^0 - T\Delta S^0$
$\qquad\qquad\qquad = 37.78$ kJ/mole $- 673$ K $(58.5 \times 10^{-3}$ kJ/mole$)$
$\qquad\qquad\qquad = -1.59$ kJ/mole

3.2 THE THERMODYNAMIC APPROACH

> At 25°C ΔG^0 is positive and ClNO will not spontaneously form Cl_2 and NO. At 400°C, however, ΔG^0 is negative and the formation of Cl_2 and NO may proceed spontaneously. The mixture of 1 mole of each of the products and the reactant is at equilibrium at a temperature of 372°C (645 K). You can verify this by setting ΔG^0 equal to zero and solving the free energy expression for T.

Often an analogy is drawn between the free energy change in a spontaneous reaction and the energetics of a boulder rolling from a mountain into a valley (see Fig. 3.2). Both systems involve potential energy changes. Potential energy in the gravitational field "drives" the mechanical system until a state of minimum potential energy is reached. Analogously, the free energy of the chemical system decreases until it reaches a minimum value. When the free energy no longer changes, a state of equilibrium exists. Later you will need to know that the free energy change in a reaction determines the maximum amount of work that can be gotten from a chemical system.

3.2.3 Concentration Changes, Free Energy, and the Equilibrium Constant

Consider as a model system for this discussion the gas phase reaction in which bromine monochloride (BrCl) is produced from bromine (Br_2) and chlorine (Cl_2) at 25°C,

$$Br_2 + Cl_2 = 2BrCl$$

When we mix 0.1 mole of Br_2 and 0.1 mole of Cl_2 in a previously evacuated 1-liter vessel, we observe a spontaneous reaction. There are changes in enthalpy and entropy such that the free energy of the system decreases until a state of equilibrium is reached. There are also changes in the concentrations of the species as the system approaches

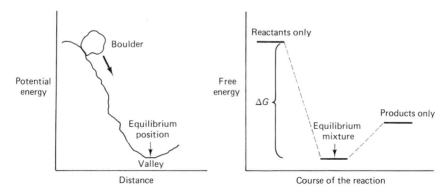

Figure 3.2 Potential energy and free energy.

equilibrium. Once the system reaches equilibrium, concentrations cease to change; any change that occurred would only increase the free energy of the system.

A plot of the free energy of the model system as a function of the concentrations of the species is shown in Fig. 3.3. A minimum free energy of about 19.2 kJ/mole occurs at concentrations of Cl_2 and Br_2 of 0.067 mole/liter and a BrCl concentration of 0.066 mole/liter.

The free energy of a species is a function of its standard free energy (G^0) and the logarithm of its concentration (C). For the species Cl_2 and Br_2 in the model system the relationships are

$$G_{Cl_2} = G^0_{Cl_2} + RT \ln(f_{Cl_2} C_{Cl_2})$$
$$G_{Br_2} = G^0_{Br_2} + RT \ln(f_{Br_2} C_{Br_2})$$

The relationship for the free energy of BrCl must take into account the fact that 2 moles of BrCl are produced for every mole of Br_2 and Cl_2 consumed:

$$2G_{BrCl} = 2G^0_{BrCl} + 2RT \ln(f_{BrCl} C_{BrCl})$$
$$= 2G^0_{BrCl} + RT \ln(f_{BrCl} C_{BrCl})^2$$

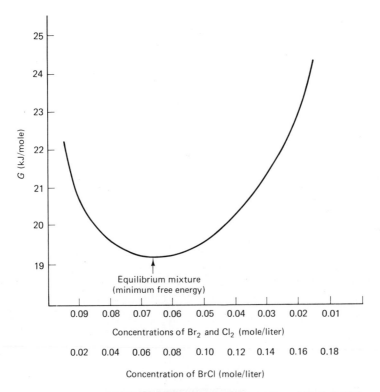

Figure 3.3 Free energy vs. composition for $Br_2 + Cl_2 = 2BrCl$ (25°C).

3.2 THE THERMODYNAMIC APPROACH

In these expressions R is the gas constant (8.314 J/mole K) and T is the absolute temperature. To simplify matters, at 298 K

$$RT \ln(X) = 5.710 \log(X) \quad \text{(units kJ/mole)}$$

The logarithmic terms in the above expressions contain the concentration and a proportionality constant, f, for each reacting species. The product fC is called the *activity* of a species and represents its effective concentration in the vessel.†

As is the case in any chemical system, for the reaction of Br_2 and Cl_2 to proceed the free energy change for the system must be negative, or

$$\Delta G_{\text{reaction}} = G_{\text{products}} - G_{\text{reactants}} < 0$$

In terms of the model system, the relationship is

$$\Delta G_{\text{reaction}} = 2G_{\text{BrCl}} - (G_{\text{Br}_2} + G_{\text{Cl}_2})$$

When we substitute into this expression the three concentration expressions shown above, we get

$$\Delta G_{\text{reaction}} = \{2G^0_{\text{BrCl}} - (G^0_{\text{Br}_2} + G^0_{\text{Cl}_2})\} + 5.710 \log(f_{\text{BrCl}} C_{\text{BrCl}})^2 \\ - 5.710 \log(f_{\text{Br}_2} C_{\text{Br}_2}) - 5.710 \log(f_{\text{Cl}_2} C_{\text{Cl}_2})$$

This should be more recognizable when terms are collected:

$$\Delta G_{\text{reaction}} = \Delta G^0_{\text{reaction}} + 5.710 \log \frac{(f_{\text{BrCl}} C_{\text{BrCl}})^2}{(f_{\text{Br}_2} C_{\text{Br}_2})(f_{\text{Cl}_2} C_{\text{Cl}_2})}$$

$$= \Delta G^0_{\text{reaction}} + 5.710 \log Q$$

where Q is called the *activity quotient* for the system. When the system is at equilibrium, $\Delta G_{\text{reaction}} = 0$, and

$$\Delta G^0_{\text{reaction}} = -5.710 \log Q$$

At equilibrium Q takes on a special meaning and is called the *equilibrium constant, K*, that is

$$Q = K \quad \text{when } \Delta G_{\text{reaction}} = 0$$

You should see that the value of K is determined by the *standard free energy change*

† This idea will be developed more fully near the end of the chapter.

for the reaction, and *not* by the concentrations of reactants or product. This is a very important idea.

EXAMPLE 3.3 Free Energy and the Equilibrium Constant

The standard free energy change for the dissociation of dichloroacetic acid, $Cl_2CHCOOH + H_2O = Cl_2CHCOO^- + H_3O^+$ is 7.18 kJ/mole at 298 K. Calculate the value of the equilibrium constant for the reaction.

Solution

At equilibrium $\Delta G_{reaction} = 0$, and $\Delta G^0 = -5.710 \log K$.
Solving for K:

$$K = \text{antilog}(-\Delta G^0/5.710) = 5.5 \times 10^{-2} = 0.055$$

Notice that we did not need to know the concentrations of any of the species at equilibrium to solve this problem. That is because K does not depend on concentrations.

Many reactions have more complex stoichiometries than the $Br_2 + Cl_2 = 2BrCl$ system. For the general reaction

$$aA + bB = cC + dD$$

the general free energy equation is

$$\Delta G_{reaction} = \Delta G^0_{reaction} + 5.710 \log \frac{a_C^c a_D^d}{a_A^a a_B^b}$$

The letter a now refers to activity. Remember that $a = fC$.

EXAMPLE 3.4 The Activity Quotient

Write the correct activity quotient for the reaction

$$2Fe^{3+} + 2I^- = 2Fe^{2+} + I_2$$

Solution

$$Q = \frac{(a_{Fe^{2+}})^2 (a_{I_2})}{(a_{Fe^{3+}})^2 (a_{I^-})^2}$$

3.2 THE THERMODYNAMIC APPROACH

It is important to notice that the free energy expression above predicts that the initial concentrations in a reacting system (and therefore the size of Q) determine the direction in which the reaction can proceed. If the log term is positive and larger than $\Delta G^0_{\text{reaction}}$, then $\Delta G_{\text{reaction}}$ is positive, and the reaction will not proceed spontaneously as it is written but will go in the *reverse* direction. This can happen when the solution contains a vast excess of "product" and a small amount of "reactant" and K is small.

EXAMPLE 3.5 Initial Activities and "Direction" of a Reaction
The reaction

$$2Fe^{3+} + 2I^- = 2Fe^{2+} + I_2$$

has an equilibrium constant of 340 at 25°C. When the initial activities of Fe^{+3} and I^- are 10^{-2} and the initial activities of Fe^{2+} and I_2 are 10^{-3}, in which direction will the reaction proceed?

Solution

The value of the equilibrium constant permits the calculation of $\Delta G^0_{\text{reaction}}$. At equilibrium $\Delta G_{\text{reaction}} = 0$, and

$$\Delta G^0_{\text{reaction}} = -5.710 \log K$$
$$= -5.710 \log 340 = -14.45 \text{ kJ/mole}$$

For the reaction to proceed to the right, that is, produce more I_2 and consume I^-,

$$\Delta G^0_{\text{reaction}} + 5.710 \log Q \quad \text{must be} <0$$

Q is calculated from initial activities:

$$Q = \frac{(a_{Fe^{2+}})^2(a_{I_2})}{(a_{Fe^{3+}})^2(a_{I^-})^2} = \frac{(10^{-3})^2(10^{-3})}{(10^{-2})^2(10^{-2})^2} = 10^{-1}$$

and $\quad \Delta G^0_{\text{reaction}} + 5.710 \log Q = -14.45 + 5.710 \log 10^{-1}$
$$= -14.45 - 5.710 \quad = -20.16 \text{ kJ/mole}$$
$$(= \Delta G_{\text{reaction}})$$

Since $\Delta G_{\text{reaction}}$ is negative, the reaction goes to the right as written, that is, to produce more I_2 at the expense of I^-.

EXAMPLE 3.6 Initial Activities and "Direction" of a Reaction

In which direction will the reaction in the previous example go if the initial activities of Fe^{3+} and I^- are 10^{-4} and those of Fe^{2+} and I_2 are 10^{-2}?

Solution

From the previous example we know that $\Delta G^0_{reaction} = -14.45$ kJ/mole and that in order to produce more I_2,

$$\Delta G^0_{reaction} + 5.710 \log Q \quad \text{must be } <0.$$

Under the initial concentration conditions in this problem,

$$Q = \frac{(10^{-2})^2(10^{-2})}{(10^{-4})^2(10^{-4})^2} = \frac{10^{-6}}{10^{-16}} = 10^{10}$$

and $\quad \Delta G^0_{reaction} + 5.710 \log Q = -14.45 + 5.710 \log 10^{10}$
$$= -14.45 + 57.10 \quad\quad = 42.65 \text{ kJ/mole}$$
$$(= \Delta G_{reaction})$$

Since this quantity is greater than zero, the reaction *consumes* I_2 and proceeds in the direction

$$2Fe^{2+} + I_2 \rightarrow 2Fe^{3+} + 2I^-$$

3.2.4 Characteristics of Chemical Equilibria

"Complete" Reactions A question often encountered in analytical chemistry is whether a particular reaction is *complete* under a given set of conditions. A reaction is generally judged to be quantitative if 99.9% of the reactants are converted to products. To arrive at a result that can be generalized, let us consider the following abstract example.

EXAMPLE 3.7 Completeness of Reactions

Given the general reaction

$$A + B = C$$

calculate the equilibrium constant needed to produce C from A and B quantitatively. Assume that A and B have the same activity.

Solution

Initial conditions: $a_A = a_B$, $a_C = 0$. If the reaction is to be 99.9% complete, then at equilibrium: $a_A = a_B = 0.001 a_C$.
The expression for K is

$$K = \frac{a_C}{a_A a_B} = \frac{a_C}{(10^{-3} a_C)^2}$$

$$= \frac{10^6}{a_C}$$

This is a perfectly general expression. But whether K is large enough for the reaction to be quantitative depends on a_C. If $a_A = 0.1$ initially, then $a_C = 0.0999$ at equilibrium, and

$$K = \frac{10^6}{a_C} = \frac{10^6}{9.99 \times 10^{-2}} \simeq 10^7$$

However, if $a_A = 0.001$ initially, then a_C must equal 9.99×10^{-4} at equilibrium, and

$$K \text{ must equal } \frac{10^6}{a_C} = \frac{10^6}{10^{-3}} = 10^9$$

for the reaction to be quantitative.

This general reaction is very much like the precipitation reactions you will study in Chapters 8 through 11.

A second common question that arises is whether a reaction with a rather small equilibrium constant can be "pushed" to completion. This question is related to LeChatelier's principle, an idea you may recall from general chemistry. According to this principle, when a stress is applied to a system at equilibrium, the system will readjust to a new equilibrium position that relieves the stress. In gas phase reactions the term stress means a change in pressure, while in solutions it means a change in activity or concentration. Remember that the equilibrium constant itself does not change. When the concentration of one species is purposely altered, the concentrations of the others will adjust to conserve K.

EXAMPLE 3.8 LeChatelier's Principle

The equilibrium constant expression for the general reaction

$$A + B = C$$

is

$$K = \frac{a_C}{a_A a_B}$$

What steps could an experimentalist take to increase the activity of species C at equilibrium?

Solution

The equilibrium constant will remain the same as long as the temperature of the system is not changed. To increase a_C, either a_A or a_B can be *increased*. If the experimentalist wants to convert species A to species C quantitatively, more B can be added to the solution.

Effect of Temperature on the Equilibrium Constant LeChatelier's principle can be used to predict the qualitative effect of a temperature change on the position of an equilibrium. For example, consider the reaction

$$H_2(g) + Cl_2(g) = 2HCl + 184.6 \text{ kJ}$$

If a mixture of H_2, Cl_2, and HCl is allowed to reach equilibrium at some temperature and is then cooled, the reaction will shift to form more HCl. On the other hand, if the original equilibrium mixture is heated, the reaction will shift to produce H_2 and Cl_2. In either case, the equilibrium position changes to relieve the "stress" created by changing the temperature of the system.

The quantitative effect of temperature on the equilibrium constant can be approximated by using an expression called the van't Hoff equation. According to this expression, the equilibrium constants K_1 and K_2 for a system at two temperatures T_1 and T_2 are related by

$$\log \frac{K_2}{K_1} = \frac{\Delta H^0}{2.303 R} \left(\frac{1}{T_1} - \frac{1}{T_2} \right)$$

where ΔH^0 is the standard enthalpy change for the reaction and R is the gas constant (8.314 J/mole K). ΔH^0 is assumed to be constant in the temperature interval from T_1 to T_2.

3.2 THE THERMODYNAMIC APPROACH

EXAMPLE 3.9 Temperature Effect

The enthalpy change for the reaction

$$H_2(g) + Cl_2(g) = 2HCl$$

is -184.6 kJ when it proceeds to the right as written. The equilibrium constant at 25°C (298 K) is

$$K = \frac{(a_{HCl})^2}{(a_{H_2})(a_{Cl_2})} = 10^{16.70}$$

Calculate the approximate value of K at 125°C (398 K).

Solution

Using the van't Hoff expression:

$$\log \frac{K_2}{K_1} = \frac{\Delta H^0}{2.303R}\left(\frac{1}{T_1} - \frac{1}{T_2}\right)$$

$$= \frac{-184.6 \times 10^3}{19.15}\left(\frac{1}{298} - \frac{1}{398}\right)$$

$$= -8.13$$

$$\frac{K_2}{K_1} = 10^{-8.13} \quad K_2 = K_1 10^{-8.13} = 10^{16.70} 10^{-8.13} = 10^{8.57}$$

Thus, increasing the temperature by 100°C decreases the equilibrium constant by a factor of about 10^8.

There are some important limitations to the use of the van't Hoff equation. First, the fact that ΔH^0 is not quite independent of temperature may result in inaccurate predictions. In general, the van't Hoff relationship is useful for reactions in which ΔH^0 is larger than about 10 kJ/mole. If ΔH^0 is much smaller than this, then changes in ΔH^0 with temperature may be relatively large, and an entirely wrong prediction could be made.

The van't Hoff equation also assumes that K for a system continues to change in the same way (monotonically) with temperature over the range T_1 to T_2. This assumption may be incorrect, especially over wide temperature ranges. For example, the equilibrium constant for the dissociation of acetic acid,

$$CH_3COOH + H_2O = CH_3COO^- + H_3O^+$$

increases and then *decreases* as temperature is changed from 0 to 100°C [$K = 1.66 \times 10^{-5}$ at 0°C, 1.75×10^{-5} at 22°C (maximum), and 1.54×10^{-5} at 60°C]. In a course in physical chemistry you will probably learn about approximating the effect of temperature on equilibrium constants in a way that more accurately reflects reality.

3.2.5 Summary of the Thermodynamic Approach

When a chemical system is in a state of equilibrium it is in a state of maximum stability. In an *isolated system* (a universe containing a reaction vessel and a large water bath), equilibrium will be achieved when the entropy of the system is as large as it can possibly be under a set of temperature, pressure, and volume conditions. Once equilibrium exists under a set of conditions, any net change in entropy decreases the stability of the system. Thus, at equilibrium $\Delta S = 0$.

When an isolated system undergoes a spontaneous change, its entropy must *increase*. The entropy change of a spontaneous *reaction* taking place in *part* of the system (the vessel) may be negative if the reaction generates enough thermal energy (*enthalpy*) to transfer to the surroundings (the water bath). The combination of changes in enthalpy and entropy in the *reaction* determine the free energy change in the reaction, according to the expression

$$\Delta G = \Delta H - T \Delta S$$

where T is the absolute temperature. When the free energy change of a *reaction* is *negative,* the reaction can proceed spontaneously. When the reaction attains equilibrium, the free energy no longer changes ($\Delta G = 0$).

The extent to which a reaction favors reactants or products depends on the size and sign of the standard free energy change (ΔG^0) and the temperature of the system. Reactions in which ΔG^0 is large and negative have large equilibrium constants, according to the relationship

$$-\Delta G^0 = RT \ln K$$

An equilibrium constant is not a function of the activities of the species involved in the reaction.

An activity quotient (Q) is equal to the equilibrium constant (K) *only* when a reaction is at equilibrium, that is, when $\Delta G = 0$. Under all nonequilibrium conditions (e.g., when reactants are mixed together at the start of a reaction), $Q \neq K$. For a reaction such as

$$A + B = C + D$$

Q is given by

$$Q = \frac{a_C a_D}{a_A a_B}$$

For the reaction to proceed to the "right" (to produce C and D),

$$\Delta G^0(\text{reaction}) + RT \ln Q \quad \text{must be} < 0$$

When there is no net reaction,

$$\Delta G^0(\text{reaction}) + RT \ln Q = 0$$
$$\text{or} \quad -\Delta G^0(\text{reaction}) = RT \ln K$$

The effect of a temperature change on the equilibrium constant for a reaction is determined (at least to a first approximation) by the sign and magnitude of the standard enthalpy change for the reaction. The van't Hoff relationship is used to calculate the effect:

$$\log \frac{K_2}{K_1} = \frac{\Delta H^0}{2.303 R} \left(\frac{1}{T_1} - \frac{1}{T_2} \right)$$

3.3 THE KINETIC APPROACH

An interesting way to describe an equilibrium condition is to examine the *rates* of processes (that is, the *kinetics*) that occur in a model system as it approaches equilibrium. Consider the reaction in which hydrogen iodide (HI) is formed from hydrogen (H_2) and iodine (I_2) in the gas phase,

$$H_2(g) + I_2(g) = 2HI(g)$$

The following facts have been established by experiment:

1. The reaction is chemically *reversible*. One can take a mixture of H_2 and I_2 and generate HI (the "forward" reaction) or take a pure sample of HI and generate H_2 and I_2 (the "reverse" reaction).
2. A condition of equilibrium will exist in the system. By definition, at this point there are no longer any net changes in concentration of H_2, I_2, or HI.
3. At room temperature a mixture of H_2 and I_2 reacts very slowly. At a temperature of 700 K (427°C), the equilibrium is established in about 3 hours.

4. The rate at which HI is produced from H_2 and I_2 depends on the concentrations of H_2 and I_2. The rate at which H_2 and I_2 are produced from HI depends on the *square* of the HI concentration.

Let us examine these facts more closely.

The rate of formation of HI depends directly on the concentrations of H_2 and I_2. This can be expressed mathematically as

$$r_f = k_f[H_2][I_2]$$

where r_f represents the rate in the forward direction (to produce HI) and k_f is the corresponding proportionality constant, called the *forward rate constant*. Let us perform an experiment in which 0.1 mole of H_2 and 0.1 mole of I_2 are placed in a 10-liter reaction flask and raised to a temperature of 700 K. The changes in concentration of the species H_2, I_2, and HI over a period of several hours are plotted in Fig. 3.4a. Notice that [HI] increases quite rapidly early in the experiment. As time goes on the rate of production of HI gradually decreases until [HI] is constant after about 5000 seconds. The H_2 and I_2 are consumed rapidly at the beginning of the experiment, but more slowly as time goes by. Their concentrations also cease to change after about 5000 seconds.

The rate of formation of H_2 and I_2 is proportional to the square of the HI concentration, or

$$r_r = k_r[HI]^2$$

where r_r is the rate of the reaction in the reverse direction (to consume HI) and k_r is the *reverse rate constant*. In a second experiment exactly 0.2 mole of HI is placed in a 10-liter reaction flask, and the concentrations of HI, H_2, and I_2 are monitored over a period of several hours. These results are also plotted in Fig. 3.4a, and show that [HI] decreases more and more slowly with time. The H_2 and I_2 concentrations increase more slowly with time until there are no longer any changes in composition. The plots show that the composition of this system after 13,000 seconds is the same as the composition of the system in the first experiment after about 5000 seconds. This is a general property of systems which exist in equilibrium: equilibrium can be approached from both reaction directions and identical distributions of species are generated.

The results of the two rate experiments, when combined, give an interesting picture of the equilibrium. The forward reaction

$$H_2 + I_2 \rightarrow 2HI$$

which is initially fast, slows down as it creates more and more HI. The production of HI "feeds" the reverse reaction

$$H_2 + I_2 \leftarrow 2HI$$

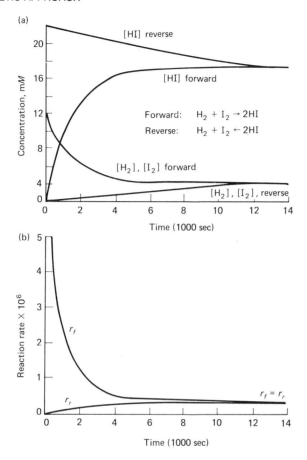

Figure 3.4 (a) Concentration-time profiles for forward and reverse reactions. (b) Reaction rates of forward and reverse processes. Conditions: Forward initially $[H_2]$ = $[I_2]$ = $0.010M$, $[HI]$ = $0.0M$; reverse initially $[HI]$ = $0.010M$, $[H_2]$ = $[I_2]$ = $0.0M$; temperature 698.6 K; k_f = 6.7×10^{-2}, k_r = 1.24×10^{-3} cc/mole sec. Data from Taylor and Crist (see text).

and makes *it* go faster and faster. Eventually a balance is struck in the system, HI is produced as fast as it is consumed, and [HI] becomes constant. We can express the condition as

$$r_f = r_r$$

Substituting from the previous expressions gives

$$k_f[H_2][I_2] = k_r[HI]^2$$

This last expression can be rearranged to give the equilibrium constant expression

$$\frac{[HI]^2}{[H_2][I_2]} = \frac{k_f}{k_r} = K$$

Using the data of Taylor and Crist (*1*) for the forward and reverse reactions at 700 K, we can calculate the value of K:

$$K = \frac{k_f}{k_r} = \frac{6.7 \times 10^{-2}}{1.24 \times 10^{-3}} = 54$$

Figure 3.4b is a plot of the forward and reverse rates. Notice that at equilibrium the rates of the forward and reverse reactions are equal, but they are *not zero*. This means that the reactions are still running at equilibrium, but what one reaction produces, the other consumes at exactly the same rate. This situation is called a *dynamic equilibrium* and is typical of chemical systems.

Before leaving the subject of reaction rates, it is important to point out that the treatment is only a device to help explain how an equilibrium state might arise. A state of equilibrium does not depend in any way on the mechanism by which it is reached; it is determined by changes in the state function, free energy. What a mechanism does influence is the rate at which a system can come to equilibrium. Generally the rate expressions for chemical reactions vaguely resemble parts of activity quotients, but only in rare cases are they the same. For example, the reaction

$$H_2 + Br_2 = 2HBr$$

which might be expected to resemble the HI reaction, has more than one reaction step, and the forward rate expression is given by

$$r_f = \frac{k_1[H_2][Br_2]^{3/2}}{[Br_2] + k_2[HBr]}$$

It is quite difficult to calculate an equilibrium constant based on so complex a rate expression. What chemists prefer to do is to let a system equilibrate and then determine as carefully as possible the concentrations of the species.

3.3.1 Catalysts and Reaction Rates

The addition of a catalyst to a system will not change the position of its equilibrium. A catalyst will, however, change the *rate* at which a system reaches equilibrium. For example, the reaction

$$H_2(g) + O_2(g) = H_2O(l)$$

has a large positive equilibrium constant (10^{57}) but is extremely slow at room temperature. A mixture of H_2 and O_2 could be placed in a sealed flask and years would go by before any water vapor was found. If a small amount of platinum metal catalyst is present in the mixture of gases, however, the reaction is rapid at room temperature. Catalysts change the rates of chemical reactions, but not the free energies of products and reactants. Free energy changes determine equilibrium constants.

3.4 SOLUTION EQUILIBRIA, ACTIVITY, AND CONCENTRATION

Gas phase reactions are well suited for examples in discussing the thermodynamic and kinetic aspects of the equilibrium condition. However, an understanding of equilibria in solutions is of greater practical importance to students of analytical chemistry. The remainder of this chapter is devoted to the solution environment and relationships between the activities of dissolved species and their concentrations.

The behavior of a chemical species depends on the nature of its environment. A solute is influenced by such solvent characteristics as polarity, dielectric constant (the insulating ability of the solvent; see below), and ability to dissociate. For example, consider the solute NaCl in the solvent H_2O. In addition to the ionic species Na^+ and Cl^-, there will be hydronium ions (H_3O^+) and hydroxyl ions (OH^-) supplied by the dissociation of the solvent, water: $2H_2O = H_3O^+ + OH^-$. A water molecule has a dipole moment that results from the uneven distribution of bonding and nonbonding electrons, with the oxygen atom being slightly more negative than the hydrogen atoms:

$$H \diagup \overset{..}{\underset{\delta^+}{O}} \diagdown_{\delta^-} H$$

We might expect many different interactions in such a solution. Positively charged hydronium ions would repel other hydronium ions and sodium ions, but attract chloride ions and hydroxyl ions (*ion-ion interactions*). The positive end of the water dipole would attract the negative ions in solution, while the negative end of the dipole would attract the positive ions (*ion-dipole interactions*). Depending on the relative importance of such interactions, sodium ion (for example) might behave as if its effective concentration were greater than, equal to, or less than its actual (analytical) concentration.

The effective concentration of a species is called its activity and is related to its analytical concentration by the simple proportionality

$$a_i = f_i C_i$$

as stated previously. The coefficient f, which was introduced in the section on free energy, is called an activity coefficient.

3.4.1 The Debye-Hückel Model

Our present understanding of the behavior of dilute solutions is based largely on the work of Debye and Hückel, whose hypothetical model of electrolyte solutions led them to mathematical expressions which can be used to calculate approximate activity coefficients.

The Debye-Hückel (DH) model of a solution is unusual because it is so idealized and yet works quite well. The solvent (water, for example) is considered to be a medium without any molecular structure, having uniform viscosity (resistance to flow) and dielectric constant. Solute ions behave as if they are simply point charges (no finite radius), and solutions are always very dilute ($10^{-3}M$ or less). Under such idealized conditions the forces between solute ions could be predicted from Coulomb's law,

$$\text{force} = \frac{1}{D}\left(\frac{q_+ q_-}{r^2}\right)$$

where q_+ and q_- are electrostatic charges of the ions, r is the distance separating them, and D is the dielectric constant of the pure solvent.† These are the kinds of forces that produce the regular structural patterns found in crystals and are responsible for some order in solutions. As in crystals, on the average a solute cation such as Na^+ can be expected to have more anions (such as Cl^-) than cations in its neighborhood. This relative order is upset, however, by thermal energy, which Debye and Hückel accounted for with the Boltzmann distribution law. The result pictured by Debye and Hückel was that the solute species behaved randomly, but that long-range electrostatic forces were always operative. Using methods of statistical thermodynamics, Debye and Hückel derived an expression for the activity coefficient of a single ion of charge z_i in a solution of ionic strength μ. The equation, known as the Debye-Hückel limiting law equation, is

$$\log f_i = -A z_i^2 \sqrt{\mu}$$

The coefficient A contains several quantities: the charge of an electron, Avogadro's number, Boltzmann's constant, the dielectric constant of the solvent, and the absolute temperature. It has the value 0.51 in water at 25°C, and the equation becomes

$$\log f_i = -0.51 z_i^2 \sqrt{\mu}$$

---

† The size of the dielectric constant of a solvent reflects its ability to insulate ions of opposite charge from each other's electrostatic charge. Highly polar solvents do this quite well. In aqueous solutions of NaCl, for example, the negative ends of the H_2O dipoles tend to cluster around Na^+ ions, and partially obscure the positive charge. Chloride ions, also surrounded by water molecules, are less strongly attracted to the insulated sodium ions. The dielectric constant of water is 78.5 at 25°C. The dielectric constant of a vacuum is 1.000.

3.4 SOLUTION EQUILIBRIA, ACTIVITY, AND CONCENTRATION

The ionic strength (μ) of a system is defined as half the sum of the analytical concentrations (C_i) multiplied by the square of the ionic charge (z_i^2) for each ionic species, that is,

$$\mu = \tfrac{1}{2} \sum C_i z_i^2$$

The units of ionic strength are the same as the units of concentration. Often the dimensions of μ are neglected in reports in the chemical literature. Example 3.10 demonstrates an ionic strength calculation.

In principle, the DH limiting law expression can be used to calculate the activity coefficient for any single ionic species. Unfortunately, the results of such a calculation cannot be checked experimentally, because it is impossible to make up a solution of a single ion. The experimentally verifiable quantity is instead the *mean activity coefficient* of a cation and anion ($f_\pm$), a geometric mean of individual ion activities. For the general salt of cation M with charge z_M and anion X with charge (absolute value) z_X,

$$(f_\pm)^{z_M+|z_X|} = (f_M)^{|z_X|}(f_X)^{z_M} \dagger$$

The Debye-Hückel equation for the mean activity coefficient, $f_\pm$, is

$$\log f_\pm = -0.51|z_+ z_-|\sqrt{\mu}$$

EXAMPLE 3.10 Ionic Strength
Calculate the ionic strength of $0.10M$ solutions of (a) NaCl and (b) $AlCl_3$. Assume that these salts dissolve and dissociate completely.

---

† This formidable looking expression is really quite simple when specific cases are considered. For a salt like NaCl ($z_M = 1$, $|z_X| = 1$),

$$(f_\pm)^{1+1} = (f_{Na^+})(f_{Cl^-})$$
$$f_\pm = [(f_{Na^+})(f_{Cl^-})]^{1/2}$$

or

For a salt like $CaCl_2$ ($z_M = 2$, $|z_X| = 1$),

$$CaCl_2 \rightarrow Ca^{2+} + 2Cl^-$$

and $C_{Cl^-} = 2C_{Ca^{2+}}$ in solution. Thus

$$(f_\pm)^{2+1} = (f_{Ca^{+2}})(f_{Cl^-})^2$$
$$f_\pm = [(f_{Ca^{+2}})(f_{Cl^-})^2]^{1/3}$$

Solution

(a) $0.1M$ NaCl:

$$\mu = \tfrac{1}{2}[C_{Na^+}(z_+)^2 + C_{Cl^-}(z_-)^2]$$
$$= \tfrac{1}{2}[0.10(1)^2 + 0.10(1)^2] = \tfrac{1}{2}(0.20)$$
$$= 0.10M$$

(b) $0.10M$ AlCl$_3$ (when dissolved, $C_{Cl^-} = 3C_{Al^{3+}}$):

$$\mu = \tfrac{1}{2}[C_{Al^{3+}}(z_+)^2 + C_{Cl^-}(z)^2]$$
$$= \tfrac{1}{2}[0.10(3)^2 + 0.30(1)^2] = \tfrac{1}{2}(1.20)$$
$$= 0.60M$$

EXAMPLE 3.11 Activity Coefficients

Calculate the mean activity coefficient for $0.01M$ NaCl using the Debye-Hückel limiting law equation.

Solution

$$\log f_\pm = -0.51(z_M|z_X|)\mu^{1/2}$$
$$= -0.51(1)(1)(0.01)^{1/2} = -0.051$$
$$f_\pm = 0.89$$

The experimental value has been found to be 0.903.

How well does the Debye-Hückel limiting law (DHLL) work? Plots of mean activity coefficients for a theoretical salt M^+X^- calculated from the Debye-Hückel limiting law appear in Fig. 3.5, along with the experimentally determined mean activity coefficients for NaCl and CaCl$_2$. It should be apparent that the DHLL calculation gives much smaller values for $f_\pm$ than are actually found as concentrations (therefore ionic strengths) increase. The limiting law does not predict at all the increase in $f_\pm$ at high concentrations. It is accurate within a few percent for ionic strengths less than about $0.03M$.

A closer approximation at the $0.1M$ level can be obtained from the extended Debye-Hückel law (EDHL). In their original paper, Debye and Hückel realized the need to consider ions of finite diameter. They modified the limiting law expression by adding a denominator term to the ionic strength

3.4 SOLUTION EQUILIBRIA, ACTIVITY, AND CONCENTRATION

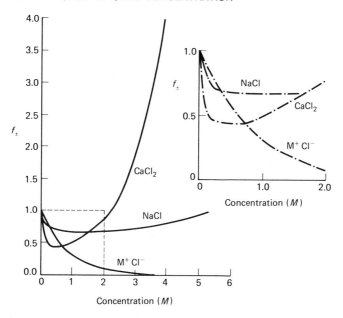

Figure 3.5 Activity coefficient vs. molar concentration for NaCl, CaCl$_2$, and a theoretical salt, MCl. Inset: expanded scale for $f_\pm$ between 0.0 and 2.0M.

$$\log f_i = -0.51 z_i^2 \frac{(\mu)^{1/2}}{1 + Ba(\mu)^{1/2}} \quad \text{(individual ion)}$$

$$\log f_\pm = -0.51 |z_+ z_-| \frac{\mu^{1/2}}{1 + Ba(\mu)^{1/2}} \quad \text{(mean)}$$

The parameter B is a function of the absolute temperature and dielectric constant of the solvent and equals 0.328 for water at 25°C. The parameter a corresponds roughly to the diameter (in angstrom units) of the *ion* surrounded by solvent. While the diameters of gaseous (free) ions are in the neighborhood of 1 to 5 Å, the diameters of hydrated ions may be in the range of 3 to 11 Å, depending on the charge and free ion diameter.†

Kielland has determined effective diameters for a variety of inorganic and organic ions in water. These appear in Table 3.1.

The extended form of the DHLL increases the upper concentration range for theoretical calculations to about 0.1M for a 1:1 electrolyte such as NaCl.

---

† The large potassium ion (free ion diameter 2.6 Å) with a low charge is assigned a = 3, while the small lithium ion (free ion diameter 1.3 Å) with low charge is assigned a = 6. Ions with +4 charges have very large a values.

TABLE 3.1 EFFECTIVE ION DIAMETERS

Inorganic ions		Organic ions	
Ion	Ion size, a (Å)	Ion	Ion size, a (Å)
H^+	9	$HCOO^-$ (formate), $H_2C_6H_5O_7^-$, (diH citrate), $CH_3NH_3^+$, (methyl ammonium ion), $(CH_3)_2NH_2^+$,	3.5
Li^+	6		
Na^+, IO_3^-, HCO_3^-, HSO_4^-, $H_2PO_4^-$, $H_2AsO_4^-$,	4	$NH_3^+CH_2COOH$ (glycinium), $(CH_3)_3NH^+$, $(C_2H_5)NH_3^+$ (ethyl ammonium)	4
K^+, Rb^+, Cs^+, Tl^+, Ag^+, NH_4^+, OH^-, F^-, SCN^-, HS^-, ClO_3^-, ClO_4^-, BrO_3^-, IO_4^-, MnO_4^-, Cl^-, Br^-, I^-, CN^-, NO_3^-	3	CH_3COO^- (acetate), chloroacetate, $C_2O_4^{2-}$ (oxalate), $NH_2CH_2COO^-$ (glycinate), $(CH_3)_4N^+$ (tetramethyl ammonium), $(C_2H_5)_2NH_2^+$ (diethyl ammonium), $HC_6H_5O_7^{2-}$ (monoH citrate)	4.5
Mg^{2+}, Be^{2+}	8		
Ca^{2+}, Cu^{2+}, Zn^{2+}, Sn^{2+}, Mn^{2+}, Fe^{2+}, Ni^{2+}, Co^{2+}	6	Dichloroacetate, trichloroacetate, triethyl ammonium, $H_2C(COO^-)_2$ (malonate), $(CHCOO^-)_2$ (succinate), $C_6H_5O_7^{3-}$, (citrate), $(CHOHCOO^-)_2$ (tartrate)	5
Sr^{2+}, Ba^{2+}, Ra^{2+}, Cd^{2+}, Pb^{2+}, Hg^{2+}, S^{2-}, CO_3^{2-}, SO_3^{2-},	5		
Hg_2^{2+}, SO_4^{2-}, $S_2O_3^{2-}$, CrO_4^{2-}, HPO_4^{2-},	4	$C_6H_5COO^-$ (benzoate), $C_6H_4OHCOO^-$ (salicylate), chlorobenzoates, phenylacetate, $C_6H_4(COO^-)_2$ (phthalate), $(CH_3)_2CHCH_2COO^-$ (isovalerate), $(C_2H_5)_4N^+$ (tetraethyl ammonium), $(CH_2)_3(COO^-)_2$ (glutarate)	6
Al^{3+}, Fe^{3+}, Cr^{3+}, Ce^{3+}, La^{3+}	9		
PO_4^{3-}, $Fe(CN)_6^{3-}$,	4	$^-OC_6H_2(NO_2)_3$ (picrate), $CH_3OC_6H_4COO^-$ (methoxybenzoate), $(CH_2)_6(COO^-)_2$ (suberate)	7
Th^{4+}, Zr^{4+}, Ce^{4+}, Sn^{4+}	11		
$Fe(CN)_6^{4-}$,	5	$(C_6H_5)_2CHCOO^-$ (diphenylacetate)	8

Source: Values from J. Kielland, *Journal of the American Chemical Society* 59, 1675 (1937).

EXAMPLE 3.12 Extended Debye-Hückel Law

Calculate an approximate mean activity coefficient for NaCl in a $0.1M$ solution (ionic strength $0.1M$).

Solution

From the Kielland values in Table 3.1, we see that $a_{Na^+} = 4$, $a_{Cl^-} = 3$. We will assume that the average of these values (3.5) can be used in the EDHL:

3.4 SOLUTION EQUILIBRIA, ACTIVITY, AND CONCENTRATION

$$\log f_\pm = -0.51(1)(1)\left[\frac{(0.1)^{1/2}}{1 + (0.328)(3.5)(0.1)^{1/2}}\right]$$
$$= -0.118$$
$$f_\pm = 0.76$$

The DHLL equation gives 0.69 for the mean activity coefficient. The experimentally determined value is 0.778. Clearly, the EDHL equation more closely approximates the experimental value.

Several empirical expressions have been proposed for the estimation of activity coefficients in solutions of relatively high ionic strength. One of the more popular is the modified Davies equation for individual activity coefficients:

$$\frac{-\log f_i}{z_i^2} = \frac{0.511(\mu)^{1/2}}{1 + 1.5(\mu)^{1/2}} - 0.2\mu$$

EXAMPLE 3.13 The Davies Equation

Calculate the approximate values for the activity coefficients of Cl^-, Ca^{2+}, and La^{3+} in an electrolyte solution with ionic strength $0.5M$.

Solution

Use the modified Davies equation to calculate a value for $-\log f_i/z_i^2$:

$$\frac{-\log f_i}{z_i^2} = \frac{(0.511)(0.5)^{1/2}}{1 + 1.5(0.5)^{1/2}} - 0.2(0.5)$$
$$= 0.0753$$

For Cl^-: $z_i^2 = 1$, $-\log f_i = 0.0753$, $f_i = 0.841$
For Ca^{2+}: $z_i^2 = 4$, $-\log f_i = 0.3012$, $f_i = 0.500$
For La^{3+}: $z_i^2 = 9$, $-\log f_i = 0.678$, $f_i = 0.210$

3.4.2 Activity Coefficients at Very High Ionic Strengths

It is not uncommon for the mean activity coefficients of electrolytes to become much larger than unity at high ionic strengths. In the case of $CaCl_2$, the mean activity coefficient is approximately 3.0 when the concentration of $CaCl_2$ is $3.6M$. This means that a $3.6M$ solution of $CaCl_2$ behaves chemically as if it were nearly $11M$. One interpretation of this behavior is that at higher concentrations of salt, as a result of hydration of ions, there is less "free" water available and the molarity of the salt is effectively increased. In the case of $CaCl_2$, each Ca^{2+} ion may tie up six water molecules in a hydration sheath. Each Cl^- ion may tie up two water molecules.† Therefore for each mole of $CaCl_2$ in solution, 10 moles of water will be held by the ions of the salt. In 1 kg of water there are about 55.5 moles of water. If 36 moles of water are tied up with 3.6 moles of the salt, then 3.6 moles of salt are dissolved in 19.5 moles of free water. The effective concentration of calcium chloride is thus (55.5/19.5)(3.6 mole/liter) = $10.25M$. This example is, of course, an oversimplification of what happens in concentrated solutions; it is intended to suggest a way of viewing such solutions.

3.4.3 Activity Coefficients of Uncharged Species

At concentrations less than about $0.1M$ the activities of uncharged species in solution are within 1 or 2% of being equal to their concentrations. Hence the activity coefficients of such species are justifiably taken as unity.

When electrolytes are present, however, the activity coefficient is more accurately calculated from

$$\log f = k\mu$$

where μ is the ionic strength and k is the *salting coefficient,* which may be either positive or negative. For small molecules such as N_2, O_2, or CO_2, salting coefficients are about equal to 0.1. In the case of larger molecules such as sugars and proteins, the salting coefficient is usually larger. When k is positive for a species, $\log f$ will be positive, and activity will exceed molar concentration. When k is negative (as it is for the acid HCN), $\log f$ is negative, and activity will be smaller than concentration.

EXAMPLE 3.14 Neutral Species

The activity coefficient of ammonia in water at 25°C is given by

$$\log f_{NH_3} = 0.12\mu$$

Calculate the activity of NH_3 in a solution in which the molar concentration of NH_3 is 0.100 and the ionic strength is 0.50.

---

† The number of water molecules tied up by ions is called a hydration number. Estimates of hydration numbers for Ca^{2+} run from 6 to 17, depending on the method of study. The Cl^- has a lower hydration number than the positive ions of groups IA and IIA, and a hydration number of 2 seems generally agreed upon.

3.4 SOLUTION EQUILIBRIA, ACTIVITY, AND CONCENTRATION

Solution

$$\log f_{NH_3} = 0.12(0.50) = 0.060$$
$$f_{NH_3} = 1.15$$
$$a_{NH_3} = f_{NH_3} C_{NH_3} = 1.15(0.100) = 0.115$$

3.4.4 Assumptions about Activity Coefficients

It is common practice in introductory courses to neglect activity coefficients in equilibrium calculations to save time and effort. At the other extreme, the detailed development of expressions that include activity coefficients can make it hard for beginners to see important underlying concepts. In this text the strategy is somewhere between the two extremes. Every time a new example of an equilibrium calculation is introduced (e.g., acid-base, solubility, redox), the connection is made between thermodynamic equilibrium constants and equilibrium constants in real systems. As you will see, at the $0.01F$ ionic strength level, activity corrections are small, and ignoring them will not lead us to draw incorrect conclusions about systems. There can be some real surprises at higher ionic strengths, however, and from time to time we will examine some of these.

It is also common practice to omit the concentration of solvent from equilibrium constant expressions in which it appears. Although the idea was not developed in the treatment in this text, the activity of a species is defined rigorously in terms of the free energy of a system and the *mole fraction* of the species. The mole fraction of solvent in most dilute systems is very nearly unity, and thus its activity is very nearly unity. Therefore the equilibrium

$$CH_3COOH + \underset{\text{(solvent)}}{H_2O} = CH_3COO^- + H_3O^+$$

has the equilibrium constant expression

$$K = \frac{(a_{CH_3COO^-})(a_{H_3O^+})}{(a_{CH_3COOH})(a_{H_2O})} = \frac{(a_{CH_3COO^-})(a_{H_3O^+})}{(a_{CH_3COOH})}$$

A similar practice is used for the activity of solid materials when present with other reacting species. For example, the solubility of silver chloride, AgCl(s), is expressed by

$$AgCl(s) = Ag^+(aq) + Cl^-(aq)$$

The mole fraction of AgCl in the solid phase is unity. The equilibrium constant expression is therefore given by

$$K = \frac{a_{Ag^+}a_{Cl^-}}{a_{AgCl}} = (a_{Ag^+})(a_{Cl^-}), \qquad a_{AgCl} = 1.0$$

REFERENCES

1. A. H. Taylor, Jr., and K. H. Crist, *Journal of the American Chemical Society,* 63, 1377–1385 (1941).

STUDY QUESTION

Complete the following statements:

A system from which matter cannot escape, but which can exchange energy with the environment, is called _____ . A system which cannot exchange energy or matter with the environment is called a _____ .

In a universe consisting of two closed systems, whatever energy one system loses must _____ . This is a statement of _____ , or the first law of thermodynamics.

Let a universe consist of only two closed systems, one of which is a reaction vessel. When the pressure and the _____ of the reaction vessel are held constant, the thermal energy lost from the reaction vessel is called the _____ change of the reaction. The sign of the change in this function is determined from the perspective of the _____ ; a negative change occurs when the reaction vessel _____ thermal energy. Such a reaction is called _____ .

A substance is in its standard state when _____ . Reference states for thermodynamic data _____ .

For a reaction to occur spontaneously the entropy change of the universe must be _____ . A spontaneous reaction in a closed system could have a negative entropy change if _____ .

The direction in which a reaction will proceed is determined by _____ . When $\Delta G_{reaction}$ is _____ , the reaction proceeds as written. Otherwise, the reaction will proceed in _____ . If $\Delta G_{reaction} = 0$, then the system _____ .

The free energy of a species depends not only on its standard free energy, but also on _____ and _____ .

The activity quotient, Q, is called an equilibrium constant, K, when _____ . The value of the equilibrium constant is determined by _____ , and not by the _____ of reactants and products.

A reaction is called quantitative if the _____, _____, and concentrations of species permit _____ % conversion of reactants to products.

According to LeChatelier's principle, when a stress is applied to a system at equilibrium, _____. Temperature is one form of stress. The direction in which an equilibrium shifts when the temperature of the system is increased depends on the sign of the _____. The quantitative change can be approximated by the _____ equation, which is written _____.

An equilibrium condition does not depend on the _____ by which it arises, but by changes in the thermodynamic function, _____. The rate at which a system reaches equilibrium does depend on _____.

A catalyst can change the _____ of a reaction, but not the position of _____. Free energies of products and reactants determine _____ and (are) (are not) influenced by a catalyst.

PROBLEMS

TABLE OF THERMODYNAMIC PROPERTIES
AT 298 K, 1 ATM, FOR PROBLEM SET

Compound	H^0 (kJ/mol)	G^0 (kJ/mol)	S^0 (J/mol)
$Ag^+(aq)$	105.75	77.09	73.38
$Ag(NH_3)_2^+(aq)$	−111.3	−17.2	245
$Ba^{2+}(aq)$	−537.64	−560.74	9.6
$BaCO_3(s)$	−1216	−1138	112
C(graphite)	0.0	0.0	5.740
CO(g)	−110.53	−137.15	197.556
$CO_2(g)$	−393.51	−394.38	213.677
$CO_2(aq)$	−413.8	−386.0	118
$H_2(g)$	0.0	0.0	130.57
$H_2O(l)$	−285.83	−237.18	69.91
$H_3O^+(aq)$	−285.83	−237.18	69.91
HCOOH(aq)	−425.43	−372	160
$HCOO^-(aq)$	−425.55	−351	92
HI(g)	26.36	1.576	206.480
I(g)	106.762	70.208	180.673
$I_2(g)$	62.421	19.360	260.567
$NH_3(aq)$	−80.29	−26.6	111
$NH_3(g)$	−45.94	−16.44	192.67
$O_2(g)$	0.0	0.0	205.037
$OH^-(aq)$	−230.025	−157.37	−10.71
S(s, rhombic)	0.0	0.0	32.054
$SO_2(g)$	−296.81	−300.10	248.11
$SO_3(g)$	−395.7	−371.1	256.6

Source: Data from T. G. Bean and N. C. Craig, "Standard State Thermodynamic Properties," Oberlin College, 1981.

3.1. Calculate the standard molar enthalpy change for the dissolution of ammonia in water:

$$NH_3(g) = NH_3(aq)$$

Is the reaction endothermic or exothermic?

3.2. Barium carbonate precipitates when a solution of barium hydroxide absorbs carbon dioxide:

$$Ba^{2+}(aq) + 2OH^-(aq) + CO_2(g) = BaCO_3(s) + H_2O$$

Is the reaction that forms $BaCO_3$ endothermic or exothermic?

3.3. What is the entropy change when a reaction vessel at a temperature of 300 K releases 10,000 joules of thermal energy to a water bath surrounding it, also at 300 K? Assume that the vessel and bath are so large that the exchange of 10,000 joules does not change their temperatures.

3.4. What is the standard molar entropy change that accompanies the formation of the diammine silver complex?

$$Ag^+(aq) + 2NH_3(aq) = Ag(NH_3)_2^+(aq)$$

3.5. The standard molar enthalpy change for the reaction

$$S(s) + O_2(g) = SO_2(g)$$

is -296.8 J/mole. Using standard entropies from the table at the beginning of the problem set, calculate the standard molar free energy change for the reaction. Does the reaction proceed to produce SO_2 spontaneously? Why?

3.6. An isolated system contains 1 mole of $I_2(g)$ and 1 mole of atomic iodine, $I(g)$. Assuming that ΔH^0 and ΔS^0 are independent of temperature and that volume is constant, will the reaction

$$I_2(g) \rightarrow 2I(g)$$

be spontaneous at 1000 K? Explain.

3.7. At what temperature will an equimolar mixture of $I_2(g)$ and $I(g)$ be at equilibrium?

3.8. Calculate the equilibrium constant for the reaction

$$C(s) + \tfrac{1}{2}O_2(g) = CO(g)$$

using data in the table at the beginning of the problem set.

3.9. Calculate the equilibrium constant for the reaction of formic acid with water:

$$HCOOH(aq) + H_2O(l) = HCOO^-(aq) + H_3O^+$$

using data in the table at the beginning of the problem set.

3.10. The reaction

$$Cl_2(g) + I_2(g) = 2ICl(g)$$

has a standard free energy change of -11.05 kJ/mole.

PROBLEMS

(a) Calculate K for the reaction.

(b) 10^{-2} mole of Cl_2 is mixed with 10^{-2} mole of ICl in a vessel and allowed to equilibrate at 25°C. At equilibrium will there be more or less Cl_2 than was present initially? Explain.

3.11. The reaction

$$I^-(aq) + I_2(aq) = I_3^-(aq)$$

has a standard free energy change of -16.28 kJ/mole. Calculate the equilibrium constant for the reaction.

(a) When 10^{-2} mole each of I^- and I_2 are added to 1 liter of solution containing 0.10 mole of I_3^-, in which direction will the reaction run to approach equilibrium?

(b) If, in another experiment, the concentrations of I^- and I_3^- are both $0.01M$ at equilibrium, what will the concentration of $I_2(aq)$ be?

3.12. The standard free energy change at 3500 K for the reaction

$$CO_2(g) + H_2(g) = CO(g) + H_2O(g)$$

is -5.23 kJ/mole. Calculate the value of ΔG for the reaction of 0.10 mole of CO_2 and 1.0 mole of H_2 with 10 moles each of CO and H_2O. In which direction will the reaction run to approach equilibrium?

3.13. The reaction

$$Cu^+ + Cl^- = CuCl(s)$$

has an equilibrium constant of 3.13×10^6 at 25°C. Enough chloride ion is added to a $10^{-3}M$ solution of Cu^+ ion so that at equilibrium the concentration of chloride ion is $0.010M$ (assume ionic strength is 0.01).

(a) Calculate the concentration of Cu^+ ion at equilibrium (assume that the activity of solid CuCl is unity).

(b) Does Cu^+ react "quantitatively" under these conditions? Explain.

3.14. How large must the equilibrium constant for the reaction

$$A + B = C + D$$

be if 99.9% of the original quantity of A is to react? Assume that initially $a_A = a_B$, and $a_C = a_D = 0$, then develop an expression for the activities at equilibrium.

3.15. Using data from the table at the beginning of the problem set, calculate an approximate value of K for the reaction

$$SO_2(g) + \tfrac{1}{2}O_2(g) = SO_3(g) \quad \text{at } 200°C$$

3.16. Hydrogen reacts with sulfur to form hydrogen sulfide according to the reaction

$$H_2(g) + S(s) = H_2S(g) \quad \Delta H^0 = -20.1 \text{ kJ/mole}$$

What will be the qualitative effect on the equilibrium of each of the following changes:

(a) a decrease in pressure
(b) a decrease in temperature
(c) addition of more hydrogen
(d) addition of more hydrogen sulfide
(e) addition of a catalyst

3.17. Explain what is meant by a dynamic equilibrium. Where are *static* equilibria found? Give an example of each type of equilibrium, and compare similarities and differences.

3.18. Calculate the ionic strength of each of the following solutions:
(a) $0.10M$ NaCl
(b) $0.05M$ BaCl$_2$
(c) $0.02M$ LaCl$_3$
(d) $0.10M$ K$_4$Fe(CN)$_6$

3.19. Calculate the mean activity coefficient for $0.05M$ AgNO$_3$ using the Debye-Hückel limiting law. Use the extended DH law relationship to find the activity coefficient when a finite ion size is considered.

3.20. Calculate the mean activity coefficient for $0.10M$ MgCl$_2$, using the DHLL. At what concentration of MgSO$_4$ will the mean activity coefficient of MgSO$_4$ equal that of $0.10M$ MgCl$_2$?

3.21. The salting coefficient of acetic acid (HOAc) is 0.0645 in $1M$ NaCl electrolyte.
(a) Calculate the activity of acetic acid in a $1M$ NaCl solution in which its concentration is $0.100M$.
(b) The equilibrium constant expression for the dissociation of acetic acid is

$$K = \frac{a_{H^+} a_{OAc^-}}{a_{HOAc}}$$

An experimentalist adds 0.100 mole of pure acetic acid to 1.00 liter of $1M$ NaCl. The activity of hydrogen ion (and therefore acetate ion, OAc$^-$, since they must be the same at equilibrium) can be measured in the laboratory with a glass electrode. If an experimentalist *assumes* that the activity coefficient of acetic acid (HOAc) is exactly unity, will she tend to overestimate or underestimate the value of K in $1M$ NaCl?

3.22. Suppose the reaction

$$2A + B = D$$

occurs in two steps:

(a) $A + B = C$
(b) $C + A = D$

Assume that the equilibrium constant for the overall process can be obtained from the forward and reverse rates of the two steps. Write expressions for the four rates in the overall process, and give the relationship between the individual rate constant (k's) and the equilibrium constant (K). Interpret your result.

Chapter 4

Concepts of Acids and Bases

4.1 INTRODUCTION

The purpose of this chapter is to outline the evolution of the concepts of acids and bases, and to describe several important contemporary theories. The section on Brønsted-Lowry theory serves as an introduction to Chapter 5 (on acid-base equilibria), while that on Lewis theory should help in understanding the material in Chapter 12 (on complex metal ion species). Section 4.5, on the solvent-system theory, should serve as an introduction to nonaqueous acid-base equilibria.

4.2 EARLY VIEWS OF ACIDS AND BASES

Many of the properties which distinguish acids from bases have been recognized for hundreds of years. In the mid-seventeenth century Robert Boyle classified compounds as "acids" or "alkalies" on the basis of tastes and colors: an acid was a substance that tasted sour, changed the color of blue plant dye (blue litmus) to red, and reacted with an alkali. An alkali, on the other hand, had a bitter taste, turned red litmus blue, and dissolved oils and sulfur. Boyle tried to explain such properties by shape. He thought acids had sharp spikes on their particles, which accounted for their sour taste and stinging effect on the skin. He thought alkalies were highly porous bodies into which the spikes of acids would penetrate and be broken off or blunted during neutralization.

G. F. Rouelle was the first to use the term "base" instead of alkali (1754). Rouelle extended Boyle's definition of a base to include several metals from what we now call groups IA and IIA of the periodic table. Furthermore, he widened the concept of salts, which had previously meant neutral and soluble compounds, by showing that when acids and bases reacted they could form acidic, neutral, and basic salts.

Antoine Lavoisier, Rouelle's student, was the first to define a general theory of the composition of acids (1777). His work on the oxidation of elements showed that the oxides of the nonmetals sulfur, carbon, nitrogen, and phosphorus produced acidic solutions when dissolved in water. Lavoisier was convinced that the acidic properties of compounds depended on the presence of oxygen. An acid which proved most troublesome for Lavoisier, however, was muriatic acid, now known as hydrochloric acid, HCl, which he was sure contained oxygen. In 1810 the English chemist Sir Humphrey Davy showed conclusively that muriatic acid was composed of hydrogen and chlorine, with no oxygen. By 1816 Davy realized that the element common to all acids was hydrogen, and not oxygen. Unfortunately, it was also known that not all hydrogen-containing compounds behaved as acids (e.g., methane and ammonia). What was missing was an explanation of what gave acids their special properties.

Justus von Liebig (1838) pointed out that the property that distinguished acids from other compounds was that their hydrogen could be *replaced* by a metal to form salts. Thus when zinc metal dissolves in a solution of sulfuric acid, hydrogen gas bubbles out and is "replaced" by zinc. Liebig's theory was based on both composition (acids contain hydrogen) and reactivity (. . . replaced by metals) and was the first theory to account for both aspects.

4.3 THE ARRHENIUS DEFINITION

The concept of acids and bases was broadened dramatically in the 1880s with the theory of electrolyte dissociation developed by Arrhenius and Ostwald. They observed that aqueous solutions of certain compounds are able to conduct electricity, suggesting that the compounds were breaking apart into positively and negatively charged species. Acids were seen to be compounds which broke apart (dissociated) to form hydrogen ions in water,

$$HA = H^+ + A^-$$
(acid)

while bases were $-OH$ containing species which dissociated to form hydroxyl ions (OH^-),

$$BOH = OH^- + B^+$$
(base)

The reaction of an acid with a base (neutralization) was simply the reaction of hydrogen ions with hydroxyl ions to form water:

$$H^+ + OH^- = H_2O$$
(from acid) (from base)

4.3 THE ARRHENIUS DEFINITION

TABLE 4.1 STRENGTH OF ACIDS
(Data of Ostwald and Arrhenius for Equimolar Solutions)

Acid	Relative conductivity	Relative ability to catalyze hydrolysis of methyl acetate	$K_a{}^a$
HCl	100	100	Very large
HNO$_3$	99.6	92	Very large
Trichloroacetic	62.3	68.2	0.20
Dichloroacetic	25.3	23.0	0.05
Monochloroacetic	4.9	4.3	0.0015
Acetic	0.42	0.35	1.8×10^{-5}

^{a} Currently accepted values. The K_a's for HCl and HNO$_3$ are too large to be measured accurately in water.

This explained why the reaction of any strong acid with a strong base produces a particular amount of heat, 58.2 kJ per formula weight.[†] All such reactions could be reduced to one reaction,

$$H^+ + OH^- = H_2O$$

The real importance of Arrhenius' theory was that it allowed chemists to deal with the strength of acids and bases in a quantitative way. The experimental data of Arrhenius and Ostwald (Table 4.1) showed that solutions of strong acids such as HCl and HNO$_3$ were not only very good conductors of electricity but also excellent catalysts for the reaction of esters such as methyl acetate with water,

$$CH_3COOCH_3 + H_2O \xrightarrow{H^+} CH_3COOH + CH_3OH \quad \text{(acid-catalyzed hydrolysis)}$$

Solutions of weak acids such as acetic acid were relatively poor conductors of electricity and poor catalysts for the hydrolysis of esters. Arrhenius said the factor common to conductance and hydrolysis was the dissociation of the acid: strong acids are highly dissociated while weak acids are only slightly dissociated.

Arrhenius expressed acid strength quantitatively by numerical constants for dissociation reactions ("dissociation constants"). For example, the extent of the dissociation of acetic acid,

$$CH_3COOH = H^+ + CH_3COO^-$$

[†] "Strong" meant completely dissociated to Arrhenius.

defines the size of the equilibrium constant,

$$K_a = \frac{[H^+][CH_3COO^-]}{[CH_3COOH]}$$

Conductance measurements show that acetic acid is about 1.35% dissociated in a $0.1M$ solution. A simple calculation can be performed to show that the dissociation constant of acetic acid is 1.8×10^{-5} at 25°C.†

The ability of Arrhenius' theory to deal with the strengths of acids and bases quantitatively tended to obscure some of its limitations as a general theory of acid-base behavior. Even though the definitions were created specifically for aqueous solutions, there remained some ambiguities in describing bases. For example, the organic amines do not contain hydroxyl ions and would therefore not be called bases by the Arrhenius definition. But amines can react with water to form hydroxyl ions by a reaction such as

$$\underset{\text{(methylamine)}}{CH_3NH_2} + H_2O = \underset{\text{(methyl ammonium ion)}}{CH_3NH_3^+} + OH^-$$

Arrhenius did not take into proper account the ability of water to act as an acid or a base or to participate in acid-base reactions. Furthermore, he tied acid strength directly to dissociation. Solutions of hydrogen chloride gas in benzene display properties characteristic of acids (neutralization reactions, ability to change the color of litmus), even though conductivity studies show that HCl is not dissociated in benzene. What was needed was a theory of acids and bases that transcended the limitations of water as a solvent.

4.4 THE BRØNSTED-LOWRY DEFINITION

A more general definition of acids and bases was proposed independently and almost simultaneously in 1923 by the Danish chemist Johannes Brønsted and the British chemist Thomas Lowry. The new definitions avoided ambiguities of the Arrhenius definitions by avoiding reference to a particular solvent. Brønsted and Lowry said that an acid is a species that has a tendency to lose a proton, and a base is a species that has a tendency to gain a proton. Notice that the word proton is used without reference to solvation. Therefore HCl can be an acid in both water and benzene. In water the base can be water itself,

† $K_a = [H^+][CH_3COO^-]/[CH_3COOH]$. For $[H^+] = [CH_3COO^-] = 1.35\% (0.1) = 1.35 \times 10^{-3}$ and $[CH_3COOH] = 0.1 - 1.35\% (0.1) = 0.1 - 1.35 \times 10^{-3}$, $K_a = (1.8 \times 10^{-6})/(0.0999) = 1.8 \times 10^{-5}$.

4.4 THE BRØNSTED-LOWRY DEFINITION

$$HCl + H_2O = H_3O^+ + Cl^-$$
$$\text{(acid)} \quad \text{(base)}$$

whereas in benzene the solvent does not react,

$$HCl + C_6H_6 \neq C_6H_7^+ + Cl^-$$

However, a base dissolved in benzene can react with HCl,

$$HCl + (CH_3)NH_2 = [(CH_3)NH_3^+ + Cl^-]\dagger$$
$$\text{(acid)} \quad \text{(base)} \quad \text{(ion pair)}$$

Once again, an acid is a proton donor and a base a proton acceptor.

Several fundamental ideas which have come from Brønsted-Lowry theory deserve close attention: the concept of conjugate pairs, autoprotolysis and the participation of amphiprotic solvents in acid-base reactions, and the ability of such solvents to limit ("level") the strengths of acids and bases.

Brønsted-Lowry acids and bases exist in what are called *conjugate pairs*. For example, in the two-step dissociation of sulfuric acid,

$$H_2SO_4 = H^+ + HSO_4^-$$
$$HSO_4^- = H^+ + SO_4^{2-}$$

bisulfate ion (HSO_4^-) is called the *conjugate base* of sulfuric acid (H_2SO_4). The second equilibrium shows that bisulfate is also the *conjugate acid* of the base, sulfate ion (SO_4^{2-}). Several conjugate pairs are listed in Table 4.2. Notice that acids and bases can be either neutral or charged. In a conjugate pair the base always has a charge less positive by one unit than the acid.

In a sense, conjugate pairs have complementary strengths: the stronger the acid, the weaker its conjugate base. For example, HCl is a very strong acid, while chloride ion is a very weak base. Ammonium ion is a weak acid, while ammonia is a strong base.

Some of the species in Table 4.2 are able to act as either acids *or* bases and are called *amphiprotic* (or *amphoteric*) compounds. The most important amphiprotic compound is water. A water molecule can lose a proton and act as an acid,

$$H_2O \rightarrow H^+ + OH^-$$

---

† Benzene, like other hydrocarbons, chloroform, and carbon tetrachloride, has a very low dielectric constant. Negligible dissociation of species occurs in such solvents, and neutralization products are found mostly as pairs of ions.

TABLE 4.2 COMMON CONJUGATE ACID-BASE PAIRS

Acid	Base
Hydrochloric acid, HCl	Chloride ion, Cl^-
Acetic acid, CH_3COOH	Acetate ion, CH_3COO^-
Sulfuric acid, H_2SO_4	Bisulfate ion, HSO_4^-
Bisulfate ion, HSO_4^-	Sulfate ion, SO_4^{2-}
Water, H_2O	Hydroxide ion, OH^-
Hydronium ion, H_3O^+	Water, H_2O
Ammonium ion, NH_4^+	Ammonia, NH_3
Ammonia, NH_3	Amide ion, NH_2^-

or gain a proton and act as a base,

$$H_2O + H^+ \rightarrow H_3O^+$$

Similarly, ammonia can act as an acid or a base,

$$NH_3 \rightarrow H^+ + NH_2^- \quad (NH_3 \text{ as an acid})$$
$$NH_3 + H^+ \rightarrow NH_4^+ \quad (NH_3 \text{ as a base})$$

In liquid ammonia as a solvent two ammonia molecules can react, one as an acid, the other as a base,

$$NH_3 + NH_3 = NH_4^+ + NH_2^-$$
$$\text{(acid)} \quad \text{(base)} \quad \text{(acid)} \quad \text{(base)}$$

This reaction of a solvent with itself is called *autoprotolysis,* or *autodissociation,* and it gives the solvent some very useful properties, which will be discussed in detail in Chapter 7 (on nonaqueous acid-base equilibria). For the time being, simply note that we can write an equilibrium constant expression for the autoprotolysis of ammonia,

$$K_s = a_{NH_4^+} * a_{NH_2^-} \quad \text{"autoprotolysis constant"}$$

taking the activity of NH_3 as the solvent to be unity. The value of K_s for ammonia is about 10^{-27}, while that of water is 10^{-14}.

The amphiprotic nature of both water and liquid ammonia makes them excellent solvents for acid-base reactions, as illustrated by the behavior of hypochlorous acid and its conjugate base, hypochlorite ion,

$$HOCl = H^+ + OCl^-$$
$$\text{(hypochlorous acid)} \qquad \text{(hypochlorite ion)}$$

4.4 THE BRØNSTED-LOWRY DEFINITION

HOCl is a weak acid in water, and protons leaving it will react with water to form hydronium ion,

$$\underset{\text{(acid)}}{\text{HOCl}} + \underset{\substack{\text{(solvent as} \\ \text{base)}}}{\text{H}_2\text{O}} = \underset{\text{(acid)}}{\text{H}_3\text{O}^+} + \underset{\text{(base)}}{\text{OCl}^-}$$

The equilibrium constant expression for this reaction is called the *acid dissociation constant* and is given by

$$K_a = \frac{a_{\text{H}_3\text{O}^+} a_{\text{OCl}^-}}{a_{\text{HOCl}} a_{\text{H}_2\text{O}}}$$

Hypochlorite ion can also react with water,

$$\underset{\text{(base)}}{\text{OCl}^-} + \underset{\substack{\text{(solvent as} \\ \text{acid)}}}{\text{H}_2\text{O}} = \underset{\text{(acid)}}{\text{HOCl}} + \underset{\text{(base)}}{\text{OH}^-}$$

with the equilibrium constant expression, called the *base dissociation constant*, given by

$$K_b = \frac{a_{\text{HOCl}} a_{\text{OH}^-}}{a_{\text{OCl}^-} a_{\text{H}_2\text{O}}}$$

Notice that the base need not have a hydroxyl group (OH$^-$), but that it may produce OH$^-$ by reacting with water.

When the expressions for K_a and K_b are multiplied together, an interesting result is obtained,

$$K_a K_b = \frac{a_{\text{H}_3\text{O}^+} a_{\text{OCl}^-}}{a_{\text{HOCl}} a_{\text{H}_2\text{O}}} \times \frac{a_{\text{HOCl}} a_{\text{OH}^-}}{a_{\text{OCl}^-} a_{\text{H}_2\text{O}}} = \frac{a_{\text{H}_3\text{O}^+} a_{\text{OH}^-}}{a_{\text{H}_2\text{O}}^2}$$

The product of K_a' and K_b' gives the autoprotolysis constant for the solvent, water,

$$K_a K_b = K_s \text{ (or } K_w\text{)} = a_{\text{H}_3\text{O}^+} a_{\text{OH}^-} \quad \text{when} \quad a_{\text{H}_2\text{O}} = 1.00$$

K_w has a numerical value of 1.0×10^{-14} at 25°C.

This kind of treatment can be made perfectly general for any proton donor-acceptor pair and an amphiprotic solvent. Brønsted-Lowry definitions do not depend on the nature of the solvent. Solvents which can accept or donate protons do, however, fit nicely into the theory.

"Strength" of an acid is a relative term in Brønsted-Lowry theory and is measured according to the ability of a species to donate protons to a particular base. So, for example, the equilibrium

$$\text{HCl} + \text{H}_2\text{O} = \text{H}_3\text{O}^+ + \text{Cl}^-$$
$$\text{(acid)} \quad \text{(base)} \quad \text{(acid)} \quad \text{(base)}$$

lies very far to the right. This means that HCl is a stronger proton donor than H_3O^+ and that H_2O is a better proton acceptor than Cl^-. However, in the case of acetic acid,

$$\text{CH}_3\text{COOH} + \text{H}_2\text{O} = \text{H}_3\text{O}^+ + \text{CH}_3\text{COO}^-$$
$$\text{(acid)} \quad \text{(base)} \quad \text{(acid)} \quad \text{(base)}$$

the equilibrium position lies toward the left. This means that acetic acid is a weaker proton donor than H_3O^+ and that H_2O is a weaker proton acceptor than CH_3COO^-. We say that in water acetic acid is a weak acid, while HCl is a strong acid. When the reference base is changed to a species which is more strongly basic than water, say liquid ammonia, both equilibria

$$\text{HCl} + \text{NH}_3 = \text{NH}_4^+ + \text{Cl}^-$$
$$\text{CH}_3\text{COOH} + \text{NH}_3 = \text{NH}_4^+ + \text{CH}_3\text{COO}^-$$

lie very far to the right. Both hydrochloric acid and acetic acid are strong acids in liquid ammonia. The only proton donor (at any appreciable concentration) in these systems at equilibrium is the ammonium ion. We say that the solvent has *leveled* the strengths of HCl and CH_3COOH to only one strength, that of NH_4^+. Evidently, the conjugate acid of a basic solvent is the strongest acid that can exist in that solvent. Similarly, the conjugate base of an acidic solvent is the strongest base that can exist in the solvent; any base which is stronger will react with the solvent to form the solvent's base form.

Brønsted-Lowry theory has proved quite successful in dealing with protonic acid-base systems. However, it is unable to explain acid-base behavior of species which do not contain protons. Acid-base reactions are observed in solvents such as thionyl chloride ($SOCl_2$) and arsenic trichloride ($AsCl_3$), even though protons play no role in their chemistry.

In summary, Brønsted-Lowry theory has been extremely popular because it extends the Arrhenius concept (and its quantitative interpretations) to a wider variety of bases and solvents. By emphasizing the donation and acceptance of protons, it removes the acid-base phenomenon from dependence on water as a solvent. There is no absolute scale of acid strength in Brønsted-Lowry theory since strength is determined by the acidity or basicity of the solvent. Brønsted-Lowry theory is quite well suited as a framework for acid-base equilibria in an introductory analytical chemistry course and is therefore the basis of the discussion in Chapter 5.

4.5 THE SOLVENT SYSTEM DEFINITION

Several definitions have been proposed for acids and bases in aprotic solvents, some of them predating Brønsted-Lowry theory. The most satisfactory definition was proposed by Cady and Elsey:

> In a general solvent an acid is a substance that releases a positive ion which is the same as the positive ion released by the solvent. A base is a substance that releases an ion which is the same as the negative ion released by the solvent.

For example, there is evidence that the aprotic solvent arsenic trichloride ($AsCl_3$)† undergoes autodissociation according to the reaction

$$2AsCl_3 = AsCl_2^+ + AsCl_4^- \qquad (K_s = 10^{-15})$$

According to the solvent-system definition, bases in $AsCl_3$ are compounds which increase the concentration of $AsCl_4^-$, while acids increase the concentration of $AsCl_2^+$. Species such as $SbCl_5$ can act as acids,

$$\underset{\text{(acid)}}{SbCl_5} + \underset{\text{(solvent, base)}}{AsCl_3} = \underset{\text{(ion pair, acid)}}{(SbCl_6^- \cdot AsCl_2^+)}$$

Species such as tetramethylammonium chloride [$(CH_3)_4NCl$] can act as bases, forming ion pairs with $AsCl_4^-$,

$$\underset{\text{(solvent, acid)}}{AsCl_3} + \underset{\text{(base)}}{(CH_3)_4NCl} = \underset{\text{(ion pair, base)}}{((CH_3)_4N^+ \cdot AsCl_4^-)}$$

Solutions of tetramethylammonium chloride in $AsCl_3$ can be titrated with $SbCl_5$ in $AsCl_3$ according to the reaction

$$\underset{\substack{\text{(ion pair)} \\ \text{(base)}}}{(CH_3)_4N^+ \cdot AsCl_4^-} + \underset{\substack{\text{(ion pair)} \\ \text{(acid)}}}{AsCl_2^+ \cdot SbCl_6^-} = \underset{\substack{\text{(ion pair)} \\ \text{("salt")}}}{(CH_3)_4N^+ \cdot SbCl_6^-} + \underset{\text{(solvent)}}{2AsCl_3}$$

Just as was the case with the Arrhenius definition, in the solvent-system definition a neutralization produces a salt and the solvent.

The solvent-system definition is by no means restricted to aprotic systems. Con-

---

† $AsCl_3$ has a wide liquid range (-13 to $130°C$). It has a far smaller dielectric constant (12.6) than water (78.3), and is less able than water to dissolve salts such as KCl and $NaCl$. The low dielectric constant of $AsCl_3$ favors the formation of ion pairs in solution. Such species are quite rare in water solutions.

sider again the solvent liquid ammonia, a protonic solvent which undergoes autodissociation as follows:

$$2NH_3 = NH_4^+ + NH_2^- \qquad (K_s = 10^{-27})$$

By the solvent-system definition an acid in liquid ammonia is a species which produces NH_4^+, while a base is a species which produces NH_2^-. Ammonium chloride (NH_4Cl) is therefore an acid in liquid ammonia and sodium amide ($NaNH_2$) a base. They react with each other in a neutralization reaction:

$$\underset{\text{(acid)}}{NH_4Cl} + \underset{\text{(base)}}{NaNH_2} = \underset{\text{(salt)}}{NaCl} + \underset{\text{(solvent)}}{2NH_3}$$

Ammonium chloride also reacts with metals in liquid ammonia to release hydrogen gas (i.e., *replace* hydrogen, according to Liebig):

$$\underset{\text{(acid)}}{2NH_4^+} + \underset{\text{(base)}}{2Na} = 2Na^+ + H_2(g) + \underset{\text{(solvent)}}{2NH_3}$$

A comparable reaction, of course, takes place in water:

$$2H_3O^+ + 2Na = 2Na^+ + H_2(g) + 2H_2O$$

A major strength of the solvent-system definition is that acidity and basicity need not be defined in terms of protons. In terms of generality, it is therefore superior to Brønsted-Lowry theory. The major weakness of the solvent-system approach is its heavy reliance on ionization. When ions can be recognized in a solvent, or better still, isolated from a solution, the solvent-system approach is quite useful in conceptualizing reactions. However, it is possible to overextend the definitions in solvents that are not conducive to forming ions. For example, for many years it was thought that liquid sulfur dioxide could autodissociate by the reaction

$$2SO_2 = SO^{2+} + SO_3^{2-}$$

Reactions of acids such as thionyl chloride ($SOCl_2$) with bases such as sodium sulfite (Na_2SO_3) were indeed found to produce a salt and SO_2,

$$SOCl_2 + Na_2SO_3 = 2NaCl + 2SO_2$$

Unfortunately, there is no experimental evidence at all that SO_2 dissociates in the liquid state, or that it participates in acid-base reactions in the way that water participates.

4.6 THE LEWIS DEFINITION

In 1923 Gilbert N. Lewis proposed a very general definition of acids and bases called the electronic theory. Lewis formulated the definitions without relying on a mechanism linked to a particular element (e.g., hydrogen) or a particular solvent interaction (e.g., dissociation). The Arrhenius, Brønsted-Lowry, and solvent-system definitions are in many respects special cases of Lewis's general theory. Lewis based his definitions on a large set of experimental results which showed that the phenomenon underlying acid-base reactions is the exchange of pairs of electrons among species, the making and breaking of *covalent bonds*. To Lewis a base was a species that *donated* a pair of electrons to form a covalent bond. An acid *accepted* a pair of electrons to form a covalent bond.

For example, the base ammonia can combine with a proton to form an ammonium ion

$$H_3N: + H^+ = (H_3N:H)^+, \quad \text{or simply } NH_4^+$$

The pair of electrons on ammonia is donated to form a covalent bond with the proton in an ammonium ion. The new N—H bond is just like the original three N—H bonds. Another example of a Lewis acid-base reaction is the reaction of the base ammonia with the Lewis acid boron trifluoride (BF_3),

$$:NH_3 + BF_3 = H_3N:BF_3$$

The electrons which form the covalent N—B bond can be pictured as being donated by the nitrogen atom.

Lewis theory has been quite popular with inorganic chemists because it provides a framework for understanding how metal ion coordination compounds form. For instance, cupric ion (Cu^{2+}) acts as a Lewis acid, accepting electron pairs from four ammonia molecules to form the very stable tetraammine copper(II) coordination complex:

$$Cu^{2+} + 4(:NH_3) = Cu(NH_3)_4^{2+} \quad (K = 4.3 \times 10^{14})$$

The term "very stable" means that the equilibrium constant for the formation of the complex is very large. Just as was the case in the Brønsted-Lowry and Arrhenius theories, the size of the equilibrium constant reflects the strength of the acid and base.

Lewis's electronic theory is the most general of any discussed so far, encompassing reactions involving hydrogen ion, metal ions, neutral compounds, and coordination compounds. However, the price paid for generality is the absence of a simple quantitative classification of acid and base strength. In the proton-based Brønsted-Lowry theory, a single constant (K_a) specifies the strength of an acid-base pair. This constant can be used to predict quantitatively the equilibrium of the pair with any other acid-base pair in the same solvent. Furthermore, the same constant allows the estimation of equilibrium positions in other solvents. Such simple relationships do not exist for Lewis acid-base pairs, since the strengths of acids and bases depend on particular reactions: a Lewis acid can seem strong in one reaction but weak in another. For example, consider the following pairs of reactions and their equilibrium constants:

$$\begin{cases} Cu^{2+} + NH_3 = Cu(NH_3)^{2+} & K = 10^{4.0} \\ Cd^{2+} + NH_3 = Cd(NH_3)^{2+} & K = 10^{2.5} \end{cases}$$
$$\begin{cases} Cu^{2+} + Br^- = CuBr^+ & K = 10^{-0.7} \\ Cd^{2+} + Br^- = CdBr^+ & K = 10^{2.2} \end{cases}$$
(acids) (bases)

In the pair of reactions in which ammonia is the base, Cu^{2+} acts as a stronger acid than Cd^{2+}; that is, K for the first reaction is larger than K for the second. However, in the reactions where bromide is the base, Cu^{2+} is a *weaker* acid than Cd^{2+}. The effect of this kind of paradox is that we cannot make a statement such as "Cu^{2+} is always a stronger acid than Cd^{2+}."

R. G. Pearson introduced a qualitative principle for deciding whether a Lewis acid will be strong or weak in reactions with Lewis bases. The idea was to divide both acids and bases into two categories, "soft" and "hard," to reflect the experimental fact that "soft acids" tend to prefer to bond with "soft bases," and "hard acids" with "hard bases." Examples of some soft and hard acids and bases are shown in Table 4.3.

It should be apparent that softer bases are those in which the electron donor atom has plenty of electrons and holds them loosely. Hard acids are generally small and positively charged, while soft acids are generally large and have a low positive or neutral charge.

TABLE 4.3 SOFT AND HARD ACIDS AND BASES[a]

Soft acids	Soft bases
Hg^+, Cu^+, Ag^+	H^-(hydride), CN^-, CO
Au^+, Cd^{2+}, BH_3	SCN^-, R_2S, $S_2O_3^{2-}$, I^-
Hg^{2+}, metals	
Hard acids	Hard bases
H^+, Na^+, Mg^{2+}, La^{3+}	NH_3, RNH_2, H_2O, OH^-
Ce^{4+}, Th^{4+}, Fe^{3+}	O^{2-}, F^-, Cl^-, CH_3COO^-
BF_3, Al^{3+}, N^{3+}, As^{3+}	CO_3^{2-}, SO_4^{2-}

[a] In addition there are acids and bases of intermediate character. Some borderline acids are Fe^{2+}, Cu^{2+}, Sn^{2+}, Pb^{2+}, and SO_2. Some borderline bases are Br^-, pyridine, azide (N_3^-), and NO_2^-.

Hard and soft acid-base (HSAB) theory is generally able to make predictions about whether or not reactions will occur. For example, Be^{2+} ion is a hard acid and readily bonds with the hard base fluoride ion,

$$Be^{2+} + 2F^- = BeF_2$$
$$\text{(hard)} \quad \text{(hard)}$$

However, Hg^{2+} ion is a soft acid and prefers a base such as iodide,

$$Hg^{2+} + 2I^- = HgI_2$$
$$\text{(soft)} \quad \text{(soft)}$$

The HSAB theory predicts that a mixture of BeI_2 and HgF_2 should react to exchange bases:

$$HgF_2 + BeI_2 \rightarrow HgI_2 + BeF_2$$
$$\text{(soft-hard)} \quad \text{(hard-soft)} \quad \text{(soft-soft)} \quad \text{(hard-hard)}$$

It is found that this reaction does indeed proceed to the right as written, with the release of nearly 400 kJ/mole.

There are several theoretical interpretations of the preference of hard acids for hard bases and soft acids for soft bases, but these are better left for a course in inorganic chemistry. You should understand that HSAB theory is a qualitative tool for predicting whether reactions will occur. The absence of a general quantitative tool is a weakness of Lewis theory.

A debate about the relative strengths and weaknesses of the acid-base theories discussed in this chapter would not be fruitful. The primary function of theories in science is to explain experimental observations. All of the theories discussed in this chapter do that, some better than others under particular circumstances. All of them have stimulated experimental work, an important function of theories. Taken as a set, they help chemists understand a wide variety of chemical reactions.

RECOMMENDED READING

Bell, R. P. *Acids and Bases,* Methuen, London, 1969. This is a brief, easy-to-read introduction to acid-base reactions and theories.

Finston, H. L. and A. C. Rychtman. *A New View of Current Acid-Base Theories,* Wiley, New York, 1982. An excellent, thorough development of the popular acid-base theories, with some new experimental insights. Literature review to mid-1970s.

Jensen, W. B. *The Lewis Acid-Base Concepts,* Wiley, New York, 1980. Chapters 7 and 8 focus on HSAB theory, using a great deal of experimental data. This book will be quite challenging for beginners.

PROBLEMS

4.1. How did the experiments of Sir Humphrey Davy affect Lavoisier's theory of the composition of acids?

4.2. In what ways were the Liebig and Arrhenius definitions of acids and bases similar? How did they differ?

4.3. According to the theory of Arrhenius, a base is a species which contains hydroxyl ions. Give examples of compounds which are bases in water but do not contain dissociable hydroxyl ions.

4.4. Unlike earlier theories, the Arrhenius theory had a quantitative foundation. For example, electrical conductance studies on monochloroacetic acid show that it is about 12% dissociated into hydrogen ions and monochloroacetate ions when 0.100 mole of the acid is dissolved in enough water to make 1 liter of solution. Calculate the dissociation constant, K_a, for monochloroacetic acid. Judging by the data given in Table 4.1, is monochloroacetic acid a stronger or weaker acid than acetic acid in water?

4.5. In what ways is the Brønsted-Lowry definition of bases more useful than the Arrhenius definition?

4.6. What is meant by the term *conjugate pair*? Give examples of conjugate pairs involving *amphiprotic* species.

4.7. When an amphoteric compound is used as a solvent, the concentrations of the acid and base forms of the compound are related by the _____ constant. In the case of the solvent water, this constant has a value of 1×10^{-14} at 25°C. The strength of an acid in some solvent is reflected in the size of the _____. The strength of the conjugate base in the solvent is reflected in the size of the _____. The product of _____ and _____ is always equal to the _____ of the solvent.

4.8. HCl and $HClO_4$ are equally strong acids in water because they are far stronger acids than H_3O^+. Their strengths are said to be _____ to that of H_3O^+ in water.

4.9. Why is acetic acid a far weaker acid than HCl in water, but as strong an acid as HCl in liquid ammonia solvent?

4.10. In what ways are the Brønsted-Lowry and solvent-system (Cady and Elsey) definitions of acids and bases similar? How are they different?

4.11. What are the strongest acids and bases in the following solvents: H_2O, CH_3COOH (acetic acid), HF, H_2SO_4, NH_3.

4.12. Arrange the following bases in order of their increasing strength relative to the acid Cd^{2+}: I^-, Cl^-, Br^-, CH_3COO^-, CN^-, NH_3, $S_2O_3^{2-}$. The following constants should help:

$$Cd^{2+} + I^- = CdI^+ \qquad K = 190$$
$$Cd^{2+} + Cl^- = CdCl^+ \qquad K = 22$$
$$Cd^{2+} + Br^- = CdBr^+ \qquad K = 56$$
$$Cd^{2+} + CH_3COO^- = CdCH_3COO^+ \qquad K = 20$$
$$Cd^{2+} + NH_3 = CdNH_3^{2+} \qquad K = 400$$
$$Cd^{2+} + CN^- = CdCN^+ \qquad K = 3.5 \times 10^5$$
$$Cd^{2+} + S_2O_3^= = CdS_2O_3 \qquad K = 8.3 \times 10^3$$

4.13. Cd^{2+} is generally considered to be a rather soft acid. Which bases in the list above are soft bases? Are there any values of K which seem unusually large or small, considering the classification scheme in Table 4.3?

4.14. Use HSAB theory to determine the direction in which each of the following reactions will proceed (to the left or right) when equimolar mixtures of all the species in each reaction are combined.
(a) $HI + AgF = HF + AgI$
(b) $CuI_2 + MgF_2 = CuF_2 + MgI_2$
(c) $HgBr_2 + 2HCN = Hg(CN)_2 + 2HBr$

Chapter 5

Acid-Base Equilibria

5.1 INTRODUCTION

This chapter will cover in detail acid-base equilibria in aqueous solutions, traditionally one of the most important subjects studied in undergraduate quantitative analysis courses. We will begin with some fundamental concepts: pH, simple equilibrium constant expressions for acids and bases, and the role of water as a solvent; much of this section will be a review of material treated in most general chemistry courses. We will then develop an algebraic and graphical approach to solving acid-base equilibrium problems and apply the methods to both simple and complex systems. The last section of the chapter is devoted to a detailed study of buffers and buffer capacity. Acid-base titrations will be treated as a separate topic in the next chapter.

5.2 REVIEW OF BASIC CONCEPTS

5.2.1 Definition of pH

The simplest definition of pH is that proposed in 1909 by Sørensen: pH is the negative logarithm of the hydrogen ion concentration,

$$\text{pH} = -\log[\text{H}^+]$$

This is the definition used in many introductory texts, and it is adequate for approximate calculations, constructing species distribution diagrams and titration curves, and selecting endpoint indicators.

The thermodynamically rigorous definition recognizes the fact that chemical reactions respond to activities rather than concentrations, and relates pH to hydrogen ion activity:

$$\mathrm{pH} = -\log a_{\mathrm{H}^+}$$

Unfortunately, it is difficult to use this definition in practical situations for two important reasons. First, as discussed in Chapter 3, the activity of a single ion

$$a_{\mathrm{H}^+} = [\mathrm{H}^+]f_{\mathrm{H}^+}$$

cannot be measured. Instead, it must be *assumed* that the activity coefficients of all singly charged species in solution are equal and given by the mean activity coefficient

$$f_\pm = (f_+ f_-)^{1/2}$$

a quantity which can be measured. A more practical definition of pH is then

$$\mathrm{pH} = -\log([\mathrm{H}^+]f_\pm)$$

The second limitation of the activity definition arises in measurement of the activity with a glass electrode (a "pH electrode") in the laboratory. A complication called "liquid junction potential," which will be explained in Chapter 14, makes the measurement unreliable. It will be sufficient at this point to say that the ideal relationship between the potential of the electrode (E_{elect}) and the pH of a test solution is given by

$$\mathrm{pH} = -\log a_{\mathrm{H}^+} = \frac{E_{\mathrm{elect}}}{0.05916} \quad \text{(ideal)}$$

but that the best measurement of E_{elect} in the laboratory will also involve some liquid junction potential ($E_{\mathrm{l.j.}}$),

$$\mathrm{pH} = -\log a_{\mathrm{H}^+} = \frac{E_{\mathrm{elect}} - E_{\mathrm{l.j.}}}{0.05916} \quad \text{(real)}$$

The size of $E_{\mathrm{l.j.}}$ varies from one electrode to the next and is difficult to assess. Given this variability, the laboratory pH scale must be an empirical (experimental) scale. Using a series of carefully prepared buffer solutions covering a wide range in acidity, one calibrates the response of the electrode and then measures the pH of the test solution. It must be assumed that $E_{\mathrm{l.j.}}$ is a constant which applies to the test solution

5.2 REVIEW OF BASIC CONCEPTS

as well as the buffer solutions.† The National Bureau of Standards specifies the pH of buffer solutions to give nominal pH values which closely follow the relationship

$$\text{pH} = -\log a_{\text{H}^+}$$

EXAMPLE 5.1 Calculating pH

Calculate the pH of a solution which has a hydrogen ion concentration of $1.05 \times 10^{-4} M$ and an ionic strength of $0.1 M$.

Solution

Using the Sørensen definition, we find

$$\text{pH} = -\log[\text{H}^+] = -\log(1.05 \times 10^{-4}) = 3.98$$

To approximate the activity we must first calculate the mean activity coefficient of a singly charged species in a solution with ionic strength $0.1 M$. The value of $f_\pm$ calculated from the modified Davies equation (see Chapter 3) is 0.81. Using the expression

$$\text{pH} = -\log[\text{H}^+]f_\pm$$

we calculate the pH to be 4.07.

The difference of 0.1 pH unit between these values is typical of the size of the error associated with using concentrations instead of activities in rather simple systems such as acetic acid and benzoic acid when the ionic strength is appreciable, as is usually the case in laboratory work.

5.2.2 Reactions with Water as Solvent

As discussed in the last chapter, water dissociates to a small extent and is able to act as either an acid or a base (it is amphiprotic). The autoprotolysis reaction for water may be written as

$$\text{H}_2\text{O} + \text{H}_2\text{O} = \text{H}_3\text{O}^+ + \text{OH}^-$$

with thermodynamic equilibrium constant given by

$$K_w = \frac{a_{\text{H}_3\text{O}^+} a_{\text{OH}^-}}{a_{\text{H}_2\text{O}}^2}$$

† If the ionic strength of the test solution is quite different from that of the buffers, this assumption may not be good.

TABLE 5.1 AUTOPROTOLYSIS CONSTANT OF WATER (AS pK_w) 0 TO 100°C

Temperature (°C)	pK_w	Temperature (°C)	pK_w
0	14.994	40	13.535
10	14.535	50	13.262
20	14.167	60	13.017
25	13.996	80	12.60
30	13.833	100	12.31

Source: Data of H. S. Harned and R. A. Robinson, *Transactions of the Faraday Society,* 36, 973 (1940) (values from 0 to 60°C are experimental, value at 80°C calculated). The 100°C value is from A. A. Noyes and Y. Kato, *Publications of the Carnegie Institution of Washington,* 63, 153 (1907).

In dilute solutions the concentration of water is large and nearly constant (55.5M), and so its activity can be considered to be unity. If we also assume that the activity coefficients of hydronium ion and hydroxide ion are unity in dilute solutions, we can replace their activities by concentrations, giving

$$K_w = [H_3O^+][OH^-] = 1.01 \times 10^{-14} \quad \text{at } 25°C$$

Once again, for convenience we will represent the hydronium ion by the proton, H^+. Using the notation $pX = -\log X$, we can then write

$$pK_w = pH + pOH = 14.0 \quad \text{at } 25°C$$

The autoprotolysis constant of water is an equilibrium constant and is therefore independent of the concentration of hydronium ion or hydroxide ion. Like all other equilibrium constants, it is, however, dependent on temperature, as shown in Table 5.1. Notice that there is a change in K_w of a factor of about 500 over the liquid range of water.

EXAMPLE 5.2 pH, pOH, and pK_w

Calculate the specified quantity in each solution. (a) What are $[H^+]$ and pH in a solution in which $[OH^-] = 2.3 \times 10^{-5} M$? (b) What are $[OH^-]$ and pOH in a solution in which $[H^+] = 9.5 \times 10^{-3} M$?

Solution

(a) Using $K_w = [H^+][OH^-]$, we calculate

$$[H^+] = (10^{-14})/2.3 \times 10^{-5}$$
$$= 4.3 \times 10^{-10} M$$
$$pH = -\log(4.3 \times 10^{-10}) = \underline{9.37}$$

(This is a basic solution.)

(b) Using $pK_w = pH + pOH$ and knowing $pH = -\log(9.5 \times 10^{-3}) = 2.02$, we calculate

$$pOH = 14.00 - 2.02 = 11.98$$

(This is an acidic solution.)

5.2.3 Acid-Base Equilibrium Constant Expressions

When a weak acid is dissolved in water a portion of its molecules dissociate to form hydronium ions (aquated protons) and conjugate base anions. For example, acetic acid (CH$_3$COOH, which we will abbreviate as "HOAc") dissociates to form hydronium ions and acetate ions,

$$HOAc = H^+ + OAc^-$$

The *thermodynamic* equilibrium constant for this equilibrium is given by the ratio of the activities of the species,

$$K_a^0 = \frac{a_{H^+} a_{OAc^-}}{a_{HOAc}}$$

and has a value calculated from the standard free energy† change for the reaction,

$$\Delta G_{dissoc}^0 = (G_{H^+}^0 + G_{OAc^-}^0) - G_{HOAc}^0 = -RT \ln K_a^0$$

As shown in Chapter 3, the dissociation constant can be expressed in terms of concentrations by using activity coefficients of the ionic species and the salting coefficient of the molecular acid,

$$K_a^0 = \frac{[H^+][OAc^-] f_\pm^2}{[HOAc] k}$$

The equilibrium constant in a solution of appreciable ionic strength (the "concentration" equilibrium constant) will be related to the thermodynamic constant in the following way:

$$K_a = K_a^0 \left(\frac{k}{f_\pm^2}\right) = \frac{[H^+][OAc^-]}{[HOAc]}$$

† The standard free energies (in kilojoules per mole) are: H$^+$, 0.00; OAc$^-$, -372.46; HOAc, -399.61. These values are used to calculate $K_a^0 = 1.746 \times 10^{-5}$, as shown in Chapter 3.

For the kinds of solutions we will deal with in this text ($0.1M$ and less concentrated) salting coefficients can be considered to be unity. Activity coefficients can be approximated by the Davies equation (Chapter 3).

EXAMPLE 5.3 Activity Coefficient Effects

Calculate the concentration equilibrium constant for acetic acid in a solution of ionic strength $0.1M$.

Solution

Using the Davies equation, we find $f_\pm = 0.81$ in a solution of this ionic strength. Taking the salting coefficient as unity, we calculate

$$K_a = \frac{K_a^0}{f_\pm^2} = \frac{1.75 \times 10^{-5}}{(0.81)^2} = 2.7 \times 10^{-5}$$

Notice that this value is considerably larger than the thermodynamic value.

A useful form of the dissociation constant for direct use in laboratory work is the "formal equilibrium constant" or formal dissociation constant. The values for these constants are determined experimentally with glass electrodes and include activity coefficients *and* liquid junction potentials in the specified electrolytes. For example, the formal dissociation constant of acetic acid in $0.2M$ KCl electrolyte is 2.3×10^{-5}. It is understood that this is the value one can expect to measure with a glass electrode and a pH meter (standardized with buffer) in a $0.2M$ solution of KCl. Since liquid junction potential is included, the value will not strictly apply to solutions of different electrolytes (e.g., NaCl or KNO_3) but will be only an approximation.

Thermodynamic, concentration, and formal equilibrium constants are all included in the appendices. Thermodynamic constants will be used in routine calculations in the remainder of this chapter.

5.2.4 Weak Base Reactions

When a weak base is dissolved in water, a portion will react to form the conjugate weak acid and hydroxide ion,

$$B + H_2O = BH^+ + OH^-$$

The thermodynamic constant for this reaction is

$$K_b^0 = \frac{a_{OH^-} a_{BH^+}}{a_B}$$

5.3 BASIC RELATIONSHIPS

and is called the *base dissociation constant*. In Chapter 4 it was shown that $K_w = K_a K_b$, a relationship you should verify. Base dissociation constants are not included in an appendix; they can always be calculated from K_w (1.01×10^{-14} at 25°C) and tabulated K_a values.

5.3 BASIC RELATIONSHIPS

The goal of this section is to develop a set of relationships with which we can accurately describe solutions of acids and bases under equilibrium conditions. When we dissolve a sample of a polyprotic acid in water, there will be several chemical species whose concentrations are related by equilibria: the weak acid (undissociated), its conjugate bases, the proton, and the hydroxide ion. In order to calculate the concentrations of these species, we need to develop a set of equations which relate them to each other and to the solvent. These equations involve solvent autodissociation (K_w) and the special algebraic relationships called *material balance* and *charge balance*. Solutions to the sets of equations arrived at are often quite complicated algebraically, and we often apply what we know about solution chemistry along with intuition to make simplifying assumptions. In many cases it will be simpler and at the same time more instructive to solve the equations by using graphical methods.

A few words of caution are needed before discussing equations. It is possible to lose sight of the goal of accurately describing a solution at equilibrium when one is deeply involved in difficult algebraic problems. While many of the derivations of the expressions are interesting problems in their own right, it should be remembered that algebra cannot substitute for a knowledge of basic chemistry. We will try to make chemically sound assumptions whenever possible to simplify the algebra and also maintain an accuracy of about 5% in the calculations.

In the discussion that follows we will often refer to the benzoic acid-benzoate ion system. Benzoic acid is a weak Bronsted acid,

$$\text{HOBz} \rightleftharpoons \text{OBz}^- + \text{H}^+ \qquad K_a = 6.17 \times 10^{-5}$$

Its gram formula weight is 122.12, and it is only slightly soluble in water (0.028M is a saturated solution at 25°C). Benzoate ion is a weak base,

$$\text{OBz}^- + \text{H}_2\text{O} = \text{HOBz} + \text{OH}^- \qquad K_b = 1.62 \times 10^{-10}$$

The salt sodium benzoate is quite soluble in water.

Let us begin with a discussion of material balances and charge balances.

5.3.1 Material Balance

The material balance is a mathematical expression of the law of conservation of matter: "in ordinary chemical reactions the number of atoms of a certain kind remains constant." If we take 0.01 mole of benzoic acid (HOBz) and dissolve it in 1 liter of water, it must be true that

$$C_{total} = 0.01F = [\text{HOBz}] + [\text{OBz}^-]$$

C_{total} is the *total concentration* of all forms of benzoic acid in solution. It is also called the *analytical concentration*. This expression reflects the chemical facts that some of the benzoic acid reacts with water to form benzoate ion and that the sum of the concentrations of acid and conjugate base *at equilibrium* must equal the total concentration. If both benzoic acid and sodium benzoate are added to water, the material balance is written as

$$C_{\text{HOBz}} + C_{\text{NaOBz}} = [\text{HOBz}] + [\text{OBz}^-]$$

A verbal statement is that the benzoic acid and benzoate ion concentrations at equilibrium must sum up to equal the total concentration of benzoic acid and sodium benzoate.

The *proton balance* is a material balance relationship among the sources and sinks of protons in a solution. For example, protons in a solution of benzoic acid can come from two sources: benzoic acid (in the dissociation, which releases an equal number of benzoate ions) or water (in the autoprotolysis reaction, which releases an equal number of hydroxide ions). The proton balance is written as

$$[\text{H}^+] = [\text{OH}^-] + [\text{OBz}^-]$$

A *hydroxide balance* is a material balance involving the sources and sinks of hydroxide ions. If the base sodium benzoate is dissolved in water, the hydroxide ion balance will be

$$[\text{OH}^-] = [\text{H}^+] + [\text{HOBz}]$$

When a water molecule dissociates, a proton is released along with the hydroxide ion. When a benzoate ion reacts with water to form a hydroxide ion, a molecule of benzoic acid is also produced,

$$\text{OBz}^- + \text{H}_2\text{O} = \text{HOBz} + \text{OH}^-$$

5.3.2 Charge Balance (Electroneutrality)

A charge balance relationship shows that a solution must have as many negatively charged species as positively charged species. All ions must be accounted for in a charge balance, even those which have no direct role in the equilibria being examined. In the case of the $0.01F$ benzoic acid solution, the charge balance is written

$$[H^+] = [OBz^-] + [OH^-]$$

Undissociated benzoic acid is not included because it has no charge.

The charge balance for a $0.01F$ solution of sodium benzoate is

$$[Na^+] + [H^+] = [OBz^-] + [OH^-]$$

EXAMPLE 5.4 Material Balances and Charge Balances

Write material balance and charge balance equations for each of the following systems:
 (a) $0.1F$ HNO_3 (a strong acid)
 (b) $0.01F$ HCN (a weak acid)
 (c) $0.1F$ NH_3 (a weak base)
 (d) $0.1F$ $NaHCO_3$ (a weak acid)

Solution

(a) At equilibrium the solution will contain H^+, OH^-, and NO_3^-. Since nitric acid is a very strong acid, we expect to find no undissociated acid. The material balances are:

$$\text{for nitrate:} \quad C_{\text{total}} = 0.1F = [NO_3^-]$$
$$\text{for protons:} \quad [H^+] = [OH^-] + [NO_3^-]$$

The charge balance is:

$$[H^+] = [OH^-] + [NO_3^-]$$

Notice that the proton balance and charge balance are identical.

(b) Since HCN is a weak acid, we expect to find the species H^+, OH^-, CN^-, *and* HCN at equilibrium. Material balance:

$$\text{for HCN:} \quad C_{\text{total}} = 0.01F = [HCN] + [CN^-]$$
$$\text{for } H^+: \quad [H^+] = [OH^-] + [CN^-]$$

Charge balance:

$$[H^+] = [OH^-] + [CN^-]$$

(Again, the proton balance and charge balance are the same.)

(c) Since NH_3 is a weak base, we expect to find some NH_3 at equilibrium as well as OH^-, H^+, and NH_4^+. Material balances:

$$\text{for ammonia:} \quad C_{total} = 0.1F = [NH_3] + [NH_4^+]$$
$$\text{for protons:} \quad [H^+] + [NH_4^+] = [OH^-]$$

(also called a hydroxide balance)
Charge balance:

$$[NH_4^+] + [H^+] = [OH^-]$$

(d) Bicarbonate ion is the conjugate base of carbonic acid, $HCO_3^- + H_2O = H_2CO_3 + OH^-$. Carbonic acid can be in equilibrium with atmospheric CO_2, $CO_2 + H_2O = H_2CO_3$. Bicarbonate ion is also the conjugate acid of carbonate ion, $HCO_3^- = H^+ + CO_3^{2-}$. We expect to find H_2CO_3, HCO_3^-, CO_3^{2-}, H^+, OH^-, Na^+, and some dissolved CO_2 at equilibrium. Material balances:

$$\text{for bicarbonate:} \quad C_{total} = 0.1F = [H_2CO_3] + [HCO_3^-] + [CO_3^{2-}] + [CO_2(aq)]$$
$$\text{for protons:} \quad [CO_2(aq)] + [H_2CO_3] + [H^+] = [OH^-] + [CO_3^{2-}]$$

Charge balance:

$$[Na^+] + [H^+] = [OH^-] + 2[CO_3^{2-}] + [HCO_3^-]$$

(It is important to point out that the proton balance can as easily be written as

$$[H^+] = [OH^-] + [CO_3^{2-}] - [H_2CO_3] - [CO_2(aq)]$$

Perhaps this more clearly indicates that a proton is *removed* from solution when a bicarbonate ion reacts to form carbonic acid. Bicarbonate ion is thus a *source* and a *sink* for protons.)

5.3.3 The Reference Level

In dealing with equilibria involving species of polyprotic acids and polybasic bases it is often useful to think in terms of a "reference level" for protons. The "zero level" of protons is the solvent and the acid or base added to it. Other species are then collected in groups of "acid-rich" or "acid-poor."

EXAMPLE 5.5 Reference Level

Identify the zero level of protons in

(a) a $0.01F$ solution of benzoic acid, and

(b) a $0.1F$ solution of potassium biphthalate (KHP).

Solution

(a) The zero level is the weak acid and water. The proton balance is written as

$$[\text{H}^+] = [\text{OH}^-] + [\text{OBz}^-]$$
"proton-rich" "proton-poor"

The ions OH^- and OBz^- are called proton-poor because they contain fewer protons than the zero-level substances H_2O and HOBz. The H^+ represents H_3O^+, a species which is rich in protons relative to H_2O.

(b) The zero level is water and biphthalate ion, HP^-. The proton balance can be written as

$$[\text{H}_2\text{P}] + [\text{H}^+] = [\text{OH}^-] + [\text{P}^{2-}].$$
"proton-rich" "proton-poor"

Notice that H_2P and H_3O^+ both contain more protons than the zero-level substances and that OH^- and P^{2-} contain fewer protons. Notice also that the zero-level substances never appear directly in the proton balance.

5.4 MONOPROTIC ACIDS AND MONOBASIC BASES

5.4.1 Solutions of Strong Acids or Strong Bases and Their Salts

When any monoprotic acid, strong or weak, is dissolved in water, two fundamental equilibria begin to operate. The first is the acid dissociation equilibrium,

$$\text{HX} = \text{H}^+ + \text{X}^-$$

with

$$K = \frac{[\text{H}^+][\text{X}^-]}{[\text{HX}]}$$

The second is the autoprotolysis of water, with the ion product expression

$$K_w = [H^+][OH^-] = 1.0 \times 10^{-14}$$

These reactions account for all of the sources of protons in the system.

In the case of a strong acid the dissociation reaction proceeds to completion and K is very large. When there is no longer any net change in the concentrations of species in solution the concentration of anion, $[X^-]$, will equal the formal concentration of the strong acid. The proton balance for the system is

$$[H^+] = [OH^-] + [X^-]$$

Substituting $K_w/[H^+]$ for $[OH^-]$ gives the quadratic equation,

$$[H^+] = \frac{K_w}{[H^+]} + [X^-]$$

While this expression accurately describes the solution, it is needed only if the concentration of strong acid is very low, less than about $10^{-6} F$. At higher concentrations $[OH^-]$ is very small and there is not a significant difference between $[H^+]$ and $[X^-]$, and so the simple expression

$$[H^+] = [X^-]$$

is used. It is important to note that when the concentration of strong acid is much smaller than $10^{-6} F$, autoprotolysis becomes an increasingly important source of protons. For example, if we were to transfer 10^{-8} mole of strong acid to 1 liter of pure water (10 µl of a milliformal solution of HNO_3 diluted to 1 liter with pure water), we would expect to find the solution slightly acidic. But if we calculate $[H^+]$ for the solution by using the simplified expression, we would find $[H^+] = [X^-] = 10^{-8} M$, pH = 8, a basic solution. We can never get a basic solution by adding an acid to water.

The strong base (MOH) case is treated in exactly the same way as the strong acid. The expression

$$[OH^-] = [M^+] = C_{total}$$

applies for solutions of monobasic bases at concentrations greater than $10^{-6} F$. The proton concentration can be found with the K_w relationship,

$$[H^+][OH^-] = K_w$$

The pH of a solution of the salt of a strong acid or base depends on the chemical nature of the salt. Solutions of the salts of monoprotic strong acids are neutral, since

5.4 MONOPROTIC ACIDS AND MONOBASIC BASES

the conjugate base of a strong acid is extremely weak. Certain salts of polyprotic acids will have dissociable protons and can be expected to produce acidic solutions. For example, sodium bisulfate ($NaHSO_4$) is a salt of the strong acid sulfuric acid (H_2SO_4) and is a weak acid,

$$HSO_4^- = H^+ + SO_4^{2-} \quad K_a = 1.2 \times 10^{-2}$$

A solution of $NaHSO_4$ will therefore be acidic. Other salts may react as weak bases to produce hydroxide ion. Since sulfate ion is a weak base,

$$SO_4^{2-} + H_2O = HSO_4^- + OH^- \quad K_b = 8 \times 10^{-13}$$

a solution of sodium sulfate will be very slightly basic.

5.4.2 Solution of a Weak Acid

In the case of a monoprotic weak acid such as benzoic acid the equilibrium

$$HOBz = H^+ + OBz^-$$

does not lie far to the right (dissociation is not favored) and the equilibrium constant

$$K_a = \frac{[H^+][OBz^-]}{[HOBz]}$$

is small (6.3×10^{-5}). The proton balance and material balance for a solution of benzoic acid are

$$\underset{\text{(proton rich)}}{[H^+]} = \underset{\text{(proton poor)}}{[OH^-] + [OBz^-]}$$

and

$$C_{\text{total}} = C_t = [HOBz] + [OBz^-]$$

respectively. The pH of a solution of benzoic acid can be calculated by taking the equilibrium constant expression and rearranging to get

$$K_a[HOBz] = [H^+][OBz^-]$$

Substituting from the material balance for [HOBz] gives

$$K_a(C_t - [OBz^-]) = [H^+][OBz^-]$$

If we assume that compared to the dissociation of HOBz, the autoprotolysis of water is only a minor source of protons and can be neglected, then we can approximate the proton balance by

$$[H^+] = [OBz^-]$$

and alter the last expression to give the quadratic equation

$$K_a(C_t - [H^+]) = [H^+]^2$$

There are several ways to deal with this last expression. First, we might assume that the acid is weak enough so that $[H^+] \ll C_t$. This assumption reduces the quadratic expression to the familiar

$$[H^+] = (K_a C_t)^{1/2}$$

Alternatively, we can rearrange the quadratic expression for $[H^+]$ to a more recognizable form,

$$[H^+]^2 + [H^+]K_a - K_a C_t = 0$$

and solve it exactly by using the quadratic formula (positive root)

$$[H^+] = \frac{-K_a + (K_a^2 + 4K_a C_t)^{1/2}}{2}$$

EXAMPLE 5.6 pH of a Solution of a Weak Acid
Calculate the pH of a $0.01F$ solution of benzoic acid, $K_a = 6.3 \times 10^{-5}$.

Solution

The quadratic equation developed in the text is

$$K_a(C_t - [H^+]) = [H^+]^2$$

This equation is used with the assumption that $[OH^-] \ll [OBz^-]$ in the proton balance expression, that is, that benzoic acid is a much more important source of hydronium ion than the autodissociation of water.

Let us assume that $[H^+] \ll C_t$ to simplify the quadratic equation. Using the simpler equation

$$[H^+] = (K_a C_t)^{1/2}$$

we find

$$[H^+] = [6.3 \times 10^{-5}(.01)]^{1/2} = 7.9 \times 10^{-4} M \quad pH = 3.10$$

When we check the assumption that $[H^+] \ll C_t$, we find that $[H^+]$ is about 8% of C_t (that is, $0.00079/0.01 = 0.079$). It is best to keep such approximations within about 5%, and so we should solve the quadratic equation exactly.

The exact solution to the quadratic equation (positive root) is $[H^+] = 7.6 \times 10^{-4} M$, or pH = 3.12. The approximate $[H^+]$ we just calculated was $7.9 \times 10^{-4} M$, about 3% larger than the exact solution value.

We should also check the assumption that $[OH^-] \ll [OBz^-]$. When $[H^+] = 7.6 \times 10^{-4}$, $[OH^-] = K_w/(7.6 \times 10^{-4}) = 1.3 \times 10^{-11}$. The concentration of benzoate is very nearly that of hydronium ion (the proton balance), or $7.6 \times 10^{-4} M$. The assumption is valid.

EXAMPLE 5.7 Rather Strong Weak Acid

Calculate the pH of a $0.01 F$ solution of trichloroacetic acid (HTCA). $K_a = 0.13$.

Solution

The simplest expression,

$$[H^+] = (K_a C_t)^{1/2}$$

gives

$$[H^+] = [0.13(0.01)]^{1/2} = 0.0361 M$$

This is an impossible result, since $[H^+]$ cannot be greater than C_t for a monoprotic acid. Apparently HTCA is such a strong acid that the assumption that $C_t \gg [H^+]$ is not valid. The quadratic equation must be solved:

$$[H^+] = \frac{-0.13 + (0.0169 + 0.0052)^{1/2}}{2} = 9.33 \times 10^{-3} M$$

$$pH = 2.03$$

This result indicates that the acid is almost completely dissociated at equilibrium. You should see that the autoprotolysis of water is an unimportant source of protons compared to the dissociation of the acid.

Whether one introduces a significant error by choosing to solve the approximate rather than the exact equation depends on the size of K_a and the total concentration

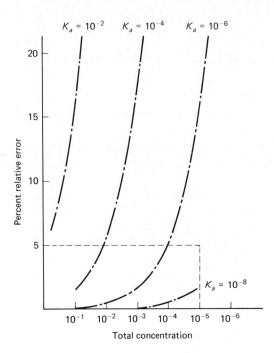

Figure 5.1 Percent relative error in calculation vs. C_{total} for several K_a's.

of acid and conjugate base. Figure 5.1 shows the percent relative error involved in choosing to solve the approximate equation rather than the exact equation as a function of these variables. A set of curves has been drawn for acids with K_a values 10^{-2}, 10^{-4}, 10^{-6}, and 10^{-8} over a wide range of total concentration values. The area blocked off contains combinations of K_a and C_t for which use of the approximate equation results in a relative error of 5% or less. Apparently, in order to remain within the 5% boundary for acids with $K_a = 10^{-3}$ one should deal with solutions more concentrated than about $0.1F$, if the approximate equation is to be used. It is permissible to use the approximate expression for weaker acids at much lower concentrations. The simplified equation can be used for $10^{-4}F$ solutions of acids with K_a values around 10^{-6}.

The vertical dashed boundary line on the right in Fig. 5.1 deserves closer attention. In dilute solutions of very weak acids both the exact and approximate equations produce absurd results. This is because neither equation correctly accounts for the autoprotolysis of the solvent.

EXAMPLE 5.8 Extremely Weak Acid
Calculate the pH of a $10^{-4}F$ solution of methylammonium chloride, $CH_3NH_3^+Cl^-$, $K_a = 2.4 \times 10^{-11}$. The dissociation reaction is

$$CH_3NH_3^+ = H^+ + CH_3NH_2$$

5.4 MONOPROTIC ACIDS AND MONOBASIC BASES

Solution

First use the approximate equation,

$$[H^+] = (K_a C_t)^{1/2} = [(2.4 \times 10^{-11})(10^{-4})]^{1/2}$$
$$= 4.9 \times 10^{-8}$$
$$pH = 7.3$$

This is an impossible result. It predicts a basic solution.

Let us begin again with the proton balance for a solution of the weak acid HA,

$$[H^+] = [OH^-] + [A^-]$$
(proton rich) (proton poor)

and substitute for both $[OH^-]$ and $[A^-]$,

$$[H^+] = \frac{K_w}{[H^+]} + \frac{K_a[HA]}{[H^+]}$$

Collecting terms, we find

$$[H^+] = (K_w + K_a[HA])^{1/2}$$

For the extremely weak acids we are dealing with (phenols, substituted ammonium salts, etc.) $[HA] = C_t$.

EXAMPLE 5.9 Extremely Weak Acid—A Second Try
Calculate the pH of a $10^{-4} F$ solution of methylammonium chloride, $K_a = 2.4 \times 10^{-11}$.

Solution

Use the expression which takes autoprotolysis into account:

$$[H^+] = (K_w + K_a C_t)^{1/2}$$
$$= [1.01 \times 10^{-14} + 2.4 \times 10^{-11}(10^{-4})]^{1/2}$$
$$[H^+] = 1.12 \times 10^{-7} M$$
$$pH = 6.95$$

It is assumed in using this expression that $[HA] = C_t$. This is indeed an excellent approximation, as we will be able to determine shortly.

5.4.3 Solution of a Monobasic Weak Base

The conjugate base of a weak acid is itself a weak base. For example, ammonia is the conjugate base of the weak acid ammonium ion, and it reacts with water in the following way:

$$NH_3 + H_2O = NH_4^+ + OH^-$$

The equilibrium constant is

$$K_b = \frac{[NH_4][OH^-]}{[NH_3]} = 1.8 \times 10^{-5} \quad \text{at } 25°C$$

We can calculate the pH of a solution of ammonia by writing material balance and charge balance expressions:

material balance: $C_t = [NH_3] + [NH_4^+]$
charge balance: $[NH_4^+] + [H^+] = [OH^-]$

The charge balance expression can also be called a hydroxide balance; it shows the sources and sinks of hydroxide ions. Starting with these expressions, we can derive the following relationships:

(approximate) $[OH^-] = (K_b C_t)^{1/2}$
(approximate, including K_w) $[OH^-] = (K_w + K_b C_t)^{1/2}$
(exact, without K_w) $[OH^-]^2 + [OH^-]K_b - K_b C_t = 0$

These expressions are totally analogous to the expressions derived for solutions of weak acids, but with K_b and $[OH^-]$ appearing wherever K_a and $[H^+]$ might appear.

EXAMPLE 5.10 Solution of a Weak Base
Calculate the pH of a 0.010F solution of ammonia. $K_b = 1.8 \times 10^{-5}$.

Solution

Since the product of K_b and C_t is much larger than 10^{-14}, we will probably not have to worry about autoprotolysis as a source of hydroxide ions. We will assume that ammonia is a weak enough base that only a small amount reacts to form ammonium ion. In other words, we will assume that $[NH_3] = C_t$. Using the simple expression for $[OH^-]$, we calculate

$$[OH^-] = (K_b C_t)^{1/2} = [1.8 \times 10^{-5}(0.010)]^{1/2} = 4.24 \times 10^{-4} M$$
$$\text{pOH} = 3.4 \quad \text{pH} = 10.6$$

5.4 MONOPROTIC ACIDS AND MONOBASIC BASES

This result shows that about 4% of the ammonia has reacted to form ammonium ion and hydroxide. The assumption that $[NH_3] = C_t$ may be too crude an assumption. The exact solution involves the quadratic equation and gives

$$[OH^-] = 4.15 \times 10^{-4} M \quad pOH = 3.38 \quad (pH = 10.6)$$

The approximate result for $[OH^-]$ is about 2% larger than this and is of acceptable accuracy.

5.4.4 Distribution of Species with Changing pH

Having determined how to calculate the pH and pOH of solutions of weak acids and weak bases, we are now in a position to ask about the concentrations of other species. What is the equilibrium concentration of undissociated benzoic acid in a solution of formal concentration 0.01? What is the concentration when the pH of the solution is adjusted to 6.0? We will begin by developing two simple expressions for quantities called the *fractions* of weak acid and conjugate base and show by diagrams how these fractions change as a function of pH. We will then deal with more complex acid-base systems.

To calculate the fraction of the species present as benzoic acid in a solution of benzoic acid, we begin with the material balance expression,

$$C_t = [HOBz] + [OBz^-]$$

and substitute in the equilibrium constant expression for $[OBz^-]$, getting

$$C_t = [HOBz] + \frac{K_a[HOBz]}{[H^+]}$$

Collecting terms and rearranging gives

$$\frac{C_t}{[HOBz]} = \frac{[H^+] + K_a}{[H^+]}$$

The inverse of this expression is the fraction of HOBz, represented by the Greek letter α:

$$\alpha_{HOBz} = \frac{[HOBz]}{C_t} = \frac{[H^+]}{[H^+] + K_a}$$

Starting with the same material balance but substituting for [HOBz] results in the expression for the fraction of benzoate ion,

$$\alpha_{OBz^-} = \frac{[OBz^-]}{C_t} = \frac{K_a}{[H^+] + K_a}$$

The concentrations of the species are related to their fractions in a very simple way,

$$[HOBz] = (\alpha_{HOBz})C_t \quad [OBz^-] = (\alpha_{OBz^-})C_t$$

Figure 5.2 shows a plot of α_{HOBz} and α_{OBz^-} as functions of pH. Notice that at any particular pH the sum of the fractions must equal unity. At a pH equal to pK_a for the weak acid (pK_a for benzoic acid is 4.21), the fractions are exactly equal ($\alpha_{HOBz} = \alpha_{OBz^-} = 0.50$). This *must* be the case, as can be seen from the K_a relationship,

$$K_a = \frac{[H^+][OBz^-]}{[HOBz]} = \frac{[H^+](\alpha_{OBz^-}C_t)}{(\alpha_{HOBz}C_t)}$$

$$= \frac{[H^+](0.50C_t)}{(0.50C_t)} = [H^+]$$

$$pK_a = pH$$

In solutions more basic than pH 4.21, the dominant species is benzoate ion. In more acidic solutions, benzoic acid dominates.

Plots of fractions as a function of pH are easily read for fractions within the range of $pK_a \pm 2$. Outside this range, however, little information can be gotten about the fractions of the species.

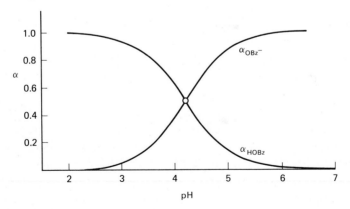

Figure 5.2 Fractions of benzoic acid species as a function of pH.

5.4 MONOPROTIC ACIDS AND MONOBASIC BASES

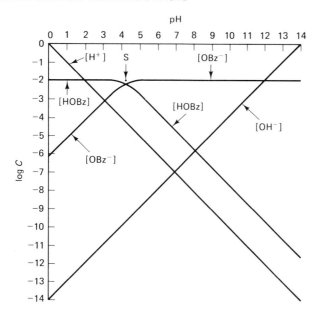

Figure 5.3 pH vs. log C for benzoic acid species.

More can be learned if, instead of plotting the fraction, we plot the logarithm of the concentration of each species as a function of pH. Figure 5.3 is a composite plot of log[H^+], log[OH^-], log[HOBz], and log[OBz^-] (generalized as "log C" on the vertical axis), all as functions of pH. This plot obviously holds a great deal of information. In fact, all the information contained in the weak acid dissociation constant expression, the autoprotolysis constant expression, and the material balances for benzoic acid species and protons is contained in one plot. Autoprotolysis is represented by the two diagonal lines (one for [H^+], the other for [OH^-]) which intersect at pH = 7, log C = -7. At any pH the sum of log[H^+] and log[OH^-] will be -14. Superimposed on the diagonal lines are the lines which represent [HOBz] and [OBz^-]. Each log C line is made up of two segments connected by a curved section. The point at which all the extrapolated segments of the [HOBz] and [OBz^-] lines intersect is called the *system point*, and it occurs where pH = pK_a (pH = 4.21 in Fig. 5.3). To the left of the system point (more acidic solutions), [HOBz] approaches $10^{-2} M$ and [OBz^-] gets smaller and smaller (log[OBz^-] decreases by one unit per unit decrease in pH). To the right of the system point, log[OBz^-] approaches -2 as log[HOBz] decreases by one unit for every unit increase in pH. The intersection of the log[HOBz] and log[OBz^-] lines occurs at pH = pK_a and where log C = -2.30, that is, 0.30 unit below the system point.† The next examples show how a logarithmic concentration plot can be used to answer questions about various solutions containing benzoic acid and benzoate ion.

† This is the result of the relationship between the concentrations when pH = pK_a. At this pH, [HOBz] = [OBz^-] = $0.5C_t$. log(0.5) = -0.30.

EXAMPLE 5.11 Graphical Determination of pH
From Fig. 5.3 determine the pH of a $0.01F$ solution of benzoic acid.

Solution

The pH of this solution can be found by applying the proton balance for the solution,

$$[H^+] = [OH^-] + [OBz^-]$$
(proton-rich)　(proton-poor)

A solution of benzoic acid will be acidic, and as a simplifying assumption we might guess that $[OH^-]$ will be much smaller than $[OBz^-]$. This gives a simpler proton balance,

$$[H^+] = [OBz^-]$$

This condition occurs where the $[H^+]$ and $[OBz^-]$ lines intersect, in this case at pH = 3.1 ($[H^+] = 8 \times 10^{-4} M$).

We can see immediately from the figure that dropping $[OH^-]$ from the proton balance was an acceptable approximation. Where $[H^+] = [OBz^-]$, $[OH^-]$ is about $10^{-11} M$.

The concentrations of the other species can be read directly from the plot,

$$\log[OBz^-] = -3.1; \quad [OBz^-] = 8 \times 10^{-4} M$$
$$\log[HOBz] = -2.0; \quad [HOBz] = 10^{-2} M$$

Notice that benzoic acid is about 8% dissociated at this pH.

EXAMPLE 5.12 Graphical Determination of pH
From Figure 5.3 determine the pH of a $0.01F$ solution of sodium benzoate.

Solution

The proton balance is

$$[H^+] + [HOBz] = [OH^-]$$
(proton-rich)　(proton-poor)

If we assume that the solution is sufficiently basic for $[H^+]$ to be unimportant, then we drop it from the proton balance and get

5.4 MONOPROTIC ACIDS AND MONOBASIC BASES

$$[HOBz] = [OH^-]$$

This occurs where $\log[HOBz] = \log[OH^-]$ in the plot, at pH = 8.1. We can see from the plot that $[H^+]$ is only about 1% of $[OH^-]$ at this point, and we were justified in dropping $[H^+]$ from the proton balance.

The concentrations of the other species are easily read from the plot:

$$\log[OBz^-] = -2.0; \quad [OBz^-] \approx 10^{-2} M$$
$$\log[HOBz] = -5.9; \quad [HOBz] = 10^{-5.9} = 1.3 \times 10^{-6} M$$

EXAMPLE 5.13 Graphical Determination of Concentrations

What are [HOBz] and [OBz$^-$] in a solution in which the total concentration of benzoic acid species is $0.01 F$ when the pH is adjusted to 6.0?

Solution

Although we have not discussed buffers yet, you may see that this question asks about the composition of a benzoic acid buffer at pH 6.0. We can find [HOBz] by moving down the pH = 6 vertical line until it meets the [HOBz] line. This occurs at $\log[HOBz] = -3.8$, or $[HOBz] = 1.6 \times 10^{-4} M$. At pH 6 the value of [OBz$^-$] is not distinguishable from $10^{-2} M$ in the figure. Nevertheless, we know that

$$C_t = 0.01 F = [HOBz] + [OBz^-]$$

and so

$$[OBz^-] = 0.010 M - 1.6 \times 10^{-4} M = 0.00984 M$$

5.4.5 Drawing a Logarithmic Concentration Diagram

The following steps describe the construction of a logarithmic concentration diagram. No calculations are necessary.

1. Draw two axes intersecting in the upper left-hand corner of a piece of graph paper ruled 10 lines/cm. The vertical axis is labeled log C and is numbered 0 to -14 moving down the page. The horizontal axis is labeled pH and is numbered from 0 to 14 from left to right.

2. Draw the solvent ion lines, $[H^+]$ and $[OH^-]$, to intersect in the middle of the plot $(7, -7)$. The $[H^+]$ line starts at $(0, 0)$, ends at $(14, -14)$, and has a slope of -1. The $[OH^-]$ line starts at $(14, 0)$, ends at $(0, -14)$, and has a slope of 1.

3. Locate the system point or points. These occur at log C = log C_t and pH = pK_a. Check Fig. 5.3 if this is unclear.

4. Draw lines for the weak acid and conjugate base. Start with horizontal lines running along the log C_t line (−2 in Fig. 5.3). Both lines bend downward as they near the system point. Curved portions intersect 0.30 unit below the system point(s). The weak acid line is extrapolated from the system point down and to the right, with a slope of −1. The conjugate base line is extrapolated from the system point down and to the left, with a slope of 1.

5. Extend the weak acid and weak base lines to the margins. As you will see later, whenever a species line crosses a vertical line containing a system point, its slope changes. This will be pointed out and explained in the section on polyprotic acids.

5.4.6 Mixture of a Weak Acid and a Strong Acid

When a strong acid such as HCl is added to a solution of a weak acid such as benzoic acid, the dissociation equilibrium, e.g.,

$$HOBz = H^+ + OBz^-$$

is driven toward the left, and dissociation is suppressed. If relatively little strong acid is added, the pH of the solution will not be changed dramatically due to the buffering effect of the weak acid. If a relatively large amount of strong acid is added, the pH will be determined mostly by the amount of strong acid.

EXAMPLE 5.14 Strong Acid–Weak Acid Mixture

Calculate the pH of $0.01F$ benzoic acid solution after it has been made $0.005F$ in the strong acid HCl.

Solution

The proton balance must reflect the presence of strong acid. The zero level of protons is water, benzoic acid, and HCl. The proton balance is thus

$$[H^+] \quad = [OH^-] + [OBz^-] + [Cl^-]$$
$$\text{(proton-rich)} \quad \text{(proton-poor)}$$

There are also two important material balances,

$$C_t = 0.01F = [HOBz] + [OBz^-]$$

and

$$C'_t = 0.005F = [Cl^-]$$

If we assume that [OH$^-$] is negligible in the proton balance, then the simplified expression is

$$[H^+] = [OBz^-] + 0.005$$

We then express the benzoate concentration in terms of K_a and the concentration of benzoic acid:

$$[H^+] = \frac{K_a[HOBz]}{[H^+]} + 0.005$$

In the presence of strong acid, it is quite likely that most of the benzoic acid species are in the form of the undissociated acid, and thus [HOBz] = 0.01M. The last relationship then becomes

$$[H^+] = \frac{0.01(K_a)}{[H^+]} + 0.005$$

When this is rearranged it looks more like a quadratic equation:

$$[H^+]^2 - (0.005)[H^+] - 0.01(K_a) = 0$$

which is solved to give

$$[H^+] = 5.12 \times 10^{-3} M \quad (pH = 2.3)$$

To check the assumptions, we can refer back to Fig. 5.3 and see first that [OH$^-$] is quite small, and second that [OBz$^-$] is much smaller than [HOBz] at pH 2.3 ([OBz$^-$] = $1.2 \times 10^{-4} M$). The pH calculated in the presence of 0.005F HCl is smaller than that calculated for 0.01F benzoic acid in pure water (pH 3.1). Notice also that benzoic acid is only about 1% dissociated in this solution, whereas in pure water it was about 8% dissociated.

A similar situation is encountered in a 0.1F solution of sulfuric acid. The first dissociation step for sulfuric acid is virtually complete in water,

$$H_2SO_4 = H^+ + HSO_4^- \quad K_{a1} \gg 1$$

The second dissociation is not complete, however,

$$HSO_4^- = H^+ + SO_4^{2-} \quad K_{a2} = 1.2 \times 10^{-2}$$

A $0.1 F$ solution of H_2SO_4 may thus be treated as if it were a $0.1 F$ solution of HSO_4^- containing $0.1 F$ strong acid. By analogy to the benzoic acid case, the proton balance for the solution will be

$$[H^+] = [SO_4^{2-}] + 0.1$$

which after substitution and rearrangement becomes

$$[H^+]^2 - 0.1[H^+] - 1.2 \times 10^{-3} = 0$$

The exact solution of this equation is

$$[H^+] = 0.17 M$$

If both dissociations were complete, the equilibrium value of $[H^+]$ would be $0.2 M$.

5.4.7 Mixture of a Weak Acid and a Weak Base

Figure 5.4 shows the logarithmic concentration diagram for a $0.01 F$ solution of the salt ammonium acetate, NH_4OAc. The ammonium ion is a weak acid,

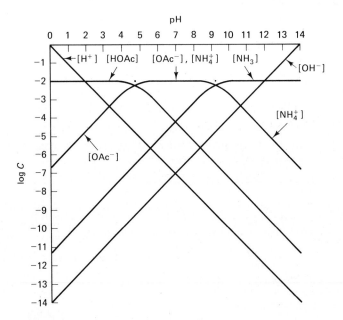

Figure 5.4 Logarithmic concentration diagram for $0.01 M$ ammonium acetate.

5.4 MONOPROTIC ACIDS AND MONOBASIC BASES

$$NH_4^+ = H^+ + NH_3 \qquad K_a = 5.7 \times 10^{-10}$$

while acetate ion is weak base,

$$OAc^- + H_2O = HOAc + OH^- \qquad K_b = 5.7 \times 10^{-10}$$

The pH of a solution containing a weak acid and a weak base depends on the relative strengths of the acid and base. In the case of ammonium acetate, the ammonium ion is as strong an acid as the acetate ion is a base, and the resulting solution is neutral.

The zero level of protons in a $0.01M$ solution of NH_4OAc is H_2O, NH_4^+, and OAc^-. The proton balance for the solution is

$$[H^+] + [HOAc] = [OH^-] + [NH_3]$$
$$\text{(proton-rich)} \qquad \text{(proton-poor)}$$

This can be simplified considerably if we can assume that $[H^+]$ and $[OH^-]$ are both small in comparison to $[HOAc]$ and $[NH_3]$. The simplified proton balance is then

$$[HOAc] = [NH_3]$$

This condition is met where the $[HOAc]$ and $[NH_3]$ lines intersect on the plot, at pH = 7.0.

To arrive at an explicit algebraic relationship for the pH of the solution, we start with the proton balance and substitute in equilibrium constant expressions involving the zero-level species, $[NH_4^+]$ and $[OAc^-]$,

$$[H^+] + \frac{[H^+][OAc^-]}{K_a} = \left(\frac{K_w}{[H^+]} + \frac{K_w[NH_4^+]}{[H^+]K_b}\right)$$

Rearranging to solve for $[H^+]$ gives

$$[H^+] = \left(\frac{K_w + K_w[NH_4^+]/K_b}{1 + [OAc^-]/K_a}\right)^{1/2}$$

Two assumptions can be made which will dramatically simplify this expression: (1) assume that $(K_w/K_b)[NH_4^+]$ is much larger than K_w, and (2) assume that $[OAc^-]/K_a$ is much greater than 1. Both of these assumptions will be valid if the acid and base are both relatively weak (that is, if K_a and K_b are small) and are present in relatively high concentrations (generally millimolar or greater). If the assumptions are valid, the complex equation can be reduced to

$$[H^+] = \left\{K_w\left(\frac{K_a}{K_b}\right)\frac{[NH_4^+]}{[OAc^-]}\right\}^{1/2}$$

EXAMPLE 5.15 pH of a Mixture of Weak Acid and Weak Base
Calculate the pH of a $0.01F$ solution of ammonium acetate.

$$K_a = 5.7 \times 10^{-10} \quad \text{(ammonium ion)}$$
$$K_b = 5.7 \times 10^{-10} \quad \text{(acetate ion)}$$

Solution

Use the simplified equation just derived.

$$[H^+] = \left\{ K_w \left(\frac{K_a}{K_b}\right) \frac{[NH_4^+]}{[OAc^-]} \right\}^{1/2} = \left\{ 10^{-14} \left(\frac{5.7 \times 10^{-10}}{5.7 \times 10^{-10}}\right) \frac{[NH_4^+]}{[OAc^-]} \right\}^{1/2}$$

$$[H^+] = \left(10^{-14} \frac{[NH_4^+]}{[OAc^-]} \right)^{1/2}$$

If the equilibrium concentrations of ammonium ion and acetate ion are equal, then

$$[H^+] = (10^{-14})^{1/2} = 10^{-7} \qquad pH = 7.0$$

The equality of $[NH_4^+]$ and $[OAc^-]$ can be checked quickly by calculating the fractions of both species at pH 7.0:

$$\alpha_{NH_4^+} = \frac{[H^+]}{[H^+] + K_a} = \frac{10^{-7}}{10^{-7} + 5.7 \times 10^{-10}} = 0.994$$

$$\alpha_{OAc^-} = \frac{K_a}{[H^+] + K_a} = \frac{1.75 \times 10^{-5}}{10^{-7} + 1.75 \times 10^{-5}} = 0.994$$

Since the fractions are the same, their concentrations will also be the same.
Check the assumptions behind the use of the simplified equation:

$$\text{Is } \frac{[OAc^-]}{K_a} \gg 1? \qquad \frac{9.94 \times 10^{-3}}{5.7 \times 10^{-10}} = 1.7 \times 10^7$$

$$\text{Is } \frac{K_w[NH_4^+]}{K_b} \gg K_w? \qquad \left(\frac{10^{-14}}{5.7 \times 10^{-10}}\right)(9.94 \times 10^{-3}) = 1.7 \times 10^{-7}$$

Both assumptions are valid.

5.5 POLYPROTIC ACIDS

This discussion will center on the polyprotic acid ascorbic acid (vitamin C),

$$\begin{array}{c}
CH_2OH \\
| \\
CHOH \\
| \quad\quad O \\
HC \diagdown \quad \diagup C=O \\
\quad C=C \\
\diagup \quad\quad \diagdown \\
OH \quad\quad OH
\end{array}$$

which has two dissociable protons in aqueous solutions,

$$H_2Asc = H^+ + HAsc^- \quad K_{a1} = 5.0 \times 10^{-5}$$
$$HAsc^- = H^+ + Asc^{2-} \quad K_{a2} = 1.5 \times 10^{-12}$$

The treatment which follows is quite general and can be extended to triprotic acids (phosphoric acid) and tetraprotic acids (ethylenediaminetetraacetic acid, EDTA) without much difficulty.

Figure 5.5 shows a logarithmic concentration diagram for ascorbic acid and its anions; it was drawn using the steps outlined previously. Several important practical questions can be answered by using this plot and various proton balance expressions.

1. What is the pH of a $0.01 F$ solution of ascorbic acid? The zero level of protons will be water and H_2Asc. The proton balance will be

$$[H^+] = [OH^-] + [HAsc^-] + 2[Asc^{2-}]$$
(proton-rich) \quad\quad (proton-poor)

The coefficient 2 on the right-hand side of this expression indicates that for every ascorbate anion which appears in solution, *two* protons must have dissociated from the starting material, H_2Asc. Since we are dealing with a weak acid, we expect the solution to be acidic and will assume that $[OH^-]$ and $[Asc^{2-}]$ are extremely small at equilibrium. This allows us to simplify the proton balance to

$$[H^+] = [HAsc^-]$$

The pH of the solution can be found where this condition is met, that is, where the $[H^+]$ and $[HAsc^-]$ lines intersect, at pH 3.1.

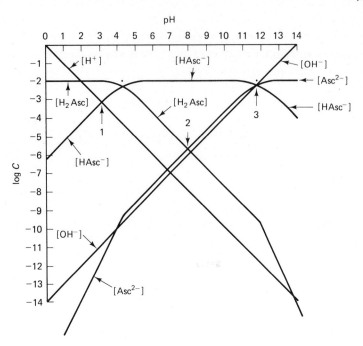

Figure 5.5 Logarithmic concentration diagram for $0.01M$ ascorbic acid.

To test the assumptions we made, we can read from the plot that $[OH^-] = 10^{-11}M$ and $[Asc^{2-}] = 10^{-12}M$. Notice that the diprotic acid behaves as if it were a simple monoprotic acid. This is a result of the great difference in size between K_{a1} and K_{a2}; K_{a2} is so small that it does not influence the pH of the solution of ascorbic acid.

2. What is the pH of a $0.01F$ solution of the salt sodium hydrogen ascorbate (NaHAsc)?

The zero level of protons will be water and $HAsc^-$. The proton balance will be

$$[H^+] + [H_2Asc] = [OH^-] + [Asc^{2-}]$$
(proton-rich)   (proton-poor)

This can be simplified if $[H^+]$ and $[OH^-]$ are small relative to $[H_2Asc]$ and $[Asc^{2-}]$, and we can write

$$[H_2Asc] = [Asc^{2-}]$$

This condition is met where the $[H_2Asc]$ and $[Asc^{2-}]$ lines cross, at pH 8.0.

The assumption that $[H^+]$ is much smaller than $[H_2Asc]$ at this pH is valid, since at pH 8 we read $[H_2Asc] = 10^{-5.7}$ from the plot. The second assumption, that

5.5 POLYPROTIC ACIDS

$[OH^-] \ll [Asc^{2-}]$, is not valid, however. Reading from the plot, we find $[OH^-] = 10^{-6} M$, $[Asc^{2-}] = 10^{-5.7} M$. We will have to examine the algebraic solution to this problem more carefully. We will take as an approximate solution $pH = 8.0$.

3. What is the pH of a $0.01 F$ solution of sodium ascorbate, Na_2Asc? The zero level of protons in this solution is water and ascorbate ion. The proton balance is written as

$$[H^+] + \underbrace{2[H_2Asc] + [HAsc^-]}_{\text{(proton-rich)}} = \underbrace{[OH^-]}_{\text{(proton-poor)}}$$

This is also called a hydroxide balance, as noted earlier. It will simplify matters considerably if we realize that the solution will probably be basic and assume that both $[H_2Asc]$ and $[H^+]$ are very small. The approximate proton balance is then

$$[HAsc^-] = [OH^-]$$

a condition which occurs at $pH = 11.7$ in Fig. 5.5, slightly below pK_{a2} (11.8). Both $[H^+]$ and $[H_2Asc]$ are very small at this pH, as we had hoped.

You should notice that sodium ascorbate is a rather strong base. The pH of a $0.01 F$ solution of the strong base NaOH is 12.0, only slightly more basic than 0.01 sodium ascorbate.

To generate the algebraic expressions needed to solve the three preceding problems, we need to know expressions and equilibrium constants for the two dissociations and K_w, material balances for ascorbic acid species, and various proton balances. We will make use of expressions for the fractions of the ascorbic acid species and will begin with their development.

Starting with the material balance expression

$$C_t = [H_2Asc] + [HAsc^-] + [Asc^{2-}]$$

we substitute for $[HAsc^-]$ and $[Asc^{2-}]$ in terms of $[H_2Asc]$ and K_{a1} and K_{a2}, getting

$$C_t = [H_2Asc] + \frac{K_{a1}[H_2Asc]}{[H^+]} + \frac{K_{a1}K_{a2}[H_2Asc]}{[H^+]^2}$$

Dividing both sides of the expression by $[H_2Asc]$ gives

$$\frac{C_t}{[H_2Asc]} = 1 + \frac{K_{a1}}{[H^+]} + \frac{K_{a1}K_{a2}}{[H^+]^2}$$

We then find a common denominator for the right-hand term and obtain

$$\frac{C_t}{[H_2Asc]} = \frac{[H^+]^2 + [H^+]K_{a1} + K_{a1}K_{a2}}{[H^+]^2}$$

The inverse of this expression is the fraction of diprotic acid,

$$\alpha_{H_2Asc} = \frac{[H_2Asc]}{C_t} = \frac{[H^+]^2}{[H^+]^2 + [H^+]K_{a1} + K_{a1}K_{a2}}$$

By completely parallel treatments the fractions of the other forms are given by

$$\alpha_{HAsc^-} = \frac{[HAsc^-]}{C_t} = \frac{[H^+]K_{a1}}{[H^+]^2 + [H^+]K_{a1} + K_{a1}K_{a2}}$$

and

$$\alpha_{Asc^{2-}} = \frac{[Asc^{2-}]}{C_t} = \frac{K_{a1}K_{a2}}{[H^+]^2 + [H^+]K_{a1} + K_{a1}K_{a2}} \dagger$$

Now we can develop the algebraic expressions for the cases solved graphically.

1. Calculate the pH of a $0.01F$ solution of ascorbic acid. The proton balance is

$$[H^+] = [OH^-] + [HAsc^-] + 2[Asc^{2-}]$$

Substituting in expressions for $[HAsc^-]$, $[Asc^{2-}]$, and $[OH^-]$ gives the expression

$$[H^+] = \frac{K_w}{[H^+]} + \frac{[H_2Asc]K_{a1}}{[H^+]} + \frac{2[H_2Asc]K_{a1}K_{a2}}{[H^+]^2}$$

which is a cubic equation. The expression can be rearranged to give

$$[H^+] = \left(K_w + [H_2Asc]K_{a1} + \frac{2[H_2Asc]K_{a1}K_{a2}}{[H^+]}\right)^{1/2}$$

If K_{a1} and $[H_2Asc]$ are sufficiently large and K_{a2} is very small, this expression will collapse to the familiar expression

$$[H^+] = (K_{a1}[H_2Asc])^{1/2} = \{K_{a1}(C_t - [H^+])\}^{1/2}$$

---

† The general relationships for an acid with n dissociable protons are

$$\alpha_{H_nA} = \frac{[H_nA]}{C_t} = \frac{[H^+]^n}{[H^+]^n + [H^+]^{n-1}K_1 + [H^+]^{n-2}K_1K_2 + \cdots (K_1K_2 \cdots K_n)}$$

$$\alpha_{H_{n-1}A^-} = \frac{[H_{n-1}A^-]}{C_t} = \frac{[H^+]^{n-1}K_1}{[H^+]^n + [H^+]^{n-1}K_1 + [H^+]^{n-2}K_1K_2 + \cdots (K_1K_2 \cdots K_n)}$$

$$\alpha_{A^{n-}} = \frac{[A^{n-}]}{C_t} = \frac{(K_1K_2 \cdots K_n)}{[H^+]^n + [H^+]^{n-1}K_1 + [H^+]^{n-2}K_1K_2 + \cdots (K_1K_2 \cdots K_n)}$$

Recall that this may be approximated by

$$[H^+] = (K_{a1}C_t)^{1/2}$$

if the acid is not too strong. If the last approximation cannot be made, the quadratic equation can be solved exactly. Using the most approximate expression, we calculate

$$[H^+] = (K_{a1}C_t)^{1/2} = 7.1 \times 10^{-4} \quad pH = 3.15$$

which is very nearly the same answer found graphically (pH = 3.1). An exact solution to the quadratic equation,

$$[H^+] = \{K_{a1}(C_t - [H^+])\}^{1/2}$$

is $[H^+] = 6.8 \times 10^{-4} M$, pH = 3.17.

Was it correct to disregard the terms K_w and $(2[H_2Asc]K_{a1}K_{a2}/[H^+])$ in the original equation? If we compare their values with the value of $K_{a1}[H_2Asc]$, we can find out. Let us begin by approximating $[H_2Asc]$ as $(C_t - [H^+])$, or $0.01 - 6.8 \times 10^{-4} = 9.3 \times 10^{-3} M$. Then we compare

$$K_{a1}[H_2Asc] = 4.7 \times 10^{-7} \quad \text{with } K_w = 10^{-14}$$

K_w is clearly much smaller. The term involving K_{a1} and K_{a2} is even smaller at pH 3.17:

$$(2[H_2Asc]K_{a1}K_{a2}/[H^+]) = 2 \times 10^{-15}$$

Therefore, we were justified in dropping both terms.

In certain cases the polyprotic acid may be weak enough or dilute enough that the other terms in the exact (cubic) equation shown above may not be neglected. The cubic equation can be solved by performing a series of approximations. First, an estimate is made of $[H^+]$, perhaps using the simplest approximation equation or a log concentration diagram. This estimate is then used to calculate an estimate for the concentration of the diprotic acid species, using the fraction expression,

$$\alpha_{H_2A} = \frac{[H^+]^2}{[H^+]^2 + [H^+]K_{a1} + K_{a1}K_{a2}}$$

The concentration of the diprotic acid is then used to calculate a second approximation for $[H^+]$. The second approximation for $[H^+]$ is then used to calculate a second approximation for the diprotic acid concentration, and so on, until successive $[H^+]$ values agree within a few percent. The process is illustrated in the next case, the pH of a $0.01F$ solution of sodium ascorbate.

2. Calculate the pH of a $0.01F$ solution of NaHAsc. The proton balance is

$$[H_2Asc] + [H^+] = [OH^-] + [Asc^{2-}]$$
(proton-rich) (proton-poor)

Substituting in expressions for $[H_2Asc]$, $[Asc^{2-}]$, and $[OH^-]$ in terms of K_{a1}, K_{a2}, and K_w gives

$$\frac{[H^+][HAsc^-]}{K_{a1}} + [H^+] = \frac{K_w}{[H^+]} + \frac{K_{a2}[HAsc^-]}{[H^+]}$$

When terms are collected and rearranged, we find the complex quadratic expression

$$[H^+] = \left(\frac{K_w + K_{a2}[HAsc^-]}{1 + [HAsc^-]/K_{a1}}\right)^{1/2}$$

The expression can be simplified by multiplying both the numerator and denominator by K_{a1}, giving

$$[H^+] = \left(\frac{K_{a1}K_w + K_{a1}K_{a2}[HAsc^-]}{K_{a1} + [HAsc^-]}\right)^{1/2}$$

There are at least two ways to proceed from this point. First, we might assume that $K_{a1}K_w \ll K_{a1}K_{a2}[HAsc^-]$ and that $K_{a1} \ll [HAsc^-]$. This leaves a remarkably simple expression for $[H^+]$,

$$[H^+] = (K_{a1}K_{a2})^{1/2}$$

which predicts that solution pH will be *independent of concentration*. In the case at hand, $0.01F$ NaHAsc, we can calculate pH = 8.06 by using this expression. Notice that these approximations are really the same ones we made in solving the problem graphically, and the result is no more accurate than that result.

The second way is to solve the exact equation by a series of successive approximations. We start by assuming that $[HAsc^-] = C_t$ and calculate an approximate value for $[H^+]$. We use this result to calculate a value for the fraction of $[HAsc^-]$, which we then use to calculate a second value for $[H^+]$, and so on, until a pair of values for $[H^+]$ and $[HAsc^-]$ are converged upon. Using $[HAsc^-] = 0.01M$, we calculate

$$[H^+] = \left(\frac{5 \times 10^{-19} + 7.5 \times 10^{-19}}{5 \times 10^{-5} + 0.01}\right)^{1/2}$$
$$= 1.12 \times 10^{-8} M \quad (pH = 7.95)$$

5.5 POLYPROTIC ACIDS

Using the expression for the fraction of HAsc⁻,

$$\alpha_{\text{HAsc}^-} = \frac{K_{a1}[\text{H}^+]}{[\text{H}^+]^2 + K_{a1}[\text{H}^+] + K_{a1}K_{a2}}$$

$$\alpha_{\text{HAsc}^-} = \frac{5.6 \times 10^{-13}}{1.25 \times 10^{-16} + 5.6 \times 10^{-13} + 7.5 \times 10^{-17}} = 1.00$$

The fraction is unity within the uncertainty of the calculation, and further approximations will not change the value of [H⁺] we have calculated.

Recall that there was some difficulty associated with the graphical solution to this problem. It was evident that [OH⁻] was not small enough to be dropped from the proton balance expression. We did not drop [OH⁻] from the proton balance in the exact algebraic solution, and so our calculation of [H⁺] is more accurate than the graphical solution. You should notice, though, that the graphical value [H⁺] = $1 \times 10^{-8} M$ was only about 12% smaller than the value found by solving the exact equation.

3. What is the pH of a 0.01F solution of Na₂Asc? Since this is a solution of a weak base, it seems sensible to deal with the weak base equilibria

$$\text{Asc}^{2-} + \text{H}_2\text{O} = \text{HAsc}^- + \text{OH}^- \qquad K_{b1} = 6.6 \times 10^{-3}$$
$$\text{HAsc}^- + \text{H}_2\text{O} = \text{H}_2\text{Asc} + \text{OH}^- \qquad K_{b2} = 2.0 \times 10^{-10}$$

The hydroxide ion balance for the solution is

$$[\text{OH}^-] = [\text{H}^+] + [\text{HAsc}^-] + 2[\text{H}_2\text{Asc}]$$

Substituting in order to express [OH⁻] only in terms of the zero-level species Asc²⁻ gives

$$[\text{OH}^-] = \frac{K_w}{[\text{OH}^-]} + \frac{K_{b1}[\text{Asc}^{2-}]}{[\text{OH}^-]} + \frac{2K_{b1}K_{b2}[\text{Asc}^{2-}]}{[\text{OH}^-]^2}$$

Rearranging, we find

$$[\text{OH}^-] = \left(K_w + K_{b1}[\text{Asc}^{2-}] + \frac{2K_{b1}K_{b2}[\text{Asc}^{2-}]}{[\text{OH}^-]} \right)^{1/2}$$

an equation entirely analogous to that used in part 1 for [H⁺]. If the product $K_{b1}[\text{Asc}^{2-}]$ is much larger than the other two terms, then this last expression collapses to the familiar approximation,

$$[\text{OH}^-] = (K_{b1}[\text{Asc}^{2-}])^{1/2}$$

The equilibrium concentration of ascorbate ion can be approximated by using the approximate pH and the expression for the fraction of ascorbate ion,

$$\alpha_{Asc^{2-}} = \frac{K_1 K_2}{[H^+]^2 + [H^+]K_1 + K_1 K_2}$$

To begin the actual calculation, we assume that $C_t = [Asc^{2-}]$ and calculate

$$[OH^-] = [0.01(6.6 \times 10^{-3})]^{1/2} = 8.1 \times 10^{-3} M$$
$$[H^+] = 1.2 \times 10^{-12} M \quad pH = 11.91$$

Next we use this value for $[H^+]$ to calculate $[Asc^{2-}]$, which we have only assumed to be $0.01M$,

$$[Asc^{2-}] = (\alpha_{Asc^{2-}})C_t = 0.556(0.01) = 5.56 \times 10^{-3} M$$

We then recalculate $[OH^-]$ using this new value for $[Asc^{2-}]$, and find

$$[OH^-] = [5.56 \times 10^{-3}(6.6 \times 10^{-3})]^{1/2} = 6.1 \times 10^{-3} M \quad (pOH = 2.2)$$
$$[H^+] = 1.6 \times 10^{-12} \quad pH = 11.79$$

Another iteration produces $[Asc^{2-}] = 4.8 \times 10^{-3} M$, which in turn produces $[H^+] = 1.8 \times 10^{-12}$, or pH = 11.75. Since the pH has not changed appreciably in the last iteration, we can take it as the pH of the solution. The graphical method gave pH = 11.7 for this solution.

Several steps back we assumed that the first and last terms of the exact equation for $[OH^-]$ could be dropped. We need to test these assumptions. First, is K_w much smaller than $K_{b1}[Asc^{2-}]$? Since 10^{-14} is very much smaller than 3.2×10^{-5}, we were justified in making the first assumption. Second, is $2K_{b1}K_{b2}[Asc^{2-}]/[OH^-]$ smaller than $K_{b1}[Asc^{2-}]$? Since 2.1×10^{-12} is much smaller than 3.2×10^{-5}, the second assumption was also valid.

In most situations we can almost guarantee that assumptions such as these will be valid. When successive dissociation constants for a polyprotic acid are separated by a factor of 1000 or more, the two equilibria act as if they are totally independent. This means that when we calculate the pH of a $0.01F$ solution of such a diprotic acid, we may comfortably treat it as if it were a *monoprotic* weak acid and neglect the second dissociation. We may treat a solution of the conjugate dibasic base as if it were a monobasic weak base.

A more difficult situation exists in solutions of acids like *trans*-fumaric acid,

```
     HOOC         H
         \       /
          C=C
         /       \
        H         COOH
```

5.6 BUFFERS

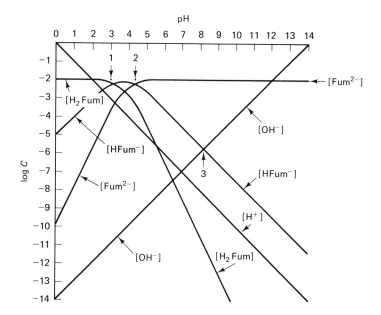

Figure 5.6 Logarithmic concentration diagram for 0.01M *trans* fumaric acid.

which has two similar dissociation constants,

$$H_2Fum = H^+ + HFum^- \qquad K_{a1} = 9.6 \times 10^{-4}$$
$$HFum^- = H^+ + Fum^{2-} \qquad K_{a2} = 4.2 \times 10^{-5}$$

A logarithmic concentration plot for a $0.01F$ solution of *trans*-fumaric acid and its conjugate bases is shown in Fig. 5.6. Notice that in the pH range 2 to 5 the concentrations of all three weak acid/weak base species are within a factor of 100 of each other. In the range of pH values bounded by the pK_a values, they are within a factor of 5 of each other. Assumptions aimed at simplifying the equations for this system must be made carefully, especially in the range between pK values.

5.6 BUFFERS

5.6.1 Composition and Applications of Buffers

An acid-base buffer is a solution that resists a change in pH. It contains compounds that are able to neutralize small added amounts of acid or base. Many of the buffers you will prepare in the laboratory contain large and nearly equal quantities of a weak acid and a salt of its conjugate base. These species are related by a weak acid dissociation equilibrium such as that between acetic acid and acetate ion:

$$HOAc = H^+ + OAc^-$$

Undissociated acetic acid in solution will act as a reservoir of hydronium ion, and can absorb extra base which might be added to solution, or be generated by another chemical reaction. Acetate ion will act as a sink for extra hydronium ion, which might be added. The pH of the solution can be kept in balance by the weak acid–weak base equilibrium as long as the amount of extra acid or base does not exceed the absorbing capacity of the buffer solution.

Buffers are used in many of the experiments described in this text, as well as in other areas of analysis. The following applications should serve to indicate how widely buffers are used:

1. Qualitative analysis (e.g., the detection of arsenic as silver arsenate in an acetic acid/sodium acetate buffer);
2. Gravimetric methods of analysis (e.g., the homogeneous precipitation of lead as lead chromate from acetic acid/sodium acetate buffer; see Chapter 10);
3. Complexometric titrations (e.g., the determination of calcium by titration with EDTA in an ammonia/ammonium chloride buffer; see Chapter 12);
4. Colorimetric methods [e.g., the determination of iron as the tris(o-phenanthroline) complex in acetic acid/sodium acetate buffers; see Chapter 13];
5. Enzyme assays [e.g., determination of the activity of lactic dehydrogenase (LDH) enzymes in serum by using a "tris" buffer; see below]; and
6. Determinations using enzymes as reagents (e.g., determination of blood urea nitrogen using the enzyme urease, buffered at pH 7.4 with a phosphate buffer).

Let us examine the properties of buffers made of acetic acid and sodium acetate. This system is chosen when slightly acidic buffers are needed ($3.8 \leq pH \leq 5.8$) and when acetate ions will not interfere with the chemical reaction being studied. The object will be to prepare a solution which contains rather large quantities of acetic acid and sodium acetate, say C_{HOAc} formal in acetic acid, and C_{OAc^-} formal in sodium acetate. The equilibrium constant expression will govern the pH of the systems at all times:

$$K_a = \frac{[H^+][OAc^-]}{[HOAc]}$$

The charge balance for the solution will be

$$[H^+] + [Na^+] = [OH^-] + [OAc^-]$$

5.6 BUFFERS

The material balance will be

$$[HOAc] + [OAc^-] = C_{HOAc} + C_{OAc^-}$$

Since sodium acetate dissociates completely in solution, it will be true that

$$[Na^+] = C_{OAc^-}$$

This may be substituted into the charge balance, giving

$$[H^+] + C_{OAc^-} = [OH^-] + [OAc^-]$$

When $[H^+]$ and $[OH^-]$ are very small compared to the concentration of sodium acetate, it will be correct to say that

$$C_{OAc^-} = [OAc^-]$$

Substituting this last equation into the material balance lets us also say that

$$C_{HOAc} = [HOAc]$$

When C_{HOAc} and C_{OAc^-} are substituted into the equilibrium constant expression, we get the simple expression for the hydronium ion concentration,†

$$[H^+] = \frac{K_a C_{HOAc}}{C_{OAc^-}}$$

This expression sometimes appears with the negative logarithms of both sides taken,

$$pH = pK_a + \log \frac{C_{OAc^-}}{C_{HOAc}}$$

Biochemists often call this expression the Henderson-Hasselbalch equation.

---

† In most practical buffers the requirement that $[H^+]$ and $[OH^-]$ be small compared to the formal concentrations of weak acid and weak base is met. If the weak acid is too strong, however, the approximate charge balance will be

$$[H^+] + C_{OAc^-} = [OAc^-]$$

and the material balance will then be

$$[HOAc] = C_{HOAc} - [H^+]$$

$[H^+]$ will be calculated more accurately from the quadratic equation

$$[H^+] = \frac{K_a(C_{HOAc} - [H^+])}{C_{OAc^-} + [H^+]}$$

The same kind of algebraic process can be used to find the pH of a solution of a weak base and a salt of its conjugate acid. For example, buffers made from the weak base "tris" [tris(hydroxymethyl)aminomethane] and its conjugate acid "trisH$^+$" are used extensively in clinical analysis. The species are related by the equilibrium

$$\underset{\text{tris}}{(HOCH_2)_3CNH_2} + H_2O = \underset{\text{trisH}^+}{(HOCH_2)_3CNH_3^+} + OH^-$$

$$K_b = \frac{[\text{trisH}^+][OH^-]}{[\text{tris}]} = 1.2 \times 10^{-6}$$

Again stipulating that [H$^+$] and [OH$^-$] are negligible compared to C_{tris} and C_{trisH^+}, the equation

$$[H^+] = \frac{K_w C_{\text{trisH}^+}}{K_b C_{\text{tris}}}$$

can be used to calculate the pH of tris/trisH$^+$ buffers.

EXAMPLE 5.16 Buffer

What is the pH of a buffer made by diluting a mixture of 0.100 mole of acetic acid and 0.045 mole of sodium acetate with enough water to make 1.00 liter of solution? $K_a = 1.8 \times 10^{-5}$.

Solution

Use the simple expression

$$[H^+] = \frac{K_a C_{\text{HOAc}}}{C_{\text{OAc}^-}} = (1.8 \times 10^{-5}) \frac{0.100}{0.045}$$

$$= 4.0 \times 10^{-5} M \qquad \underline{pH = 4.40}$$

We are able to use this simple expression because

$$0.1 M \gg [H^+](4 \times 10^{-5} M) \quad \text{or} \quad [OH^-](2.5 \times 10^{-10} M)$$

5.6.2 Preparing Buffer Solutions

Three alternative strategies are used to make buffers. The following descriptions apply to acetic acid/sodium acetate buffers in the pH range $pK_a \pm 1$ (3.8 ≤ pH ≤ 5.8). In each case a calibrated glass electrode should be used to check the pH of the solutions.†

---

† The electrode should be calibrated with a standard buffer solution of the same ionic strength to avoid errors arising from activity changes and liquid junction potential. Compositions of several standard buffers and problems associated with liquid junction potentials are discussed in Chapter 15.

5.6 BUFFERS

1. Add sodium acetate (NaOAc) to a $0.1F$ solution of acetic acid (HOAc). When the resulting solution is $0.01F$ in NaOAc and $0.1F$ in HOAc, its pH will be about 3.8. When it is $1.0F$ in NaOAc and $0.1F$ in HOAc, its pH will be about 5.8. Notice that these solutions have very different ionic strengths.
2. Add HOAc to a $0.1F$ solution of NaOAc. When the solution is $0.1F$ in NaOAc and $0.01F$ in HOAc, its pH will be 5.8. When the solution is $0.1F$ in NaOAc and $1.0F$ in HOAc, its pH will be 3.8. Notice that the pH range is the same as that spanned by the solutions in recipe 1, but that the ionic strength is kept constant over the range because the amount of the salt NaOAc is kept constant.
3. Add strong base to a solution of acetic acid or add strong acid to a solution of sodium acetate. (This method is useful if only one of the buffer compounds is available. It is commonly used in preparing tris buffers.)

EXAMPLE 5.17 Tris Buffer

An enzyme assay in serum requires a tris buffer of pH 7.4, the pH of normal blood. How would you prepare a liter of this buffer, knowing that the ionic strength of serum is about 0.15 and might be important in the assay? The thermodynamic K_b for tris is 1.20×10^{-6}. In a solution of $0.15M$ ionic strength, K_b is about 1.8×10^{-6} (see Example 5.3 for the method of correcting K).

Solution

The tris/trisH$^+$ equilibrium constant expression is

$$K_b = 1.8 \times 10^{-6} = \frac{[\text{trisH}^+][\text{OH}^-]}{[\text{tris}]}$$

The simple relationship for the [H$^+$] of a buffer of tris will be

$$[\text{H}^+] = \frac{K_w C_{\text{trisH}^+}}{K_b C_{\text{tris}}}$$

Solving for the ratio of the formal concentration of trisH$^+$ to tris, we find

$$\frac{C_{\text{trisH}^+}}{C_{\text{tris}}} = \frac{[\text{H}^+]K_b}{K_w} = \frac{4.0 \times 10^{-8}(1.8 \times 10^{-6})}{1 \times 10^{-14}}$$
$$= 7.2$$

This means that the solution requires a 7.2-fold excess of trisH$^+$ to tris, as well as an ionic strength of 0.15. A combination of concentrations which would give this would be

$$C_{\text{trisH}^+} = 0.15M \qquad C_{\text{tris}} = 0.021M$$

To prepare the solution one could start with 1 liter of $0.171F$ tris in deionized water and add 0.15 mole of a strong acid such as HCl (12.5 mL of $12F$ HCl).† The pH of the resulting solution would then be checked with a calibrated glass electrode.

5.6.3 Measures of the Effectiveness of a Buffer

For a buffer to be effective at regulating solution pH it must meet two criteria. First, its pH ought not to change very much when small amounts of strong acid or strong base are added. Second, its pH ought not to change appreciably when the solution is diluted. We will begin this discussion with a description of buffer capacity, then define it in quantitative terms using the buffer index. We will then discuss the buffer dilution factor of Bates.

Buffer Capacity The ability of a buffer solution to absorb strong acid or base is called buffer capacity and depends on several factors:

1. The greater the concentrations of weak acid and weak base in the buffer, the greater is its ability to absorb strong acid and strong base. The buffer capacity of pure water is very small but not zero.
2. If there is far more weak acid in a buffer than weak base, the buffer will have a greater capacity to absorb strong base than strong acid without undergoing a large change in pH.
3. Buffer capacity is a maximum when the concentrations of weak acid and conjugate base are the same. In such a solution pH = pK_a for the weak acid.

These factors can be illustrated with the following examples.

EXAMPLE 5.18 Buffer Capacity

A liter of buffer solution contains 0.1 mole of acetic acid and 0.01 mole of sodium acetate. Calculate the pH of the solution as it stands and then (a) after the addition of 0.01 mole of NaOH; (b) after the addition of 0.01 mole of HCl. Use $K_a = 1.8 \times 10^{-5}$.

Solution

The original solution will have a pH found by using the simple relationship

$$[H^+] = \frac{K_a C_{HOAc}}{C_{OAc^-}} = \frac{1.8 \times 10^{-5}(0.1)}{0.01}$$
$$= 1.8 \times 10^{-4} M$$
$$pH = 3.74$$

---

† The chloride added must be accounted for in the ionic strength. The hydronium ion added is tied up in trisH⁺.

5.6 BUFFERS

(a) When 0.01 mole of strong base is added, it will react with an equivalent amount of acetic acid to produce an equivalent amount of acetate ion,

$$[H^+] = \frac{K_a(C_{HOAc} - 0.01)}{(C_{OAc^-} + 0.01)} = \frac{K_a(0.09)}{0.02}$$
$$= 8.1 \times 10^{-5} M$$
$$pH = 4.09$$

The solution has become more basic by about 0.4 pH unit.

(b) When 0.01 mole of strong acid is added to the original buffer solution, it will react with all the acetate ion to form acetic acid. The resulting solution of acetic acid has a pH given approximately by

$$[H^+] = (K_a C_{HOAc})^{1/2}$$
$$= [1.8 \times 10^{-5}(0.11)]^{1/2} = 1.4 \times 10^{-3} M$$
$$pH = 2.85$$

The solution has become more acidic by about 0.9 pH unit. The original buffer solution was less able to absorb strong acid than strong base; that is, the capacities for strong acid and strong base were not equal.

It should be noted that when 0.01 mole of the strong acid HCl is added to 1 liter of pure water, the pH drops from 7 to 2. When 0.01 mole of the strong base NaOH is added to 1 liter of water, the pH rises from 7 to 12. Pure water has a relatively low capacity to absorb acid or base.

It will be profitable to repeat the calculation in this example with a buffer made from an equimolar mixture of acetic acid and sodium acetate. The pH of such a buffer is the same as the pK_a, and the capacities to absorb strong acid and strong base are the same.

When we take a sample of an acid and add small increments of strong base to it, the rate of change of solution pH with added base can be used as a measure of the capacity of the solution for base. Van Slyke (Ref. *1*) proposed that the inverse of this response be called the "buffer index,"

$$\beta = \frac{dC_b}{dpH}$$

and be used as a quantitative measure of buffer capacity. The larger β, the more resistant the solution to an increase in pH with added strong base. The buffer index for the addition of strong acid is defined as

$$\beta = \frac{-dC_a}{d\text{pH}}$$

and since pH *decreases* with added strong acid, β is always a positive number.

It is possible to arrive at an algebraic expression for β of a weak acid/weak base buffer beginning with a charge balance, material balances, and the K_a expression for the weak acid. Let us use a mixture of acetic acid and acetate ion at a total concentration C_t to derive the general expression. For the most general case we will also have available the strong acid HCl at concentration C_a and the strong base NaOH at concentration C_b. The charge balance of the solution will be

$$[\text{H}^+] + [\text{Na}^+] = [\text{OH}^-] + [\text{Cl}^-] + [\text{OAc}^-]$$

To simplify this expression, recall that the concentration of acetate ion will be given as a function of $[\text{H}^+]$ and K_a by

$$[\text{OAc}^-] = C_t \alpha_{\text{OAc}^-} = \frac{C_t K_a}{K_a + [\text{H}^+]}$$

The only source of sodium ion is assumed to be NaOH, and so

$$C_b = [\text{Na}^+]$$

Similarly, the only source of chloride ion is assumed to be HCl, and so

$$C_a = [\text{Cl}^-]$$

As is always the case, $[\text{OH}^-]$ can be expressed as $K_w/[\text{H}^+]$. Making these substitutions in the charge balance and rearranging to solve for C_b gives

$$C_b = \frac{K_w}{[\text{H}^+]} - [\text{H}^+] + C_a + \frac{C_t K_a}{K_a + [\text{H}^+]}$$

We now differentiate this expression with respect to $[\text{H}^+]$ and find

$$\frac{dC_b}{d[\text{H}^+]} = -\frac{K_w}{[\text{H}^+]^2} - 1 + 0 - \frac{C_t K_a}{(K_a + [\text{H}^+])^2}$$

Since the buffer index is defined as

$$\beta = \frac{dC_b}{d\text{pH}}$$

5.6 BUFFERS

we must multiply $dC_b/d[H^+]$ by $d[H^+]/dpH$. Since

$$\frac{dpH}{d[H^+]} = \frac{d(-\log[H^+])}{d[H^+]} = \frac{-1}{2.303[H^+]}$$

the expression for β becomes

$$\beta = \frac{dC_b}{d[H^+]} \cdot \frac{d[H^+]}{dpH} = -2.303[H^+]\left\{-\frac{K_w}{[H^+]^2} - 1 - \frac{C_t K_a}{(K_a + [H^+])^2}\right\}$$

or

$$\beta = 2.303\frac{K_w}{[H^+]} + 2.303[H^+] + \frac{2.303 C_t K_a [H^+]}{(K_a + [H^+])^2}$$

It should be noted that β is the sum of the β's of the components of the solution:

$$\beta = \beta_{OH^-} + \beta_{H^+} + \beta_{HOAc}$$

Figure 5.7 is a plot of buffer index vs. pH for solutions of acetic acid and acetate ion in which $C_t = 0.1M$. The maximum buffer index is found in an equimolar mixture of acetic acid and acetate ion (pH = pK_a = 4.6 at $0.1M$ ionic strength). If $[H^+] = K_a$ is used in the buffer index equation, the buffer capacity is given by

$$\beta = \beta_{max} = 0.576 C_t$$

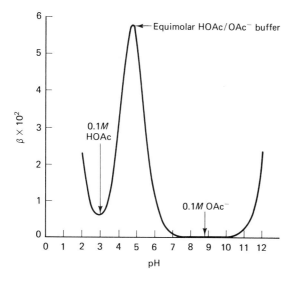

Figure 5.7 Buffer index as a function of pH.

Considerably smaller buffer indices are found for a solution containing $0.1F$ acetic acid alone ($\beta = 0.61 \times 10^{-2}$ at pH 2.9) and $0.1F$ sodium acetate alone ($\beta = 0.034 \times 10^{-2}$ at pH 8.8). The plot also shows that buffer solutions of acetic acid and acetate ion which are more acidic than the equimolar mixture are better buffered against the addition of base than they are against the addition of acid. Similarly, acetic acid/acetate ion buffers which are more basic than the equimolar mixture are better buffered against the addition of acid than against the addition of base.

The large buffer index values at very high and very low pH deserve special comment. The buffer index of a strong acid is given by

$$\beta_{H^+} = 2.303[H^+]$$

and that of a strong base by

$$\beta_{OH^-} = 2.303[OH^-]$$

These values are large (but finite) in concentrated solutions of strong acid or strong base. In acidic solutions β_{OH^-} is negligibly small, and the solution is apparently well buffered against the addition of base. Notice that the buffer indices of strong acids and bases are about four times greater than that of an equimolar mixture of weak acid and conjugate base. In a neutral solution without a weak acid-conjugate base pair, the buffer index is a minimum,

$$\beta_{min} = \beta_{H^+} + \beta_{OH^-} = 2\beta_{H^+}$$
$$= 2(2.303[H^+]) = 4.6 \times 10^{-7}$$

It is possible to extend the range of buffering by mixing two or more monoprotic acids and the salts of their bases, or by using a polyprotic acid and its salts. A plot of buffer index vs. pH for a mixture of acetic acid/acetate ion and tris/trisH$^+$ ion is shown in Fig. 5.8. Notice that in comparison with the buffer index vs. pH plot for acetic acid/acetate ion, the buffer index in this solution is considerably greater in the basic pH range. By mixing in several more acid-base pairs, it is possible to prepare a "universal buffer," a solution with a large buffer capacity over the entire pH range.

Buffer capacities for solutions of a diprotic acid and its salts are quite similar to those of mixed monoprotic acids and their salts if K_1 and K_2 for the diprotic acid are several orders of magnitude different. If K_1 and K_2 are relatively close, however, the buffer index between the β_{max} points will be somewhat greater than in the mixture of two monoprotic acids and their salts of the same total concentration. A detailed discussion of this difference can be found in Butler's text (Ref. 2).

Effect of Dilution on Buffer pH In an idealized world the pH of a buffer would depend only on the *ratio* of the concentrations of weak acid and conjugate base forms. For

5.6 BUFFERS

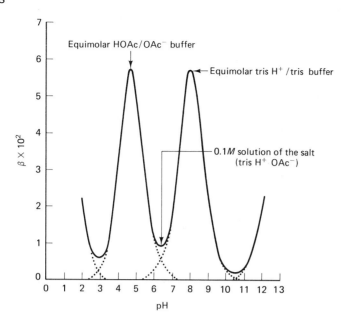

Figure 5.8 Buffer index vs. pH. Mixture of HOAc/OAc⁻ (C_t = 0.1M) and trisH⁺/tris (C_t = 0.1M).

example, the following set of acetic acid/acetate ion buffers would all have the same pH:

$$[\text{H}^+] = \frac{K_a[\text{HOAc}]}{[\text{OAc}^-]} = \frac{K_a(1.0M)}{1.0M} = K_a\frac{(0.01M)}{(0.01M)} = \frac{K_a(0.001M)}{0.001M}$$

However, this is not found to be the case. As a system containing a weak acid and conjugate base is diluted, the pH *increases* slightly. When a buffer containing a weak base and a salt of its conjugate acid is diluted, its pH is found to *decrease* slightly. Bates (Ref. 3) suggested that a unit called the "dilution value," $\Delta \text{pH}_{1/2}$, be used to express the effect of dilution for a buffer. $\Delta \text{pH}_{1/2}$ is defined as the increase in pH which results from diluting a solution with an equal volume of water (halving the concentration):

$$\Delta \text{pH}_{1/2} = \text{pH}_{C/2} - (\text{pH})_C$$

Bates found that $\Delta \text{pH}_{1/2}$ depends on both the strength of the buffer's acid (or base) form and the molar concentrations of the solution species. For example, for buffers in the pH range 5 to 9 made up from monoprotic acids or monobasic bases (HOAc/OAc⁻ or tris/trisH⁺, among others), $\Delta \text{pH}_{1/2}$ was about 0.03 pH unit at the 0.1M concentration level. For buffers of polyprotic acid species in the same pH range (e.g., $H_2PO_4^-/HPO_4^{2-}$), $\Delta pH_{1/2}$ was about 0.1 pH unit at the 0.1M concentration level.

The stronger the weak acid in a buffer system, the more dependent the pH is on dilution. A weak acid like acetic acid (pK_a about 5) has a dilution value of only about 0.01 at the 0.005M concentration level, while a weak acid with pK_a = 3 has a dilution value of about 0.08 at the same level.

$\Delta pH_{1/2}$ values of 0.03 or less can be accounted for largely by changes in activity coefficients during dilution (see Chapter 3). Larger dilution values reflect a shift in equilibrium which results from the acid and base (amphiprotic) character of water as a solvent.

While dilution values seem to be relatively minor factors, they are experimental details which must be accounted for in careful analytical work. Generally, it should never be simply *assumed* that a buffer can be diluted by a factor of 100 and still have the same experimental pH.

REFERENCES

1. D. D. Van Slyke, *Journal of Biological Chemistry,* 52, 525 (1922).
2. J. N. Butler, *Ionic Equilibrium—A Mathematical Approach,* Addison-Wesley, Reading, MA, 1964.
3. R. G. Bates, *Analytical Chemistry,* 26, 871 (1954).

RECOMMENDED READING

Bates, R. G. "Concept and Determination of pH," in *Treatise on Analytical Chemistry,* 2nd ed., part 1, vol. 1, I. M. Kolthoff and P. J. Elving, eds., Wiley, New York, 1978, chapter 14.

Bates, R. G. *Determination of pH,* 2nd ed., Wiley, New York, 1973.

Bruckenstein, S., and I. M. Kolthoff, "Acid-Base Strength and Protolysis Curves in Water," in *Treatise on Analytical Chemistry,* 1st ed., I. M. Kolthoff and P. J. Elving, eds., Wiley, New York, 1959, chapter 12.

Good, N. E., G. D. Winget, W. Winter, T. N. Connolly, S. Izawa, and R. Singh, "Hydrogen Ion Buffers for Biological Research," *Biochemistry,* 5, 467–477 (1966).

Laitinen, H. A., and W. E. Harris, "Acid-Base Equilibria in Water," in *Clinical Analysis,* 2nd ed., McGraw-Hill, New York, 1975, chapter 3.

Rosenthal, D., and P. Zuman, "Acid-Base Equilibria, Buffers, and Titrations in Water," in *Treatise on Analytical Chemistry,* 2nd ed., part 1, vol. 2, I. M. Kolthoff and P. J. Elving, eds., Wiley, New York, 1978, chapter 18.

PROBLEMS

5.1. Calculate [H$^+$], [OH$^-$], pH, and pOH for each of the following solutions. Correct for effects of ionic strength to calculate pH.
  (a) $0.05F$ HCl
  (b) $0.05F$ NaOH
  (c) $0.05F$ Ba(OH)$_2$

5.2. The thermodynamic equilibrium constant for the dissociation of chloroacetic acid is 1.36×10^{-3}. Calculate the concentration equilibrium constant in a solution in which the concentration of chloroacetic acid is $0.10F$.

5.3. The thermodynamic equilibrium constant for the second dissociation of o-phthalic acid, represented by HP$^-$ = H$^+$ + P^{2-}, is 3.908×10^{-6} at 25°C. Calculate the concentration equilibrium constant in a solution in which the formal concentration of potassium biphthalate (KHP) is $0.10F$.

5.4. A careful measurement of pH with a calibrated glass electrode indicates that the pH of a $0.0500F$ solution of o-fluorobenzoic acid is 2.25. What is the equilibrium constant for the dissociation of the acid?

5.5. The thermodynamic pK_a for propionic acid is 4.874. Sufficient potassium nitrate is added to a 50-mL solution of $0.01F$ propionic acid to make the ionic strength $0.01M$. Calculate the concentration equilibrium constant for propionic acid.

5.6. Use the Davies equation (Chapter 3) to calculate concentration equilibrium constants for a monoprotic acid at several ionic strengths between 0.01 and $1.0M$. Does the weak acid become weaker or stronger as ionic strength increases?

5.7. Write material balance and charge balance expressions for each of the following systems:
  (a) 0.05 mole of ammonium chloride diluted to 500 mL with water.
  (b) 20 mmole of ammonium chloride and 10 mmole of HCl diluted to 250 mL with water.
  (c) 0.03 mole of potassium biphthalate and 0.01 mole of sodium hydroxide diluted to 500 mL with water.
  (d) 10 mmole of sodium dihydrogen phosphate (NaH$_2$PO$_4$) diluted to 1.0 liter with water.

5.8. Identify the reference level (zero level) of protons and write correct proton balance expressions for each of the following solutions:
  (a) $0.10F$ formic acid
  (b) $0.03F$ sodium formate
  (c) $0.01F$ citric acid (triprotic)
  (d) $0.10F$ sodium bisulfite, NaHSO$_3$

5.9. Calculate the pH of a solution obtained by adding 5.00 mL of concentrated acetic acid to enough water to make 1 liter of solution. Concentrated acetic acid is 99.5% acetic acid and has a density of 1.051 g/mL at 20°C.

5.10. A $0.100F$ solution of a weak monoprotic acid has a pH of 3.8. Calculate K_a for the weak acid.

5.11. A $0.0735F$ solution of a weak monoprotic acid has a pH of 5.9. Calculate K_a for the weak acid.

5.12. A $0.0500F$ solution of a monobasic weak base has a pH of 8.2. Calculate K_b for the weak base.

5.13. What must be the pK_a of a weak acid if a $0.0105F$ solution of its conjugate base has a pH of 9.10?

5.14. What must be the pK_b of a weak base if a $0.0050F$ solution of its conjugate acid has a pH of 5.62?

5.15. Calculate the pH of a $0.01F$ solution of each of the following species. You need not correct for ionic strength.
  (a) ammonia
  (b) pyridinium chloride
  (c) periodic acid
  (d) aniline
  (e) ethylamine
  (f) triethanolammonium chloride

5.16. Calculate the pH of a $0.01F$ solution of each of the following species. You need not correct pK_a values for ionic strength.
  (a) anilium chloride
  (b) sodium bisuccinate (salt of succinic acid)
  (c) *o*-vanillin
  (d) sodium *p*-nitrobenzoate
  (e) cystein hydrochloride
  (f) sodium chromate

5.17. Calculate the concentrations of formic acid and formate anion at equilibrium when 5.00 mmole of sodium formate are dissolved in 100 mL of water.

5.18. Calculate the concentrations of *p*-chlorobenzoic acid and *p*-chlorobenzoate ion at equilibrium if 10.0 mmole of *p*-chlorobenzoic acid are dissolved in 100 mL of water and the pH is adjusted to 5.0.

5.19. Nicotinic acid (3-pyridinecarboxylic acid) has a molecular weight of 123.11. The pK_a of the neutral compound is 4.70 in a solution of $0.05M$ ionic strength. 10.00 grams of pure nicotinic acid are dissolved in enough water to make 1 liter of solution, and the pH is adjusted to 4.60 with sodium hydroxide. Calculate the concentrations of nicotinic acid and nicotinate anion at pH 4.60.

5.20. 5.00 grams of potassium binoxalate, KHC_2O_4 (gram formula weight 128.13), are dissolved in enough water to make 500 mL of solution. Calculate the pH and pOH of the solution and the fractions of all the oxalate-containing species. Use $pK_{a1} = 1.27$ and $pK_{a2} = 4.27$.

PROBLEMS

5.21. Draw a log concentration diagram for the HNO_2/NO_2^- system at a total concentration of $0.01F$. Assume $pK_a = 3.35$.
  (a) What is the pH of a $0.01F$ solution of HNO_2? Write and simplify a proton balance expression. Perform a calculation to check your answer.
  (b) What is the pH of a $0.01F$ solution of NO_2^-? Write and simplify a proton balance expression. Check your answer with a calculation.
  (c) What is the pH of an equimolar mixture of sodium nitrite and nitrous acid?

5.22. Draw a log concentration diagram for a $0.10F$ solution of ammonium formate. What is the pH of the solution? What are the concentrations of ammonia and formic acid in the solution?

5.23. Write and run a program in either BASIC or Pascal which will calculate and plot the log concentration diagram for acetic acid (use $pK_a = 4.76$). Plot points every 0.1 pH unit between pH 0 and pH 12.

5.24. Draw a log concentration diagram for the $0.1F$ malonic acid system (H_2Mal, $HMal^-$, Mal^{2-}, $pK_{a1} = 2.86$, $pK_{a2} = 5.70$).
  (a) Write a proton balance expression for a $0.1F$ solution of H_2Mal. Make appropriate assumptions and read the pH of the solution from the diagram. Check your assumptions.
  (b) Write a proton balance expression for a $0.1F$ solution of $HMal^-$. Make the appropriate assumptions, find the pH of the solution, and check the validity of the assumptions.
  (c) What are the concentrations and fractions of all the species in a $0.1F$ solution of $HMal^-$?
  (d) Write a hydroxide balance expression for a $0.1F$ solution of Na_2Mal. Find the pH of the solution from the diagram. Test any assumptions you make.
  (e) Describe the solution you get by mixing exactly equal molar quantities of H_2Mal and Na_2Mal and diluting with water until the solution is $0.10F$ (total concentration of malonic acid species).

5.25. Draw a log concentration diagram for the $0.01F$ phosphoric acid system. Use $pK_{a1} = 2.15$, $pK_{a2} = 7.20$, $pK_{a3} = 12.38$. Check the validity of any assumptions you make in answering the following questions.
  (a) What is the pH of a $0.01F$ solution of H_3PO_4? Is a simplified proton balance expression adequate to calculate this pH? Explain.
  (b) What is the pH of a $0.01F$ solution of NaH_2PO_4?
  (c) What is the pH of a $0.01F$ solution of Na_3PO_4?
  (d) Describe at least two recipes for the preparation of a buffer at pH 7.2.
  (e) The pK_{a2} value given and plotted is the thermodynamic value. Calculate the ionic strength of a mixture which is $0.005F$ in NaH_2PO_4 and $0.005F$ in Na_2HPO_4 and a more appropriate value for pK_{a2} for this real solution.

5.26. Draw a log concentration diagram for the $0.01F$ citric acid system (H_2Cit, $HCit^-$,

Cit^{3-}, $pK_{a1} = 3.13$, $pK_{a2} = 4.76$, $pK_{a3} = 6.40$). Check the validity of any assumption you make in answering the following questions.

(a) What is the pH of a 0.01F solution of NaH_2Cit?

(b) What is the pH of a 0.01F solution of Na_3Cit? What are the concentrations of the other citric acid species at this pH?

(c) Is the concentration of $HCit^{2-}$ greater than 1/1000 that of Cit^{3-} at the pH found in part (b)?

(d) A 0.01F solution of citric acid has its pH lowered to 1.0 by the addition of the strong acid HCl. Is 99.9% of the citric acid in the form of H_3Cit at pH 1.0?

(e) 4.5 mmole of NaH_2Cit and 0.5 mmole of Na_2HCit are mixed together and diluted to 500 mL with water. What is the pH of the resulting solution?

5.27. A 100-mL solution is prepared to be 0.05F in potassium biphthalate (KHP). The pH is of the solution is adjusted to 5.5. (H_2P, HP^-, P^{2-}, $pK_{a1} = 2.95$, $pK_{a2} = 5.41$.)

(a) Calculate the fractions of all phthalate species at this pH.

(b) How would you increase the fraction of phthalate ion in this solution by a factor of 1.2?

(c) What would be the effect of adding 0.005 mole of H_2P to the original solution?

5.28. Calculate the pH of a solution made by mixing 6.00 g of pure ammonium chloride and 20.00 mL of concentrated ammonia with enough water to make exactly 1 liter of solution. Concentrated ammonia is 28.0% NH_3 and has a density of 0.898 g/mL at 20°C.

5.29. One liter of buffer solution is prepared with 0.125 mole of NaH_2PO_4 and 0.125 mole of Na_2HPO_4. Use $pK_a = 7.20$ for the following calculations.

(a) What is the pH of this buffer solution?

(b) What is the pH after adding 0.05 mole of HCl?

(c) What is the pH of 1 liter of pure water before and after the addition of 0.05 mole of HCl?

(d) Compare your answers to (b) and (c) and explain the behavior.

5.30. Using any of the phosphoric acid system species, describe how you would prepare buffers at pH 4.0, 7.0, and 9.0.

5.31. Pyridine is a base which will react with HCl to form the weak acid pyridinium ion ($pK_a = 5.22$). What molar ratio of pyridine to HCl is necessary to prepare a buffer at pH 7.2?

5.32. Ethylmalonic acid is a diprotic acid with $pK_{a1} = 2.96$ and $pK_{a2} = 5.90$. Using a logarithmic concentration diagram, discuss the pH range over which useful buffers could be prepared from ethylmalonic acid and its conjugate base forms.

5.33. Calculate the pH and equilibrium concentration of HSO_3^- in a 0.04F solution of sodium bisulfite, $NaHSO_3$.

PROBLEMS

5.34. How would you prepare a "tris" (THAM) buffer at pH 7.5 with total concentration of tris-containing species equal to $0.10F$?

5.35. A trisH$^+$/tris buffer is prepared by mixing 0.10 mole of trisH$^+$Cl$^-$ with 0.10 mole of tris in enough water to make 1 liter of solution.
  (a) Calculate the pH of the buffer, knowing that pK_b for tris is 5.92.
  (b) What is the pH of the buffer solution after adding 10 mL of $0.1F$ HCl?
  (c) What is the pH of the buffer after adding 10 mL of concentrated HCl ($12M$)?

5.36. Calculate the buffer index of a $0.1F$ solution of HCl. Does adding NaCl increase the buffer index? Explain.

5.37. Calculate the buffer index of pure water.

5.38. What is the buffer index of a $0.1F$ solution of NaOH? Explain with a specific example how the solution acts as a buffer.

5.39. Calculate the buffer index of a buffer made by adding 0.010 mole of NaOH to 0.030 mole of 2,4-dichlorophenol and diluting to 100 mL with water. Use $pK_a = 7.9$.

5.40. Plot buffer index as a function of pH for a solution containing ammonia and ammonium chloride (total concentration $0.10F$) and benzoic acid and sodium benzoate (total concentration $0.10F$).

5.41. Write a computer program in either BASIC or Pascal which will solve Problem 5.40. Plot buffer index at 0.1 pH unit intervals between pH 2 and pH 13.

5.42. The dilution value for a $0.05M$ solution of potassium biphthalate buffer is 0.05. If the pH of this solution is 4.00, what will the pH of the solution be when an equal volume of water is added to it?

5.43. The pH of a buffer made by mixing equal volumes of $0.20F$ KH$_2$PO$_4$ and $0.20F$ Na$_2$HPO$_4$ is found to be 6.90. If the average dilution value is 0.08 for a buffer containing these two compounds in the concentration range 0.10 to $0.005F$, what will be the approximate pH of the solution obtained when 5 mL of the buffer is diluted to 100 mL with pure water?

5.44. What are the analytical implications of the answer to Problem 5.43? Is it a safe practice to keep a large supply of concentrated buffer solution and take small portions of it for dilution?

Chapter 6

Acid-Base Titrations

6.1 INTRODUCTION

A detailed discussion of acid-base equilibria is now behind us in Chapter 5, and we have enough information to really understand some important applications that rely on acid-base titrations. Most students perform an acid-base titration in general chemistry laboratory. In a course in analytical chemistry students go on to learn how to plan a titration, how to prepare and standardize solutions, how to decide if an indicator can be used or a pH electrode will be needed, and how to interpret results. The use of titration curves to determine the number of dissociable protons, molecular weight, and proton dissociation constants of an acid is also a part of most analytical course laboratory programs. This chapter is a mixture of information important in the practice of analytical chemistry and material of a more theoretical flavor important in understanding why titration curves appear as they do.

Let us begin by defining some terms. In a titrimetric determination of an acid, a measurable volume of base *titrant* of accurately known concentration (a "standard solution") is added to a solution of the sample being tested until the sample has reacted completely. This point is called the *equivalence point*, the point at which chemically equivalent amounts of acid and base have reacted. Several methods are used to tell us when the equivalence point has been reached, among them a color change in an acid-base indicator dye or a sudden voltage change of an electrode sensitive to solution pH. The signal gives us an *endpoint* for the reaction. Under the best of circumstances the endpoint and equivalence point occur at exactly the same volume of titrant. It may not always be possible to make the endpoint and equivalence point coincide, but we will discuss later some decisions we can make to minimize error resulting from their noncoincidence.

For a titration to have promise as an analytical tool it must meet all of the following criteria:

1. It must involve a rapid reaction. After the addition of a small volume increment of titrant, equilibrium should be reached within a few seconds. Almost all Brønsted acid-base reactions and many complexation reactions behave this way. Precipitation reactions, however, may be sluggish because precipitates often form slowly.
2. Its reactions must be stoichiometric and complete. If there are any secondary or side reactions, they cannot affect the analytical results. The criterion for completeness is that the reaction be 99.9% complete at the equivalence point.
3. It must be free of interferences. Reactions of the titrant with the air or sample impurities may cause determinate error. Atmospheric carbon dioxide can be an important interference in many acid-base titrations, as you will see shortly.
4. It must have a discernible endpoint. Abrupt color changes and pH electrode voltage changes help the accuracy and precision of acid-base titrations.

6.2 STANDARD SOLUTIONS

Most standard solutions of acid and base titrants are prepared as *secondary standards,* that is, by diluting concentrated stock solutions to approximately the desired concentration and then titrating the resulting solution against a *primary standard* substance. Primary standards must either be available in pure form or at least be purified with ease, dried and stored without decomposing, and have high gram formula weights (to minimize weighing error).

6.2.1 Standard Base Solutions

Sodium hydroxide is by far the most popular base for use when water is the solvent, although potassium hydroxide and barium hydroxide are used occasionally. None of these compounds is used as a primary standard because the commercially available materials contain silica, carbonates, chloride, sulfate, and water. Furthermore, these bases all absorb water from the air and are therefore quite difficult to handle and weigh. The presence of carbonate in particular may prove to be quite troublesome.

The Carbonate Nuisance Atmospheric carbon dioxide reacts with hydroxide to produce carbonate ion according to the following reaction:

6.2 STANDARD SOLUTIONS

$$CO_2 + 2OH^- = CO_3^{2-} + H_2O$$

Carbonate ion is soluble in $0.1F$ NaOH and $0.1F$ KOH, but precipitates from even dilute solutions of $Ba(OH)_2$ as $BaCO_3$.

Whether or not carbonate changes the effective concentration of the base in which it dissolves depends on the nature of the titration in which the base is used. When carbonate-containing base is added as a titrant to an acidic solution, the carbonate impurity reacts with two protons,

$$2H^+ + CO_3^{2-} = CO_2 + H_2O$$

Since each CO_2 molecule reacts with two hydroxides when absorbed by the base, there is no net error introduced in the titration:

$$\begin{aligned} CO_2 + 2OH^- &= CO_3^{2-} + H_2O \\ 2H^+ + CO_3^{2-} &= CO_2 + H_2O \\ \hline \text{net}\quad 2H^+ + 2OH^- &= 2H_2O \end{aligned}$$

A problem arises, however, if carbonate-containing base titrant is added to a neutral or slightly basic solution. Under these conditions carbonate will react with only one proton, forming bicarbonate ion:

$$CO_3^{2-} + H^+ = HCO_3^-$$

One of the hydroxides consumed by an absorbed molecule of CO_2 is effectively lost from solution when the base is added to neutral solutions, and the concentration of the base is lower than expected.

All of this chemistry translates into titration error in the following way: whenever a titration of an acid with a base can be terminated at pH 7 or less (strong acids), carbonate error will be insignificant.

Several strategies have been suggested for avoiding contamination with atmospheric CO_2. For example, the addition of barium chloride to a stock solution of sodium hydroxide will precipitate $BaCO_3$, although barium ions may precipitate many ions (sulfate, oxalate) which may be present in samples being titrated. It is probably a better idea to avoid carbonate by taking advantage of the low solubility of Na_2CO_3 in 50% NaOH, which is commercially available or prepared without much difficulty.†
A $0.1F$ solution of NaOH can be prepared by adding about 6 mL of 50% NaOH (density 1.525 g/mL) to 1 liter of deionized water which has been boiled for a few minutes to remove dissolved CO_2, then cooled in a stoppered flask. The $0.1F$ solution

† Dissolve 50 g of NaOH in 50 g of water in a Pyrex flask. Either centrifuge the syrupy solution in a stoppered test tube and draw off clear solution or filter it through a Gooch crucible in a nitrogen-filled air bag.

may be stored for short periods of time in a polyethylene bottle, but for longer-term storage borosilicate glass (Pyrex) vessels are recommended.†

Absorption of CO_2 by standard base solutions during short exposures to the air (e.g., transferring solution to a buret) is not a serious problem. In extremely careful or slow work a tube containing sodium oxide (soda lime) should be attached to the top of a buret to remove CO_2 from air that enters the buret as solution is dispensed. More complicated apparatus can be assembled when a great many titrations are run routinely (see Ref. 1) but such elaborate equipment is not practical for instructional laboratories.

Standardizing a Solution of Base A solution of base is standardized by using it to titrate replicate portions of a primary standard acid. Many fine primary standard acids are available:

Potassium Biphthalate (KHP)

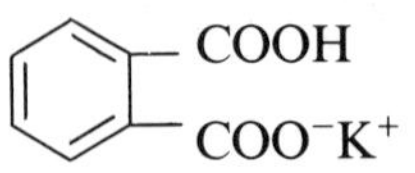

[structure: benzene ring with COOH and COO^-K^+ groups]

Formula weight: 204.23

pK_a: 5.51

KHP is by far the most popular primary standard acid. It is available from many suppliers in 99.95 to 100.05% purity and from at least one supplier in 99.99% purity (Baker Ultrex). The National Bureau of Standards can supply 99.97% material. The principal impurity is usually phthalic acid, which can be assayed by ether extraction (Ref. 2). KHP should be dried at 110°C for 2 hours to remove adsorbed water. At temperatures above 125°C KHP may decompose to phthalic anhydride.

A minor disadvantage of KHP as a titrimetric standard is that it is a rather weak acid and as a result exhibits a more limited pH change at the equivalence point than does a strong acid (see below). Phenolphthalein is the recommended indicator for titrations of KHP.

Benzoic Acid

[structure: benzene ring with COOH group]

Formula weight: 122.12

pK_a: 4.21

---

† Polyethylene is permeable to CO_2, but is preferred over glass if the solution needs to be stored for a few days. Strongly alkaline solutions dissolve soft glass (producing soluble silicates) and will not maintain their original concentration. Borosilicate glass is more resistant to attack than is soft glass. Soluble silicate produces an error similar to that caused by soluble carbonate, since silicic acid is a very weak acid.

This standard has been popular for decades. It is available commercially in very high purity (99.95–100.05%) and as a standard reference material (National Bureau of Standards, SRM 350a, 99.9958 ± 0.0027%). The acid can be dried by melting at 130°C, decanting into a platinum crucible, and crushing the cooled melt. The acid is not very hygroscopic and is therefore easily weighed.

A disadvantage of benzoic acid is its limited solubility in water (approximately 0.25 g per 100 mL at 25°C). Alcohol may be added to increase the solubility, but the acid tends to creep up vessel walls in alcohol solutions, a minor source of annoyance. Since many alcohols contain acidic impurities, pretitration of the solvent to a phenolphthalein endpoint may be necessary.

Potassium Sulfosalicylate (3:2 Salt) (KHS·K₂S·H₂O)

Formula weight: 550.655
pK_a: 2.85

This compound has recently been recommended as a primary standard acid by Butler and Bates (Ref. *3*) It has an extremely high gram formula weight and is hygroscopic only in extreme humidity. It should be dried for 1 hour at 110°C. While it is not yet commercially available, it is easily prepared and purified (see Ref. *3*). The acid is rather strong and shows a large equivalence point break when the titration is monitored with a glass electrode. Phenolphthalein can be used as an indicator.

Potassium Biiodate, $KH(IO_3)_2$ Formula weight: 389.912; strong acid.

This compound is a very versatile primary standard, suitable for acid-base studies and redox titrations (see Chapter 16). It has a very high gram formula weight and gives very sharp equivalence point breaks.

6.2.2 Standard Acid Solutions

Like standard base solutions, standard acid solutions are prepared as secondary standards. For example, an analyst wishing to make 1 liter of $0.1F$ HCl to be used as a standard solution will take about 8 mL of concentrated HCl (37%, concentration approximately $12.4M$) and dilute it to about 1 liter with deionized water. This stock solution will then be standardized by using it to titrate samples of a primary standard base such as sodium carbonate. Commercially available concentrated acids should never be used as primary standards.

Table 6.1 gives concentrations and densities of commercially available acids.

EXAMPLE 6.1 Preparing a Stock Solution to Be Standardized

A bottle of perchloric acid is labeled "70% perchloric acid," but its molarity is not indicated. How would you prepare 1 liter of a $0.10F$ solution of $HClO_4$ in water?

Solution

Table 6.1 shows that the density of 70% $HClO_4$ is 1.668 g/mL at 20°C. Exactly 1 mL of 70% perchloric acid will hold 0.70(1.668) = 1.17 g of perchloric acid, or 1.17/100.47 = 0.0116 mole. To prepare 1 liter of $0.10F$ acid, you would need to take 0.10/0.0116 = 8.6 mL of 70% $HClO_4$ and dilute to 1.00 liter with water.

Standardizing a Solution of Acid Several primary standard bases are available. Sodium carbonate has long been very popular, although tris(hydroxymethyl)aminomethane (THAM) if properly dried is a good alternative.

Sodium Carbonate (Na_2CO_3) Formula weight: 106.00; equivalent weight: 53.00.

The full reaction of sodium carbonate with a strong acid is

$$CO_3^{2-} + 2H^+ = CO_2 + H_2O$$

and so the equivalent weight of Na_2CO_3 is half its formula weight. Sodium carbonate

TABLE 6.1 COMMERCIALLY AVAILABLE MINERAL ACIDS

Acid	Gram formula weight	Density, 20°C	Concentration (wt %)	Molarity
Acetic	60.05	1.051	99.5–99.9	17.4
Hydrochloric	36.46	1.179–1.189	36.5–38.0	11.9–12.4
Hydrofluoric	20.01	1.16	48–51	27.8–29.6
Nitric	63.02	1.41–1.42	69–71	15.5–16.0
Perchloric	100.47	1.668	70	11.6
Phosphoric	98.00	1.69–1.71	85–87	14.8–15.2
Sulfuric	98.08	1.84	95–98	17.8–18.4

is commercially available in primary standard grade. The major impurity is sodium bicarbonate, which is converted to Na_2CO_3 by heating to temperatures above 160°C for 3 or 4 hours. The only potential drawback in the use of Na_2CO_3 is its rather low equivalent weight. Only about 200 mg of Na_2CO_3 is required for a titration involving about 40 mL of $0.1F$ strong acid.

Tris(hydroxymethyl)aminomethane (THAM or Tris), $(HOCH_2)_3$-C-NH_2 Formula weight: 121.14; K_b: 1.2×10^{-6} [also known as 2-amino-2-(hydroxymethyl)-1,3-propanediol].

THAM is a weak base and can be thought of as a derivative of ammonia. Many authors recommended drying at a temperature of 100 to 103°C for 1 or 2 hours, but decomposition is a risk even at these temperatures. It is probably safer to grind the material and dry it in a vacuum oven at 70 or 80°C. Diehl and co-workers (Ref. *4*) found that samples of reagent grade THAM contained up to 0.7% water entrained in crystals, and drying was quite difficult even with grinding. Since the purity of commercial THAM seems somewhat undependable, it is a good idea to compare each batch with primary standard sodium carbonate. The National Bureau of Standards sells primary standard THAM (SRM 723a) with certified purity of 99.9703 ± 0.0028 wt%.

Borax (Sodium Tetraborate Decahydrate, $Na_2B_4O_7 \cdot 10H_2O$)   Formula weight: 381.37.

Tetraborate reacts with two protons to form the weak acid H_3BO_3,

$$B_4O_7^{2-} + 2H^+ + 5H_2O = 4H_3BO_3$$

At the equivalence point of the titration of sodium tetraborate with strong acid, the solution is a dilute solution of boric acid (pK_a 9.2). At the $0.1F$ concentration level, the pH of the solution is about 5.1.

Borax is available in highly pure form and has a high gram formula weight. The only important disadvantage is that special precautions must be taken to make sure that the material remains a decahydrate. Rather than oven dry at a low temperature, one should keep a supply of borax in a desiccator over a saturated solution of sodium bromide dihydrate.

4-Aminopyridine

Formula weight: 94.117

K_b: 1.3×10^{-5}

This material has recently become available commercially in primary standard grade purity. It can be purified by sublimation under vacuum, and was used by Diehl and co-workers as a standard for the determination of the Faraday constant (Ref. 5). It appears to be a cancer-causing agent and should be handled carefully.

Sodium Oxalate ($Na_2C_2O_4$)   Formula weight: 134.00.

This material was used for many years as a source of highly pure sodium carbonate. It is converted by heating to sodium carbonate:

$$Na_2C_2O_4 = Na_2CO_3 + CO$$

which is then used to standardize acids. Sodium oxalate can also be used as a primary standard reducing agent in redox titrations (see Chapter 16).

EXAMPLE 6.2   Standardizing an Acid Solution

The $0.10F$ solution of $HClO_4$ prepared in the last example is to be standardized with primary standard sodium tetraborate decahydrate. A 475.2-mg sample of sodium tetraborate decahydrate is dissolved in 100 mL of water. Perchloric acid solution is added to the tetraborate solution from a buret, and the progress of the titration is monitored with a pH electrode (see below). 26.46 mL of perchloric acid is required to reach the endpoint at pH 5. Calculate the molarity of the perchloric acid.

Solution

We begin by noting that 1 mmole of tetraborate reacts with 2 mmole of perchloric acid, a monoprotic acid. The millimoles of tetraborate are calculated from the gram formula weight and amount of material weighed:

$$\text{millimoles ``borax''} = 475.2/381.37 = 1.246 \text{ mmole}$$

This will react with 2(1.246 mmole) = 2.492 mmole of perchloric acid. The concentration of the perchloric acid solution is therefore 2.492 mmole/26.46 mL or $0.09418M$.

6.3 EQUIVALENT WEIGHT AND NORMALITY

Most of the calculations in this text involve concentrations expressed in terms of molarity, that is, moles (gram formula weights) of solute per liter of solution. Another

6.3 EQUIVALENT WEIGHT AND NORMALITY

way to express concentrations, mentioned briefly in Chapter 1, is with *normality,* and the measurement of *equivalent weights* of solute. To help you understand the relationships between normality and molarity, consider the following three acids:

Acetic acid: monoprotic weak acid, CH_3COOH, gram formula weight 60.05

Succinic acid: diprotic weak acid, $(HOOC)CH_2CH_2(COOH)$, gram formula weight 118.09

Citric acid: triprotic acid, H_2CCOOH, gram formula weight 192.12
$$\begin{array}{c} | \\ HOCCOOH \\ | \\ H_2CCOOH \end{array}$$

Let us take 1-mmole portions of strong base (NaOH) and see how much of each acid they will consume. We find that 1 mmole of acetic acid will react completely with 1 mmole of NaOH. The 1 mmole of acetic acid weighs 60.05 mg, an amount which is *chemically equivalent* to 1 mmole of NaOH. We next find that 0.5 mmole of succinic acid is enough to react with 1 mmole of NaOH; this is because 0.5 mmole of succinic acid provides 1.0 mmole of protons. The 0.5 mmole of diprotic succinic acid weighs 59.05 mg, an amount which is chemically equivalent to 1 mmole of NaOH and 60.05 mg of acetic acid. Finally, we find that only 0.333 mmole of citric acid (64.04 mg) is needed to react with 1 mmole of NaOH. Each molecule of citric acid produces three protons; $\frac{1}{3}$ mmole of citric acid is therefore chemically equivalent to 1 mmole of NaOH (and 1 mmole acetic acid and 0.5 mmole of succinic acid). Notice that if we knew nothing of the molecular structures of these three acids and the fact that they have different numbers of reactive protons, we might conclude that they all had about the same gram formula weight. This apparent gram formula weight is called the equivalent weight. It is the weight of acid which will release 1 mole of protons in an acid-base reaction. The equivalent weight of a base is the weight of base which will react with 1 mole of protons. Just as there are milligram formula weights, there are also milliequivalent weights.

The normality (N) of a solution is defined as the number of equivalents of solute per liter of solution. A $0.1N$ solution of acetic acid contains 0.1 equivalent per liter and is also $0.1M$. A $0.1N$ solution of succinic acid, however, is prepared by dissolving 0.05 mole of succinic acid in enough water to make 1 liter of solution. A $0.05M$ solution of a diprotic acid such as succinic acid is $0.1N$.

The normal concentration system was developed at a time when chemists were much less sure of molecular structures. Using the normality system, one can always say that at the equivalence point of a titration

Equivalents of acid = equivalents of base

or Normality of acid * volume of acid = normality of base * volume of base

An unknown acid can be characterized by the amount of it needed to react with an equivalent of base (a mole of a monobasic base). For day-to-day assay work this quantity can be every bit as useful as the amount needed to react with a mole of base. The only problem is that the normality of a solution depends on the reaction for which it is used. An often cited example is phosphoric acid, H_3PO_4. Since it is a triprotic acid, its equivalent weight ought to be one-third its gram formula weight. However, the third proton is so difficult to remove ($HPO_4^{2-} = H^+ + PO_4^{3-}$, $pK_a = 12.4$) that phosphoric acid is almost never used as a triprotic acid, but rather as a diprotic acid. You must be careful when using a solution whose concentration is expressed as a normality. If you cannot be sure of the intended chemical reaction, you should restandardize it.

EXAMPLE 6.3 Normality

A 25.00-mL sample of oxalic acid is titrated with $0.1073N$ NaOH. 17.37 mL is required to reach the equivalence point. What is the normality of the oxalic acid solution, and how could it be prepared? The reaction is

$$H_2C_2O_4 + 2NaOH = 2H_2O + Na_2C_2O_4$$

Solution

Since there must be chemically equivalent amounts of acid and base at the equivalence point, we can write

$$N_{acid} * vol_{acid} = N_{base} * vol_{base}$$

and solve for N_{acid};

$$N_{acid} = N_{base} * (vol_{base}/vol_{acid}) = 0.1073(17.37/25.00)$$
$$N_{acid} = 0.07455$$

The molarity of the oxalic acid solution is one-half this value, or $0.03728M$. Remember that 2 moles of protons (i.e., 2 equivalents) are released by every mole of oxalic acid.

We can prepare a liter of this oxalic acid solution by taking 0.07455 equivalent weight of oxalic acid and diluting to 1 liter with water. 1.000 equivalent weight of oxalic acid is (90.03/2) or 45.02 g, and so we could weigh out 3.356 g.

6.4 ACID-BASE INDICATORS

6.4.1 Single Indicators

When an acid-base reaction is well known and many routine samples must be titrated quickly, an analyst uses an indicator to signal the equivalence point. Acid-base indicators are organic dyes whose color depends quite critically on the presence of one or two dissociable protons. These dyes have such intense colors that only a few drops of a 0.1% indicator solution are needed to color a solution being titrated. While indicators are themselves acids and bases, they consume very little titrant in most applications.†

Let us discuss indicator color range, using bromcresol green as an example. The acid form of bromcresol green (HBG^-) is yellow, while its base form (BG^{2-}) is blue-green. Structures are shown in Fig. 6.1 along with those for phenolphthalein and methyl red.

We can represent the dissociation of HBG^- by the reaction

$$HBG^- = H^+ + BG^{2-}$$

and the weak acid dissociation constant (the "indicator constant") by

$$K_{ind} = \frac{a_{H^+} a_{BG^{2-}}}{a_{HBG^-}}$$

Taking negative logarithms of both sides and rearranging gives

$$pH = pK_{ind} + \log \frac{a_{BG^{2-}}}{a_{HBG^-}}$$

The pH of a solution in which the activities of HBG^- and BG^{2-} are the same is numerically the same as pK_{ind}, 4.7 in this case. As we raise the pH by adding a strong base, the solution turns blue-green (color of the anion). At some point an analyst with normal color vision will decide that the addition of more base produces no further color change. The pH at this point is the basic end of the indicator range, and for bromcresol green is pH 5.4. By the same kind of process, but adding acid instead of base, the acid end of the bromcresol green range (yellow) is found to be pH 3.8. The

---

† Five drops of a 0.1% solution of bromcresol green (gram formula weight 698) in 100 mL of solution is a concentration of about $3.6 \times 10^{-6} M$. This amount of acid will consume 0.0036 mL of a $0.1F$ solution of strong base. In more dilute systems an "indicator blank" may be needed to correct for the titration of the indicator. In ordinary laboratory situations the same amount of indicator is used in standardizing the titrant as is used in titrating samples, and this compensates for the indicator.

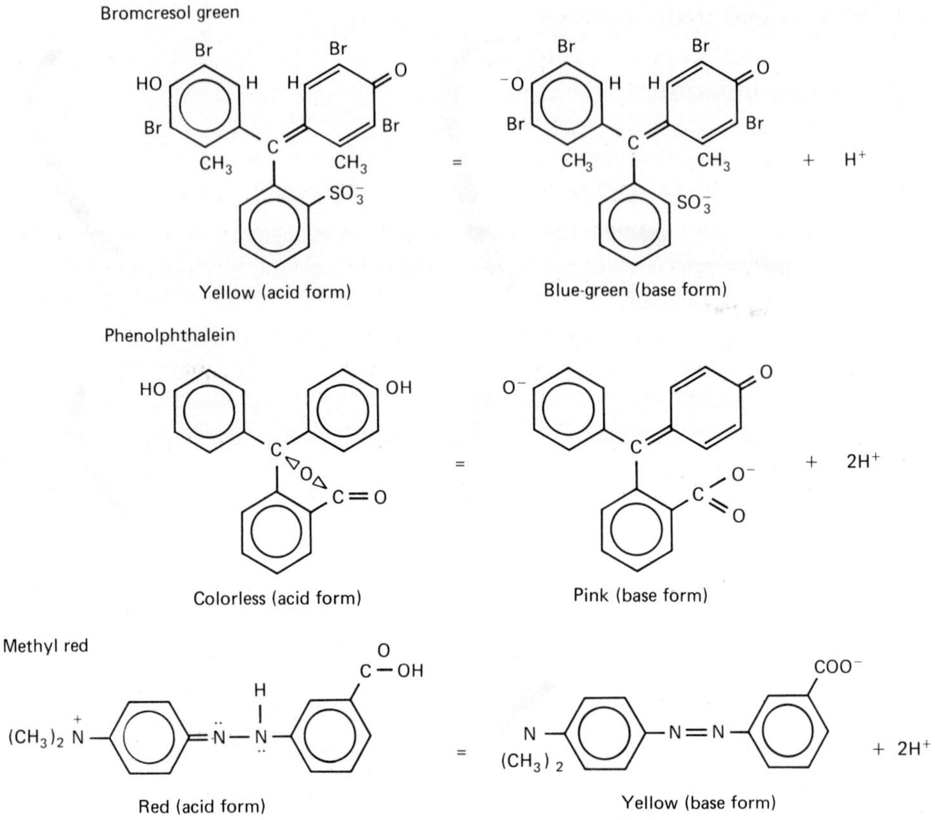

Figure 6.1 Indicator structures.

ratios of $a_{BG^{2-}}$ to a_{HBG^-} at these points can be calculated from the last algebraic expression,

$$\text{acid end: at pH } 3.8, \quad \frac{a_{BG^-}}{a_{HBG}} = 0.10$$

$$\text{base end: at pH } 5.4, \quad \frac{a_{BG^-}}{a_{HBG}} = 5.0$$

Why is the color range not symmetrical about the pK_{ind}? Why does the basic end occur at pH 5.4 instead of pH 5.7? Asymmetry of a range is quite common and can be attributed to two causes. First, the color of the anion, BG^{2-}, is inherently more intense† than that of HBG^-; a 0.01 mM solution of BG^{2-} will be more deeply colored

---

† Intensity and hue are different properties. The intensive property called "molar absorptivity" is responsible for the effect just described and will be discussed in detail in Chapter 13.

6.4 ACID-BASE INDICATORS

than a 0.01 mM solution of HBG$^-$. The second cause, although not very important for this particular pair of species, is the sensitivity of the human eye to different colors. Our eyes are far more sensitive to green light (middle of the visible spectrum) than to either blue (short-wavelength end) or red (long-wavelength end).

In the absence of specific information about the range of an indicator, it is generally safe to assume that a 10-fold excess of base form over acid form will produce the base color and that a similar excess of acid form over base form will produce the acid color. We should then expect to see indicator ranges of about plus or minus one pH unit around pK_{ind}.

The most popular and useful acid-base indicators are shown in Fig. 6.2 along with their color ranges. Some indicators, notably cresol red and thymol blue, have

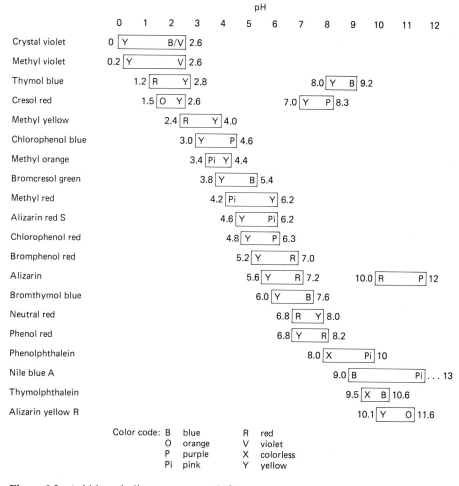

Figure 6.2 Acid-base indicator ranges at 25°C.

two color ranges. These indicators have two proton sites which have quite different pK_a values.

The pH ranges for the indicators in Fig. 6.2 are for aqueous solutions at 25°C. Ranges for uncharged or negatively charged species (bromcresol green and phenolphthalein, for example) undergo very little change with temperature. Positively charged indicators (methyl red, methyl orange, and methyl yellow) shift their entire range to lower pH values with increased temperature. At 25°C the range for methyl red is pH 4.2 to 6.3, while at 100°C it is pH 4.0 to 6.0.

6.4.2 Mixed and Screened Indicators

By mixing two indicators ("mixed") or an indicator and a neutral dye ("screened"), it may be possible to shorten the color range and sharpen the color change at the endpoint of a titration. Examples of mixed and screened indicators are given in Table 6.2, along with their formulations and the colors of their acid and base forms. Methylene

TABLE 6.2   MIXED AND SCREENED INDICATORS

Indicator[a,b]		pH for color change	Acid color	Base color
Mixed indicators				
Bromcresol green[c]/methyl orange		4.3	Orange	Blue-green
0.1% alc	0.02% aq		(pH 3.5)	(pH 4.3)
Bromcresol green[c]/chlorophenol red[c]		6.1	Yellow-green	Blue-violet
0.1% aq	0.2% alc		(pH 5.4)	(pH 6.2)
Bromthymol blue[c]/phenol red[c]		7.5	Yellow	Violet
0.1% aq	0.1% aq	(pH 7.2)	(pH 7.6)	
Phenolphthalein/thymolphthalein		9.9	Colorless	Rose
0.1% alc	0.1% alc			(pH 9.6)
Screened indicators				
Dimethyl yellow/methylene blue		3.25	Blue-violet	Green
0.1% aq	0.1% aq		(pH 3.2)	(pH 3.4)
Methyl red/methylene blue		5.4	Red-violet	Green
0.2% alc	0.1% alc		(pH 5.2)	(pH 5.6)
Neutral red/methylene blue		7.0	Violet-blue	Green
0.1% alc	0.1% alc		(pH 6.8)	(pH 7.3)

[a] Concentrations are percent wt/vol in ethanol (alc) or water (aq). Color changes observed with 5 to 10 drops in 100 ml.
[b] All screened and mixed indicators should be stored in dark bottles as a precaution against photodecomposition.
[c] Denotes sodium salt.
Source: Data from I. M. Kolthoff and V. A. Stenger, *Volumetric Analysis,* vol. II, 2nd ed., Interscience, New York, 1947, pp. 58–60.

6.5 ACID-BASE TITRATION CURVES

blue is a very common screening dye. The screen produces a neutral gray color in the middle of the transition range.

6.5 ACID-BASE TITRATION CURVES

There are two important kinds of information we can get from acid-base titrations. As you have already seen, the quantity of acid or base present in a sample can be determined by carefully measuring the volume of titrant needed to reach the equivalence point, signaled, perhaps, by a color change of an acid-base indicator. In addition, it may also be possible to measure the dissociation constant of an acid or base if solution pH is monitored throughout the titration. In this section we will discuss how to make these measurements and how accurate we can expect them to be. The chapter will conclude with the application of some simple principles to the analysis of complex acid-base systems and a description of some acid-base methods used routinely in analysis.

6.5.1 Experimental Considerations

Measuring pH with a Glass Electrode pH is measured in the laboratory with a glass membrane electrode (see Fig. 6.3) and a voltmeter (a "pH meter"). The active element

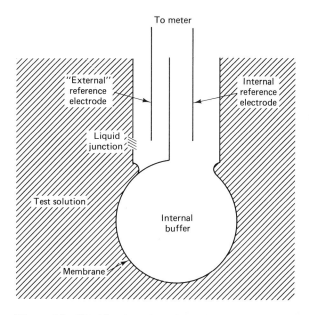

Figure 6.3 Combination glass electrode.

of the electrode is a thin glass membrane sphere which contains a solution of fixed hydronium ion activity (0.1M HCl, for example). An electrical potential can develop across this membrane when the hydronium ion activity in a solution outside the membrane is different from that of the internal solution.

In order to get useful data with the glass electrode, we must calibrate the electrode potential with solutions of standard buffers of known pH and ionic strength.[†] A properly functioning glass pH electrode will give a potential, E, which changes linearly with changes in the logarithm of the hydronium ion activity in solutions over a wide range of hydronium activities. This means that an equation of the following type is obeyed:

$$E_{\text{measured}} = m * \log a_{H^+} + \text{constant}$$

where m is a slope. In the laboratory we use at least two standard buffer solutions to calibrate the slope and constant for a glass electrode and pH meter. First we dip an electrode in pH 7.0 buffer and adjust the meter "zero" control until the meter reads 7.0. We are adjusting the "constant" term in the equation when we do this. We then remove the electrode from the pH 7.0 buffer, wash it with deionized water, and dip it into pH 4.0 buffer. We then turn the "slope" dial on the pH meter until it reads 4.0. This adjusts the slope term in the equation above to be 0.05916 volt/pH unit. This value is predicted by the Nernst equation for the glass electrode and will be explained in Chapter 15. The calibration process is shown diagrammatically in Fig. 6.4. The object of calibrating the electrode and meter is to make the response line labeled "uncalibrated" coincide with the line labeled "calibrated." Note that there are two vertical axes showing the units of pH and the electrical potential which corresponds to the pH if the meter is set to read 0.000 volt at pH 7.0 and −0.1775 volt at pH 4.0. More information about the use and calibration of pH electrodes can be found in Chapter 15.

With a properly calibrated electrode one can expect measured pH values to be within about ±0.05 pH unit of the true value over a pH range of about 3 to 11, if test solutions and buffers are of similar ionic strength (0.05M). However, if the ionic strength of a test solution is, say, 1M, the accuracy of pH measurements with a pH electrode calibrated in much more dilute solutions is doubtful. The reasons for this will be discussed in Chapter 15. For the time being you should think of such pH measurements as only rough approximations.

To create realistic expectations about titration curves, we make some departures from the traditional treatment of this material. First, we will discuss real experimental titration data obtained with the kind of equipment available in most undergraduate laboratories.[‡] Second, because we will be comparing results for several acid-base systems

---

[†] The National Bureau of Standards has defined pH values to within ±0.005 pH unit for several buffers at a number of temperatures. A list of these primary buffers is given in Chapter 15.

[‡] In the examples a Bower and Haack Digital pH Analyzer model S5000 (3.5 digit) and a model S200C combination glass electrode were used. Their responses were calibrated with pHydrion pH buffers of pH 7.00 ± 0.02 and pH 4.00 ± 0.02 at 25°C.

6.5 ACID-BASE TITRATION CURVES

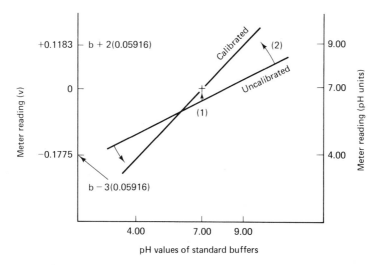

Figure 6.4 Calibration of a pH measuring system. The calibration of an electrode and meter is a two-step process. First, the meter is set to 0.00 V (pH 7.00) with a pH 7.00 standard buffer. Second, the meter slope response is set to 0.05916 V/pH unit with either a pH 4.00 or pH 9.00 standard buffer.

at different formal concentrations, we will convert the traditional "pH vs. volume of titrant" format for titration curves to "pH vs. the *fraction titrated.*" By definition, the equivalence point is reached at a fraction titrated of 1.00, or 100% titrated. If a titration requires 41.52 mL of titrant for complete reaction, after 30.00 mL has been added the fraction titrated is 30.00/41.52, or 0.7346. This can also be stated as 73.46% titrated. With a little practice you will find this format easy to use.

The third departure from traditional treatments is in the use of activities rather than concentrations in the discussion of titration curves and equilibrium constants. In Chapter 3 the extended Debye-Hückel and Davies equations were presented for use in calculating individual ion and mean activity coefficients from concentrations. The careful treatment of activities is not intended to be an obstacle; rather, it is done to show you how to get meaningful, reliable results when characterizing acids and bases and performing analyses.

6.5.2 Titration of a Strong Acid with a Strong Base

Figure 6.5 shows the pH throughout the course of titration of the strong acid HCl with the strong base NaOH at the $0.1F$ concentration level. Notice the extreme sharpness of the equivalence point break and the symmetry of the curve. Strong acids and bases are completely dissociated and all their reactions can be reduced to the simple solvent reaction

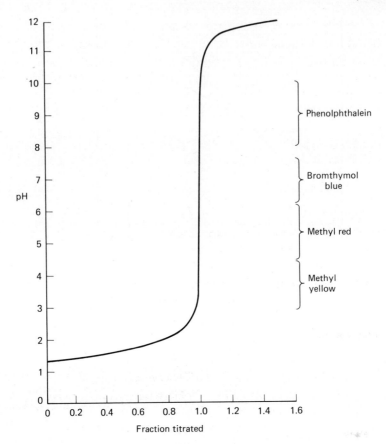

Figure 6.5 Titration of HCl with NaOH, 0.1F concentration level. Actual values were: 0.0923F HCl, 50-mL volume; 0.1427F NaOH.

$$H^+ + OH^- = H_2O$$

Before the equivalence point solution pH is determined by the activity of unreacted hydronium ion. After the equivalence point pH is determined by the activity of excess hydroxide ion. At the equivalence point the autoprotolysis of water determines the pH:

$$[H^+] = [OH^-] = K_w^{1/2} = 10^{-7.00} \quad \text{at } 25°C$$

The pH at the equivalence point in a strong acid/strong base (SA/SB) titration is *always* 7.0 at 25°C.

The SA/SB titration curve gives us no exact information about acid or base strength, since strong acids and bases are completely dissociated.

6.5 ACID-BASE TITRATION CURVES

The most striking characteristic of the SA/SB titration curve is the sharpness of the pH change near the equivalence point. Experimental data from this portion of the curve are given in Table 6.3. The addition of 0.10 mL of base (about two drops) from a buret results in a pH change of 2.8 units at the equivalence point.

Superimposed on the titration curve in Fig. 6.5 are the color transition ranges for four popular acid-base indicators. The best possible indicator would change color abruptly (with one drop of titrant, if possible) right at the equivalence point. Of the four indicators the best choice would be bromthymol blue; its transition range (pH 6.2 to 7.6) straddles the equivalence point pH. If the titration were stopped at the first permanent blue color the volume would be within 0.1% of the equivalence point volume, and there would be no appreciable indicator error.

Phenolphthalein indicator merits special attention because of its great popularity. Phenolphthalein is colorless until the solution pH increases to about 9, where it turns faintly pink. In the titration being discussed, pH 9 is achieved at 32.45 mL (100.3% titrated), about 0.1 mL past the equivalence point. Carbon dioxide contamination (the "carbonate nuisance") can further increase the volume of base needed to reach pH 9 and can be a source of determinate error in titrations using phenolphthalein as the indicator.

Methyl red can be used in an SA/SB titration at the $0.1M$ concentration level, but its color transition range occurs before the equivalence point (pH 4.4 to 6.2, about 99.7 to 99.9% titrated). The color change occurs with the addition of about two drops of base and is therefore a less abrupt change than that of bromthymol blue. Titration to a permanent yellow (pH > 6) should result in a negligible titration error at the $0.1F$ concentration level.

Methyl yellow is not a satisfactory indicator for this titration because its transition range begins and ends well before the equivalence point (pH 2.8 to 4, about 98 to 99.7% titrated). Furthermore, it will not show an abrupt color change, since about 0.5 mL of base must be added to change the pH from 2.8 to 4. An analyst would have

TABLE 6.3 pH vs. PERCENT TITRATED AND VOLUME OF TITRANT FOR THE SA/SB TITRATION IN FIG. 6.5

Percent titrated	Volume delivered	pH
98	31.70	3.12
99.3	32.10	3.40
99.6	32.20	3.62
99.9	32.30	5.92
100.2	32.40	8.74
100.5	32.50	9.20

trouble seeing subtle color differences during the addition of several drops in the color transition range, and poor precision is likely to result. Stopping at the first permanent yellow would result in an error of about 0.3% in the titration used as an example.

Figure 6.6 shows the effect of dilution on an SA/SB titration curve. The curve obtained from the titration of HCl with NaOH at the 0.001F ("milliformal") concentration level is drawn over that shown in the last figure. Notice that the milliformal curve appears to be compressed on the pH axis. The equivalence point still occurs at pH 7.0, but the initial plateau spans a pH range of 3 to 4, instead of 1.5 to 2.5 as was the case in the titration at the 0.1F level. Also notice that the upper plateau of the milliformal curve is in the pH range 10 to 10.5, rather than 11 to 12. The equivalence point break is thus much less sharp at the milliformal level than at the 0.1F level; a

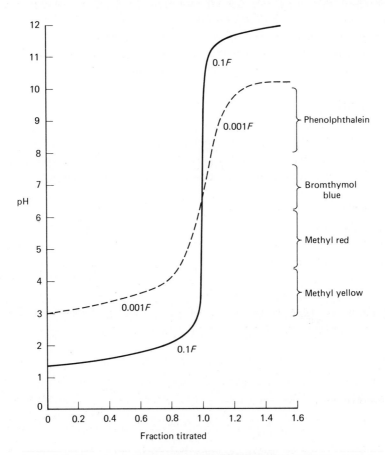

Figure 6.6 Titrations of HCl with NaOH, 0.1F and 0.001F levels. Actual values were: 0.0923F HCl, 0.1427F NaOH, ionic strength 0.092 at start of titration; and $1.205 \times 10^{-3}F$ HCl in 0.095F KCl, $1.427 \times 10^{-3}F$ NaOH, ionic strength 0.095 at start of titration.

change in pH from 6 to 8 requires almost 3 mL of base in the milliformal titration. The practical effect of this is to rule out single indicators for dilute acid-base titrations because they will not change color abruptly. Even screened indicators are of little value. Although a mixture of neutral red and methylene blue will change color in the narrow pH range 6.8 to 7.3, the delivered volume over this range at the milliformal level is more than 1 mL. Phenolphthalein will not begin to turn pink until the 120% titrated point has been reached, while methyl red shows a very gradual color change between 85 and 96% titrated. Methyl yellow begins its color change almost at the start of the titration and is not at all suitable as an endpoint indicator.

It might be possible to overcome the problem of the gradual color changes of indicators in dilute SA/SB titrations by following the titration with a glass electrode. Unfortunately, regardless of the method used to monitor the endpoint, extreme care must be exercised to avoid contaminating dilute solutions with dissolved CO_2, a very serious source of error at low concentrations. Generally it is inadvisable to work at concentrations below about $0.01F$ in acid-base titrations.

In summary, for an SA/SB titration endpoint (at the $0.1F$ level), bromthymol blue is the best choice of indicator for high accuracy and precision. Both phenolphthalein and methyl red can be used if errors of about 0.2% can be tolerated. Methyl yellow, however, is not suitable; not only is the color change premature, it occurs over several drops of base titrant.

6.5.3 Titration of a Weak Acid with a Strong Base

From a weak acid/strong base titration curve we can determine not only the equivalence point volume, but also a quantitative measure of the acid's strength. Figure 6.7 is the titration curve for acetic acid (HOAc) with sodium hydroxide at the $0.1F$ concentration level. Notice that the equivalence point break is sharp, but not as sharp as that in the SA/SB titration curve in Fig. 6.5. The lack of sharpness is a consequence of the weakness of acetic acid as an acid. Since it is only slightly dissociated, the pH during the titration is never as low as in the case of the strong acid HCl at the same fraction titrated. Up to the equivalence point the pH of the solution is governed by the equilibrium

$$HOAc = H^+ + OAc^-$$

and is determined by the fraction of unreacted HOAc and the dissociation constant, K_a,

$$K_a = \frac{[H^+][OAc^-]}{[HOAc]}$$

The relationship between pH, pK_a, and fraction of HOAc (α_{HOAc}) can be obtained by

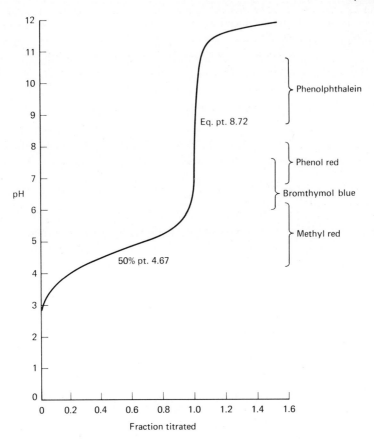

Figure 6.7 Titration of acetic acid with NaOH, 0.1F concentration level. Actual values were 0.102F HOAc, 50-mL volume, 0.1427F NaOH ($\mu = 0.03M$ at 50% point).

substituting $C_{total}(\alpha_{HOAc})$ for [HOAc], and $C_{total}(1 - \alpha_{HOAc})$ for [OAc$^-$] into the Henderson-Hasselbalch equation to get

$$pH = pK_a - \log \frac{\alpha_{HOAc}}{1 - \alpha_{HOAc}}$$

pK_a is equal to solution pH at the point in the titration where $\alpha_{HOAc} = 1 - \alpha_{HOAc}$, that is, where $\alpha_{HOAc} = 0.50$, at the 50% titrated point. From Fig. 6.7 the experimental pH at the 50% point is 4.67. When corrected for ionic strength the value is 4.75, in good agreement with the accepted value of 4.76. The following example shows how to correct the titration curve data for the effect of ionic strength.

6.5 ACID-BASE TITRATION CURVES

EXAMPLE 6.4 Calculating pK_a

The pK_a of acetic acid was set equal to the pH at the 50% point in Fig. 6.7 and was 4.67. What is the thermodynamic value for pK_a? The initial concentration of acetic acid was 0.102F (50.0 mL volume), and the titrant was 0.1427M NaOH. To reach the equivalence point 5.80 mL of titrant was used.

Solution

The approach is to express the experimental pH in terms of pK_a and the activity coefficients of acetic acid and acetate ion. We use the extended Debye-Hückel equation to calculate the activity coefficient of acetate, assume that that of acetic acid is 1.00, and calculate pK_a.

The Henderson-Hasselbalch equation can be written in terms of activities:

$$\text{pH} = \text{p}K_a - \log \frac{a_{\text{HOAc}}}{a_{\text{OAc}^-}}$$

Substituting in $f_{\text{HOAc}} C_{\text{HOAc}}$ and $f_{\text{OAc}^-} C_{\text{OAc}^-}$ for their respective activities gives

$$\text{pH} = \text{p}K_a - \log \frac{C_{\text{HOAc}}}{C_{\text{OAc}^-}} - \log \frac{f_{\text{HOAc}}}{f_{\text{OAc}^-}}$$

At the 50% titrated point, $C_{\text{HOAc}} = C_{\text{OAc}^-}$, and so

$$\text{pH} = \text{p}K_a - \log \frac{f_{\text{HOAc}}}{f_{\text{OAc}^-}}$$

At the 50% point in the titration the solution is 0.50(0.102)(50 mL/67.9 mL) or 0.0376M in both acetic acid and acetate ion. The solution is also 0.0376M in sodium ion (from NaOH, the titrant). The ionic strength is therefore

$$\mu = \tfrac{1}{2}(C_{\text{Na}^+} z_{\text{Na}^+}^2) + \tfrac{1}{2}(C_{\text{OAc}^-} z_{\text{OAc}^-}^2)$$
$$= \tfrac{1}{2}(0.0376) + \tfrac{1}{2}(0.0376) = 0.0376M$$

Using the extended Debye-Hückel law (EDHL) equation,

$$-\log f_i = \frac{0.511 z^2 (\mu)^{1/2}}{1 + 0.329 \mathring{a}(\mu)^{1/2}}$$

with Kielland radius $\mathring{a} = 4.5$ for acetate ion, we find $-\log f_{\text{OAc}^-} = 0.077$, $f_{\text{OAc}^-} = 0.84$.

The activity coefficient for the neutral species, acetic acid (the "salting coefficient" discussed in Chapter 3), can be taken as unity at ionic strengths below $0.1M$. The pK_a corrected for ionic strength is thus

$$pK_a = pH + \log \frac{1.00}{0.84} = 4.67 + 0.08 = 4.75.$$

The accepted value is 4.76.

In Fig. 6.8 the experimental data from the titration of acetic acid with sodium hydroxide at the milliformal level are superimposed on the curve shown in the last

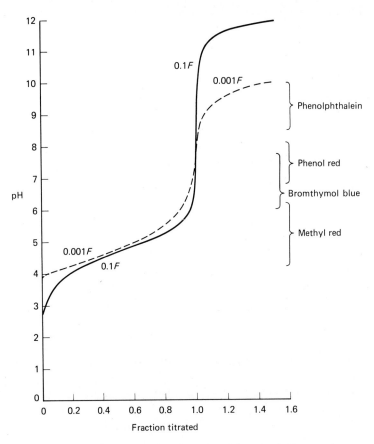

Figure 6.8 Titrations of acetic acid with NaOH, $0.1F$ and $0.001F$ levels. Actual values: $0.102F$ HOAc, 50-mL volume, $0.1427F$ NaOH ($\mu = 0.03M$ at 50% point); and $0.608 \times 10^{-3}F$ HOAc in $0.054F$ KCl, $0.921 \times 10^{-3}F$ NaOH ($\mu = 0.04M$ at 50% point).

6.5 ACID-BASE TITRATION CURVES

figure. There are only slight differences between the two curves in the range of 20 to 80% titrated, in contrast with the dramatic effect of dilution on an SA/SB titration in the same range (see Fig. 6.6). Notice that the expression just used to calculate pK_a contains no explicit concentration dependence, but rather a dependence on *fractions*. In fact, in the region of the lower plateau of a WA/SB titration, the system is a buffer composed of untitrated weak acid and the conjugate base produced in the titration. Dilution has only a slight effect on the pH of such buffers, as discussed at the end of Chapter 5. Once past the equivalence point, dilution markedly decreases the sharpness of the endpoint break. Remember that after the equivalence point, solution pH is determined by the quantity of strong base added in excess.

To evaluate pK_a from titration data for relatively strong weak acids ($pK_a < 3$), we must extend the simple expression used in the last example to account for greater dissociation of the acid. A more satisfactory expression is much like the one derived near the end of Chapter 5,

$$pH = pK_a - \log \frac{C_{HA} - [H^+]}{C_{A^-} + [H^+]}$$

EXAMPLE 6.5 Calculating pK_a

A 50.0-mL portion of a $0.500M$ solution of 2,3-dinitrobenzoic acid is titrated with $0.0500M$ NaOH. After exactly 25 mL of NaOH has been added the pH is measured as 2.20. Calculate the pK_a of 2,3-dinitrobenzoic acid.

Solution

We can use the last equation with estimates of $[H^+]$ from the experimental pH and of f_{H^+} from the EDHL equation. The concentrations of HA and A^- are determined from stoichiometry. The activity of A^- can be approximated when a value for f_{A^-} is calculated from the EDHL equation. f_{HA} is taken as unity.

To solve for pK_a we write

$$pK_a = pH + \log \frac{f_{HA}[C_{HA} - (a_{H^+}/f_{H^+})]}{f_{A^-}[C_{A^-} + (a_{H^+}/f_{H^+})]}$$

After 25 mL of NaOH solution has been added (50% titrated), $C_{HA} = C_{A^-} = 0.01667M$.

The activity coefficients of H^+ and A^- are estimated as follows

$$-\log f_{A^-} = \frac{0.51(\mu)^{1/2}}{1 + 0.329(\mathring{a})(\mu)^{1/2}} = 0.052 \quad (\mathring{a} = 6, \mu = 0.01667)$$

$$f_{A^-} = 0.89$$

$$-\log f_{H^+} = 0.048 \quad (\mathring{a} = 9, \mu = 0.01667)$$

$$f_{H^+} = 0.90$$

We then use the experimental pH (2.20) ($a_{H^+} = 6.31 \times 10^{-3}$) to calculate pK_a:

$$pK_a = 2.20 + \log \frac{1.0[0.01667 - (6.31 \times 10^{-3}/0.90)]}{0.89[0.01667 + (6.31 \times 10^{-3}/0.90)]}$$
$$= 2.20 - 0.34 = \underline{1.86}$$

If we had ignored the strength of the weak acid and assumed that the formal and equilibrium concentrations of the acid and base forms were identical, we would have substantially overestimated the pK_a.

6.5.4 Titration of a Weak Base with a Strong Acid

Weak base/strong acid (WB/SA) titrations can be viewed as mirror images of WA/SB titrations; in a WA/SB titration a weak acid is converted to its conjugate base, while in a WB/SA titration a weak base is converted to its conjugate acid. Figure 6.9 shows the experimental curve for the titration of the weak base tris(hydroxymethyl)amino methane (THAM), with HCl at the $0.1F$ level. The mirror-image relationship to the WA/SB case can be appreciated if one holds a mirror on a vertical line at 100% titrated. The initial solution in a WB/SA titration contains only weak base; its pH can be calculated from a hydroxide ion balance and equations developed in Chapter 5. Up to the equivalence point, the solution contains a mixture of weak base and conjugate acid produced by reaction with HCl. At the 50% titrated point K_a can be calculated, just as in the WA/SB case. Since K_w is always known, K_b can also be calculated. The equivalence point pH is determined by K_a and the concentration of the conjugate weak acid.

An indicator for a particular WB/SA titration must have a pH color transition range close to the pH of a solution of the conjugate weak acid. This pH depends on dilution which occurs during the titration. Bromcresol green is a fine indicator for the titration of a base with $pK_b = 8$ with HCl at the $0.1F$ concentration level.

Think about how dilution will affect the shape of the curve in Fig. 6.9. How would dilution affect your choice of an indicator?

6.5.5 Titrations of Polyprotic Weak Acids

The ideas just presented for the treatment of data from WA/SB and WB/SA titrations are easily extended to polyprotic WA/SB and polybasic WB/SA titrations. Once again, it will be to our advantage to think of titrations as sequences of changes in the concentrations of solutions of acids or bases.

6.5 ACID-BASE TITRATION CURVES

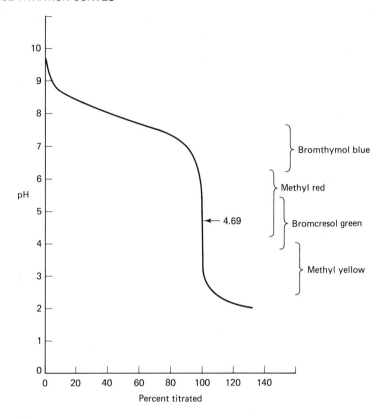

Figure 6.9 Titration of THAM with hydrochloric acid: $(HOCH_2)_3$-$CNH_2 + H^+ \rightarrow (HOCH_2)_3CNH_3^+$. Initial concentrations: $0.0635F$ THAM (50-mL volume), $0.1120M$ HCl. Ionic strength at 50% = $0.0253M$, at 100% = $0.0419M$.

Figure 6.10 shows the experimental curve for the titration of 50 mL of $0.0432F$ malonic acid with $0.1427F$ NaOH. Two distinct endpoint breaks are seen. The first break (pH $\cong 4.1$)

$$\underset{HO}{\overset{O}{\underset{\|}{C}}}-CH_2-\underset{OH}{\overset{O}{\underset{\|}{C}}}$$

malonic acid

(GFW 104.06, $pK_{a1} = 2.84$, $pK_{a2} = 5.66$)

corresponds to the titration of the diprotic weak acid ("H_2Mal") to form the conjugate

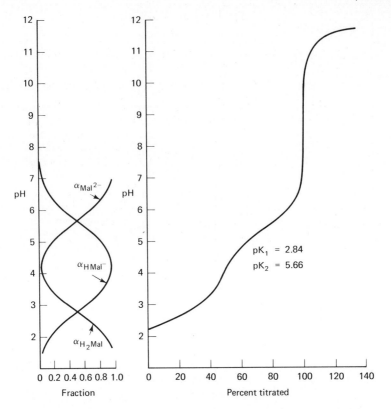

Figure 6.10 Titration of malonic acid with sodium hydroxide. Initial concentrations: 0.0432F malonic acid (volume 50 mL) and 0.1427M NaOH. The plot of pH vs. fraction of species shows that the fraction of HMal$^-$ is about 0.95 at the 50% titrated point and the fractions of the other two forms are small. The dominance of the intermediate species is characteristic of diprotic acids for which K_1/K_2 is greater than about 500.

base (also a weak acid, "HMal$^-$"). The second break (pH $\cong$ 8.6) signals the completion of the reaction of HMal$^-$ to form the weak base "Mal^{2-}." Distinct endpoint breaks are observed in systems in which successive dissociation constants differ by a factor of about 500, that is, when successive pK_a values differ by a factor of about 2.7.

There are four points of interest in Fig. 6.10. For simplicity, let us agree that a fraction titrated value of unity corresponds to the completion of the full two-step reaction,

$$2OH^- + H_2Mal \rightarrow Mal^{2-} + 2H_2O$$

and that the fraction 0.5 corresponds to the completion of only the first step,

6.5 ACID-BASE TITRATION CURVES

$$OH^- + H_2Mal \rightarrow HMal^- + H_2O$$

The solutions at the four points of interest are described as follows:

1. At 25% titrated (7.57 mL of 0.1427M NaOH). Here the solution is an equiformal mixture of H_2Mal and $HMal^-$. If H_2Mal is a weak enough acid, its first pK_a can be approximated just as was the pK_a of acetic acid. The concentration is given by

$$C_{H_2Mal} = C_{HMal^-} = \frac{(0.0432)(50.0)(.500)}{50.0 + 7.57} = 0.0188F$$

2. At 50% titrated (15.14 mL of 0.1427M NaOH). At this point the principal species in solution is $HMal^-$, at a formal concentration of

$$\frac{(0.0432)(50.0)}{50.0 + 15.14} = 0.0332F$$

In the previous chapter the ascorbic acid system was used to show that at this point the experimental pH can be approximated by the average of the successive pK_a values of the weak acid.

3. At 75% titrated (22.71 mL of 0.1427M NaOH). Here the solution is an equiformal mixture of $HMal^-$ and Mal^{2-},

$$C_{HMal^-} = C_{Mal^{2-}} = \frac{(0.0432)(50.0)(.500)}{50.00 + 22.71} = 0.0149F$$

If $HMal^-$ is a sufficiently weak acid, pK_{a2} can be calculated from the experimental pH by the simple method used to calculate pK_a for acetic acid from the experimental pH of an acetic acid buffer solution.

4. At 100% titrated (30.27 mL of 0.1427M NaOH). The solution at this point contains only the weak base, Mal^{2-},

$$C_{Mal^{2-}} = \frac{(0.0432)(50.0)}{50.00 + 30.27} = 0.0269F$$

The calculations of pK_{a1} and pK_{a2} from the experimental data are left for you to do as an exercise. The literature values are pK_{a1} = 2.84 and pK_{a2} = 5.66. Since H_2Mal is a rather strong weak acid, the corrected form of the Henderson-Hasselbalch equation used in Example 6.3 is recommended. Use the 25 and 75% titrated data points and be sure to correct for ionic strength. The author's calculations gave pK_{a1} = 2.78, pK_{a2} = 5.66.

Determination of the successive pK_a values of a polyprotic acid presents special computational problems when the pK_a values are close. In the case of malonic acid, the pK_a's differ by about 2.8 units, and so the two dissociation steps can be dealt with independently: when studying the first equilibrium step, $H_2Mal = HMal^- + H^+$, we can neglect $[Mal^{2-}]$, a part of the second equilibrium step. However, in the case of succinic acid, the pK_a values differ by only 1.34 units, and the effect of succinate ion should be considered in the region where succinic acid and bisuccinate ion dominate (around $pH = pK_{a1}$). Figure 6.11 shows the effect that close pK_a values have on the appearance of a titration curve. The titration curves of malonic acid and succinic acid are shown along with plots of species fraction vs. pH, tilted on end to show how the loss of a curve inflection point follows the increasing complexity of solutions when pK_1 and pK_2 are similar.

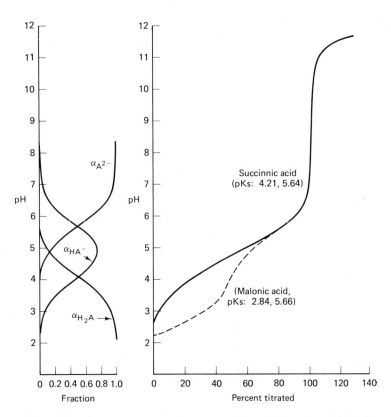

Figure 6.11 Titration of succinic acid with sodium hydroxide. Initial concentrations: $0.0454F$ succinic acid (volume 50 mL) and $0.1427M$ NaOH. The malonic acid titration curve break is drawn with dashes. The plot of pH vs. fraction of species shows that all three forms of succinic acid (H_2A, HA^-, A^{2-}) are present in substantial quantities at the 50% titrated point. In the case of malonic acid, the intermediate (monoprotonated) species was the only form present in substantial quantities at this point.

6.5 ACID-BASE TITRATION CURVES

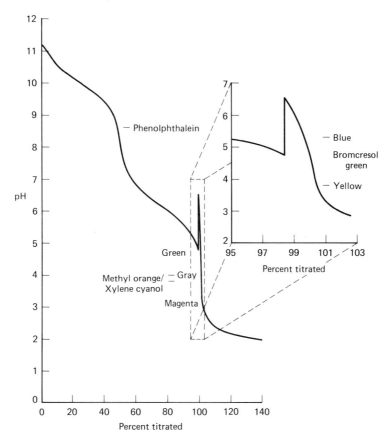

Figure 6.12 Titration of sodium carbonate with hydrochloric acid. Initial concentrations: $0.0439F$ Na_2CO_3 (50 mL volume) and $0.1220M$ HCl.

The best strategy to use in calculating close pK_a values is once again to use experimental pH values from two parts of the curve (25 and 75% titrated, as recommended for malonic acid), but to relate the values to equations which involve *both* K_1 and K_2. The equations can be derived from proton balance expressions and appear in Appendix V.

6.5.6 Titration of a Polybasic Weak Base with a Strong Acid

Figure 6.12 shows an experimental curve for the titration of sodium carbonate (Na_2CO_3) with hydrochloric acid at the $0.1F$ concentration level. In this titration, the first endpoint break occurs when the reaction

$$CO_3^{2-} + H^+ = HCO_3^-$$

is complete. The second break signals the final conversion of bicarbonate ion to carbonic acid [and to $CO_2(g)$]

$$HCO_3^- + H^+ = H_2CO_3 = H_2O + CO_2(g)$$

The fact that the concentration of carbonic acid depends on the solubility of gaseous CO_2 creates an interesting complication not found in other systems. In the region between the two endpoint breaks, the chemical system governing pH is the weak acid dissociation,

$$H_2CO_3 = H^+ + HCO_3^- \qquad pK_{a1} = 6.35\dagger$$

As more and more acid is added, the reservoir of H_2CO_3 builds, and the solubility of CO_2 is gradually exceeded. Carbon dioxide gas begins to bubble out of solution as the system tries to reequilibrate. The pH of the solution thus falls when HCl titrant is added and then rises gradually as CO_2 bubbles form. The acid-base indicator bromcresol green changes color from blue to green at the endpoint but then reverts to blue as CO_2 leaves the solution. To avoid the slow color change, we can simply boil the solution when the titration is just about complete (about pH 5) to drive off CO_2. The pH jumps to about 6.5, and the titration can be completed with a sharp endpoint. With bromcresol green indicator, one titrates to a green color (pH $\cong$ 4.8), boils the solution (the indicator turns blue again), and titrates to permanent yellow.‡ An alternative is to use a mixed (screened) indicator such as 0.10% methyl orange/0.14% xylene cyanole. The HCl titrant is added until the solution color changes from green to gray (pH 3.8). The solution is boiled to expel CO_2, then cooled to room temperature to continue the titration to the acid color, magenta.

Mixtures of Na_2CO_3, $NaHCO_3$, and NaOH We can create some interesting variations on the titration of Na_2CO_3 with HCl by adding either strong base (e.g., NaOH) or weak base (e.g., $NaHCO_3$). We can determine the components of the mixture of bases by observing the two endpoint breaks separately. An indicator such as phenolphthalein might be used to monitor the first break (pH $\cong$ 8.3), while bromcresol green can be used for the second. The volume of titrant used to reach each endpoint could then be used to determine the composition of the sample. The possible experimental outcomes are sketched in Fig. 6.13.

---

† Notice that carbonic acid is a rather weak acid, far weaker than either acetic or benzoic acid.

‡ The boiled solution need not even be cooled because the pH range of bromcresol green is not very temperature-sensitive. The ranges of methyl yellow, methyl orange, and the screened indicator methyl orange/xylene cyanole are far more temperature-sensitive, and solutions must be cooled when they are used.

6.5 ACID-BASE TITRATION CURVES

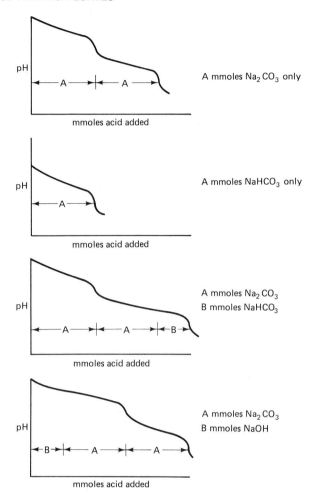

Figure 6.13 Titration curves for possible mixtures of Na_2CO_3, $NaHCO_3$, and NaOH. Other combinations react chemically to produce one of these mixtures. For example, A moles of NaOH plus B moles of $NaHCO_3$ produce A moles of Na_2CO_3 and leave $(A - B)$ moles of $NaHCO_3$ (if $A < B$).

EXAMPLE 6.6

A 10.00-mL portion of a sample containing Na_2CO_3 and either NaOH or $NaHCO_3$ requires 15.05 mL of $0.1010M$ HCl to reach the phenolphthalein endpoint. A second 10.00-mL portion requires 42.22 mL of the acid to reach the bromcresol green endpoint. How many millimoles of each species were present in a 10.00-mL portion?

Solution

More than twice as much acid is needed to reach the bromcresol green endpoint than was needed to reach the phenolphthalein endpoint. The titration curve is like case 3 in Fig. 6.13.

If the sample were pure Na_2CO_3, 2(15.05) = 30.10 mL of HCl would be needed to reach the bromcresol green endpoint. The extra acid, 42.22 − 30.10 = 12.12 mL, is used to titrate the $NaHCO_3$ present at the outset. Therefore

$$\text{millimoles } Na_2CO_3 = (15.05)(0.1010) = 1.520 \text{ mmole}$$
$$\text{millimoles } NaHCO_3 = (12.12)(0.1010) = 1.224 \text{ mmole}$$

If NaOH had been present, the bromcresol green endpoint would have required *less* than twice the volume needed to reach the phenolphthalein endpoint.

In actual laboratory work, the first equivalence point break is not large enough to give a sharp color change with phenolphthalein or any other indicator. It helps to titrate until the color matches that of the indicator in a solution of sodium bicarbonate of about the same concentration as the sample being titrated. Even so, errors on the order of 1% must be expected.

A more reliable method for determining carbonate and bicarbonate in a sample involves a back titration. The first step is to titrate the sample with HCl all the way to the pH 4 endpoint. Next, add an excess of standard strong base to a second (identical) portion of the sample to convert the original bicarbonate to carbonate. Then add a large excess of barium chloride to precipitate barium carbonate ($BaCO_3$), and titrate the excess strong base to the phenolphthalein endpoint.

EXAMPLE 6.7

A 0.8768-g sample containing Na_2CO_3 and $NaHCO_3$ is dissolved in 100 mL of deionized water. A 25.00-mL aliquot requires 29.18 mL of 0.1010M HCl for titration to the bromcresol green endpoint. Next, 25.00 mL of 0.09182M NaOH is pipetted into a second 25-mL aliquot of the sample and 1 g of $BaCl_2 \cdot 2H_2O$ is added to precipitate the carbonate. For back titration of excess base to a phenolphthalein endpoint, 11.85 mL of 0.1010M HCl is required. Calculate the millimoles of bicarbonate and carbonate present in the original sample.

Solution

The first step converts all the carbonate *and* bicarbonate to carbonic acid. Therefore,

$$2 \text{ mmole } Na_2CO_3 + \text{mmole } NaHCO_3 = \text{mmole HCl}$$
$$= 29.18 \text{ mL } (0.1010 \text{ mmole/mL})$$
$$= 2.947 \text{ mmole}$$

Next, all the bicarbonate in a second *identical* sample is converted to carbonate. The millimoles of bicarbonate is given by the difference between the millimoles of standard base added and the millimoles of HCl used in the back titration:

$$\text{mmole NaHCO}_3 = \text{mmole NaOH} - \text{mmole HCl}$$
$$= (25.00 \text{ mL})(0.09182 \text{ mmole/mL}) - (11.85 \text{ mL})(0.1010)$$
$$= 2.296 - 1.197 = \underline{1.099} \text{ mmole NaHCO}_3$$

Knowing this, we can calculate the millimoles of Na_2CO_3:

$$\text{mmole Na}_2\text{CO}_3 = \frac{2.947 - 1.099}{2} = 0.924 \text{ mmole}$$

The original 100-mL sample contained

$$4(1.099) = 4.396 \text{ mmole NaHCO}_3$$
$$4(0.924) = 3.70 \text{ mmole Na}_2\text{CO}_3$$

6.6 APPLICATIONS

A great many compounds can be determined by titration with an acid or base. Analyses for carbon, nitrogen, sulfur, phosphorus, and the halogens can be performed with a quantitative acid-base titration as the last step. Table 6.4 lists the most important examples.

TABLE 6.4 ELEMENTAL ANALYSES USING ACID-BASE TITRATIONS

Element	Form	Trap reactions	Titrate	Titrant
N	NH_3[a]	$NH_3 + HCl = NH_4^+ + Cl^-$	Excess HCl	Standard NaOH
S	SO_2[b]	$SO_2 + H_2O_2 = H_2SO_4$	H_2SO_4	Standard NaOH
C	CO_2[b]	$CO_2 + Ba(OH)_2 = BaCO_3 + H_2O$	Excess $Ba(OH)_2$	Standard HCl
Cl(Br)	HCl	$HCl + H_2O = H_3O^+ + Cl^-$	H_3O^+	Standard NaOH
P	PO_4^{3-}	$(NH_4)_3PO_4 \cdot 12MoO_3$ ppt[c]	Excess NaOH	Standard HCl

[a] See Kjeldahl procedure for details.
[b] Combustion in stream of oxygen.
[c] Precipitate PO_4^{3-} with MoO_3 as ammonium phosphomolybdate at 40°C. Wash crystals, dissolve in excess NaOH, titrate with standard HCl. Reaction:

$$(NH_4)_3PO_4 \cdot 12MoO_3 + 26OH^- = 12MoO_4^{2-} + HPO_4^{2-} + 3NH_3 + 11H_2O$$

A wide variety of organic compounds can also be determined by acid-base titrimetry. Carboxylic acids, amino acids, sulfonic acids, and many amines can all be determined directly, while esters, ketones, aldehydes, and alcohols require intervening chemical reactions to form acids or bases. Saponification (hydrolysis) of esters and acetylation of alcohols are slow reactions, and so a good strategy is to perform the reactions in a known (excess) amount of reagent, and then back titrate with standard acid or base after the reactions are complete.

EXAMPLE 6.8 Back Titration

A 10.00-mL portion of a solution containing the ester ethyl acetate is pipetted into a round-bottom flask containing exactly 50 mL of $0.2378M$ KOH. A reflux condenser is attached and the flask is heated for 30 minutes to hydrolyze the ester completely:

$$CH_3CH_2OCOCH_3 + OH^- = CH_3CH_2OH + CH_3COO^-$$

The unreacted hydroxide is then titrated to the phenolphthalein endpoint with 32.75 mL of $0.3172M$ HCl. Calculate the number of millimoles of ethyl acetate in the 10.00 mL sample.

Solution

There were 50.00 mL(0.2378 mmole/mL) = 11.89 mmole of base present initially. After the reaction there were 32.75 mL(0.3172 mmole/mL) = 10.38 mmole of base unreacted. The difference, 1.51 mmole, corresponds to the amount of ethyl acetate present in the sample.

Do you understand why phenolphthalein indicator was used rather than bromcresol green?

Table 6.5 lists the most popular acid-base methods for organic groups.

6.6.1 Determination of Nitrogen by the Kjeldahl Method

The Kjeldahl method for nitrogen is an acid-base method which deserves special attention because of its great importance in the routine analysis of protein in grain, flour, animal feed, and meat. The general scheme is first to oxidize an organic sample in hot sulfuric acid, converting nitrogen to ammonium ion, carbohydrates to CO_2 and H_2O, and sulfur to SO_2 or sulfate. Next an excess of strong base is added to the reaction flask, the solution is heated, and ammonia is distilled off. The ammonia is trapped in a receiving flask and titrated.

6.6 APPLICATIONS

TABLE 6.5 FUNCTIONAL GROUP ANALYSES USING ACID-BASE TITRATIONS[a]

Group		Reaction	Titrant
Acid	RCOOH	Direct neutralization	NaOH (phenolphthalein)
	RSO$_3$H(sulfonic)	Direct neutralization	NaOH (phenolphthalein)
Amine	RNH	Direct neutralization[b] (nonaqueous solvent if pK_b > 10)	HCl
Ester	RCOOR'	Saponify (hydrolyze) with excess base	HCl (excess base)
Ketone aldehyde	RR'CO (R' = H)	$R_2CO + NH_2OH \cdot HCl \xrightarrow{\text{reflux}}$ $R_2C=NOH + HCl$ (oxime) (excess)	Standard NaOH
Alcohol	ROH	$ROH + (OAc)_2O \rightarrow ROAc + HOAc$[c] $(OAc)_2O + H_2O = 2HOAc$	Standard NaOH

[a] A variety of classical methods are discussed in Ref. 7.
[b] Aromatic amines are generally too weak to be titrated in water. Glacial acetic acid, a stronger acid than water, can be used as a solvent for many amines.
[c] An alcohol is heated with a known excess of acetic anhydride in pyridine solvent. Water is then added to hydrolyze the unreacted anhydride to acetic acid, which is then titrated with NaOH in ethanol. Alcohol is determined by difference. Amines will interfere.

The digestion process is a critical part of the analysis. Special care must be taken to make sure that all the nitrogen in the sample is converted to ammonium ion. Nitrogen present in low-oxidation-state forms (such as amines and amides) is quickly and easily converted to ammonium ion. However, nitrogen in the +5 state (nitrate) or +3 state (nitrite), as well as hydrazines (R—NHNH$_2$) and azo (R—N=N—R) compounds, is apt to be converted into either N$_2$ or nitrogen oxides. Since these products are volatile gases, they will not be held by the sulfuric acid and will be lost, yielding low results. This source of error can be avoided if, prior to digestion with sulfuric acid, the sample is treated with a reducing agent to convert the nitrogen to lower-oxidation-state forms. Nitrate and nitrite are reduced readily by a powdered metal alloy called Devarda's alloy (50% Cu, 45% Al, 5% Zn), while azo compounds can be reduced by heating with stannous chloride in hydrochloric acid. Hydrazines can be reduced with formic acid, zinc metal dust, and hydrochloric acid.

Digestion in sulfuric acid may be quite slow, particularly if pyridines or other oxidation-resistant compounds are present. A variety of catalysts have been proposed to speed the digestion: mercuric oxide, mercury metal, copper, and selenium are all effective. If mercury compounds are used, sodium sulfide must be added to the solution after digestion to precipitate HgS; mercuric ions complex ammonia and may give low results. Another strategy to hasten digestion is to add potassium sulfate to raise the boiling point of sulfuric acid and thus the temperature of the oxidation reaction. It is possible to make the temperature too high, however, and loss of nitrogen in higher oxidation states may be the cost of greater speed in digestion.

The distillation process, while involving much simpler chemistry, requires great care. A typical distillation apparatus is shown in Fig. 6.14. The long-necked reaction flask, called a **Kjeldahl flask**, is used for both digestion and distillation. When the digestion is complete the flask and solution are cooled, water is added, and concentrated sodium hydroxide is trickled down the inside of the neck. The dense hydroxide forms a layer on the bottom of the flask, and very little ammonia is evolved. The contents of the flask are then swirled gently and the flask is attached quickly to the spray trap of the distillation apparatus. The solution is then heated to drive out ammonia, which is trapped in a receiving solution. The spray trap keeps sodium hydroxide from contaminating the receiving solution.

There are two methods used to trap and determine ammonia. In the first method the receiver contains a known amount of standard acid (HCl). After the distilled ammonia has been trapped the excess standard acid is back-titrated with standard base. Since the solution in the trap contains both a strong acid (HCl) and a weak acid (NH_4^+), an indicator such as methyl red is used to detect the HCl endpoint before NH_4^+ is titrated. Notice that this method requires two standard solutions.

A more convenient method is to trap the ammonia in a solution containing an unknown excess of boric acid. The trap reaction is

$$HBO_2 + NH_3 = NH_4^+ + BO_2^-$$

The borate ion produced in this reaction is a weak base (pK_b 4.8) and is titrated with standard hydrochloric acid. The solution at the equivalence point contains ammonium ion and boric acid; the endpoint is conveniently indicated by bromcresol green.

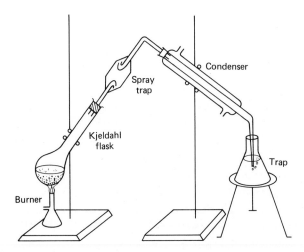

Figure 6.14 Apparatus for Kjeldahl determination of nitrogen. Apparatus shown for distillation step. During digestion the Kjeldahl flask is upright and not connected to the spray trap and condenser.

REFERENCES

1. W. J. Blaedel and V. W. Meloche, *Elementary Quantitative Analysis,* 2nd ed., Harper & Row, New York, 1963, Fig. 18.7, p. 360.
2. A. K. Covington and A. J. Utting, *Analyst,* 105, 470 (1980).
3. R. A. Butler and R. G. Bates, *Analytical Chemistry,* 48, 1669 (1976).
4. W. F. Koch, D. L. Biggs, and H. Diehl, *Talanta,* 22, 637 (1975).
5. W. F. Koch, W. C. Hoyle, and H. Diehl, *Talanta,* 22, 717 (1975).
6. R. W. Ramette, *Chemical Equilibrium and Quantitative Analysis,* Addison-Wesley, Reading, Mass., 1981, chapter 11.
7. J. Mitchell, Jr., I. M. Kolthoff, E. S. Proskauer, and A. Weissberger, *Organic Analysis,* vols. 1–4, Interscience, New York, 1953–1960.

PROBLEMS

6.1. Describe how you would prepare each of the following solutions:
  (a) 100 mL of $0.07537 M$ HCl from a stock of $1.5073 M$ HCl
  (b) one liter of $0.10 M$ NaOH from NaOH pellets
  (c) one liter of $0.10 M$ NaOH from 50% NaOH [specific gravity (sp. gr.) 1.5253]
  (d) 500 mL of $0.10 M$ $HClO_4$ from 70% $HClO_4$ (sp. gr. 1.674)
  (e) 250 mL of a solution whose hydroxide concentration is $0.0105 M$, from solid $Ba(OH)_2$

6.2. How would you prepare 500 mL of $0.10 F$ solutions of the following acids and bases, given the specified starting forms:
  (a) HCl (conc. HCl, 38% wt/wt, sp. gr. 1.189)
  (b) acetic acid (glacial CH_3COOH, 100%, sp. gr. 1.0498)
  (c) ammonia (20% wt/wt, sp. gr. 0.9229)
  (d) NaOH (50% wt/wt, sp. gr. 1.5253)

6.3. What weight of each of the following acids should be taken so that 40 mL of $0.1000 M$ NaOH will be required for titration?
  (a) potassium biphthalate (GFW 204.23)
  (b) potassium sulfosalicylate (GFW 550.655)
  (c) sulfuric acid (conc. $1.55 M$, sp. gr. 1.095)
  (d) citric acid (GFW 192.12, triprotic)

6.4. What weight of each of the following bases should be taken so that 40 mL of $0.1000 M$ HCl will be required for titration?
  (a) sodium benzoate (GFW 144.11)
  (b) sodium carbonate (GFW 106.00)
  (c) potassium binoxalate (GFW 128.13)
  (d) sodium tetraborate

6.5. Calculate the equivalent weights of the following acids in reactions with sodium hydroxide titrant:
  (a) $HClO_4$
  (b) H_2SO_4
  (c) $(COOH)_2$ (oxalic acid)
  (d) H_3PO_4

6.6. Calculate the equivalent weights of the following bases in reactions with hydrochloric acid:
  (a) $CaCO_3$
  (b) NaH_2PO_4
  (c) Na_2HPO_4
  (d) $CH_2ClCOO^-Na^+$
  (e) $NH_2CH_2COO^-Na^+$ (sodium glycinate—will form protonated cation)

6.7. How would you prepare each of the following solutions?
  (a) 100 mL of approximately $0.12N$ H_2SO_4 from $1.20M$ H_2SO_4
  (b) 250 mL of $0.250N$ KOH from KOH pellets
  (c) 500 mL of $0.173N$ citric acid (GFW 192.12)
  (d) 100 mL of $0.05N$ Na_2HAsO_4 for titration with strong acid
  (e) 50 mL of $0.010N$ $Ba(OH)_2$ from anhydrous $Ba(OH)_2$
  (f) 1.00 liter of $0.05N$ Na_2CO_3 for titration with strong acid

6.8. 208.2 mg of primary standard grade sodium carbonate react with 21.72 mL of a solution of HCl. Calculate the normality of the HCl solution.

6.9. A $0.03375N$ solution of HCl is used to determine the concentration of a solution of barium hydroxide. A 13.31-mL portion of the HCl solution is required to titrate 25.00 mL of the $Ba(OH)_2$ solution.
  (a) Calculate the normality of the $Ba(OH)_2$ solution.
  (b) How would you prepare the barium hydroxide solution from $Ba(OH)_2 \cdot 8H_2O$?

6.10. Sodium tetraborate, discussed in the text, is to be used to standardize a solution of HCl. What is the equivalent weight of sodium tetraborate for reactions with strong acids? How much would you weigh out to use about 40 mL of $0.10N$ HCl solution in a titration?

6.11. Calculate the pH at the 50% titrated point and the equivalence point in each of the following titrations (account for dilution):
  (a) $0.01F$ HCl with $0.01F$ NaOH
  (b) $0.01F$ chloroacetic acid ($pK_2 = 2.87$) with $0.01F$ NaOH
  (c) $0.01F$ sodium chloroacetate with $0.01F$ HCl
  (d) $0.05F$ succinic acid ($pK_1 = 4.21$, $pK_2 = 5.64$) with $0.075F$ NaOH
  (e) $0.05F$ Na_2CO_3 with $0.05F$ HCl (to HCO_3^- only) ($pK_{a1} = 6.35$, $pK_{a2} = 10.33$)

6.12. What indicators would you choose to signal the equivalence points in the titrations in Problem 6.11?

PROBLEMS

6.13. The titration of a diprotic weak acid with NaOH is monitored with a glass electrode and only one equivalence point break is observed. The pH at 50% titrated is 3.25. Within what limits will the pK_1 and pK_2 lie?

6.14. How large an error is introduced by using a methyl red endpoint as opposed to a phenolphthalein endpoint in the titration of 50.00 mL of 0.0200M acetic acid ($pK_a = 4.76$) with 0.0200M NaOH?

6.15. Give examples of titrations in which each of the following indicators could be used:
  (a) methyl violet
  (b) bromthymol blue
  (c) methyl yellow
  (d) alizarin yellow

6.16. A 0.1% solution of methyl yellow in 90% ethanol has an indicator color range of 2.9 (red) to 4.0 (yellow). The pK_a of methyl yellow is 3.2. What do these facts suggest about the "inherent intensities" (molar absorptivities) of the acid and base forms?

6.17. A 0.1257-g sample of pure aniline ($C_6H_5NH_2$, $pK_b = 9.38$) is dissolved in 50.0 mL of 0.03947M HCl. Describe the titration curve you would expect if you titrated the solution with 0.05M NaOH.

6.18. Adipic acid, $HOOC(CH_2)_4COOH$, is a weak diprotic acid, $pK_1 = 4.41$, $pK_2 = 5.30$. Describe the titration curve obtained when 5.00 mmole of sodium adipate ("Na_2Adp") is titrated with 0.2000M HCl. What indicator would you choose to signal the endpoint?

6.19. Draw a log concentration diagram for the 0.033F adipic acid system, using pK_a values from Problem 6.18.
  (a) Locate the point on the log concentration diagram which corresponds to the equivalence point in the titration of 0.100F adipic acid with 0.100F NaOH. Is 99.9% of the adipic acid species in the form of the anion, Adp^{2-}, by this point?
  (b) Can you see why there is only one break in the titration curve? Calculate the fractions of various species at the point which should correspond to the first equivalence point break.
  (c) If you were to titrate 0.100F Na_2Adp with 0.100F HCl, at what pH would the equivalence point break(s) be? Recommend a possible indicator.
  (d) If you were to titrate 0.100F Na_2Adp with 0.100F HCl, would the titration be complete (99.9%) at the equivalence point?

6.20. Anilium ion has a pK_a of 4.60. Draw a log concentration diagram for the 0.05F anilium chloride system.
  (a) Using the diagram, locate the pH at the equivalence point in the titration of 0.100F anilium chloride with 0.100M NaOH.

(b) Locate or calculate the fractions of anilium ion and aniline at the equivalence point pH. Is the reaction quantitative at the equivalence point?

6.21. There are several possible titrimetric methods for the determination of the amino acid glycine. The protonated form of glycine, "H_2Gly^+," is a rather strong acid, $pK_a = 2.35$. Neutral glycine, "HGly," is much weaker, $pK_a = 9.78$.
(a) Sketch a curve for the titration of $0.100F$ glycine (HGly) with NaOH. Judging by the appearance of the curve, would you want to use this titration in an analysis?
(b) Confirm your judgment in part (a) by calculating the fraction of unreacted glycine remaining at the equivalence point pH.
(c) Sketch a curve for the titration of sodium glycinate ("NaGly") with NaOH at the $0.1F$ concentration level. Would this titration be suitable for an analytical method?

6.22. The diprotic acid suberic acid (1,6-hexanedioic acid) has successive pK_a values of 2.52 and 5.33. Consider a titration of $0.010F$ suberic acid with $0.0100M$ NaOH.
(a) Sketch the titration curve. Be sure to justify the number of equivalence point breaks you draw.
(b) Calculate the pH when the reaction

$$OH^- + HSu^- = Su^{2-} + H_2O$$

is 99.9% complete. Is this point the same as the equivalence point?
(c) Suggest an appropriate indicator for the titration reaction.

6.23. Piperidine ($C_5H_{11}N$) can be produced by electrochemical reduction of pyridine (C_5H_5N). Both pyridine and piperidine are weak bases, with pK_b values of 8.78 and 2.88, respectively. A 100-mL sample containing pyridine and no reducible impurities is partially reduced electrochemically. A 10-mL portion of the solution is removed by pipet and titrated with $0.05125M$ HCl titrant. 4.32 mL of HCl solution is required to reach the first equivalence point. An additional 12.78 mL is required to reach the second equivalence point. What is the composition of the solution after electrochemical reduction?

6.24. Sketch the titration curve you would expect to observe when 50.0 mL of a $0.0250M$ solution of sodium sulfite (Na_2SO_3) is titrated with $0.05000M$ HCl.
(a) What is the pH of the solution of sodium sulfite?
(b) What is the pH of the solution at the first equivalence point break?
(c) What is the pH of the solution when it is an equiformal mixture of HSO_3^- and SO_3^{2-}? What volume of titrant is required to reach this point?
(d) What is the pH of the solution when it is an equiformal mixture of HSO_3^- and H_2SO_3? What volume of titrant is required to reach this point?
(e) Is it possible to add enough HCl to convert 99.9% of the sulfite-containing species to H_2SO_3? Explain.

PROBLEMS

6.25. Bicarbonate ion is a very weak acid ($pK_a = 10.33$). Why would the determination of bicarbonate by titration with $0.1M$ NaOH be a poor method? Be sure to include indicators in your discussion.

6.26. Bicarbonate and carbonate in a sample of seawater are to be determined by titration with standard HCl solution. A 100-mL sample of water requires 1.55 mL of $0.04530M$ HCl for titration to the phenolphthalein endpoint and 18.34 mL of the same acid to titrate all the way to the bromcresol green endpoint. Calculate the concentrations of carbonate and bicarbonate in milligrams per 100 mL.

6.27. 15.25 mL of $0.100F$ H_3PO_4 is added to 50.00 mL of a $0.1250F$ solution of Na_3PO_4. Describe the curve you would expect to obtain when titrating the resulting solution with $0.2000M$ HCl. (H_3PO_4: $pK_1 = 2.15$, $pK_2 = 7.20$, $pK_3 = 12.38$)

6.28. 1.50 mmole of NaOH is added to 1.25 mmole of $NaHCO_3$, diluted to 50 mL with deionized water and titrated with $0.1M$ HCl. Describe the titration curve you would expect to observe.

6.29. A 25.00-mL aliquot of a 500-mL solution containing Na_2CO_3 and either NaOH or $NaHCO_3$ requires 21.63 mL of $0.1023M$ HCl for titration to the phenolphthalein endpoint. A second 25.00-mL aliquot of the solution requires 35.34 mL of the same acid titrant for titration to the bromcresol green endpoint. What is the composition of the original solution?

6.30. A 10.00-mL aliquot of a solution containing H_3PO_4 and either HCl or NaH_2PO_4 is titrated with $0.1005M$ NaOH. The titration is monitored with a glass electrode, and two equivalence point breaks are seen. The first break, at pH 4.6, occurs when 8.75 mL of base has been added. The second break (pH 9.9) occurs when 24.67 mL has been delivered. How many millimoles of each species are present in the 10-mL aliquot?

6.31. A 100-mL solution containing Na_3PO_4 and either NaOH or Na_2HPO_4 was titrated with $0.2042M$ HCl. The first endpoint break was observed at pH 9.9 after 22.72 mL of titrant was added. The second break, at pH 4.6, was observed after an additional 30.95 mL of titrant was added. What is the composition of the solution?

6.32. Nitrilotriacetic acid (NTA) is a triprotic acid: $pK_1 = 1.65$, $pK_2 = 2.95$, $pK_3 = 10.28$. Describe the titration curve you would observe if you titrated 25 mL of $0.05F$ NTA with $0.05F$ NaOH:
  (a) How many equivalence point breaks would you observe?
  (b) At what points in the titration would you observe breaks?
  (c) Describe in detail the calculation you would perform to determine the purity of a sample of NTA.

6.33. The percentage of crude protein in flour and wheat is estimated by multiplying the percentage of nitrogen by the factor 5.7. (Different materials have different factors.) A 2.453-g portion of flour is digested in a Kjeldahl flask. After treatment with NaOH, ammonia is distilled and collected in a trap containing 100.0 mL of 0.01120M HCl. The solution in the trap requires 12.37 mL of 0.01600M NaOH for back-titration. Calculate the percentage of crude protein in the flour.

6.34. A sample containing NH_4Cl and NH_4NO_3 and inert material is to be analyzed. A 2.450-g portion of the sample is dissolved in deionized water, and a one-tenth aliquot is transferred to a distillation apparatus. Concentrated NaOH is added, the solution is heated, and NH_3 is driven off into a trap containing 1.800 meq of HBO_2. The contents of the trap require 23.08 mL of 0.04892M HCl for titration to a methyl orange endpoint. A second one-tenth aliquot of the original sample is treated with Devarda's alloy in a Kjeldahl flask. After distillation of NH_3 into a trap containing 1.920 meq of HBO_2, 34.81 mL of the standard HCl solution is required to titrate to the methyl orange endpoint. Calculate the percentages of NH_4Cl and NH_4NO_3 in the sample.

6.35. A 1.784-g sample of pure ester is saponified by heating in 25.0 mL of an ethanol-KOH solution. After an hour of heating, the solution requires 13.24 mL of 0.4132M H_2SO_4 solution for titration to a bromthymol blue endpoint. A 25.0-mL ethanol-KOH blank requires 29.84 mL of the same standard acid. What is the equivalent weight of the ester?

6.36. A 5.0-g sample containing ethanol is acetylated in a closed container with acetic anhydride. The acetic acid produced in the reaction is subsequently titrated with 0.1020M NaOH. If 12.35 mL of base is required to titrate to a phenolphthalein endpoint, what is the concentration of ethanol in the sample in percent?

6.37. A sample contains ethanol, acetic acid, and acetaldehyde. Using acid-base titrations, design a scheme to analyze the mixture for each of the components.

6.38. Theophylline (GFW 180.17) is a drug used to relieve symptoms of asthma. It is too weak an acid to be determined by direct titration with sodium hydroxide. However, silver ion will displace hydrogen ion and precipitate silver theophyllinate:

$$C_7H_8N_4O_2 + Ag^+ = AgC_7H_7N_4O_2 + H^+$$

The hydronium ion formed can then be titrated with standard NaOH to a phenolphthalein endpoint.

A 327.2-mg tablet containing theophylline is dissolved in water, and excess silver nitrate is added. The acid released requires 14.62 mL of 0.03720M NaOH for titration. Calculate milligrams of theophylline present in the tablet.

6.39. Mercaptans (thioalcohols) can be determined in solvents such as benzene by equilibration with aqueous solutions of mercuric chloride. The reaction for ethyl mercaptan is

$$C_2H_5SH + HgCl_2 = C_2H_5SHgCl + H^+ + Cl^-$$

PROBLEMS

A 50-mL benzene solution containing ethyl mercaptan is transferred to a separatory funnel containing a similar volume of 1% mercuric chloride in water. The benzene and mercuric chloride solution are shaken and then allowed to separate. Three 50-mL portions of deionized water are then equilibrated with the benzene to strip out remaining HCl. The combined aqueous layers are titrated to a methyl red endpoint with 23.12 mL of 0.0537M NaOH. What is the molar concentration of ethyl mercaptan in the benzene?

6.40. Chloral hydrate will react with strong base to form chloroform and formate ion (HCOO$^-$), as follows:

$$Cl_3CCH(OH)_2 + OH^- = CHCl_3 + HCOO^- + H_2O$$

Because the reaction is too slow for direct titration, a measured excess of standard NaOH solution is added, the reaction is allowed to run its course, and finally unreacted base is back-titrated with standard hydrochloric acid solution.

A 25.00-mL portion of 0.09420M NaOH is added to a vessel holding a sample containing chloral hydrate. The vessel is sealed and set aside for 10 minutes. The back titration of unreacted OH$^-$ requires 12.43 mL of 0.1020M HCl. Calculate the number of milligrams of chloral hydrate originally present in the vessel.

6.41. The contents of an old bottle labeled "3% hydrogen peroxide" are to be assayed for hydrogen peroxide using a reaction with bromine and subsequent titration of hydronium ion:

$$H_2O_2 + Br_2 = 2H^+ + 2Br^- + O_2$$

Exactly 5 mL of the solution in the bottle is taken by pipet, treated with bromine, and allowed to react for 10 minutes. Excess bromine is then driven off, and the hydronium ion produced in the reaction is titrated with 0.2578M NaOH. To reach the endpoint 22.25 mL are required. Calculate the millimoles of peroxide present in the 5-mL aliquot and percent H$_2$O$_2$ (wt/vol) in the bottle.

EXPERIMENT 6.1: Titration of a Pure Weak Acid

Determination of Equivalent Weight and Dissociation Constant

A weak acid is characterized by two important properties, its equivalent weight and dissociation constant. Both properties can be determined from a titration in which solution pH is monitored as a function of the volume of standard base titrant. A hydronium ion-sensitive glass electrode is used to measure the pH of the solution during titration. The electrical potential developed between a glass electrode and a reference electrode (housed in the same body—a "combination electrode") is translated to a pH value by a pH meter and displayed to three significant figures. A titration curve is drawn by plotting the pH observed after each addition of standard base solution. Chapter 6 discusses the theory and practice of acid-base titrations, and Chapter 15 contains a detailed description of the glass electrode.

The equivalent weight of an acid is found from the volume of base required to reach the equivalence point in the titration. This point lies in the region of the titration curve where the pH changes most rapidly, and is approximated by the inflection point in the curve. The equivalent weight is related to the molecular weight of the acid in a very simple way: the equivalent weight is equal to the molecular weight divided by the number of hydrogen ions involved in the titration reaction. Since you will be working with an unknown weak acid, you will not know if it has one, two, or three reacting hydrogens. Therefore the only real information you can obtain from this particular titration is the equivalent weight. If you are given a diprotic weak acid, its dissociation constants will be so close that only one equivalence point break will be observed.

The dissociation constant of the weak acid can be determined from the pH at the midpoint of the titration (the 50% point). At this point the acid ("HA") and its conjugate base (A^-) will be the same concentration (50% of the total concentration, C_t, of weak acid species). Using the expression for the thermodynamic dissociation constant, we can substitute concentrations for activities as explained in the text:

$$K_a^0 = \frac{a_{H^+} a_{A^-}}{a_{HA}} = a_{H^+} * \frac{[A^-] f_{A^-}}{[HA] f_{HA}}$$

Since $[A^-] = 0.50\, C_t$ and $[HA] = 0.50 C_t$ at 50% titrated, the concentration terms divide out, leaving a ratio of activity coefficients,

$$K_a^0 = a_{H^+} * \frac{f_{A^-}}{f_{HA}} \quad \text{or} \quad pK_a = pH - \log \frac{f_{A^-}}{f_{HA}}$$

We use the glass electrode and pH meter to measure hydronium ion activity. Therefore,

EXPERIMENT 6.1: TITRATION OF A PURE WEAK ACID

if we can find a way to approximate the activity coefficients of A⁻ and HA, we can evaluate the thermodynamic constant.

There will be two main tasks in performing the experiment: standardizing the titrant (sodium hydroxide) and titrating the samples of unknown. Prepare your base as described below; then titrate samples of your unknown while the primary standard potassium biphthalate dries in the oven. Near the end of the lab period (or the next week) standardize the sodium hydroxide solution.

Procedure

1. Prepare and standardize a $0.1M$ solution of sodium hydroxide. Boil about 1 liter of deionized water to remove dissolved carbon dioxide. Carefully pour the hot water into a liter polyethylene bottle. Cap the bottle and chill the water to room temperature with ice. At some point during the cooling process add about 7 mL of clear 1:1 NaOH to the water, using a 10-mL graduated cylinder. Be very careful, as 50% NaOH is quite caustic and will cause burns. Mix the resulting solution well and keep the bottle capped. Do not use this titrant solution until it is at room temperature.

Dry about 4 g of primary standard potassium biphthalate (KHP, GFW 204.23) in a weighing bottle at 110°C for 2 hours. Cool and store the KHP in your desiccator. Weigh out as carefully as possible three samples of KHP (800 to 900 mg each) into 250-mL Erlenmeyer flasks. When you are ready to standardize, dissolve each KHP sample in about 50 mL of deionized water. Add about two or three drops of 0.1% phenolphthalein indicator, and titrate to the first permanent pink color with your NaOH titrant.

If you find yourself waiting for something to dry or cool, look up the pK_a of KHP and the color range of phenolphthalein, and check the wisdom of choosing phenolphthalein as the indicator.

2. Calibrate a pH meter with two buffer solutions available in the lab. Set the meter to 7.00 with the electrode dipped in the pH 7.00 buffer using the "STD" or "CALIBRATE" control on the front of the meter. Next, with the electrode dipped in the pH 4.01 buffer, set the meter to read 4.01 using the "SLOPE" control. If there is no slope control ask your instructor for help. Be sure to rinse off the electrode with deionized water whenever you move it from one solution to another.

3. Get an unknown from the stockroom. DO NOT DRY THE UNKNOWN. Do a trial titration to find out what weight of weak acid sample will be needed to consume between 30 and 40 mL of your titrant. Start by weighing out about 100 mg of your unknown and dissolving it in about 100 mL of deionized water. Place the freshly rinsed pH electrode in the solution; fill the buret to the 0.00-mL mark with standard base solution. Titrate using 1-mL volume increments until you see the sudden jump in pH which signals your trial endpoint. Read the volume delivered from the buret, and use it to calculate the approximate weight of weak acid samples you need.

Weigh by difference three samples of unknown weak acid into 250-mL Erlenmeyer flasks and dissolve each in 100 mL of deionized water. The volume of water used to dissolve each sample should be measured in a graduated cylinder; it will be important to know that the volume is 100 mL. Titrate each sample by adding base in 1-mL increments until you see the pH begin to move rather rapidly with each addition (about pH 6 for most samples). Near the equivalence point add smaller increments, say 0.2 mL. Increments of equal size are recommended; treatment of experimental data may be easier with equal small volumes (see below). Once you are 1 to 2 mL past the equivalence point you may add base in 1-mL increments again. Stop titrating when you are about 5 mL past the equivalence point. The object is to collect many data points close to the equivalence point. Fewer data points are needed away from the equivalence point.

4. Plot your titration curves on high-quality graph paper (10 lines/cm) and attach them permanently to your notebook. In Chapter 15 there is a BASIC program for a Savitsky-Golay derivative function. Use this program to take the derivative of your titration curve in order to see the equivalence point more clearly. Read the discussion of digital filter programs. Calculate the equivalent weight of your unknown weak acid from the equivalence point volume.

Locate the midpoints of the titration curves and calculate the ionic strength of each solution at that point. Use one of the activity coefficient equations in Chapter 3 to calculate f_{A^-}. If you use the extended Debye-Hückel equation, assume a Kielland radius of 6 for the anion. You may assume that the weak acid is *monoprotic*, and that the anion is singly charged. Calculate K_a^0 and report pK_a^0.

EXPERIMENT 6.2: The Titrimetric Determination of Sodium Carbonate (Soda Ash)

The determination of sodium carbonate by titration with standard acid solution is of interest for several reasons. Carbonate is an important component of limestone, natural waters, and the commercial material "soda ash" and is frequently encountered in commercial analyses. In addition, some experience with primary standard grade sodium carbonate is important to students of analytical chemistry because it is so widely used for preparing standard acid solutions.

Carbonate ion is a weak base ($K_b = 2.1 \times 10^{-4}$) and can accept a proton to form bicarbonate ion:

$$CO_3^{2-} + H^+ = HCO_3^-$$

Bicarbonate ion is an even weaker base than carbonate ion ($K_b = 2.2 \times 10^{-8}$) and can combine with another proton to form carbonic acid:

$$HCO_3^- + H^+ = H_2CO_3$$

EXPERIMENT 6.2: THE TITRIMETRIC DETERMINATION OF SODIUM CARBONATE

Carbonic acid can be in equilibrium with CO_2 gas:

$$H_2CO_3 = H_2O + CO_2(g)$$

In the titration of sodium carbonate with hydrochloric acid, carbonate is titrated all the way to CO_2 (two protons per carbonate). You will use the indicator bromcresol green, which is blue at pH values greater than 5.5, yellow at pH values below 4.0, and green in the region of pH 4.8. Direct titration to the yellow color would be possible if it were not for unfortunate fact that the solubility reaction for CO_2 in water is very slow to reach equilibrium, and a fading endpoint color change is observed along with bubbles of CO_2. Equilibration is hastened by heating samples to boiling just before the yellow color is reached. The CO_2 boils out, the color reverts to blue, and the subsequent blue-yellow color change is quite sharp. The change is so sharp that the intermediate green color may not be seen. A carbonate titration curve monitored with a pH electrode will be given out during the laboratory lecture; it shows the effect of dissolved CO_2 on solution pH near the equivalence point.

In this experiment you will have your choice of primary standard bases, sodium carbonate or tris(hydroxymethyl)aminomethane (THAM). THAM reacts as follows:

$$H_2NC(CH_2OH)_3 + H^+ = {}^+H_3NC(CH_2OH)_3$$

The gram formula weights of sodium carbonate and THAM are 105.99 and 121.14, respectively. The equivalent weights are 53.00 and 121.14.

Procedure

Place a little less than 1 liter of deionized water in a clean glass-stoppered 1-liter bottle. Add a predetermined volume of concentrated hydrochloric acid (12M HCl, in the fume hood) with a small graduated cylinder and mix the solution thoroughly.

> CAUTION: 12M HCl is dangerous. Wear eye protection and work in a fume hood. Do not breathe HCl vapors.

Dry about $1\frac{1}{2}$ g of primary standard grade sodium carbonate in a weighing bottle for 2 hours at 140°C. At this temperature any bicarbonate in the sample will be converted to carbonate, carbon dioxide, and water:

$$2NaHCO_3 = Na_2CO_3 + H_2O(g) + CO_2(g)$$

After cooling the material in a desiccator, weigh as accurately as possible three 0.2-g samples into 250-mL flasks. Dissolve each sample in about 50 mL of deionized water. Add about two or three drops of bromcresol green to each flask. Titrate with 0.1M hydrochloric acid until the solution color changes from blue to green. Be careful not to overshoot to a yellow solution. Boil the solution to expel CO_2, cool it to room temperature (use ice), and complete the titration by adding acid until the solution is yellow.

As an alternative you may wish to standardize your hydrochloric acid solution with THAM. Calculate the weight of THAM which will require 30 to 40 mL of 0.1M hydrochloric acid solution for titration. Do not dry the THAM provided; it has already been dried in a vacuum oven. Weigh and titrate three samples of THAM to the bromcresol green endpoint. Is boiling necessary?

Dry your unknown sample for 2 hours at 140°C. Weigh out three or more samples, each containing 0.25 to 0.35 g, into 250-mL flasks, and dilute as before. Titrate with standard acid to the endpoint as before.

Report the results as percent sodium carbonate, % Na_2CO_3 (range 20 to 50% Na_2CO_3).

Chapter 7

Titrations in Nonaqueous Solvents

7.1 INTRODUCTION

Up to this point the discussion of acid-base equilibria has involved only reactions in the solvent water. Water is by far the most common solvent used in analysis due to its abundance, the relative ease with which it can be purified, and its limited reactivity with atmospheric gases. Water has two additional properties which make it a highly desirable solvent: its ability to react as either an acid or a base (it is an amphiprotic compound) and its high dielectric constant. Water is a weaker acid than mineral acids (e.g., hydrochloric acid and sulfuric acid) and carboxylic acids (e.g., acetic acid). It is also a far weaker base than ammonia or substituted amines. As a result of its intermediate acidity, water can participate in a great number of acid-base reactions. In addition, the large dielectric constant of water allows it to dissolve ionic compounds and keep their ions separated in solution. While this property might not seem vital at first glance, it has allowed chemists largely to ignore such complicating factors as dimer and ion-pair formation when performing acid-base equilibrium calculations for aqueous systems.

There are certain limitations to the use of water as a solvent, particularly when used as a medium for the titration of very weak acids or very weak bases. As you will see shortly, many acids and bases simply cannot be determined quantitatively in aqueous solution. In this brief chapter we will discuss solvents which are useful alternatives to water. A quick review of the section of Chapter 4 which deals with Brønsted-Lowry theory and the "leveling effect" would be helpful before reading further.

7.2 THE CASE OF A VERY WEAK ACID

Phenol is a very weak acid in water,

[phenol structure: benzene ring with —OH]

$pK_a = 10.0$, solubility $\sim 0.9M$ at 15°C

Is it possible to titrate phenol in water quantitatively with the strong base sodium hydroxide?

EXAMPLE 7.1

In a titration of $0.05F$ phenol with $0.10M$ NaOH, is the reaction 99.9% complete at the equivalence point?

Solution

The pH at the equivalence point can be approximated using the simple expression for $[OH^-]$ in a $0.033F$ solution of the weak base, phenolate ion:

$$[OH^-] = \sqrt{K_b C_b} = 2.2 \times 10^{-3}$$
$$[H^+] = 4.5 \times 10^{-12}$$
$$pH = 11.3$$

Now at pH 11.3 the fraction of unreacted phenol is found to be about 0.05:

$$\alpha_{HA} = \frac{[H^+]}{[H^+] + K_a} = \frac{10^{-11.3}}{10^{-11.3} + 10^{-10}} = 0.05$$

If the titration were quantitative, the fraction of unreacted weak acid would be less than 0.001.

How could we make the titration of phenol quantitative? First, we might look for a stronger base than hydroxide ion. Unfortunately, as long as we use water as a solvent, hydroxide will be the strongest base we can use (due to the leveling effect). Stronger bases than hydroxide can exist, however, in solvents which are more basic than water. At the same time a more basic solvent enhances the dissociation of phenol, making it a stronger acid than it is in water.

Figure 7.1 should help you understand these ideas. Two titration curves are sketched in this figure. Curve B is the titration of phenol with NaOH in water. The pH on the upper plateau of the curve is determined by the strength of the base, OH^-, and its concentration. The size of the equivalence point break (its sharpness) is determined both by the upper pH limit and by the strength and concentration of the

7.3 CLASSIFICATION SCHEME FOR SOLVENTS

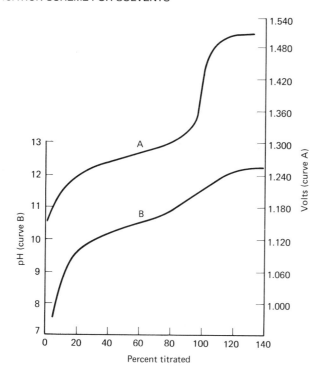

Figure 7.1 Titration of phenol in water and ethylenediamine. Curve A: titration of phenol with 0.2M sodium aminoethoxide in ethylenediamine; glass electrode voltage vs. SCE. Curve B: titration of phenol with 0.5M NaOH in water; glass electrode voltage converted to pH scale. [Sketched from composite of Figs. 3 and 5 in M. L. Moss, J. H. Elliott, and R. T. Hall, *Analytical Chemistry,* 20, 784 (1948). Reprinted with permission of the American Chemical Society.]

weak acid. Curve A shows the effect of titrating phenol in the basic solvent ethylenediamine and using aminoethoxide anion, a stronger base than OH$^-$, as titrant. The equivalence point break is accentuated, the titration curve becomes sharp, and the reaction becomes quantitative (see reference cited in figure). Phenol becomes about as strong an acid in ethylenediamine as acetic acid is in water (pK_a = 4.76).

Other solvents might be suitable alternatives to water as a solvent for the titration of a very weak acid. What are the criteria one uses to select a suitable solvent?

7.3 CLASSIFICATION SCHEME FOR SOLVENTS

While there are several characteristics by which solvents can be classified, the most important in acid-base studies are acidity and dielectric constant. A classification scheme proposed by Kolthoff (Ref. *1*) and based on earlier ideas of Brønsted provides a good framework for discussion.

Kolthoff first grouped solvents according to the extent to which their reactions can involve protons. Solvents which can either accept or donate protons are called *amphiprotic,* while those which interact with protons only slightly (or act only as proton acceptors, but not as donors) are called *aprotic.* Within the amphiprotic class are the *protogenic* (proton-making) and the *protophilic* (proton-taking) solvents. Protogenic solvents are much stronger acids than water, while protophilic solvents are much stronger bases than water. Solvents which are both acceptors and donors of protons, like water, are called *neutral* solvents.

Within each of Kolthoff's three acidity groups there are solvents with dielectric constants which are either greater or less than 20. Those with dielectric constants greater than 20 are called high dielectric solvents; the rest are called low dielectric solvents. A dielectric constant of 20 roughly marks the value above which equilibria involve the interactions of solvated single ions. Larger clusters of ions are much more important in solvents with dielectric constants below 20.

Kolthoff divides the aprotic solvents into *protophilic, protophobic* (literally "proton fearing"), and *inert.* Protophilic aprotic solvents are proton acceptors (bases) and do not behave as acids. Protophobic solvents are both very weak acids and very weak bases. Inert solvents such as benzene, carbon tetrachloride, and hexane do not participate in acid-base reactions. Kolthoff's scheme is shown in Table 7.1, along with example systems.

The following general recommendations can be made in light of the foregoing discussion and the information in Table 7.1:

1. To titrate a very weak acid ($pK_a > 9$ in water) with strong base, try a more basic amphiprotic solvent such as ethylenediamine or formamide or a basic aprotic solvent such as pyridine or dioxane.
2. To titrate a very weak base ($pK_b > 9$ in water) with a strong acid, try a more acidic (protogenic) amphiprotic solvent than water, such as acetic acid or formic acid, or an aprotic solvent such as methyl isobutyl ketone or acetonitrile.

7.4 A CLOSER LOOK AT FOUR CRITICAL FACTORS

Let us now take the discussion a step further and examine more carefully the properties of solvents which might allow us to titrate pure acids or bases, or determine the components of a mixture. The properties we will discuss are the inherent acidity (related to leveling ability), autoprotolysis constant, dielectric constant, and solvating ability. The first two factors combine to produce the "differentiating range" of a solvent, that is, the pH range over which acids can be distinguished from the protonated form of

7.4 A CLOSER LOOK AT FOUR CRITICAL FACTORS

TABLE 7.1   SOLVENT CLASSIFICATION SCHEME[a]

Solvent type	Dielectric constant	Relative acidity	Relative basicity	Examples[b]
Amphiprotic				
Neutral	+	(comparable to water)		Water, methanol *t*-BuOH, cyclohexanol
	−			
Protogenic	+	(++)	(−−)	Sulfuric acid, formic acid
	−	(++)	(−−)	Acetic acid
Protophilic	+	(−−)	(++)	DMSO, formamide
	−	(−−)	(++)	EN, NH$_3$, TMG
Aprotic				
Protophilic	+	(−)	(+)	DMF
	−	(−)	(++)	Pyridine, dioxane, THF, ethers
Protophobic	+	(−)	(−)	AN, ketones
	−	(−)	(−)	MIBK, MEK
Inert	−	(−)	(−)	benzene, alkanes

[a] In the dielectric constant column (+) and (−) mean greater than and less than 20, respectively. Elsewhere (+) and (−) mean somewhat stronger and somewhat weaker than water; (++) and (−−) mean much stronger and much weaker than water.

[b] Key: AN (acetonitrile), CH$_3$CN

dioxane, O(CH$_2$CH$_2$)$_2$O (ring)

EN (ethylenediamine), H$_2$N(CH$_2$)$_2$NH$_2$

DMF (dimethylformamide), HC(=O)N(CH$_3$)$_2$

DMSO (dimethylsulfoxide), H$_3$CS(=O)CH$_3$

Formamide, HC(=O)NH$_2$

MEK (methyl ethyl ketone), H$_3$CCCH$_2$CH$_3$ (with C=O)

MIBK (methyl isobutyl ketone) H$_3$C−C(=O)−CH$_2$CH(CH$_3$)$_2$

pyridine, HC(N)CHCHCHCH (ring)

THF (tetrahydrofuran), H$_2$C−CH$_2$−CH$_2$−CH$_2$−O (ring)

TMG (tetramethylguanidine), HN=C(N(CH$_3$)$_2$)$_2$

the solvent (lyonium ion) and bases can be distinguished from the deprotonated form of the solvent (lyate ion).

7.4.1 Inherent Acidity

It has been stated several times that the acidity of a solvent determines its ability to level the strengths of acids and bases. It may be helpful to describe the leveling effect

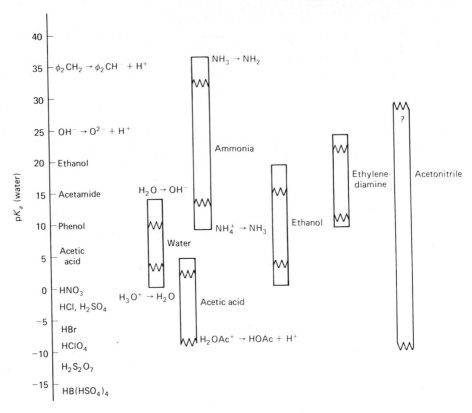

Figure 7.2 Approximate leveling and differentiating pH ranges for several solvents.

graphically with Fig. 7.2. In this figure the pK_a values of several acids are located on a vertical pH scale running from -15 to 40. These values, along with the pH range limits for the solvents, were determined potentiometrically with glass electrodes.† Remember that very weak acids in water have large pK_a values (e.g., the pK_a of ethanol is 19), while very strong acids have small or negative pK_a values (e.g., HBr in acetic acid, $pK_a = -7$). Several bars are drawn in the figure to describe the pH ranges over which acids or bases can be differentiated in several solvents. The smooth ends of the bars correspond to approximate pK_a values of lyonium ions and their neutral forms. For example, experiments indicate that the pK_a of hydronium ion is about 0 and that of neutral water is about 14. The bar for water thus extends from 0 to 14. The pK_a of

---

† The experiments are similar to that described in the previous chapter for measuring the pH of a solution of a weak acid. The pH range for ethanol can be measured by dipping a glass electrode into ethanol solutions of the base sodium ethoxide and the acid, perchloric acid. The response of the electrode in millivolts can then be converted to pH units, knowing that 60 mV is about equivalent to one pH unit. Many of these measurements are difficult to make, and the ranges are only approximations. Some special precautions for studying nonaqueous systems potentiometrically are given in Chapter 15.

ammonium ion is about 10, while that of ammonia is about 25, and so the bar for ammonia extends from 10 to 25. The jagged boundaries are drawn to show the differentiating boundaries for each solvent; an acid whose pK_a lies below a jagged line cannot be distinguished from the solvent's lyonium ion. You have already seen that a wide differentiating zone can be very helpful in titrating a wide variety of acids and bases.

Consider the behavior of the mixture of perchloric acid, hydrochloric acid, acetic acid, and phenol in several solvents. In water, perchloric and hydrochloric acids are strong acids, while acetic acid and phenol are weak acids. The pK_a values of acetic acid and phenol lie within the differentiating zone of water, so these acids can be distinguished from hydronium ion (and each other, because their pK_a values are sufficiently different).

Now if we take the same four acids dissolved in glacial acetic acid (a solvent more acidic than water) and titrate with the base sodium acetate, we find that perchloric acid is a stronger acid than HCl. This means that the equilibria

$$HClO_4 + HOAc = H_2OAc^+ClO_4^- \quad \text{(ion pair)}$$
$$HCl + HOAc = H_2OAc^+Cl^- \quad \text{(ion pair)}$$

do not lie equally far to the right, as they did in water. Hydrochloric acid is inherently weaker as an acid than is $HClO_4$, but we could not have established that fact in a solvent as basic as water. Acetic acid is a "neutral" species in glacial acetic acid, just as water is a "neutral" species in water.†

Phenol is such an extremely weak acid in glacial acetic acid that no base is strong enough to react with it, and it appears to be inert. On the other hand, the phenolate anion is such a strong base in acetic acid solvent that its strength is lowered to that of acetate ion. Notice in Fig. 7.2 that the pK_a of phenol lies above the differentiation range of acetic acid. It will always be true that an acid whose pK_a lies above the differentiation zone of a solvent cannot be determined by titration with a base in that solvent.

If we next dissolve the four acids in ethylenediamine (EN), a far more basic solvent than water or acetic acid, and titrate with a base like lithium ethylenediaminate (EN^-), we find that the strengths of $HClO_4$, HCl, and HOAc are all leveled to that of protonated ethylenediamine. As you saw earlier in the chapter, phenol is about as strong an acid in ethylenediamine as acetic acid is in water. Since the pK_a values of $HClO_4$, HCl, and HOAc lie below the differentiating zone of EN, their strengths cannot be differentiated.

† Remember that in Brønsted theory the strength of an acid depends on the base (or solvent) to which it is donating protons. There is really no compound which is defined as neutral outside a particular solvent frame of reference. Water is, of course, not neutral relative to acetic acid, and acetic acid is not neutral relative to sulfuric acid. Some chemists, Brønsted among them, have objected to the use of the term neutral in any discussion of acid-base equilibria.

All four acids are differentiable, however, in the solvent methyl isobutyl ketone (MIBK), an aprotic, protophobic solvent with a small dielectric constant (~ 12). The lyonium ion (MIBKH$^+$) is quite unstable and is thus a very strong acid. Figure 7.3 shows an often-cited example of the resolution of a strong-weak-very weak acid mixture by titration with tetrabutylammonium hydroxide in MIBK.

7.4.2 Autoprotolysis Constant

The autoprotolysis constant of a solvent has an important influence on the size of the differentiation range. The position of the lower boundary of each band in Fig. 7.2 indicates the strength of the lyonium ion of each solvent as an acid. For the general protonic solvent SH we can write the equilibrium

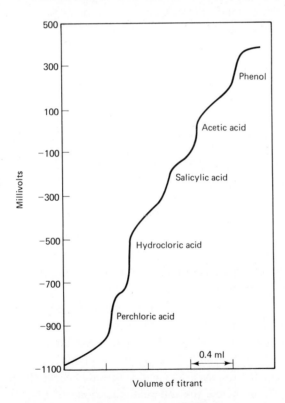

Figure 7.3 Resolution of acids in methyl isobutyl ketone. The titrant was $0.2M$ tetrabutylammonium hydroxide. Vertical axis shows voltage of glass electrode vs. platinum reference electrode. [Reproduced from Fig. 5 of D. B. Bruss and G. E. A. Wyld, *Analytical Chemistry*, 29, 232 (1957) with permission of the American Chemical Society.]

7.4 A CLOSER LOOK AT FOUR CRITICAL FACTORS

$$SH_2^+ = SH + H^+$$

to represent the boundary. The lower the boundary in the figure, the stronger is the lyonium ion as an acid. The position of the upper boundary of a band indicates the strength of the lyate ion as a base. Again, for the general solvent the boundary reaction can be written as

$$S^- + H^+ = SH$$

The higher an upper boundary in the figure, the stronger is the lyate ion (S^-) as a base.

The reaction between the lyonium and lyate ions is found by simply adding the boundary reactions:

$$SH_2^+ + S^- = 2SH$$

The equilibrium quotient expression for this autoprotolysis reaction is

$$K_s = \frac{a_{SH}^2}{a_{SH_2^+} a_{S^-}}$$

and is the inverse of the "autoprotolysis constant" when the solvent is at unit activity. When the lyonium ion is a very strong acid and the lyate ion a very strong base, the equilibrium quotient will be very large and the autoprotolysis constant very small. Solvents with very small autoprotolysis constants show very large bands in Fig. 7.2.

It should be clear from Fig. 7.2 that water, ethanol, acetic acid, and ethylenediamine have relatively small differentiation ranges. Ammonia and acetonitrile have much larger differentiation ranges. The ammonium ion is a weak acid in water; amide ion must be an extremely strong base, because K_s is so small. Protonated acetonitrile is a very strong acid; the acetonitrile anion must be an extremely strong base, again because K_s is so small.

7.4.3 Dielectric Constant

Think of the dissociation of an uncharged acid, HA, as taking place in two steps. In the first step solvent molecules react with HA to form a solvated ion pair in a process called "ionization":

$$HA + \underset{\text{(solvent)}}{2SH} = \underset{\text{(ion pair)}}{(SH_2^+)(ASH^-)}\dagger$$

† The notation (ASH^-) means a solvated anion.

We will call the equilibrium constant for this reaction K_i. In the second step the ion pair separates (dissociates) with the help of solvent molecules to insulate the electrostatic charges:

$$(SH_2^+)(ASH^-) = SH_2^+ + ASH^-$$

We will call the equilibrium constant for this reaction K_d.

The measured protolysis constant for HA should be thought of as an "overall dissociation constant," which is related in a simple way to the equilibrium constants for the two steps:

$$K_a = K_i K_d = \frac{a_{(SH_2^+)(ASH^-)}}{a_{HA} a_{SH}^2} \cdot \frac{a_{SH_2^+} a_{ASH^-}}{a_{(SH_2^+)(ASH^-)}} = \frac{a_{SH_2^+} a_{ASH^-}}{a_{HA} a_{SH}^2}$$

Solvents which have large dielectric constants are able to interfere with the attractive forces holding the ion pairs together, and in such solvents the dissociation step is essentially complete. In solvents with small dielectric constants, even very good potential proton donors may show small experimental K_a values because ion pairs (or larger clusters) cannot be broken apart by solvent interactions.

The effect that solvent dielectric constant has on measured dissociation constants depends critically on the charges of the acid and conjugate base forms. Three cases are most interesting:

1. Neutral acid, negative conjugate base,

$$HA + SH = A^- + SH_2^+$$

2. Positive acid, neutral conjugate base,

$$HB^+ + SH = B + SH_2^+$$

3. Negative acid, doubly negative conjugate base,

$$HC^- + SH = C^{2-} + SH_2^+$$

(Conjugate bases are shown unsolvated in the interest of clarity.)

Let us say for the sake of discussion that in a high dielectric solvent such as water, HA, HB$^+$, and HC$^-$ are of equal strength (their pK_a values are identical). As we add a solvent which has a lower dielectric constant, say isopropanol (dielectric constant 19.9), the strengths of the three acids respond differently. The acid HA donates a proton to the solvent and a pair of ions must be separated by the solvent. Therefore

HA becomes a weaker acid as isopropanol is added to its solution. Similarly, HC$^-$ donates a proton to the solvent, and the resulting ion pair must be separated. Since C^{2-} is a doubly charged species, it is more strongly paired with the solvated proton than was the singly charged species A$^-$, and so the acid HC$^-$ becomes weaker than HA as isopropanol is added to its solution. The positive species HB$^+$ is perhaps the most interesting case, because its reaction with solvent involves no net change in the charges of the interacting species. When isopropanol is added to an aqueous solution of an acid of this type, there should be no observable change in acidity due to dielectric constant.

Figure 7.4 shows a set of curves obtained from the titration of a diprotic weak acid, succinic acid (pK_{a1} = 4.21, pK_{a2} = 5.64 in water), with tetrabutylammonium hydroxide in a series of solutions containing increasingly large amounts of isopropanol. In pure water the pK_a values of succinic acid are so close that only one equivalence

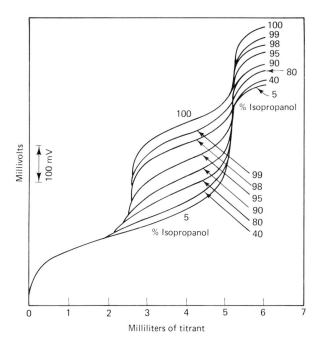

Figure 7.4 Titration of succinic acid in isopropanol/water mixtures. The results of several titrations of equal quantities of succinic acid in isopropanol/water mixtures of various concentrations with tetrabutylammonium hydroxide. Vertical axis shows voltage of glass electrode vs. SCE. The curves are plotted to make the first plateaus coincide. [Reproduced from Fig. 3 in G. A. Harlow and G. E. A. Wyld, *Analytical Chemistry*, 30, 69 (1958) with permission of the American Chemical Society.]

point break can be seen. As isopropanol is added both pK_{a1} and pK_{a2} increase, but pK_{a2} increases more dramatically. As more and more isopropanol is added, a distinct second equivalence point break appears. By the time the solution is 95% isopropanol, the second break is quite sharp. This behavior is predictable if we notice that the second dissociation involves separating the doubly negative succinate anion from the solvated proton, which is a more difficult task for a solvent with a low dielectric constant.

7.4.4 Solvating Ability

The extent to which solvent molecules associate with both charged and uncharged species (that is, "solvate" them) can dramatically affect the degree of dissociation of acids and bases. Solvents which associate strongly with protons or anions, for example, will promote the dissociation of uncharged acids. Protophilic solvents like water, ethylenediamine, and ammonia thus enhance the strength of such acids. Protophobic solvents like acetonitrile do not associate strongly with protons, so they suppress the dissociation of weak acids.

A complication that arises in titrations performed in weakly solvating, low or medium dielectric constant solvents, is that of self-condensation or *homoconjugation*. The anionic conjugate base of a weak acid may have a greater affinity for the undissociated weak acid than for molecules for the solvent, and the equilibrium

$$A^- + HA = AHA^-$$

may become important. The effect in a titration would be to generate an unexpected acid species which is a weaker acid than the original uncharged acid. The titration curve plateau would, in the best of cases, pitch upward slightly, or, in the worst of cases, actually show a second inflection. This, of course, will make it quite difficult to differentiate a mixture of acids whose dissociation constants are similar.

7.5 APPLICATIONS

7.5.1 Neutral Solvents: Alcohols

The simpler alcohols, methanol (MeOH), ethanol (EtOH), isopropanol (*i*-POH), and tertiary butanol (*t*-BuOH), have all been used extensively as nonaqueous solvents in analytical chemistry. All four solvents can be obtained in very pure form from commercial sources and can be used without further purification for many applications.

7.5 APPLICATIONS

Table 7.2 contains dielectric constants and autoprotolysis constants for these solvents and water, and selected weak acid dissociation constants obtained in them.

Methanol is the most waterlike solvent of the simple alcohols. While the large differentiation range of methanol ($pK_s = 16.7$) would seem to make it a good choice for titrating weak acids, its moderate dielectric constant discourages full dissociation when the acid-base reaction products are charged species. Curves for the titration of acetic acid with hydroxide in water and methoxide ion (OCH_3^-) in methanol are sketched in Fig. 7.5. Acetic acid is apparently more effectively titrated in water than in methanol; while the greater basicity of methoxide ion raises the pH on the upper plateau of the curve (relative to the curve in water), the extreme weakness of acetic acid in methanol also raises the lower plateau, offsetting any anticipated advantage in endpoint sharpness. This behavior is quite general among uncharged weak acids. Notice in Table 7.2 that the pK_a values of the uncharged organic acids (the first group) are about 4.7 units higher in methanol than in water.

There is a definite advantage to titrating positively charged acids in methanol rather than water, as indicated in Fig. 7.6, where curves for the titration of ammonium ion in water (with hydroxide) and in methanol (with methoxide) are compared. As noted before, the pK_a values of singly charged cationic acids are much less influenced by solvent dielectric constant, and the additional endpoint sharpness gained by using

TABLE 7.2 SOLVENT PROPERTIES AND DISSOCIATION CONSTANTS FOR FIVE NEUTRAL AMPHIPROTIC SOLVENTS

	Solvent				
Property	Water	Methanol	Ethanol	i-Propanol	t-Butanol
e (dielectric constant)	78	32.6	24.3	19.9	12.5
pK_s	14	16.7	19.5	20.8	—
pK_a of					
Acetic	4.75	9.7	10.4	11.4	14.3
Benzoic	4.2	9.3	10.0	—	15.0
Salicylic	3.0	7.9	8.6	—	—
p-Nitrobenzoic	3.5	8.3	8.8	9.6	12.0
Phenol	9.9	14.3	—	—	—
pK_a of					
Anilium	4.6	6.0			
Pyridinium	5.2	5.5			
Ammonium	9.2	10.0			
4-Methylanilium	4.4	5.9			
3-Methylpyridinium	5.7	6.5			

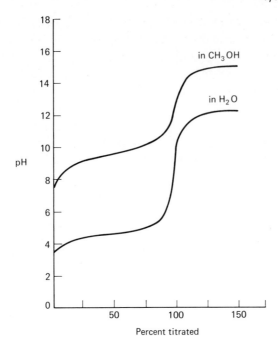

Figure 7.5 Titration of acetic acid in water and methanol. Curves calculated from simple expressions (see Chapter 6), assuming titrant and initial acid concentrations are both $0.1F$. Titrant in methanol is sodium methoxide; titrant in water is NaOH. Methanol solvent raises and compresses the titration curve; water is the preferred solvent.

a solvent with a large differentiating range is not offset by the the effects of a low dielectric constant. The slight differences observed in pK_a values in the two solvents (about one unit) can be attributed to methanol's slightly greater acidity.

Weak base/strong acid titrations may also be performed in methanol, although once again dielectric effects limit the solvent's usefulness. Weak anionic bases which react to form uncharged conjugate acids are profitably titrated in methanol. For example, acetate ion, too weak a base to titrate with strong acid in water ($pK_b = 9.25$), is a stronger base in methanol ($pK_b = 7.4$).

Ethanol has a smaller dielectric constant than methanol but a far larger differentiating range ($pK_s = 19.5$). Uncharged weak acids are weaker in ethanol than in water by about 5.5 pK units. Titrations of cationic weak acids give sharper endpoints in ethanol than in water, as they do in methanol.

Isopropanol and *t*-butanol are weaker acids than either ethanol or methanol and thus offer some advantages as solvents for the titration of weak acids. The very low dielectric constant of *t*-butanol means that acids weaker than phenol cannot be titrated

7.5 APPLICATIONS

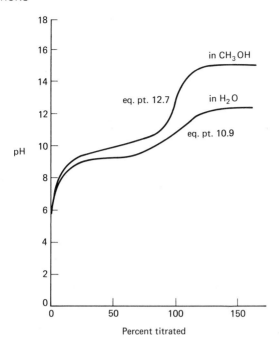

Figure 7.6 Titration of ammonium ion in water and methanol. Curves calculated from pK_a values, using simple expressions for pH (see Chapter 6), assuming concentrations of ammonium ion and base are 0.1F. Sodium hydroxide is the titrant in water; NaOCH$_3$ is the titrant in methanol.

satisfactorily, even though t-butanol has a larger differentiating range than ethanol. Hydrochloric acid, a fully dissociated acid in methanol, is only slightly dissociated in t-butanol (pK_a = 5.0).†

The most popular base titrants in neutral amphiprotic solvents have been sodium methoxide (CH$_3$O$^-$Na$^+$), potassium ethoxide (CH$_3$CH$_2$O$^-$K$^+$), and tetraalkylammonium hydroxides, most notably tetrabutylammonium hydroxide (TBAH). TBAH is commercially available as a 1M solution (25%) and is easily diluted to 0.1M with alcohol. Solutions of TBAH gradually decompose and may be unstable in organic solvents if too little water is present. Like all strong bases, TBAH solutions absorb atmospheric CO$_2$ and must be kept from contact with air. The methods described in Chapter 6 for preventing CO$_2$ absorption may not be adequate for TBAH solutions, and it may be necessary to perform titrations in an inert atmosphere in a glove box.

Sodium methoxide is also a strong base in alcohol solutions. It is prepared by treating methanol with sodium metal,

† See Ref. 2 for experimental studies with t-butanol.

$$2Na + 2CH_3OH = 2NaOCH_3 + H_2(g)$$

Solutions of base may be standardized against primary standard grade benzoic acid.

The most popular acidic titrant is perchloric acid. Solutions of perchloric acid can be standardized against potassium biphthalate or tris(hydroxymethyl)amino methane (see Chapter 6). Reference 3 describes the preparation of titrant solutions in detail.

7.5.2 Acidic Solvents: Acetic Acid

A great deal of experimental work has been done on titrations in glacial acetic acid. In fact, it may be the most thoroughly studied nonaqueous solvent. The earliest systematic experiments were done by Hall (Ref. 4), who titrated 50 weak bases with perchloric acid in glacial acetic acid. In an important series of papers in the 1950s, Kolthoff and Bruckenstein (Ref. 5) interpreted the behavior of acids and bases in this solvent, determined the autoprotolysis constant, and calculated curves for the titration of bases. Studies up to 1970 are summarized in a review by Popov (Ref. 6), and biennial reviews in the journal *Analytical Chemistry* provide useful updates.

Acetic acid is an acidic solvent with a very low dielectric constant (6.2 at 18°C) and a pK_s (14.5) comparable to that of water. This combination of properties allows an analyst to do two kinds of titrations that at first might seem incompatible: differentiate mixtures of strong acids and determine very weak bases. Table 7.3 gives values for overall dissociation constants of several acids and bases in glacial acetic acid. Corresponding values in water are given to provide some perspective. You should notice

TABLE 7.3 OVERALL DISSOCIATION CONSTANTS IN GLACIAL ACETIC ACID AT 25°C

Acids	pK_a (HOAc)	pK_a (H$_2$O)
Perchloric	4.87	≪0
Sulfuric	7.24	<0
Hydrochloric	8.55	<0
Bases	pK_b (HOAc)	pK_b (H$_2$O)
Sodium acetate	6.68	9.24
Potassium acetate	6.15	9.24
Tribenzylamine	5.4	9.3
Ammonia	6.40	4.75
Urea	10.2	13.8
Salts	pK_d (HOAc)	
Sodium perchlorate	5.48	
Tribenzylamine hydrochloride	6.71	
Urea hydrochloride	6.96	
Potassium chloride	6.88	

immediately that there are no completely dissociated acids or bases in glacial acetic acid solvent. Perchloric acid ($HClO_4$) is the most dissociated acid, but is really no more dissociated in acetic acid than is acetic acid in water. Weak bases are stronger in glacial acetic acid than in water by about 3 pK units. An exception is ammonia, which seems to be slightly weaker.

Perchloric acid is by far the most popular acidic titrant in glacial acetic acid. Stock solutions can be prepared by adding a calculated amount of 72% perchloric acid to glacial acetic acid. The water added with the perchloric acid can be tolerated except in the titration of very weak bases. In such cases acetic anhydride can be added to react with the water and form acetic acid,

$$(CH_3CO)_2O + H_2O = 2CH_3COOH$$
(acetic anhydride)

A slight excess of acetic anhydride can be added, except when primary or secondary amines are to be determined. Recall from the discussion at the end of Chapter 6 that these compounds will react to form amides. Since perchloric acid is not a primary standard, stock solutions of titrant must be standardized against a base like potassium biphthalate.

Since acetate ion is the strongest base that can exist in glacial acetic acid, solutions of sodium acetate can be used for the titrations of mixtures of strong acids. A stock solution of sodium acetate can be standardized with a previously standardized solution of perchloric acid.

7.5.3 Basic Solvents: Ammonia and Amines

The most important solvents in this class are ethylenediamine (EN) and liquid ammonia. Both solvents are strong bases and are excellent levelers of acids, except for the very weakest. Ammonia has a larger dielectric constant than does EN (23 vs. 12.5) and so can dissolve a wider range of solutes. The autoprotolysis constant of ammonia is quite a bit smaller than that of EN (pK_s about 33 vs. 15.3), and so it should have a very large differentiating range for the titration of weak acids. Were it not for the experimental difficulties associated with working at temperatures below $-33°C$, ammonia would doubtless be a very popular solvent. Herlem and Thiebault (Ref. 7) investigated the titration of weak acids in ammonia. Amide ion (NH_2^-) is a strong base in ammonia.

Ethylenediamine has been an excellent solvent for the titration of phenols, as mentioned at the beginning of this chapter. The lithium salt of the ethylenediamine lyate ion (EN^-) and sodium aminoethoxide ($H_2NCH_2CH_2O^-Na^+$) have both been useful as strong base titrants. Comprehensive studies and interpretations of equilibria in EN were made by Schaap et al. (Ref. *8*) and Bruckenstein and Mukherjee (Ref. *9*).

The concepts of these interpretations have been useful in studies of other low-dielectric media, particularly liquid ammonia (Ref. *10*).

7.5.4 Aprotic Solvents

The most important aprotic solvents are the protophilic compounds dimethyl sulfoxide (DMSO) and N,N-dimethylformamide (DMF) and the protophobic compounds acetonitrile (AN) and methyl isobutyl ketone (MIBK). Structures of these compounds are shown in Fig. 7.1. Dielectric constants and pK_s values for these solvents are presented in Table 7.4, along with some representative weak acid dissociation constants determined in them.

Although DMSO has been classified for years as an aprotic solvent, it is now known that its protons can be involved in acid-base reactions. While DMSO is a very weak acid, it is a reasonably strong base, slightly stronger than water. It is not nearly as strong a base as ethylenediamine, however, and it can be used in titrations in which mixtures of acids must be determined. Price and Whiting (Ref. *11*) were the first to prepare the lyate form of DMSO, which they called "DMSYL," and used it to titrate over 25 acids, including lower alcohols (methanol, isopropanol), ethylene glycol, glycerol, and even cyclopentadiene and indene. They used triphenylmethane as a visual indicator (colorless to red in strong base). DMSYL solutions are very strongly basic and react with atmospheric moisture, oxygen, and carbon dioxide.

DMF has also been a popular solvent for the titration of mixtures of acids. Like DMSO, DMF has a moderate dielectric constant and dissolves many salts. Its differrentiating range is somewhat more limited than that of DMSO due to its smaller pK_s.

TABLE 7.4   APROTIC SOLVENTS

Property	Solvent			
	DMSO	DMF	AN	MIBK
e (dielectric constant)	46	37	36	12
pK_s	~33	~18	~33	—
pK_a of				
Ammonium	10.5	9.4	16.5	—
Anilinium	3.6	4.4	10.7	—
Pyridinium	3.4	—	12.3	—
p-Nitrophenol	—	12.6	20.7	—

Unlike DMSO, DMF has no stable lyate form and is a genuine aprotic solvent. Tetrabutylammonium hydroxide in isopropanol solution is used as a basic titrant. A major disadvantage of DMF is that it is extremely reactive with water, forming dimethylamine and formic acid, and is therefore difficult to handle in the laboratory.

Acetonitrile is a protophobic solvent and can be used as a solvent for the titration of either acids or bases. Solutions of perchloric acid in glacial acetic acid or dioxane are used as acid titrants. Solutions of pure perchloric acid in AN are unstable. Acetonitrile is both a weaker acid and a weaker base than water; in fact, both the lyonium and lyate forms of AN are unstable. As a result, the mineral acids are not leveled in AN. Unfortunately, the potentially tremendous differentiating power of the solvent may be difficult to realize in practice as a result of its rather poor solvating ability. Homoconjugation in AN solutions may result in poorly shaped titration curves, particularly when weak acids are titrated with strong bases. Weak bases are less susceptible to homoconjugation.

Methyl isobutyl ketone is an excellent solvent for the titration of acid mixtures and bases. The remarkably large range of acid strengths that can be differentiated in MIBK was discussed above and illustrated in Fig. 7.3. The paper of Bruss and Wyld referred to in the figure legend should be consulted for details.

REFERENCES

1. I. M. Kolthoff, "Acid-Base Equilibria in Dipolar Aprotic Solvents," *Analytical Chemistry,* 46, 1992 (1974).
2. L. W. Marple and J. S. Fritz, *Analytical Chemistry,* 35, 1223 (1963).
3. J. S. Fritz, *Acid-Base Titrations in Nonaqueous Solvents,* Allyn & Bacon, Boston, 1973.
4. N. F. Hall, *Journal of the American Chemical Society,* 52, 5115 (1930).
5. I. M. Kolthoff and S. Bruckenstein, *Journal of the American Chemical Society,* 79, 5915 (1959) (the last paper in a series of five).
6. A. I. Popov, in *The Chemistry of Nonaqueous Solvents,* vol. III, J. J. Lagowski, ed., Academic, New York, 1970, Chapter 5.
7. M. Herlem and A. Thiebault, *Bulletin de la Societe Chimique de France,* 1970, p. 383.
8. W. B. Schaap, R. E. Bager, J. R. Seifker, J. Y. Kim, P. W. Brewster, and F. C. Schmidt, *Record of Chemical Progress,* 22, 197 (1961).
9. S. Bruckenstein and L. M. Mukherjee, *Journal of Physical Chemistry,* 66, 2228 (1962).
10. R. E. Cuthrell, E. C. Fohn, and J. J. Lagowski, *Inorganic Chemistry,* 4, 1002 (1965); 5, 111 (1966).
11. G. G. Price and M. C. Whiting, *Chemical Industry,* 1963, p. 775.

RECOMMENDED READING

Covington, A. K. and T. Dickenson, eds. *Physical Chemistry of Organic Solvents,* Plenum, New York, 1973. Review of solvent properties with tabulated data.

Fisher, R. and D. G. Peters. *Quantitative Chemical Analysis,* 3rd ed., Saunders, Philadelphia, 1968. Chapter 9 gives an excellent treatment of equilibria and titrations in ethylenediamine.

Fritz, J. S. *Acid-Base Titrations in Nonaqueous Solvents,* Allyn & Bacon, Boston, 1973. An excellent, easily read introduction to the subject, along with practical material for laboratory work.

Huber, W. *Titrations in Nonaqueous Solvents,* Academic Press, New York, 1967. An introductory text also containing practical material.

Kolthoff, I. M. and P. J. Elving, eds. *Treatise on Analytical Chemistry,* 2nd ed. part 1, vol. 2, Wiley, New York, 1978. Chapter 19 is a good compilation of recent developments in acid-base studies in nonaqueous solvents.

PROBLEMS

7.1. Phenylurea ($C_6H_5CONH_2$) is a very weak base in water ($pK_b = 14.3$). Which solvents would you investigate if you wished to determine phenylurea by titration with strong acid?

7.2. Formic acid (HCOOH) is a protogenic solvent, with $pK_s = 6.2$ and a dielectric constant of 58. What would you expect to be some advantages and disadvantages of formic acid relative to acetic acid as a solvent?

7.3. Explain why solvent dielectric constant is an important factor in considering nonaqueous solvents for acid-base titrations.

7.4. Biphosphate ion, HPO_4^{2-} ($pK_a = 12.3$), is too weak an acid to titrate with a strong base in water. On the basis of acidity alone, could biphosphate be titrated in ethylenediamine? What might pose a serious limitation?

7.5. What is homoconjugation? What solvent characteristics make homoconjugation a problem in acetonitrile, but not in water or DMSO?

7.6. What advantage might isopropanol have over water as a solvent for the titration of a mixture of anilium hydrochloride ($C_6H_5NH_3^+Cl^-$, $pK_a = 4.6$ in water), and 4-aminobenzoic acid ($H_2NC_6H_4COOH$, $pK_a = 4.9$ in water)?

7.7. The pK_s of acetonitrile is quite difficult to determine, but probably has a value of about 30. The overall dissociation constants of phenol and benzoic acid are 26.6 and 20.7, respectively, in acetonitrile. Calculate pK_b for phenolate and benzoate ion, and predict which would be a better titrant for the determination of weak acids. What other factors would you consider?

PROBLEMS

7.8. Tribenzylamine (TBA) can be titrated with perchloric acid in glacial acetic acid solvent. The titration reaction is

$$HClO_4 + TBA = TBAH^+ClO_4^- \text{ (ion pair)}$$

The following dissociation constants have been determined in glacial acetic acid:

$$TBA + HOAc = TBAH^+ + OAc^- \qquad pK_b = 5.36$$
$$TBAH^+ClO_4^- = TBAH^+ + ClO_4^- \qquad pK_d = 6.71$$
$$HClO_4 + HOAc = H_2OAc^+ + ClO_4^- \qquad pK_a = 4.87$$
$$2HOAc = H_2OAc^+ + OAc^- \qquad pK_s = 14.45$$

(a) Show that K for the reaction of $HClO_4$ with TBA is given by

$$K = K_a K_b / K_d K_s$$

and calculate a numerical value for K.

(b) Sodium acetate would appear to be a weaker base than TBA, with $pK_b = 6.58$. The value of pK_d for sodium perchlorate is 5.48. Is the reaction of sodium acetate with perchloric acid any more complete than the reaction of TBA with perchloric acid?

Chapter 8

Solubility Equilibria

8.1 INTRODUCTION

Imagine an experiment in which a gram of finely divided silver chloride is stirred vigorously in a liter of pure water until an equilibrium condition exists. When the stirred solution is allowed to stand and the suspended AgCl settles out, the analysis of a portion of the solution will show the presence of three important species: solvated silver ions, Ag^+(aq), solvated chloride ions, Cl^-(aq), and solvated but undissociated silver chloride, AgCl(aq). If the experiment is performed again in a liter of $0.1M$ NaCl rather than pure water, the anionic complexes $AgCl_2^-$, $AgCl_3^{2-}$, and $AgCl_4^{3-}$ will also be important species at equilibrium. The purpose of this chapter is to investigate the relationships between the amount of some slightly soluble material which actually dissolves (its *solubility*) and the concentrations of its many forms at equilibrium. Once we have gained some appreciation of the equilibria which accompany dissolution, we will examine in the next chapter the process of *precipitation* and its applications in analytical chemistry.

8.2 SIMPLE SOLUBILITY RELATIONSHIPS

Consider again the silver chloride systems described above. The activity of solvated, undissociated silver chloride, AgCl(aq) at equilibrium is called the *intrinsic solubility* of AgCl. The activity of AgCl(aq) is related to the activity of solid AgCl by the simple equilibrium quotient,

$$K_{s1} = \frac{a_{AgCl(aq)}}{a_{AgCl(s)}}$$

As long as there is solid AgCl present in the system there will always be AgCl in the solution to the extent permitted by the intrinsic solubility.

The activities of $Ag^+(aq)$ and $Cl^-(aq)$ depend on the extent to which solvated AgCl dissociates,

$$AgCl(aq) = Ag^+(aq) + Cl^-(aq)$$

The equilibrium quotient expression for this reaction is

$$K_{s2} = \frac{a_{Ag^+(aq)} a_{Cl^-(aq)}}{a_{AgCl(aq)}}$$

When the expressions for K_{s1} and K_{s2} are multiplied together, we obtain the expression for the *thermodynamic solubility product constant* K_{sp}^0,

$$K_{sp}^0 = K_{s1} K_{s2} = \frac{a_{Ag^+(aq)} a_{Cl^-(aq)}}{a_{AgCl(s)}}$$

K_{sp}^0 is an equilibrium constant and is determined by the standard free energies of the species.

EXAMPLE 8.1 The Thermodynamic K_{sp}

Calculate the thermodynamic solubility product constant for silver chloride knowing the following standard free energies:

Species	G^0 (kJ/mole)
AgCl(s)	−109.7
Ag^+(aq)	77.11
Cl^-(aq)	−131.2

Solution

The reaction is

$$AgCl(s) = Ag^+(aq) + Cl^-(aq)$$

with a standard free energy change given by

$$\begin{aligned}\Delta G^0 &= G^0 \text{ (products)} - G^0 \text{ (reactants)} \\ &= \{G^0[Ag^+(aq)] + G^0[Cl^-(aq)]\} - G^0[AgCl(s)] \\ &= \{77.11 + (-131.2)\} - (-109.7) \text{ kJ/mole} \\ &= 55.6 \text{ kJ/mole}\end{aligned}$$

8.2 SIMPLE SOLUBILITY RELATIONSHIPS

Recall from Chapter 3 that ΔG^0 and K are related by

$$\Delta G^0 = -RT \ln K = -5.707 \log K$$

Substitution gives

$$\log K_{sp}^0 = -55.6/5.707 = -9.74$$
$$K_{sp}^0 = 1.80 \times 10^{-10}$$

For simplicity, the activity of solid material is assumed to be unity,† and the solubility product relationship becomes

$$K_{sp}^0 = a_{Ag^+(aq)} a_{Cl^-(aq)}$$

This kind of expression is called an *ion product*.

The solubility product constant of a compound determines the activities of the ions of the compound at equilibrium. K_{sp}^0 can be determined experimentally by careful measurement of the activities of the ionic species at equilibrium.

EXAMPLE 8.2 K_{sp}^0 and Activities

A sample of solid silver chloride is equilibrated with pure water. The activities of silver ion and chloride ion in the solution are each found to be 1.34×10^{-5}. Calculate K_{sp}^0 for silver chloride.

Solution

$$K_{sp}^0 = a_{Ag^+} a_{Cl^-} = (1.34 \times 10^{-5})^2$$
$$= 1.80 \times 10^{-10}$$

Notice that the (aq) notation has been dropped for simplicity.

The actual solubility of a compound is the sum of the ionic concentration and the intrinsic solubility, as shown in the following example.

EXAMPLE 8.3 Intrinsic Solubility

The thermodynamic solubility product for silver chloride is 1.80×10^{-10}. If 1.35×10^{-5} mole of AgCl actually dissolves in a liter of pure water, what is the intrinsic solubility of AgCl?

---

† The presence of major impurities will lower the activity of a solid phase. Remember that activity is a function of mole fraction. Throughout this text we will assume that the activity of every solid phase is unity.

Solution

The activity of silver ion must equal that of chloride ion since the solution must be electrically neutral, that is,

$$a_{Ag^+} = a_{Cl^-}$$

Using the K_{sp}^0 expression,

$$K_{sp}^0 = a_{Ag^+}a_{Cl^-} = a_{Ag^+}^2$$

this tells us that the activity of silver ion at equilibrium must be given by the square root of K_{sp}^0, or

$$a_{Ag^+} = (K_{sp}^0)^{1/2} = 1.34 \times 10^{-5} M$$

Since the ionic strength of this solution is very low ($1.35 \times 10^{-5} M$), let us assume that the concentration of silver ion is the same as its activity. The actual solubility of AgCl is equal to the sum of the concentrations of silver ion and solvated AgCl, or

$$\text{Actual solubility} = [Ag^+] + [AgCl(aq)]$$

The actual solubility was given as $1.35 \times 10^{-5} M$. Therefore the intrinsic solubility, [AgCl(aq)], is given by

$$[AgCl(aq)] = 1.35 \times 10^{-5} M - 1.34 \times 10^{-5} M$$
$$= 1 \times 10^{-7} M$$

The intrinsic solubility of AgCl is about 1% of the actual solubility.

The intrinsic solubilities of most slightly soluble inorganic salts are on the order of 0.1 to 2% of their actual solubilities. Ordinarily we can closely approximate solubility products and simplify calculations by neglecting contributions from intrinsic solubility. Some caution is advised, however, in dealing with organic ligand-metal ion compounds, whose intrinsic solubilities lie in the range 10^{-6} to $10^{-9} M$. For example, the solubility of the bis(dimethylglyoxime)nickel(II) complex, based on K_{sp}^0 calculations is $1 \times 10^{-8} M$. However, the actual solubility of the complex is 100 times greater than this due to its high intrinsic solubility in water. In such a case the solubility calculation is of theoretical interest, but the total solubility is the quantity of practical interest to the analyst isolating the compound and trying to minimize the amount left in solution. You should remember that the intrinsic solubility guarantees a certain minimum solubility of a slightly soluble salt.

8.3 CONCENTRATIONS: K_{sp}^0 vs. K_{sp}

The solubility product constant for a compound can be determined by careful measurement of the activities of the cation and anion of the compound by techniques such as potentiometry (see Chapter 15). However, solubility involves quantities of material, and concentrations rather than activities are frequently of interest to an analyst. It was shown in Chapter 3 that concentrations are proportional to activities and that the proportionality constant (activity coefficient) depends on the ionic strength of a solution. Therefore the thermodynamic solubility product constant for silver chloride is expressed in terms of concentrations in the following way:

$$K_{sp}^0 = a_{Ag^+} a_{Cl^-} = f_+[Ag^+]f_-[Cl^-]$$
$$= f_{\pm}^2 [Ag^+][Cl^-]$$

Since solutions of slightly soluble salts are usually quite dilute, the Debye-Hückel limiting law can often be used to approximate the mean activity coefficient, $f_{\pm}$. When the thermodynamic solubility product constant is divided by the mean activity coefficient raised to the appropriate power (2 in the case of AgCl), the result is K_{sp}, simply the *solubility product,* that is,

$$K_{sp} = \frac{K_{sp}^0}{f_{\pm}^2} = [Ag^+][Cl^-]$$

When slightly soluble salts such as the silver halides are dissolved in pure water, the ionic strength is so low ($10^{-5} M$ or less) that $f_{\pm}$ is nearly equal to unity.

The next example is intended to show the difference between K_{sp}^0 and K_{sp}.

EXAMPLE 8.4 Effects of Ionic Strength
Calculate the solubility of silver chloride in $0.01 M$ KNO_3, given $K_{sp}^0 = 1.80 \times 10^{-10}$.

Solution

The relationship between K_{sp}^0 and K_{sp} in the case of AgCl is

$$K_{sp} = \frac{K_{sp}^0}{f_{\pm}^2}$$

To approximate $f_\pm$ we use the Debye-Hückel limiting law,

$$\log f_\pm = -0.510 z_+ z_- (\mu)^{1/2}$$
$$= -0.510(1)(1)(0.01)^{1/2} = -0.0510$$
$$f_\pm = 0.89$$

Using this value for $f_\pm$, we calculate

$$K_{sp} = \frac{1.80 \times 10^{-10}}{(0.89)^2} = 2.27 \times 10^{-10}$$

Since for every AgCl unit that dissolves one Ag^+ is released, it is convenient to represent the solubility as the silver ion concentration. Using the K_{sp} relationship and the condition of electroneutrality ($[Ag^+] = [Cl^-]$),

$$K_{sp} = [Ag^+][Cl^-] = [Ag^+]^2$$
$$\text{Solubility} = [Ag^+] = (K_{sp})^{1/2} = (2.27 \times 10^{-10})^{1/2}$$
$$= 1.51 \times 10^{-5} M$$

or $(1.51 \times 10^{-5} M)(143.3 \text{ g AgCl/mole}) = 2.16 \times 10^{-3}$ g AgCl/liter
$= 2.16$ mg AgCl/liter.

When the mean activity coefficient is unity, the solubility is 1.9 mg/liter. The intrinsic solubility of AgCl is about 0.03 mg/liter and has been neglected in these calculations.

It is important to know about the effect of ionic strength on the solubility of slightly soluble compounds, because precipitations in analytical procedures are invariably made from solutions containing rather large concentrations of foreign electrolytes. For example, in order to analyze a sample containing Na_2SO_4 for sulfate ion, an analyst would add an excess of $BaCl_2$ to precipitate $BaSO_4$. The quantity of $BaSO_4$ collected and weighed would be used to determine the percentage of Na_2SO_4 in the original sample. Precipitation is done not in water, but in a solution containing at least millimolar quantities of sodium and chloride ion.

Generally speaking, when a precipitate does not react with an electrolyte to form a soluble complex species, its solubility can be expected to increase with ionic strength up to about $\mu = 1M$ and then decrease again at higher electrolyte concentrations (see Fig. 3.5, in which $f_\pm$ is plotted as a function of the concentrations of NaCl and $CaCl_2$). The activity coefficients of ions with multiple charges are more sensitive to changes in ionic strength than those of singly charged ions. Therefore it might be expected that the solubilities of precipitates of highly charged ions will be much more influenced by ionic strength than will those of species such as AgCl or AgBr. Barium sulfate (Ba^{2+}, SO_4^{2-}) is about 1.7 times more soluble in $0.01M$ KNO_3 than in pure water, whereas

8.4 SOLUBILITY CALCULATIONS FOR MORE COMPLEX SPECIES

silver chloride (Ag^+, Cl^-) is only about 1.4 times more soluble in $0.01 M$ KNO_3 than in pure water.

8.4 SOLUBILITY CALCULATIONS FOR MORE COMPLEX SPECIES

When intrinsic solubility is neglected, the solubility of a simple 1:1 salt such as AgCl is given by either the cation or anion concentration at equilibrium. In the cases of 2:1, 1:2, 3:2, or 2:3 salts, however, it may be difficult to decide which ionic concentration gives the solubility. In the case of barium iodate, $Ba(IO_3)_2$, for example, the solubility equilibrium is represented by

$$Ba(IO_3)_2 = Ba^{2+} + 2IO_3^-$$

If we adopt the rather artificial view that solid barium iodate is made up of a collection of $Ba(IO_3)_2$ units ("formula units"), then we can imagine that the solubility of barium iodate is determined by the number of units that leave the solid (dissolve). Each barium iodate formula unit which dissolves produces one barium ion and two iodate ions,

$$Ba(IO_3)_2 = Ba^{2+} + 2IO_3^-$$

Therefore there is a one-to-one correspondence between the solubility of barium iodate and the barium ions at equilibrium. Since the solution must be electrically neutral, it must be true that

$$2[Ba^{2+}] = [IO_3^-]$$

Substituting this into the expression for K_{sp} gives,

$$K_{sp} = [Ba^{2+}][IO_3^-]^2 = [Ba^{2+}](2[Ba^{2+}])^2$$
$$= 4[Ba^{2+}]^3$$

Solving for $[Ba^{2+}]$ gives the solubility,

$$\text{solubility} = [Ba^{2+}] = \left(\frac{K_{sp}}{4}\right)^{1/3} = \left(\frac{1.5 \times 10^{-9}}{4}\right)^{1/3}$$
$$= 7.3 \times 10^{-4} M$$

The iodate ion concentration at equilibrium will be twice that of barium ion, or $1.5 \times 10^{-3} M$. The next example shows how to calculate the solubility of the 2:3 salt Bi_2S_3.

EXAMPLE 8.5 Using K_{sp} to Calculate Solubility

Calculate the molar solubility of Bi_2S_3 neglecting intrinsic solubility. $K_{sp} = 1 \times 10^{-96}$.

Solution

For every formula unit of Bi_2S_3 that dissolves, two bismuth ions and three sulfide ions are released, that is,

$$Bi_2S_3(s) = 2Bi^{3+} + 3S^{2-}$$

The solubility of Bi_2S_3 is therefore given by *one-half* the Bi^{3+} ion concentration at equilibrium.

Electroneutrality requires that

$$3[Bi^{3+}] = 2[S^{2-}]$$

Substituting this condition into the solubility product relationship

$$K_{sp} = [Bi^{3+}]^2[S^{2-}]^3$$

gives

$$K_{sp} = [Bi^{3+}]^2(\tfrac{3}{2}[Bi^{3+}])^3$$
$$1 \times 10^{-96} = \tfrac{27}{8}[Bi^{3+}]^5$$
$$[Bi^{3+}] = (3 \times 10^{-97})^{1/5} = 4.9 \times 10^{-20} M$$
$$\text{solubility} = \frac{[Bi^{3+}]}{2} = 2.5 \times 10^{-20} M\dagger$$

The same result can be obtained if the solubility is set equal to $[S^{2-}]/3$, and it is again realized that $[Bi^{3+}] = (2/3)[S^{2-}]$.

8.5 USING K_{sp} TO PREDICT PRECIPITATION

An important question is often whether or not an ion can be precipitated from solution by the addition of a certain amount of some reagent. What must be considered is

---

† This is an interesting theoretical result, but unfortunately it is not very useful for laboratory work with sulfide precipitates. Equilibrium conditions are almost never attained in sulfide precipitations. Complex sulfides (BiS_2^-, BiS_3^{3-}, etc.), soluble hydroxides, and halide complexes are formed and all contribute to increasing the solubility of metal sulfides by factors of 10^5 or more. Slow equilibration makes the measurement of species activities quite difficult. For this reason solubility products for metal sulfides are calculated from free energy data.

8.5 USING K_{sp} TO PREDICT PRECIPITATION

whether the product of the concentrations of the ions which form the insoluble compound exceeds the solubility product. If the product of the concentrations (the "ion product") is greater than K_{sp}, the system is not at equilibrium, and precipitation will occur until equilibrium is reached.

EXAMPLE 8.6 Exceeding K_{sp}

A liter of solution contains 11.0 mg of dissolved silver. How many milligrams of potassium chromate (K_2CrO_4) can be added to the solution before silver chromate precipitates? The K_{sp} of Ag_2CrO_4 is 1.3×10^{-12}. The atomic weight of silver is 107.8, and the formula weight of K_2CrO_4 is 194.2.

Solution

Begin by calculating the molar concentration of silver. Next use the solubility product relationship for Ag_2CrO_4 to calculate the concentration of chromate ion needed to exceed K_{sp} in the presence of the silver ion in the solution. Finally, convert the chromate concentration to a weight of potassium chromate.

$$[Ag^+] = \frac{11.0 \times 10^{-3} \text{ g/liter}}{107.8 \text{ g/mole}} = 1.02 \times 10^{-4} \text{ mole/liter}$$

Precipitation of Ag_2CrO_4 should begin when the solubility product is just exceeded by the ion product,

$$K_{sp} \leq [Ag^+]^2[CrO_4^{2-}]$$

Rearranging and solving for chromate ion concentration gives

$$[CrO_4^{2-}] \geq \frac{K_{sp}}{[Ag^+]^2} = \frac{1.3 \times 10^{-12}}{(1.02 \times 10^{-4})^2} = 1.25 \times 10^{-4} M$$

This concentration corresponds to

$$1.25 \times 10^{-4} \text{ mole} \times 194.2 \text{ g } K_2CrO_4/\text{mole} = \underline{24.3 \text{ mg } K_2CrO_4}$$

Under certain circumstances it may be possible to separate two or more ions from each other by precipitation. In fact, this technique forms the basis of qualitative analysis schemes studied by most students in general chemistry. The completeness of separation depends on the relative solubilities of the compounds precipitating, as shown in the next example.

EXAMPLE 8.7 Separating Ions by Precipitation

Solid sodium chloride is added in small increments to a solution containing $0.10F$ Ag^+ and $0.10F$ Pb^{+2}. Predict which compound will precipitate first, $PbCl_2$ or $AgCl$. Solubility products are 1.6×10^{-5} and 1.8×10^{-10}, respectively.

Solution

The compound whose ion product more greatly exceeds its solubility product will precipitate first. The following parallel calculations are used to determine the concentration of chloride ion needed to begin precipitation of $AgCl$ and $PbCl_2$:

AgCl	PbCl$_2$
$K_{sp} = [Ag^+][Cl^-]$	$K_{sp} = [Pb^{2+}][Cl^-]^2$
$[Cl^-] = K_{sp}/[Ag^+]$	$[Cl^-] = (K_{sp}/[Pb^{2+}])^{1/2}$
$\quad = 1.8 \times 10^{-10}/0.10$	$\quad = (1.6 \times 10^{-5}/0.10)^{1/2}$
$\quad = 1.8 \times 10^{-9} M$	$\quad = 1.3 \times 10^{-2} M$

If a micromole (10^{-6} mole) of chloride were added to a liter of the solution in this example, silver chloride would precipitate, since $10^{-6} M$ chloride ion exceeds the critical concentration of chloride in $0.10 M$ Ag^+. A micromole of chloride would not precipitate lead chloride, however, since the solubility product of lead chloride would not be exceeded. Silver chloride will precipitate before lead chloride.

EXAMPLE 8.8 Separating Ions by Precipitation

Solid sodium chloride is added to 1 liter of a solution containing 0.10 mole Ag^+ and 0.10 mole Pb^{2+} until $PbCl_2$ just begins to precipitate. Calculate the concentration of silver ion left in solution at this point.

Solution

The concentration of chloride ion needed to start the precipitation of lead chloride can be calculated from the K_{sp} relationship for $PbCl_2$:

$$K_{sp} = [Pb^{2+}][Cl^-]^2$$
$$[Cl^-] = (K_{sp}/[Pb^{2+}])^{1/2} = [(1.6 \times 10^{-5})/0.10]^{1/2}$$
$$= 1.3 \times 10^{-2} M$$

The concentration of silver ion in solution can be calculated from the K_{sp} relationship for AgCl and the chloride ion concentration just calculated:

$$K_{sp} = [Ag^+][Cl^-]$$
$$[Ag^+] = K_{sp}/[Cl^-] = 1.8 \times 10^{-10}/1.3 \times 10^{-2}$$
$$= 1.4 \times 10^{-8} M$$

This concentration of silver ion is about $(1.4 \times 10^{-8}/0.10)(100) = 10^{-5}$% of the original silver ion concentration.

EXAMPLE 8.9 Quantitative Separation

Given the solution used in the previous two examples, calculate the range in chloride ion concentrations over which Ag^+ can be separated quantitatively (99.9%) from solution without the precipitation of $PbCl_2$.

Solution

If 99.9% of the silver ion is to be precipitated, the lower chloride ion concentration limit can be calculated from the K_{sp} relationship for AgCl, with $[Ag^+] = 0.1$% $(0.1\ M) = 10^{-4}\ M$:

$$[Cl^-] = K_{sp}/[Ag^+] = 1.8 \times 10^{-10}/10^{-4} = 1.8 \times 10^{-6} M$$

The upper limit for chloride ion concentration is reached just before $PbCl_2$ begins to precipitate, that is,

$$[Cl^-] = (K_{sp}/[Pb^{2+}])^{1/2} = (1.6 \times 10^{-5}/0.10)^{1/2} = 1.3 \times 10^{-2} M$$

Therefore the range in chloride ion concentration for the quantitative separation of Ag^+ from Pb^{2+} is

$$1.8 \times 10^{-6} M \leq [Cl^-] < 1.3 \times 10^{-2} M$$

8.6 THE COMMON ION EFFECT

The position of a solubility equilibrium will, of course, depend on the relative concentrations of the species which form the insoluble salt. As pointed out in Chapter 3, an equilibrium constant is independent of concentrations; when the concentration of one species is changed, the concentrations of the other species increase or decrease to conserve the equilibrium constant. So, for example, if we take a saturated solution of barium iodate and add to it more iodate ion, the equilibrium

$$Ba(IO_3)_2(s) = Ba^{2+}(aq) + 2IO_3^-(aq)$$

will shift to form more solid barium iodate. The solubility of barium iodate is therefore suppressed by the addition of iodate ion, the ion "common" to both solution and precipitate. In the following example the solubility of barium iodate is calculated for solutions in pure water and $0.01F$ potassium iodate.

EXAMPLE 8.10 Common Ion Effect

The solubility product of barium iodate is 1.5×10^{-9}. Calculate the solubility in moles per liter of barium iodate in pure water and then in $0.01F$ KIO_3. Assume that the intrinsic solubility of $Ba(IO_3)_2$ is negligible and that concentrations can be used instead of activities.

Solution

(a) In pure water, electroneutrality requires that

$$2[Ba^{2+}] = [IO_3^-]$$

For every barium iodate formula unit which dissolves, one barium ion is released. Therefore the barium ion concentration at equilibrium will equal the solubility of barium iodate. (The iodate ion concentration will equal twice the solubility.)
Using the K_{sp} relationship,

$$K_{sp} = [Ba^{2+}][IO_3^-]^2$$

and expressing it in terms of barium ion only gives

$$K_{sp} = [Ba^{2+}](2[Ba^{2+}])^2$$

Solving for $[Ba^{2+}]$, we find

$$4[Ba^{2+}]^3 = 1.5 \times 10^{-9}$$
$$\text{solubility} = [Ba^{2+}] = \underline{7.2 \times 10^{-4} M}$$

(b) In $0.01F$ KIO_3, electroneutrality requires that

$$2[Ba^{2+}] + [K^+] = [IO_3^-]$$

8.6 THE COMMON ION EFFECT

Since KIO_3 is completely soluble and dissociated, $[K^+] = 0.01 M$. The solubility product expression for barium iodate is

$$K_{sp} = [Ba^{2+}][IO_3^-]^2$$

Substituting the electroneutrality condition for $[IO_3^-]$ gives

$$K_{sp} = [Ba^{2+}](2[Ba^{2+}] + 0.01)^2$$

This is a cubic equation, and its exact solution is difficult and time-consuming. Rather than try to solve it exactly, let us simplify it by *assuming* for the moment that

$$2[Ba^{2+}] \ll 0.01$$

The simplified expression is then more easily solved:

$$K_{sp} = [Ba^{2+}](0.01)^2$$
$$[Ba^{2+}] = K_{sp}/10^{-4} = 1.5 \times 10^{-9}/10^{-4} = 1.5 \times 10^{-5} M$$

This is the molar solubility of barium iodate in $0.01 F\ KIO_3$. The simplifying assumption must be tested:

$$2[Ba^{2+}] \ll 0.01?$$
$$2(1.5 \times 10^{-5} M) = 3 \times 10^{-5} M \ll 0.01 M$$

and the assumption was valid. For this kind of assumption to be valid the equilibrium concentration should be less than about 5% of the initial common ion concentration. Barium iodate is almost 50 times less soluble in $0.01 F\ KIO_3$ than in pure water. (For practice you may wish to calculate the solubility of barium iodate in $0.01 F\ KIO_3$ taking into account the effect of KIO_3 electrolyte on the activities of the ions. Is $Ba(IO_3)_2$ 50 times less soluble when a Debye-Hückel calculation is used?)

In the event that a simplifying assumption about the concentration of common ion is invalid, the equation for solubility must be solved exactly. The following example shows such a situation.

EXAMPLE 8.11 Common Ion Effect

Calculate the solubility of TlCl in $0.01 F\ KCl$. $K_{sp} = 1.9 \times 10^{-4}$. KCl is completely dissociated.

Solution

The electroneutrality condition requires that

$$[Tl^+] + [K^+] = [Cl^-]$$

The solubility product expression for thallium chloride is

$$K_{sp} = [Tl^+][Cl^-]$$

Substituting the electroneutrality condition for $[Cl^-]$ gives

$$K_{sp} = [Tl^+]([Tl^+] + 0.010)$$
$$= [Tl^+]^2 + 0.010[Tl^+]$$

The last expression is a quadratic equation. To approximate a solution we might assume that $[Tl^+] \ll 0.010M$, and so

$$[Tl^+] = \frac{K_{sp}}{0.010} = 1.9 \times 10^{-2} M$$

Clearly this concentration is *larger* than $0.010M$, and the simplifying assumption is not valid.

At this point it is appropriate to solve the expression exactly using the quadratic formula,

$$x = \frac{-b \pm (b^2 - 4ac)^{1/2}}{2a}$$

In this example $x = [Tl^+]$ and $a = 1$, $b = 0.010$, $c = K_{sp}$.

$$[Tl^+] = \frac{-0.010 \pm (10^{-4} + 4K_{sp})^{1/2}}{2}$$
$$= 9.7 \times 10^{-3} M \text{ (positive root)}$$
$$= \text{solubility of TlCl}$$

8.7 COMPLEXATION

In this section we will carefully examine equilibria which produce soluble "complex ion" species when Lewis bases compete for the cation of a slightly soluble salt. The

8.7 COMPLEXATION

goal will be to understand such systems well enough to be able to calculate the solubility and the concentrations of all soluble forms of a salt when complexation occurs. The treatment you are about to study is an extension of the methods described in Chapter 5 for dealing with equilibria involving Brønsted-Lowry acids and bases, and it is extremely process-oriented. At this point it is more important to learn how to deal with complex equilibria than to focus on applications. The practical significance of complexation reactions (beyond their effects on solubility) and some interesting descriptive chemistry will be taken up in Chapter 12.

Let us begin by describing what is known about the silver chloride system. There are four soluble AgCl species, uncharged AgCl [AgCl(aq)] and three anionic species, $AgCl_2^-$, $AgCl_3^{2-}$, and $AgCl_4^{3-}$. These species are related to solid AgCl by the following equilibria:

(0) $\quad AgCl(s) = AgCl(aq) \quad K_0 = [AgCl(aq)] = 2.7 \times 10^{-7}$

(1) $\quad Ag^+ + Cl^- = AgCl(aq) \quad K_1 = \dfrac{[AgCl(aq)]}{[Ag^+][Cl^-]} = 7.1 \times 10^2$

(2) $\quad AgCl(aq) + Cl^- = AgCl_2^- \quad K_2 = \dfrac{[AgCl_2^-]}{[AgCl(aq)][Cl^-]} = 7.4 \times 10^2$

(3) $\quad AgCl_2^- + Cl^- = AgCl_3^{2-} \quad K_3 = \dfrac{[AgCl_3^{2-}]}{[AgCl_2^-][Cl^-]} = 2.1$

(4) $\quad AgCl_3^{2-} + Cl^- = AgCl_4^{3-} \quad K_4 = \dfrac{[AgCl_4^{3-}]}{[AgCl_3^{2-}][Cl^-]} = 7.2$

(5) $\quad AgCl(s) = Ag^+ + Cl^- \quad K_{sp} = [Ag^+][Cl^-] = 3.80 \times 10^{-10}$
(Notice that $K_0 = K_1 K_{sp}$.)

Reactions (1) through (4) are written as *formation reactions* rather than *dissociation reactions* as you became used to seeing in Chapter 5. The equilibrium constants for reactions written in this way are called *formation constants* or *stability constants*. The larger the formation constant, the more thermodynamically stable the complex. These are not thermodynamic constants. They were determined in $0.20M$ $NaClO_4$ and are "concentration constants."

The complex species all serve as sinks of solid silver chloride and explain why its solubility increases in the presence of excess chloride. The solubility is expressed as a material balance for silver:

$$\text{Solubility} = C_{Ag} = [Ag^+] + [AgCl(aq)] + [AgCl_2^-] + [AgCl_3^{2-}] + [AgCl_4^{3-}]$$

This expression should remind you of the expressions used in Chapter 5 to tally the various forms of ascorbate ion in solutions of ascorbic acid. It is an extension of an expression used earlier in this chapter when complexes were not being considered

$$\text{Solubility} = [Ag^+] + [AgCl(aq)]$$

The concentration of free silver ion at equilibrium is given by the product of C_{Ag} and the fraction of free silver ion, α_{Ag^+},

$$[Ag^+] = (\alpha_{Ag^+})C_{Ag}$$

We can then relate the solubility product constant to C_{Ag} and the chloride ion concentration by substitution,

$$K_{sp} = (\alpha_{Ag^+})C_{Ag}[Cl^-]$$

We are now able to define a special equilibrium constant called the *conditional constant* in terms of the actual solubility (the sum of the concentrations of all the soluble forms of silver ion) in a solution containing chloride:

$$K'_{sp} = K_{sp}/\alpha_{Ag^+} = C_{Ag}[Cl^-]$$

The conditional constant describes the solubility of AgCl at any particular concentration of chloride ion. When using the conditional constant, we treat all the silver-containing species as if they were a single species.

The fraction of free silver ion can be calculated from the material balance

$$\alpha_{Ag^+} = \frac{[Ag^+]}{C_{Ag}} = \frac{[Ag^+]}{[Ag^+] + [AgCl(aq)] + [AgCl_2^-] + [AgCl_3^{2-}] + [AgCl_4^{3-}]}$$

To evaluate the fraction numerically we use the strategy of expressing all terms in the denominator as products of formation constants and the equilibrium concentration of chloride ion. For example, this is done for the species $AgCl_2^-$ by combining the expressions for K_2 and K_0,

$$K_2 = \frac{[AgCl_2^-]}{[AgCl(aq)][Cl^-]} \qquad K_0 = [AgCl(aq)] = K_1 K_{sp}$$

Rearranging the expression for K_2 gives

$$[AgCl_2^-] = K_2[AgCl(aq)][Cl^-]$$

Substitution for [AgCl(aq)] gives

$$[AgCl_2^-] = K_{sp}K_1K_2[Cl^-] = K_{sp}\beta_2[Cl^-]$$

By similar processes we find that

$$[AgCl_3^{2-}] = K_{sp}K_1K_2K_3[Cl^-]^2 = K_{sp}\beta_3[Cl^-]^2$$
$$[AgCl_4^{3-}] = K_{sp}\beta_4[Cl^-]^3$$

8.7 COMPLEXATION

The new term, β, has been introduced to help simplify the task of writing such long expressions. It is the product of two or more stepwise formation constants. The subscript describes the number of formation constants multiplied together; for example, $\beta_3 = K_1 K_2 K_3$. Some references use a different subscript code, and you should always check a reference key.

Using the beta notation we can write the expression for the fraction of silver ion as

$$\alpha_{Ag^+} = \frac{K_{sp}/[Cl^-]}{K_{sp}/[Cl^-] + K_{sp}K_1 + K_{sp}\beta_2[Cl^-] + K_{sp}\beta_3[Cl^-]^2 + K_{sp}\beta_4[Cl^-]^3}$$

or, dividing through by $K_{sp}/[Cl^-]$,

$$\alpha_{Ag^+} = \frac{1}{1 + K_1[Cl^-] + \beta_2[Cl^-]^2 + \beta_3[Cl^-]^3 + \beta_4[Cl^-]^4}$$

The following example illustrates the calculation.

EXAMPLE 8.12 Formation of Soluble Complexes
Calculate the molar solubility of silver chloride in $0.20F$ NaCl.

Solution

The solubility can be calculated from the conditional K_{sp} by assuming that the equilibrium concentration of chloride ion is $0.20F$:

$$\text{solubility} = C_{Ag} = K'_{sp}/[Cl^-] = K_{sp}/\alpha_{Ag}[Cl^-]$$

The conditional K_{sp} can be calculated once the fraction of free silver ion is calculated. We must remember to test the assumption that the initial and equilibrium concentrations of chloride are the same.

Substituting values for the formation constants and formal concentration of chloride ion into the equation derived above for the fraction of free silver gives a value of 2.35×10^{-5}. The conditional formation constant is then calculated to be $3.80 \times 10^{-10}/2.35 \times 10^{-5} = 1.62 \times 10^{-5}$. The solubility is found by dividing the conditional constant by the chloride ion concentration,

$$\text{solubility} = C_{Ag} = K'_{sp}/[Cl^-] = 8.10 \times 10^{-5} M$$

The concentrations of the other soluble forms of silver can be calculated from their respective fraction equations. In $0.20 M$ Cl^- these concentrations are:

$$[Ag^+] = 1.9 \times 10^{-9} M$$
$$[AgCl(aq)] = 2.7 \times 10^{-7} M$$
$$[AgCl_2^-] = 4.0 \times 10^{-5} M$$
$$[AgCl_3^{2-}] = 1.7 \times 10^{-5} M$$
$$[AgCl_4^{3-}] = 2.4 \times 10^{-5} M$$

As a quick check on the calculations, notice that the sum of these concentrations equals 8.1×10^{-5} M, as of course it must.

As a final detail we must check the validity of the assumption that the equilibrium concentration of chloride ion is not significantly smaller than its initial value. We can do this by evaluating a material balance expression for chloride ion:

$$\text{total Cl}^- = 1.0 M = [Cl^-] + [AgCl(aq)] + 2[AgCl_2^-] + 3[AgCl_3^{2-}] + 4[AgCl_4^{3-}]$$

In this expression the initial chloride ion is divided among free chloride and the soluble chloride species. Since two chlorides are tied up in the species $AgCl_2^-$, the concentration of $AgCl_2^-$ at equilibrium is multiplied by the factor 2. Similar reasoning applies to the factors 3 and 4 used for $AgCl_3^{2-}$ and $AgCl_4^{3-}$, respectively.

The free chloride ion concentration can be found by rearranging the Cl^- material balance and substituting the concentrations of the species just calculated. The result is $[Cl^-] = 0.1998 M$. Since this is only 0.1% smaller than the initial concentration, we can be comfortable accepting the assumption that the initial and equilibrium concentrations of chloride ion are the same. However, had the free chloride concentration turned out to be less than about 95% of the initial concentration, in the interest of accuracy we would substitute the calculated value back into the equations used to determine the concentrations of the individual species (and the solubility) and go through them again. The new concentration values would then be used to calculate a new free chloride ion concentration, and so on and so on, until concentrations converged on correct values.

Complexation equilibria such as these often produce high-order equations (e.g., $[Cl^-]^4$). Appendix 6 describes an iterative method which is very useful for solving such problems.

8.8 GRAPHICAL REPRESENTATIONS

The equilibria of complex ion systems are generally more difficult to understand than the equilibria of acid-base systems simply because there are so many more species to think about and keep track of. There are two graphical representations that may help in learning about the ways in which the common ion effect and the formation of soluble complexes simultaneously influence solubility. The first involves a plot of total solubility as a function of common ion concentration. For example, Lieser (Ref. *1*) studied the solubilities of AgI, AgCl, and AgBr as functions of their respective halide

8.8 GRAPHICAL REPRESENTATIONS

ion concentrations. His data for AgI and AgCl are plotted in Fig. 8.1. Notice that both compounds show a minimum in solubility; AgCl is least soluble in a solution in which $[Cl^-] = 2.5 \times 10^{-3} M$ while AgI is least soluble in a solution in which $[I^-] = 1 \times 10^{-6} M$. The solubilities of these compounds in pure water are *greater* than these minimum values and lie to the left of the curve minima. Both AgCl and AgI become more soluble as halide ion concentrations are increased (to the right of the minima) and soluble complexes form. While the figure does show the total concentration of all forms of each salt, we cannot tell anything about the relative importance of the various forms.

Much more can be learned if we plot formation constant and solubility product data in the manner that we developed in Chapter 5 for acid-base systems. Remember that in those systems all species were related by hydronium ions. The oxalic acid species system is an example:

$$H_2C_2O_4 = H^+ + HC_2O_4^- \qquad pK_{a1} = 1.27$$
$$HC_2O_4^- = H^+ + C_2O_4^{2-} \qquad pK_{a2} = 4.27$$

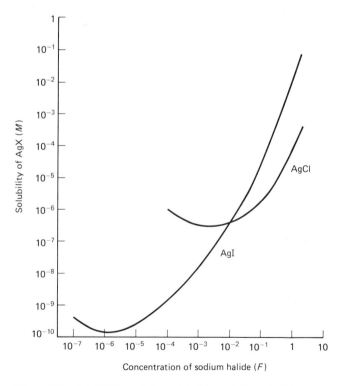

Figure 8.1 Solubilities of silver chloride and silver iodide as functions of sodium halide concentration at 18°C. [Data from K. H. Lieser, *Zeitschrift für Anorganische und Allgemeine Chemie*, 292, 97 (1957).]

In solutions in which [H$^+$] is large, H$_2$C$_2$O$_4$ and HC$_2$O$_4^-$ are the dominant species. In solutions in which [H$^+$] is small, C$_2$O$_4^{2-}$ is the dominant species. Recall that we drew diagrams in Chapter 5 in which logarithms of the concentrations of the various acid species were plotted as a function of the negative logarithm of the concentration of the species that relates them all, [H$^+$].

In an analogous manner, the soluble complex species in the silver chloride system are all interrelated by chloride ions. In solutions with high chloride ion concentrations, the species AgCl$_4^{3-}$ and AgCl$_3^{2-}$ are dominant, while in solutions containing very little chloride ion, AgCl(aq) and Ag$^+$ are dominant. Rather than use $-\log[\text{H}^+]$ as the plotting variable, we will use $-\log[\text{Cl}^-]$ and draw lines representing the changes in concentrations of the complex species as shown in Fig. 8.2. Starting with the simplest relationship,

$$K_{sp} = [\text{Ag}^+][\text{Cl}^-]$$

we arrange to solve for [Ag$^+$],

$$[\text{Ag}^+] = K_{sp}/[\text{Cl}^-]$$

and take logarithms of both sides,

$$\log[\text{Ag}^+] = \log K_{sp} - \log[\text{Cl}^-] = -9.75 - \log[\text{Cl}^-]\dagger$$

In Fig. 8.2 we plot a line representing [Ag$^+$] which intersects the log C axis at -9.75 and has a slope of -1 (note the sign of the vertical axis).

Next consider [AgCl(aq)]:

$$[\text{AgCl(aq)}] = K_{sp}K_1 \quad \text{where} \quad K_1 = \frac{\text{AgCl(aq)}}{[\text{Ag}^+][\text{Cl}^-]} = 10^{3.30}$$

Again, take logarithms of both sides,

$$\log[\text{AgCl(aq)}] = \log K_{sp} + \log K_1 = -9.75 + 3.30 = -6.45$$

Since [AgCl(aq)] is independent of [Cl$^-$], a line in Fig. 8.2 for [AgCl(aq)] is drawn as a horizontal line intersecting the log C axis at -6.45.

The expression for [AgCl$_2^-$] derived earlier was

$$[\text{AgCl}_2^-] = K_{sp}K_1K_2[\text{Cl}^-]$$

Taking logarithms of both sides gives

$$\log[\text{AgCl}_2^-] = \log(K_{sp}K_1K_2) + \log[\text{Cl}^-]$$
$$= -4.45 + \log[\text{Cl}^-]$$

---

† It will simplify matters if we use thermodynamic constants in the graphical treatment. Ionic strength changes dramatically across the diagram, and corrections are tedious.

8.8 GRAPHICAL REPRESENTATIONS

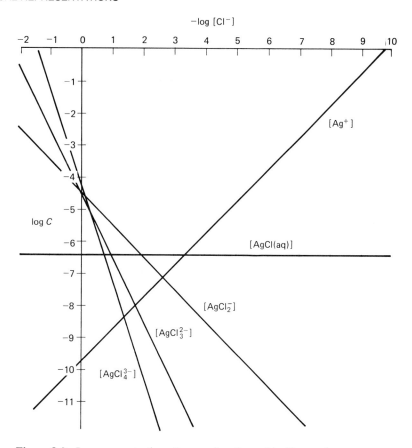

Figure 8.2 Log concentration diagram for silver chloride species.

This is plotted in Fig. 8.2 as a straight line of slope $+1$ intersecting the log C axis at -4.45.

By exactly the same procedure, the relationships for $\log[\text{AgCl}_3^{2-}]$ and $\log[\text{AgCl}_4^{3-}]$ are found:

$$\log[\text{AgCl}_3^{2-}] = -4.45 + 2\log[\text{Cl}^-]$$
$$\log[\text{AgCl}_4^{3-}] = -4.19 + 3\log[\text{Cl}^-]$$

Let us now answer some simple questions using the diagram.

EXAMPLE 8.13 Graphical Methods
Use Fig. 8.2 to answer the following questions.
(a) What is the equilibrium constant for the following reaction?

$$\text{AgCl}_3^{2-} = \text{Ag}^+ + 3\text{Cl}^-$$

(b) What is the solubility of AgCl in 1.0F chloride ion (Example 8.12), and what are the concentrations of all species in a solution in which [Cl$^-$] = 1.0F?

(c) At what chloride ion concentration is the solubility of AgCl minimal?

Solutions

(a) Using $K = [\text{Ag}^+][\text{Cl}^-]^3/[\text{AgCl}_3^{2-}]$, take logarithms of both sides and rearrange:

$$\log K - 3\log[\text{Cl}^-] = \log[\text{Ag}^+] - \log[\text{AgCl}_3^{2-}]$$

Look for the point on the diagram where $\log[\text{Ag}^+] = \log[\text{AgCl}_3^{2-}]$, that is, where the [Ag$^+$] and [AgCl$_3^{2-}$] lines intersect. At this point $\log[\text{Cl}^-]$ is read as -1.77. Therefore

$$\log K = 3\log[\text{Cl}^-] = 3(-1.77) = -5.31$$
$$K = 4.9 \times 10^{-6}$$

(b) The solubility is given by the sum of all the concentrations. When [Cl$^-$] = 1.0M, the most important species is AgCl$_4^{3-}$, as can be seen in the diagram. AgCl$_3^{2-}$ and AgCl$_2^-$ are only slightly less important. [Ag$^+$] and [AgCl(aq)] are clearly too small to be considered important in this system. Therefore

$$\text{solubility} = [\text{AgCl}_4^{3-}] + [\text{AgCl}_3^{2-}] + [\text{AgCl}_2^-]$$
$$= 10^{-4.19} + 10^{-4.45} + 10^{-4.45} = 1.4 \times 10^{-4} M$$

(c) Once again, the solubility is given by the sum of the concentrations of all the species. We can see the point of minimum solubility as the minimum in the sum of the logarithms of [Ag$^+$] and [AgCl$_2^-$], occurring at $\log[\text{Cl}^-] = -2.65$ or [Cl$^-$] = 2.5 $\times 10^{-3} M$. Refer back to the data of Lieser, plotted in Fig. 8.1.

Figure 8.3 is a log concentration diagram for the cuprous chloride system. A question in the problem set involves the equilibria in this system.

8.9 FOREIGN LIGAND COMPETITION

The solubility of a compound such as AgCl can be increased dramatically by the addition of a species (a foreign ligand) which forms a soluble complex with the metal ion. For example, ammonia is known to form two complexes with silver ion, the silver ammine complexes,

8.9 FOREIGN LIGAND COMPETITION

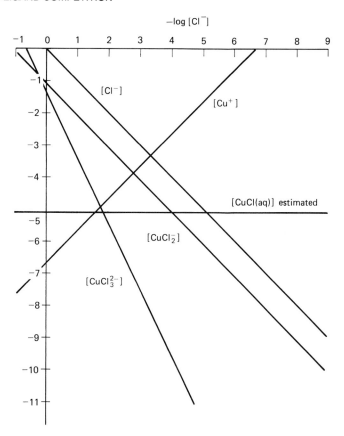

Figure 8.3 Log concentration diagram for cuprous chloride species.

$$Ag^+ + NH_3 = Ag(NH_3)^+ \qquad K_1 = \frac{[Ag(NH_3)^+]}{[Ag^+][NH_3]} = 2.34 \times 10^3$$

$$Ag(NH_3)^+ + NH_3 = Ag(NH_3)_2^+ \qquad K_2 = \frac{[Ag(NH_3)_2^+]}{[Ag(NH_3)^+][NH_3]} = 6.92 \times 10^3$$

When ammonia is added to a solution saturated with AgCl, it will react with free silver ion and pull the solubility equilibrium

$$AgCl(s) = Ag^+ + Cl^-$$

to the right, increasing solubility. In such a situation the solubility of silver chloride can be expressed by the silver material balance:

$$\text{solubility} = C_{\text{total Ag}} = [Ag^+] + [AgCl(aq)] + [Ag(NH_3)^+] + [Ag(NH_3)_2^+]$$

The intrinsic solubility of AgCl is about $2 \times 10^{-7} M$ and will be neglected in this treatment. Therefore, whatever AgCl dissolves will free one Ag^+ and one Cl^-; the chloride ion concentration at equilibrium will also equal the solubility of AgCl. Let us assume that there will always be solid AgCl present.

The solubility product is related to the total silver concentration and the fraction of free silver by

$$K_{sp} = (\alpha_{Ag^+} C_{\text{total Ag}})[Cl^-]$$

The conditional constant is given by

$$K'_{sp} = \frac{K_{sp}}{\alpha_{Ag^+}} = [Cl^-] C_{\text{total Ag}}$$

We calculate the fraction of free silver ion from the relationship

$$\alpha_{Ag^+} = \frac{[Ag^+]}{[Ag^+] + [Ag(NH_3)^+] + [Ag(NH_3)_2^+]}$$

Substituting in equilibrium constants gives

$$\alpha_{Ag^+} = \frac{\cancel{[Ag^+]}}{\cancel{[Ag^+]}(1 + K_1[NH_3] + K_1 K_2[NH_3]^2)}$$

Using a value for the equilibrium concentration of ammonia and numerical values for K_1 and K_2, we can calculate K'_{sp} and therefore the solubility.

EXAMPLE 8.14 Competing Equilibrium

Calculate the solubility of silver chloride in a solution in which the equilibrium concentration of ammonia is $0.10 M$. Assume that complex chlorides and hydroxides of silver are unimportant.

Solution

First we calculate the fraction of free silver ion:

$$\alpha_{Ag^+} = \frac{1}{1 + (2.34 \times 10^3)(0.1) + (2.34 \times 10^3)(6.92 \times 10^3)(0.01)}$$
$$= 6.17 \times 10^{-6} \text{ (very little silver is uncomplexed)}$$

Next we find K'_{sp},

$$K'_{sp} = \frac{K_{sp}}{\alpha_{Ag^+}} = \frac{1.8 \times 10^{-10}}{6.17 \times 10^{-6}} = 2.9 \times 10^{-5}$$

Since we know that $[Cl^-] = C_{total\ Ag}$ = solubility, then

$$\text{solubility} = (K'_{sp})^{1/2} = 5.4 \times 10^{-3} M$$

Recall that the solubility of AgCl in pure water is only $1.34 \times 10^{-5} M$. Ammonia substantially increases the solubility of AgCl.

We can calculate the concentrations of the other species at equilibrium by using the solubility, $[Cl^-] = 5.4 \times 10^{-3} M$.

$$[Ag^+] = K_{sp}/[Cl^-] = 3.3 \times 10^{-8} M$$
$$[Ag(NH_3)^+] = K_1[Ag^+][NH_3] = 2.34 \times 10^3 (3.3 \times 10^{-8})(0.1) = 7.8 \times 10^{-6} M$$
$$[Ag(NH_3)_2^+] = K_2[Ag(NH_3)^+][NH_3] = 5.4 \times 10^{-3} M$$

Notice that the only species with a significant concentration is the diammine complex. It is almost totally responsible for the solubility of AgCl in $0.1M$ ammonia. To keep the problem manageable, we set the *equilibrium* concentration (not the initial concentration) of free ammonia at $0.1M$. We do not have to make any assumptions about its size or test any assumptions after the calculation. Presumably the initial concentration of NH_3 would have been about $0.1 + [Ag(NH_3)^+] + 2[Ag(NH_3)_2^+] = 0.1 + 8 \times 10^{-6} + 0.0108 = 0.111 F$.

8.10 ACID-BASE COMPETITION

Often the anion of a slightly soluble compound is a weak base and is able to react with water (solvent) to form a weak acid. The anions sulfate, chromate, acetate, iodate, and fluoride, among others, can be expected to react with water as weak bases. The general effect of the acid-base reaction is to increase the solubility of the metal salt.

Consider what happens when calcium oxalate, $Ca(C_2O_4)$, is equilibrated with pure water. The solubility equilibrium is

$$Ca(C_2O_4)(s) = Ca^{2+} + C_2O_4^{2-} \qquad K_{sp} = 1.3 \times 10^{-9}$$

Oxalate ion is a base and reacts with water to form the weak acid species $HC_2O_4^-$ and $H_2C_2O_4$. These acids dissociate with the following equilibrium constants:

$$H_2C_2O_4 = H^+ + HC_2O_4^- \quad K_1 = 8.8 \times 10^{-2}$$
$$HC_2O_4^- = H^+ + C_2O_4^{2-} \quad K_2 = 5.1 \times 10^{-5}$$

These reactions compete with Ca^{2+} for oxalate ions. The fact that oxalate is a base means that the fraction of free oxalate ion will be kept small by the solvent, and solubility will be increased. In this case the conditional solubility product is

$$K'_{sp} = \frac{K_{sp}}{\alpha_{ox^{2-}}} = C_{\text{total oxalate}}[Ca^{2+}]$$

The expressions for the fraction of weak acid forms were derived in Chapter 5. Recall that the fraction of a base like oxalate will be given by

$$\alpha_{ox^{2-}} = \frac{K_1 K_2}{[H^+]^2 + [H^+]K_1 + K_1 K_2}$$

EXAMPLE 8.15 Acid-Base Competition

Calculate the solubility of calcium oxalate in a solution buffered at pH 5.0.

Solution

In this problem and others like it, we will assume that the buffer has enough capacity to maintain its initial pH while the system reaches equilibrium.
Begin by calculating the fraction of free oxalate at pH 5.0.

$$\alpha_{ox^{2-}} = \frac{4.49 \times 10^{-6}}{1 \times 10^{-10} + 8.8 \times 10^{-7} + 4.49 \times 10^{-6}}$$
$$= 0.84$$

Now we can solve for K'_{sp}:

$$K'_{sp} = \frac{K_{sp}}{\alpha_{ox^{2-}}} = \frac{1.3 \times 10^{-9}}{0.84} = 1.5 \times 10^{-9}$$

Solubility will equal $[Ca^{2+}]$ and $C_{\text{tot.ox.}}$, and so

$$\text{solubility} = (K'_{sp})^{1/2} = 3.9 \times 10^{-5} M$$

8.10 ACID-BASE COMPETITION

This is slightly larger than the value predicted by the K_{sp} without taking into account competition (solubility = $3.6 \times 10^{-5} M$).

Now consider what happens if we add to the solution in the last example an agent such as ethylenediaminetetraacetic acid (EDTA) to complex free Ca^{2+}. The conditional constant will now become

$$K''_{sp} = \frac{K_{sp}}{(\alpha_{ox^{2-}})(\alpha_{Ca^{2+}})} = C_{tot.ox.} C_{tot.Ca}$$

EDTA combines with Ca^{2+} according to the reaction

$$Ca^{2+} + EDTA^{4-} = Ca(EDTA)^{2-} \qquad K' = 1.26 \times 10^4 \text{ in pH 5 buffer†}$$

The fraction of free calcium ion can be found from the material balance for calcium,

$$C_{tot.Ca} = [Ca^{2+}] + [Ca(EDTA)^{2-}]$$

$$\alpha_{Ca^{2+}} = \frac{[Ca^{2+}]}{[Ca^{2+}] + [Ca(EDTA)^{2-}]} = \frac{1}{1 + K'[EDTA^{4-}]}$$

EXAMPLE 8.16 Complex and Acid-Base Competition

Calculate the solubility of calcium oxalate in a solution buffered at pH 5 containing $0.05 M$ EDTA.

Solution

We calculated $\alpha_{ox^{2-}} = 0.84$ in the pH 5 buffer in the last example. Now calculate the fraction of free calcium ion from the relationship just given:

$$\alpha_{Ca^{2+}} = \frac{1}{1 + 1.26 \times 10^4 (0.05)} = 1.58 \times 10^{-3}$$

The solubility is given by $C_{tot.Ca} = C_{tot.ox.} = (K''_{sp})^{1/2}$. Now

$$(K''_{sp})^{1/2} = \left(\frac{K_{sp}}{\alpha_{Ca^{2+}} \alpha_{ox^{2-}}}\right)^{1/2} = 9.9 \times 10^{-4} M$$

---

† Only one EDTA combines with each calcium ion. This equilibrium constant is itself a conditional constant for a pH 5 solution. EDTA reactions will be discussed in detail in Chapter 12.

We have assumed that the equilibrium concentration of EDTA will be the same as the initial concentration of EDTA. The solubility tells us that the equilibrium concentration of EDTA will be at least $0.05 - 0.00099 = 0.049 M$.

Neglecting both kinds of competition, the solubility of calcium oxalate is calculated to be $3.6 \times 10^{-5} M$. Clearly the addition of EDTA has increased the solubility of calcium oxalate.

PROBLEMS

8.1. Using the standard free-energy data provided, calculate the thermodynamic solubility product constants for AgCl, AgI, and Ag_2CrO_4. (You may wish to refer to Chapter 3.)

Species	G^0 (kJ/mole)
AgCl(s)	−109.7
AgI(s)	−66.32
Ag_2CrO_4(s)	−647.2
Ag^+(aq)	77.11
Cl^-(aq)	−131.2
I^-(aq)	−51.67
CrO_4^{2-}(aq)	−736.8

8.2. Calculate the solubility in moles per liter of each of the following compounds. Use solubility product constants from Appendix II.
(a) $BaSO_4$ (b) $CaSO_4$ (c) CuI (d) $AgBr$

8.3. Using solubility product constants from Appendix II, calculate the solubilities of the following compounds in moles per liter and milligrams per 100 mL of pure water.
(a) PbI_2 (b) Ag_2SO_4 (c) $Ce(IO_3)_3$ (d) $Ba_3(AsO_4)_2$

8.4. The solubility of silver iodide is $5.74 \times 10^{-9} M$ at 18°C. The solubility product constant of silver iodide is 3.16×10^{-17} at the same temperature. What is the intrinsic solubility of silver iodide at 18°C?

8.5. 0.4310 gram of $Ag(IO_3)$ is added to 1 liter of pure water at 25°C. After the solution comes to equilibrium the remaining solid silver iodate is filtered, dried, and found to weigh 0.3810 g. Calculate the solubility product of silver iodate. The formula weight of silver iodate is 282.8.

8.6. 0.9520 gram of strontium iodate, $Sr(IO_3)_2$, is added to 100 mL of pure water. After the solution comes to equilibrium, the remaining precipitate is filtered, dried, and found to weigh 0.7610 g. Calculate the solubility product constant of $Sr(IO_3)_2$. The formula weight is 437.4.

PROBLEMS

8.7. The solubility of silver chloride in $0.01M$ KNO_3 is $1.49 \times 10^{-5}M$. If the mean activity coefficient of Ag^+/Cl^- is 0.896 in this solution, calculate the thermodynamic solubility product constant for AgCl. Neglect the intrinsic solubility of silver chloride.

8.8. Calculate the concentration of anion needed to just start the precipitation of metal ion from each of the following solutions.
   (a) $0.01M$ Ag^+ precipitated as AgCl
   (b) $10^{-3}M$ Ba^{2+} precipitated as $BaSO_4$.
   (c) $10^{-2}M$ Zn^{2+} precipitated at $Zn_3(PO_4)_2$
   (d) $10^{-4}M$ Ag^+ precipitated as AgOH

8.9. Calculate the concentration of anion needed to decrease the metal ion concentration in each of the following solutions to 0.1% its initial value.
   (a) $0.01M$ Ba^{2+}, precipitated as $BaCO_3$
   (b) $0.10M$ Fe^{2+}, precipitated as $Fe(OH)_2$
   (c) $0.10M$ Cu^{2+}, precipitated as $Cu_3(AsO_4)_2$
   (d) $0.001M$ Ag^+, precipitated as Ag_2S

8.10. Calculate the molar solubility of silver iodide in each of the following solutions:
   (a) pure water
   (b) $0.10M$ KNO_3
   (c) $0.10M$ $AgNO_3$
   Compare the solubilities and interpret the results.

8.11. A solution contains $0.010M$ Cu^+ and $0.010M$ Ag^+. Solid NaBr is added to the solution in very small increments.
   (a) Calculate the bromide ion concentration needed to start the precipitation of CuBr.
   (b) Calculate the concentration of bromide ion needed to start the precipitation of AgBr.
   (c) Which compound will precipitate first, CuBr or AgBr? Explain.
   (d) What is the concentration of the metal ion which precipitates first when the second metal ion begins to precipitate?

8.12. Two compounds, $A(OH)_2$ and $B(OH)_3$ have K_{sp} values which are numerically the same. Which compound is more soluble (molar concentration) in $0.10M$ NaOH?

8.13. 0.0100 mole of $AgNO_3$ is added to 100.0 mL of a solution containing 0.0100 mole of Cl^- and 0.0100 mole of IO_3^-. Using K_{sp} of $AgIO_3 = 3.00 \times 10^{-8}$ and K_{sp} of $AgCl = 1.80 \times 10^{-10}$,
   (a) Calculate the sum of the concentrations of chloride ion and iodate ion after the addition of silver nitrate.
   (b) Calculate the individual concentrations of chloride ion and iodate ion after the addition of silver nitrate.

(c) Calculate the chloride ion concentration at the moment when silver iodate begins to precipitate.

8.14. Calculate the concentrations of metal ion *and* anion in each of the following solutions:
(a) solid $Pb(IO_3)_2$ added to 100 mL of $0.01M$ KIO_3
(b) solid $Pb(BrO_3)_2$ added to 100 mL of $0.5M$ $KBrO_3$
(c) solid $Pb(BrO_3)_2$ added to 100 mL of $0.5M$ $Pb(NO_3)_2$

8.15. If 1.200 g of silver sulfate is added to 100 mL of $0.500M$ $AgNO_3$ solution, how many grams of Ag_2SO_4 will remain undissolved when the solution has reached equilibrium?
$K_{sp} = 1.2 \times 10^{-5}$. (Formula weight of $Ag_2SO_4 = 311.8$.)

8.16. Consider the following data for silver iodide complexes:

Reaction	K (18°C) (notation in text)
$AgI(s) = Ag^+ + I^-$	3.16×10^{-17} (K_{sp})
$Ag^+ + I^- = AgI(aq)$	3.8×10^6 (K_1)
$AgI(aq) + I^- = AgI_2^-$	1.47×10^5 (K_2)
$AgI_2^- + I^- = AgI_3^{2-}$	86 (K_3)
$AgI_3^{2-} + I^- = AgI_4^{3-}$	0.27 (K_4)

The object of this group of problems will be to calculate the solubility of AgI in $0.100M$ KI and examine complex ion equilibria.
(a) Write a material balance for silver.
(b) Express the material balance in terms of iodide ion as the only ion.
(c) Calculate the concentration of each complex silver iodide species assuming that the equilibrium concentration of iodide ion is the same as the initial concentration, $0.100M$.
(d) Identify the complex species which is in highest concentration.
(e) What is the intrinsic solubility of AgI?
(f) What is the solubility of AgI in $0.100M$ KI?
(g) What is the actual concentration of iodide ion at equilibrium?

8.17. Ethylenediaminetetraacetic acid (EDTA) forms strong 1:1 complexes with many metal ions. The equilibrium constant for the reaction EDTA anion, $EDTA^{4-}$, with barium ion,

$$Ba^{2+} + EDTA^{4-} = Ba(EDTA)^{2-}$$

is $K = 5.75 \times 10^7$.
Calculate the molar solubility of barium sulfate ($K_{sp} = 1.08 \times 10^{-10}$) in a solution which is made up to contain $0.05M$ $EDTA^{4-}$ before any barium sulfate dissolves.

PROBLEMS

(Hint: the initial and equilibrium concentrations of $EDTA^{4-}$ may be quite different.)

8.18. The following formation constants have been determined for cadmium iodide complexes:

$$Cd^{2+} + I^- = CdI^+ \quad K_1 = 190$$
$$CdI^+ + I^- = CdI_2 \quad K_2 = 44$$
$$CdI_2 + I^- = CdI_3^- \quad K_3 = 12$$
$$CdI_3^- + I^- = CdI_4^{2-} \quad K_4 = 13$$

Calculate the solubility of cadmium arsenate, $Cd_3(AsO_4)_2$, in $0.1M$ KI solution.

8.19. Calculate the solubility of calcium fluoride (CaF_2) in a solution in which the hydrogen ion concentration is maintained at exactly $1 \times 10^{-2}M$. Fluoride ion is the conjugate base of the weak acid, HF.

$$HF = H^+ + F^- \quad K_a = 6.7 \times 10^{-4}$$
$$CaF_2 = Ca^{2+} + 2F^- \quad K_{sp} = 4 \times 10^{-11}$$

8.20. Calculate the solubility of CaF_2 in 1 liter of solution in which hydrogen ion concentration is maintained at $10^{-2}M$, and to which 0.05 mole of $CaCl_2$ has been added. $CaCl_2$ dissociates completely.

8.21. Lead chromate, $PbCrO_4$, has a solubility product of 1.8×10^{-14} at 25°C. Chromate ion can react with water to form the weak acid $HCrO_4^-$. This acid dissociates as follows:

$$HCrO_4^- = CrO_4^{2-} + H^+ \quad K = 3.3 \times 10^{-6}$$

0.3000 g of $PbCrO_4$ is added to 1 liter of solution which is $0.10M$ in HNO_3. How many grams of $PbCrO_4$ could be recovered by filtering the solution? (formula weight $PbCrO_4 = 323.2$)

8.22. Answer these questions by using Fig. 8.3.
(a) What are the equilibrium constants for the following reactions?

$$Cu^+ + Cl^- = CuCl$$
$$Cu^+ + 2Cl^- = CuCl_2^-$$
$$CuCl_3^{2-} = CuCl + 2Cl^-$$

(b) What is the minimum solubility of CuCl in the presence of Cl^-?
(c) What is the solubility of CuCl in a solution in which $[Cl^-] = 0.01M$?

8.23. Given the following information about the thallium(I)-bromide system, prepare a diagram like Figs. 8.2 and 8.3 for TlBr:

$$\text{Tl}^+ + \text{Br}^- = \text{TlBr(s)} \qquad K_{sp} = 10^{-4.81}$$
$$\text{Tl}^+ + \text{Br}^- = \text{TlBr(aq)} \qquad K_1 = 10^{0.32}$$
$$\text{TlBr(aq)} + \text{Br}^- = \text{TlBr}_2^- \qquad K_2 = 10^{-0.17}$$
$$\text{TlBr}_2^- + \text{Br}^- = \text{TlBr}_3^{2-} \qquad K_3 = 10^{-0.45}$$
$$\text{TlBr}_3^{2-} + \text{Br}^- = \text{TlBr}_4^{3-} \qquad K_4 = 10^{-0.75}$$

Calculate the intrinsic solubility of TlBr and the bromide concentration (up to $1F$) at which the solubility of TlBr is a minimum.

8.24. Do Problem 8.23 using data for the lead(II)-bromide system:

$$\text{Pb}^{2+} + 2\text{Br}^- = \text{PbBr}_2(\text{s}) \qquad K_{sp} = 10^{-4.56}$$
$$\text{Pb}^{2+} + \text{Br}^- = \text{PbBr}^+ \qquad K_1 = 10^{1.56}$$
$$\text{PbBr}^+ + \text{Br}^- = \text{PbBr}_2(\text{aq}) \qquad K_2 = 10^{0.46}$$
$$\text{PbBr}_2(\text{aq}) + \text{Br}^- = \text{PbBr}_3^- \qquad K_3 = 10^{0.23}$$
$$\text{PbBr}_3^- + \text{Br}^- = \text{PbBr}_4^{2-} \qquad K_4 = 10^{0.41}$$

Chapter 9

Precipitation

9.1 INTRODUCTION

Analytical chemists use the phenomenon of precipitation both as a method of separation and as the basis for methods of analysis called *gravimetric methods.* The goal of a precipitation separation might be simply the removal of a compound which might interfere with the determination of another species; collecting the precipitate is not the goal, but removing the interference is. Alternatively, it might be possible to analyze a sample for one species by precipitating it quantitatively, collecting it carefully by filtration, purifying it by washing, then drying it and weighing it.

In this chapter we will look at precipitation as both a chemical and a physical phenomenon. We will discuss some important properties of precipitates and how some of these properties (such as surface area) can make analysis and separations more difficult. We will see how to control some of these properties to make analyses more accurate. We will also make use of some of the ideas on competing solubility equilibria presented in the last chapter to illustrate how precipitation can be used in separations.

9.2 NUCLEATION AND GROWTH OF PRECIPITATE CRYSTALS

Imagine a solution of sodium chloride at equilibrium with solid sodium chloride at the bottom of its container. Since the solution is at equilibrium, there is no net change in the amount of solid NaCl in the container; individual crystals are dissolving and reforming constantly, but there is no net change. Now imagine that we quickly evaporate 20 mL of water. Temporarily, the solution contains more sodium chloride than it can hold at equilibrium. We say that the solution has become "supersaturated" and is unstable. Solid NaCl will begin to appear as the solution tries to return to equilibrium. A process called *nucleation* starts to form the smallest crystalline precipitate particles. The number of crystal nuclei and the rate at which they form are critical factors in determining the size and purity of the precipitating crystals. If the precipitation process produces a large number of nuclei, precipitation will occur very quickly and the resulting crystals will be very small. Small crystals will tend to gather impurities (as you will see later), and they are also very hard to collect and wash with filters. If, however, the precipitation begins with very few nuclei, it will proceed much more slowly, and the crystals which form will be larger and more pure. The number of nuclei is important in determining the properties of the precipitate.

Although the exact mechanism by which nuclei form depends on the characteristics of each particular system, there are two models which are appealing in their simplicity. According to the first, nuclei can form "homogeneously" within the liquid phase. Local concentration fluctuations in solutions bring together aggregates of ions pairs (or "formula units"). These aggregates are in equilibrium with isolated ions in solution and continually form and disperse. As more solute is added the equilibrium shifts to favor the aggregates, some of which reach a critical size and do not disperse. Crystals form around these aggregates. According to the second model, nuclei are formed "heterogeneously" (i.e., in two phases) around foreign particles such as dust or pollen present in any solution, or even on the microscopically rough surfaces of a beaker. It is likely that a precipitation will involve both of these mechanisms.

The forces which draw ions from solution to nucleation sites are probably both electrostatic (attractions of opposite charges) and chemical in nature. For example, any site with an extra negative charge in a solution containing silver ions will attract silver ions to it. A cluster of silver ions will in turn attract chloride ions, which will attract more silver ions, and so on. The most effective nucleation sites will be those with specific chemical attractions for the ions being precipitated, what we might call "seed crystals." If the goal is simply to remove an impurity from solution by precipitating it, it might help to add some seed crystals to the solution to stimulate precipitation. However, if the goal is a quantitative assay of the precipitate, seed crystals would contribute weight to the product.

As mentioned above, precipitation occurs spontaneously from a solution which is supersaturated. There are varying degrees of supersaturation, depending on the concentration of solute and temperature of the solution. The behavior of a hypothetical solute is sketched in Fig. 9.1.

Referring to Fig. 9.1, an equilibrium condition exists when solid and solution are present at any pair of concentration and temperature values on the line labeled E. Solutions along this line are called *saturated* solutions. All solutions with (C, T) values above line E are *supersaturated* solutions. Line D is the line of critical supersaturation: solutions having (C, T) values that lie above line D are always unstable.

9.2 NUCLEATION AND GROWTH OF PRECIPITATE CRYSTALS

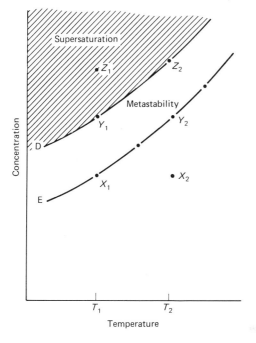

Figure 9.1 Plot of concentration vs. temperature for a solute.

Precipitation will occur from such solutions, lowering the system to line E. The region between lines D and E is the region of metastability. Solutions whose (C, T) values fall in this region may appear to be stable for short periods of time, but will eventually return to line E by precipitation.

Consider the points shown in Fig. 9.1. When a solution at point X_2 is cooled from T_2 to T_1, the concentration of solute remains constant, but the solution becomes saturated. Saturation is also achieved if enough solute is added at constant temperature to reach point Y_2. If solution Y_2 is cooled from T_2 to T_1, precipitation *may* not occur for some period of time, because the solution, while supersaturated, is metastable. Physical shock or a speck of dust falling into such a solution may stimulate sudden precipitation. Solution Z_2 lies on a line of critical supersaturation; if Z_2 is cooled only slightly, precipitation will occur and the (C, T) values for the solution will return to line E. Solution Z_1 is never a stable solution.

The degree of supersaturation in a solution influences the rate at which a precipitate forms. The proportionality between rate and degree of supersaturation is expressed in a semiquantitative way by the so-called von Weimarn equation,

$$\text{Rate} \propto \frac{Q - S}{S}$$

In this expression Q is the *actual* concentration of solute when precipitation begins, and S is the equilibrium solubility of coarse crystals of the solute. In terms of the plot in Fig. 9.1, $Q - S$ corresponds to the vertical distance between points X_1 and Y_1. The von Weimarn relationship is at best semiquantitative because the equilibrium solubility (S) is a function of particle size (small particles are more soluble than large ones) and ionic strength. Nevertheless, by changing $(Q - S)/S$ for a given precipitate it is possible to change the rate of precipitation and the size of the precipitate particles quite dramatically. When $(Q - S)/S$ is small, nuclei form slowly, and those that form grow slowly into relatively large crystals. When $(Q - S)/S$ is large, however, a large number of nuclei form at the very start of precipitation, and dissolved solute is used up so quickly that crystals cannot grow very large before the solution returns to a condition of saturation. In some cases small crystals formed quickly can be increased in size by secondary processes such as digestion, a topic to be discussed later.

Once a nucleus has been formed, a crystal can grow by a two-step process involving *diffusion* of solute ions to the nucleus surface and their subsequent *deposition*. The rate of growth can be limited by either process. The rate of diffusion depends on the concentration of ions near the surface and in the bulk of the solution and the temperature, whereas the rate of deposition depends on the nature of the surface and the pattern of crystal growth.

9.3 COLLOIDS

An understanding of the properties of colloidal dispersions is important to an analytical chemist, because at some point during precipitation every insoluble compound passes through the colloidal state. An analyst tries to avoid stabilizing the colloidal state in a quantitative precipitation because colloids are hard to trap in filters. On the other hand, there are several methods of analysis which depend on the formation of stable colloids. These are light-scattering methods (nephelometry and turbidimetry) and titrimetric methods which rely on adsorption indicators (Chapter 11).

A colloidal state exists when particles with sizes in the range of 1 nanometer (nm; 10^{-9} meter) to about 250 nanometers are dispersed in a medium. *Sols* are the most common colloids in inorganic analysis. They are called hydrophobic ("water-hating") colloids and can be brought out of dispersion ("flocculated") by the addition of small amounts of electrolyte (see below). *Gels* are called hydrophilic ("water-loving") colloids and are quite difficult to work with. They hold water tenaciously and are viscous, like jellies. They are also quite difficult to flocculate, requiring large amounts of electrolyte.† A number of complex organic molecules such as proteins and carbohydrates tend to form gels. A few inorganic precipitates form gels, notably silicic acid [a polyacid simply represented by $Si(OH)_4$] and hydrous aluminum and stannic oxides.

---

† Gels can be turned into sols by heating. There are several familiar sols: a dispersion of a solid in water is called a *hydrosol*. A dispersion of a solid in a gas is called an *aerosol*. *Emulsions*, perhaps the most important colloids in the chemical industry, are dispersions of one liquid in another.

Colloidal dispersions have peculiar properties. Like true solutions, they pass through ordinary filter paper and do not settle out by gravity. They can, however, be filtered by dialysis,† or be removed by ultracentrifugation. Colloids do not affect the freezing point, boiling point, or osmotic pressure of a solvent in the same ways as solutes in true solutions. Rather, colloids act as if they were solutions of solutes of very high molecular weight.

The most interesting properties of colloids are their optical and surface properties. A true solution of ions is said to be "optically empty"; that is, when a strong beam of light is shone through the solution no light is scattered at right angles to the light path. If, however, particles larger than about 10 nm are dispersed in a solvent, light is scattered at right angles at high intensity (the Tyndall effect). Like a true solution, a colloidal dispersion appears perfectly clear to the naked eye or even under a microscope. However, images of colloids are visible in an ultramicroscope, an instrument based on the Tyndall effect.‡

Perhaps the most important property of colloids is their immense surface area. Imagine a cube of silver chloride, 1.00 cm on a side, weighing about 5.6 g. It has a surface area of 6 cm^2 (six faces of the cube, each with area 1.00 cm^2). If this cube is divided into 1000 cubes, 1.0 mm on a side, each weighing 5.6 mg, the total surface area of the sample would be 6000 mm^2, or 60 cm^2. Further division of each 1.0-mm cube into 10^9 cubes 1 micrometer (μm) on a side, each weighing 5.6×10^{-12} g, produces a total surface area for the sample of 60,000 cm^2. To be within the size range for colloidal particles, each cube of silver chloride would have to be about 0.1 μm on a side, with a total sample surface area of 6×10^5 cm^2, or 60 m^2. This corresponds to the floor space of a room 24 feet square! The process of subdivision just described is summarized in Table 9.1. The immense surface area of samples of colloidal particles allows them to adsorb impurities with high efficiency. The next section deals with adsorption.

9.4 ADSORPTION

The contamination of colloidal surfaces involves both physical and chemical interactions. The physical force is electrostatic: positive and negative centers on the particle surface attract oppositely charged species in solution. There are also four specific

---

† Two substances can be separated by dialysis if they diffuse through the pores of a semipermeable material such as cellophane at different rates. In an experiment a colloidal dispersion is placed in a sack of cellophane, which is immersed in a vessel of pure water. Dissolved ions and colloidal particles try to diffuse from the sack to the water, but the colloidal particles diffuse through the membrane much too slowly to leave the sack.

‡ Ordinary microscopes view objects along the line of light propagation. The ultramicroscope, invented at the turn of the twentieth century, is used to view objects at right angles to the light path. The image is a diffraction image from light scattered by colloidal particles; small points of light appear to move against a dark background.

TABLE 9.1 DIVIDING A 5.6-GRAM CUBE OF AgCl TO PRODUCE COLLOIDAL PARTICLES

Number of cubes	Edge dimension	Total surface area
1	1 cm	6 cm^2
10^3	10^{-1} cm (1 mm)	60 cm^2
10^9	10^{-4} cm (1 μm)	6×10^4 cm^2
10^{12}	10^{-5} cm (0.1 μm)	6×10^5 cm^2

chemical factors which favor the adsorption of one ion rather than another of the same charge at a surface site:

1. *Solubility.* If there are two adsorbable ions in a solution in contact with a precipitate, the ion which forms the *less soluble* compound with one of the ions of the precipitate will be preferentially adsorbed. This is known as the Paneth-Fajans-Hahn rule, and is quite useful in predicting the natures of the adsorbed layers (see below).
2. *Concentration.* When there are two ions with the same tendency to adsorb on a precipitate, the one present at higher concentration will be adsorbed preferentially.
3. *Ionic charge.* Other factors being the same, the higher the charge on an ion, the more readily it is adsorbed. This results from the electrostatic component of the surface interaction mentioned above. The La^{3+} ion has been found to adsorb on colloidal As_2S_3 to a much greater extent than does Na^+.
4. *Ionic size.* Generally speaking, precipitates prefer to adsorb ions which are similar in size to those making up the precipitate crystal lattice. For example, radium ion is strongly adsorbed on barium sulfate, but not on calcium sulfate. The ionic radii of Ra^{2+} (162 pm) and Ba^{2+} (156 pm) are similar; the radius of Ca^{2+} is much smaller (126 pm).

Figure 9.2 shows a two-dimensional picture of a particle of silver chloride in a solution of NaCl. The surface of the particle is covered with a layer of solute ions called the *primary adsorbed-ion layer*. Ions in this layer are so close to the particle surface that chemical bonding forces favor the attraction of chloride ions to positive Ag^+ surface sites over the attraction of sodium ions to negative Cl^- surface sites. An excess of chloride builds up in the primary layer, and the particle develops a negative charge. The negative charge increases until additional chloride ions entering the layer are repelled by the excess negative charge, and an equilibrium condition is reached.

The negatively charged particle draws positively charged sodium ions into the

9.4 ADSORPTION

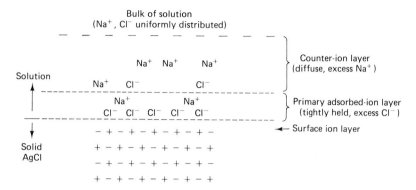

Figure 9.2 Silver chloride particle in sodium chloride solution.

counter-ion layer. This layer is somewhat more diffuse than the primary layer, and the sodium ions, held by electrostatic forces, are relatively free to exchange places with ions in the bulk of the solution. Taken together, the primary and counter-ion layers make up the so-called *electrical double layer* between colloidal particles and the solution. It is this double layer that is responsible for the stability of colloidal dispersions. The particles in a dispersion of AgCl in dilute NaCl electrolyte will all have the same charge (negative) and will repel each other when they collide, rather than stick together. The result is rapid motion called "Brownian motion," which prevents settling out of the dispersion.

If the electrical double layer can be destroyed, particles will no longer repel one another, and the dispersion will no longer be stable. The particles will then stick together, or flocculate. Flocculation can be stimulated by gentle heating or by adding a few millimoles of an electrolyte to the solution. The particular electrolyte chosen depends on the sign of the surface charge of the colloidal particles; negatively charged particles are most easily flocculated by cations of high charge. Thus aluminum chloride ($AlCl_3$, Al^{3+}) is a far more effective flocculator for negatively charged silver chloride particles than is NaCl (Na^+). On the other hand, ferricyanide ion [$Fe(CN)_6^{4-}$] is a far more efficient flocculator of positively charged colloidal particles than is chloride ion.

Experiments have shown that the ions of flocculating agents actually replace the original counterions. Therefore when colloidal hydrous ferric oxide (precipitated with excess ammonium hydroxide; see Chapter 10) is flocculated with sodium sulfate, sulfate ions replace many of the hydroxide ions in the double layer. Weaker flocculating agents such as chloride and nitrate are less tightly adsorbed and replace hydroxide ion less completely than do sulfate ions.

Sometimes a problem arises in a gravimetric analysis procedure when a flocculated precipitate must be washed with deionized water. If the concentration of flocculating agent is lowered too far in washing, part of the precipitate may pass back into the colloidal state and be lost through the filter paper. This phenomenon is called *peptization* and can usually be avoided by washing precipitates with solutions of electrolytes. As you will see later, silver chloride is washed with dilute nitric acid to avoid peptization.

9.5 IMPURITIES IN PRECIPITATES

The exact mechanism by which a precipitate may become contaminated may be highly complex. To help in understanding contamination, it is convenient to describe it in two basic categories: *coprecipitation* and *postprecipitation*. Coprecipitation involves either the adsorption of impurities on the surfaces of precipitates as they form, the incorporation of a contaminant in the precipitate lattice, or the physical entrapment of soluble species in aggregates of small precipitate particles. Postprecipitation is a delayed precipitation of a slightly soluble impurity as a second solid phase on top of a pure precipitate. We will examine each of these mechanisms and then describe some strategies that analysts use to avoid error resulting from them.

9.5.1 Coprecipitation by Adsorption

The nature of adsorption was discussed earlier. The adsorption of foreign ions can be of practical importance when an analyst must deal with finely divided precipitates, for example, in the precipitation of the silver halides, metal sulfides, and hydrous oxides of aluminum and tin.

9.5.2 Coprecipitation by Occlusion

In this form of contamination impurities (foreign ions and solvent) are trapped either in cavities within crystals or in cavities within aggregates of small crystals. It is particularly troublesome in precipitates which form and flocculate rapidly. Occluded impurities may make up as much as 0.1 to 0.2% of the mass of a precipitate and result in significant error.

9.5.3 Coprecipitation by Isomorphous Replacement

Certain pairs of compounds have the ability to form mixed crystals. Isomorphous replacement occurs when one compound is able to fit into the crystal lattice of another compound and form a mixed crystal with little or no distortion of the lattice. Generally it is necessary that the ions making up the two compounds have about the same ionic radius (within about 15%) and that the chemical bonds of the two compounds have

9.5 IMPURITIES IN PRECIPITATES

TABLE 9.2 ISOMORPHOUS PAIRS OF COMPOUNDS

	Radii (picometer)		
Compound pairs	Cation	Anion	Mixed crystal?
[BaSO$_4$ [PbSO$_4$	156 143	230 230	Yes
[BaSO$_4$ [BaCrO$_4$	156 156	230 240	Yes
[AgCl [AgBr	129 129	167 182	Yes
[KCl [PbS	152 133	167 170	No

about the same degree of covalency. Table 9.2 is a list of pairs of compounds along with cation and anion radii. Compounds in each bracket have the same basic crystal structure.† Mixed crystals form in the cases of BaSO$_4$/PbSO$_4$, BaSO$_4$/BaCrO$_4$, and AgCl/AgBr. When, for example, sulfate ion is added to a solution containing Ba^{2+} with a small amount of Pb^{2+}, lead ion will replace some of the barium ion sites in the barium sulfate lattice. The replacement tends to be rather uniform throughout the precipitate, in contrast to the highly localized contamination that results from occlusion.

The pair of compounds KCl/PbS is included in the table to show the importance of the nature of the chemical bonds in a pair of compounds. These compounds both pack in a cubic lattice and have similar ionic radii, but do not form mixed crystals. The +2 charge on lead ion and the −2 charge on sulfide produce a chemical bond in PbS which is far more covalent than the highly ionic bond in KCl.

9.5.4 Postprecipitation

This kind of contamination occurs when a precipitate separates in a form more or less free from impurities, and a second insoluble impurity phase then forms. The second phase develops *after* the first precipitation and is therefore postprecipitated rather than coprecipitated. Postprecipitation can be an important analytical problem in the precipitation of calcium as calcium oxalate in the presence of magnesium. Calcium oxalate ($K_{sp} = 1.3 \times 10^{-9}$) is less soluble than magnesium oxalate ($K_{sp} = 8.6$

---

† Two crystal structures are the same type if they belong to the same symmetry class. There are 32 symmetry classes grouped into seven crystal systems. You will learn more about these ideas in a course in inorganic chemistry.

$\times 10^{-5}$). When ammonium oxalate is added to a solution containing both Ca^{2+} and Mg^{2+}, calcium oxalate precipitates and coprecipitates some magnesium oxalate. If calcium oxalate is allowed to remain in contact with the solution for a long period, magnesium oxalate gradually precipitates on the surface of the original precipitate.† Once postprecipitation occurs, double or even triple reprecipitation (see below) may not be effective in removing magnesium ion.

Postprecipitation can also be a problem in the separation of metal ions by means of hydrogen sulfide. In most qualitative analysis schemes metals in group II (Hg, Cu, Cd, and others) precipitate as sulfides from acidic solutions, while those in group III (Zn, Fe, Mn, Ni) precipitate as sulfides only in basic solution, where sulfide ion concentrations are high.‡ Zinc sulfide, which precipitates very slowly in the absence of other metal ions in acidic solutions of hydrogen sulfide, postprecipitates with copper sulfide. Metal sulfides are known to be very strong adsorbers of sulfide ion, and presumably adsorption stimulates the postprecipitation of ZnS.

Postprecipitation is sometimes desirable. For example, traces of lead can be removed from solution by postprecipitation on strontium sulfate. This is the basis of the determination of traces of radioactive lead. The $SrSO_4$ is said to act as a "carrier" for lead.

Contamination by postprecipitation can be largely avoided if the initial precipitate can be made filterable and removed from the solution as quickly as possible. Heating a solution in an attempt to flocculate a colloidal precipitate may stimulate postprecipitation, particularly if boiling occurs, and make the situation worse. In some cases it may be possible to add a small amount of some water-immiscible organic liquid immediately after precipitation to coat the precipitate and prevent postprecipitation. The organic liquid would have to be volatile so that it would be removed when the filtered precipitate was dried in an oven or heated in a crucible over a flame.

9.5.5 Methods Used to Reduce Contamination

Precipitates may become contaminated by any or all of the mechanisms just described. If the analyst has some idea of the mechanism of contamination, measures can be taken to avoid or lessen the problem. The following measures may be useful.

---

† Supersaturated solutions of magnesium oxalate tend to be metastable at room temperature, and precipitation of magnesium oxalate in the absence of calcium oxalate is quite slow. The surface of calcium oxalate seems to stimulate the precipitation of magnesium oxalate, perhaps by localizing a high concentration of oxalate ion in the primary adsorbed layer.

‡ Remember that sulfide ion exists in equilibrium with H_2S and HS^-:

$$H_2S = H^+ + HS^- \quad K_{a1} = 9 \times 10^{-8}$$
$$HS^- = H^+ + S^{2-} \quad K_{a2} = 1 \times 10^{-15}$$

The presence of excess hydroxide ion in a solution of hydrogen sulfide will pull these equilibria to the right, raising the sulfide ion concentration.

Washing the Precipitate This is probably the simplest method of removing surface contamination, especially when contaminants are in the counter-ion adsorbed layer.† Washing is not effective in reducing other forms of contamination. Solutions chosen for washing should neither dissolve appreciable quantities of the precipitate, nor peptize it. Peptization can be avoided if wash solutions contain dissolved electrolytes. The electrolytes chosen should be volatile at drying-oven temperatures, so as not to increase the final weight of the precipitate. Dilute nitric acid, ammonia, or ammonium salts are often chosen as washing solution electrolytes because of their volatility.

Washing a precipitate with a solution of an ion in common with the precipitate might seem desirable, but may result in serious solubility losses if complex ions can form. For example, washing silver chloride precipitate with $0.1M$ HCl would wash out species such as $AgCl_2^-$ and might ruin an analysis. Washing with pure water should be avoided when the precipitate is known to react with water. The Fe^{3+} salts, for example, react with water to form soluble hydrous oxide complexes.

Whether a wash solution should be cold or hot depends on the solubility of the precipitate at higher temperatures. Hot solutions should be used if possible, since impurities will probably be more soluble in them. Hot solutions also speed filtration by lowering solution viscosity.

Maintenance of Low Supersaturation A low degree of supersaturation encourages the growth of large, coarse crystals. These are desirable for several reasons. They have relatively small surface areas (per unit weight) on which to adsorb impurities, are less soluble than fine crystals, settle out of solution quickly, and are easy to filter and wash. More soluble compounds such as $PbCl_2$, $KClO_4$, and $MgNH_4PO_4 \cdot 6H_2O$ tend to precipitate slowly and form large crystals. Less soluble compounds which precipitate quickly can be encouraged to form large crystals by taking the following steps to keep the degree of supersaturation low:

1. Mix reagents slowly and stir vigorously.
2. Precipitate from hot, dilute solutions.
3. If possible, generate the precipitating reagent by a chemical reaction in the presence of the ion to be precipitated ("*in situ*"). This technique is called *homogeneous precipitation* and will be discussed in detail at the end of this chapter.

Digestion A precipitate is digested by allowing it to stand in contact with the solution from which it has formed. Digestion promotes the purity of precipitates in the following ways:

1. Small crystals dissolve and their ions are redeposited on the surfaces of larger crystals in a process called *Ostwald ripening*. Impurities trapped in aggregates of small crystals are returned to the solvent when the small crystals dissolve.

† Ions in the primary layer may be too tightly held to be washed away.

2. Equilibrium may be reached among adsorbed impurities and species in solution. For example, when barium chloride is added to a solution of sodium sulfate, barium sulfate begins to precipitate immediately. At the beginning of the precipitation, when sulfate ion is in excess, coprecipitation of sodium sulfate is quite important.[†] If, however, the precipitation is completed and the precipitate is digested in an excess of $BaCl_2$, coprecipitated sodium sulfate reacts to form $BaSO_4$. Barium chloride adsorbed during digestion can be volatilized after the precipitate is filtered.

3. Precipitates can undergo structural changes to more stable forms. For example, when calcium oxalate is precipitated from cold solutions, mono-, di-, and trihydrate compounds form: $CaC_2O_4 \cdot H_2O$, $CaC_2O_4 \cdot 2H_2O$, and $CaC_2O_4 \cdot 3H_2O$. On standing in contact with solution, the less stable di- and trihydrates are transformed to the monohydrate. As a result the solubility of calcium oxalate appears to decrease during digestion. The transformation occurs more quickly in warm solutions.

Digestion should be avoided where postprecipitation is a possibility. It is of little value when contamination occurs by isomorphous replacement.

Reprecipitation In this process a precipitate is filtered, washed, dissolved to make a new solution, and precipitated a second time. If the original impurities are mostly adsorbed or occluded, the second precipitate can be expected to be much purer. If contamination has occurred by only slight postprecipitation, reprecipitation may increase purity. However, if the precipitate is a good carrier for an impurity and is grossly contaminated, reprecipitation will be of little help. Reprecipitation is not very helpful in removing isomorphous contaminants.

Because of the extra manipulations involved, reprecipitation may increase the likelihood of error. In addition, where time is important, reprecipitation may not be desirable.

Elimination of the Contaminant Problems of contamination can be avoided if a contaminant can either be removed from solution before precipitation or made less objectionable by some chemical process. Removing a contaminant from solution can be quite difficult, especially if it is chemically similar to the substance of interest. Additional steps in an analysis present opportunities for error, and actually removing an impurity should be a tactic of last resort.

Changing a contaminant by a complexation or redox reaction may sometimes be of benefit and may not compromise accuracy. For example, in the determination of sulfate as $BaSO_4$, Fe^{3+} is much more strongly adsorbed than Fe^{2+}. Hydroxylamine can be used to reduce Fe^{3+} to Fe^{2+} prior to precipitation of $BaSO_4$.

[†] If the precipitate were isolated at this point, dried, and weighed, the determination of percent sulfate in the original sample would be low. This is because the precipitate is assumed to be pure $BaSO_4$, and the sodium sulfate impurity has a lower formula weight (142 g/mole) than barium sulfate (233.4 g/mole). This point may seem vague, but it should be more clear in the section on gravimetric calculations in Chapter 10.

9.6 PRECIPITATION FROM HOMOGENEOUS SOLUTION

A summary of the methods described above and the kinds of contamination they are generally effective in reducing appears in Table 9.3.

9.6 PRECIPITATION FROM HOMOGENEOUS SOLUTION

In the previous section it was stated that limited supersaturation in a solution promotes the growth of large, pure crystals. Generating a precipitating agent in the solution from which precipitation occurs ("homogeneously") is an effective way to limit supersaturation in the solution. The technique requires a chemical reaction which produces a precipitating agent at a rate about equal to that at which crystals form. A few classic examples will illustrate the principles involved.

9.6.1 Precipitation of Hydroxides with Urea

Urea reacts slowly with water in hot solutions to produce carbon dioxide and ammonia,

$$(NH_2)_2CO + H_2O \rightarrow CO_2 + 2NH_3$$

The ammonia gradually produced in an acidic or neutral solution increases the basicity of the solution by reactions such as

$$NH_3(aq) + H_3O^+ = NH_4^+ + H_2O \text{ (acid solution)}$$

and
$$NH_3(aq) + H_2O = NH_4^+ + OH^-$$

The production of hydroxide ion by urea has been applied to the precipitation of the hydrous oxides of aluminum, chromium(III), and iron. When ammonia is added

TABLE 9.3 IS A METHOD OF REDUCING CONTAMINATIONS EFFECTIVE?

Method	Form of contamination			
	Adsorption	Occlusion	Isomorphous	Postprecipitation
Washing	Yes	(Some)	No	No
Maintenance of low supersaturation	Yes	Yes	No	No
Digestion	Yes	Yes	No	No
Reprecipitation	Yes	Yes	No	(Some)
Elimination	Yes	Yes	Yes	Yes

directly to acidic solutions of Al^{3+}, Cr^{3+}, or Fe^{3+}, jellylike colloidal precipitates of the metal hydroxides form. These precipitates have immense surface areas, are easily contaminated through adsorption, and are quite difficult to filter. Such undesirable qualities can be avoided if precipitation is carried out homogeneously. In the case of aluminum, if urea is added to a solution containing Al^{3+} and a succinic acid buffer,† and the solution is heated, a coarse, nongelatinous precipitate containing a mixture of aluminum oxides and aluminum succinate forms in the pH range 4.2 to 4.6. The precipitate is filtered and ignited to convert the succinate/oxide mixture to aluminum oxide, Al_2O_3, an excellent weighing form.

9.6.2 Precipitations with Sulfate Ion

Ions such as barium, calcium, strontium, and lead can be precipitated from solutions in which sulfate is generated homogeneously. Two popular sources of sulfate ion have been sulfamic acid (reaction I) and dimethyl sulfate (reaction II):

(I) $\quad\quad\quad HSO_3NH_2 + 2H_2O \rightarrow H_3O^+ + SO_4^{2-} + NH_4^+$

(II) $\quad\quad\quad (CH_3)_2SO_4 + 4H_2O \rightarrow 2H_3O^+ + SO_4^{2-} + 2CH_3OH$

(NOTE: Dimethyl sulfate is a suspected carcinogen and should be used with care.)
A novel method of generating sulfate is by reduction of persulfate ion,

$$\underset{\text{(persulfate)}}{S_2O_8^{2-}} + 2e = \underset{\text{(sulfate)}}{2SO_4^{2-}}$$

by ethylenediaminetetraacetic acid (EDTA). An excess of EDTA is added to a solution containing Ba^{2+} to form the soluble complex species $Ba(EDTA)^{2-}$. Next, an excess of ammonium persulfate is added and the solution is heated. Persulfate ion oxidizes the excess EDTA quite rapidly. As the complexed EDTA is oxidized more slowly, Ba^{2+} is released and precipitates as $BaSO_4$. Alternatively, hydrogen peroxide has been used to oxidize EDTA in solutions of $Ba(EDTA)^{2-}$ in the presence of sulfate ion. These reactions are commonly called "cation release" reactions.

9.6.3 Precipitation with Sulfide Ion

Many metal ions form insoluble sulfides. When sulfides are precipitated by bubbling H_2S into solutions, precipitations occur quite rapidly, and the precipitates are quite

---

† The buffer keeps the solution from becoming too basic and thus controls supersaturation. See Willard and Tang (Ref. 1).

9.6 PRECIPITATION FROM HOMOGENEOUS SOLUTION

hard to handle. Metal ion sulfides can be precipitated from homogeneous solution by the acid- or base-catalyzed hydrolysis of thioacetamide or thioformamide,

$$CH_3CSNH_2 + H_2O \rightarrow CH_3CONH_2 + H_2S$$
$$HCSNH_2 + H_2O \rightarrow HCONH_2 + H_2S$$

Thioacetamide has been used in the homogeneous precipitation of copper, cadmium, zinc, silver, mercury, arsenic, bismuth, and antimony. Thioformamide has been used in the separation of copper and arsenic and in the separation of arsenic, antimony, and tin.

9.6.4 Precipitation of Ni(II) with Homogeneously Generated Dimethylglyoxime

Nickel(II) can be precipitated as the neutral bis(dimethylglyoximate) complex,

$$Ni^{2+} + 2 \begin{bmatrix} CH_3 & OH \\ C=N & \\ C=N & \\ CH_3 & OH \end{bmatrix} \rightarrow [\text{Ni(DMG)}_2] + 2H^+$$

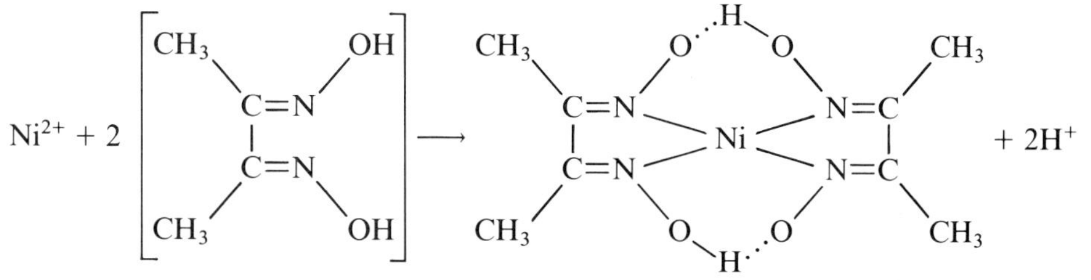

DMG

Precipitation can be carried out directly by adding dimethylglyoxime to a solution of Ni^{2+} with a pH in the range 5 to 9.

Dimethylglyoxime can be synthesized in solution by the reaction of biacetyl with hydroxylamine,

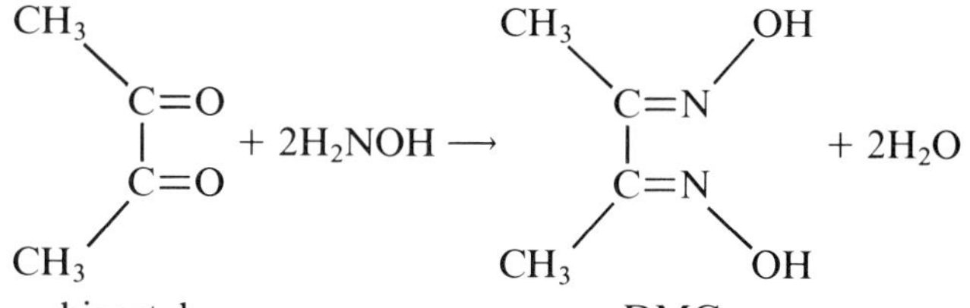

$$\begin{array}{c} CH_3 \\ C=O \\ C=O \\ CH_3 \end{array} + 2H_2NOH \rightarrow \begin{array}{c} CH_3 \quad OH \\ C=N \\ C=N \\ CH_3 \quad OH \end{array} + 2H_2O$$

biacetyl   DMG

The mechanism by which the bis(dimethylglyoximate) complex of nickel precipitates is apparently quite complicated. There is evidence for the formation of nickel complex with an intermediate species in the biacetyl-hydroxylamine reaction (Ref. 2). The ultimate product of the reaction, Ni(DMG)$_2$, forms slowly as large, easily filtered crystals.

9.6.5 Limitations of Homogeneous Precipitation Methods

The principal disadvantage of homogeneous precipitation methods is the relatively long time required for chemical reactions to occur. This need not be a serious disadvantage, however, since the time required for the actual handling of solutions is about the same as that required for a conventional precipitation. Furthermore, the time required for digestion of precipitates obtained by conventional methods may be as great as that required for the homogeneous generation of the precipitating agent.

A second potential disadvantage of homogeneous methods is that the analyst needs a rough idea of the composition of the sample prior to analysis. In conventional methods, the precipitating reagent is added until precipitation stops and then a predetermined excess is added. In homogeneous methods the analyst must know beforehand the approximate amounts of reagents required to avoid using either too little reagent or an excess so large as to produce other errors. This disadvantage would not be terribly serious in, for example, quality control work, where ranges of sample composition are well established. Preliminary rough assay can also be used to determine the approximate composition of a sample.

The major advantage of precipitations from homogeneous solution is, of course, larger, more easily filtered crystals. In many gravimetric procedures this advantage far outweighs the disadvantages mentioned above.

9.7 PRECIPITATION METHODS OF SEPARATION

It may be possible to use precipitation to separate elements or compounds in a mixture. Separation by precipitation involves a change in phase. We can add a solution of some precipitating agent to a solution containing a mixture of ions and, by a chemical reaction, form a solid compound containing only one (we hope!) of the elements in the original mixture. We then separate the solid and liquid phases by a mechanical or physical process such as filtration.

We can describe the completeness of a precipitation and of a separation in quantitative terms with the help of two ratios, the *recovery factor*, R, and the *separation factor*, α (alpha). The recovery factor is the fraction of a particular species or element recovered in a precipitation process,

$$R = Q_{\text{recovered}}/Q_{\text{initial}} = Q_r/Q_i$$

where the Q terms are masses in grams or milligrams, or in moles. A recovery factor

9.7 PRECIPITATION METHODS OF SEPARATION

of 0.999 is a quantitative precipitation. In Chapter 17 (separations) you will see a term very similar to this called a distribution coefficient (K), which is the ratio of concentrations of a species at equilibrium in two immiscible (mutually insoluble) solvents which are brought together in, for example, a separatory funnel.

The separation factor, α, is defined as a ratio of the recovery factors for two elements or compounds which are to be separated by precipitation,

$$\alpha = R_1/R_2$$

Two elements which have nearly identical recovery factors will be very hard to separate by precipitation. For such a pair α will be very close to unity. If two elements have quite different recovery factors, α will be either very large or very small (depending on which species you assign numbers 1 and 2, above), and separation by precipitation will be much more feasible. The challenge to the analytical chemist is then to find the proper solution conditions to make the recovery factor for one of the elements in a mixture very nearly 1.0, and at the same time make the separation factor very large or very small. The following example illustrates these factors.

EXAMPLE 9.1 Separation

A 100-mL solution contains 0.0010 mole of Ba^{2+} and 0.0010 mole of Mg^{2+}. Enough oxalic acid is added to make the formal concentration of oxalic acid 0.20, and the pH is adjusted to 3.0 with sodium hydroxide. Calculate the recovery factors for barium ion and magnesium ion, knowing that the fraction of oxalate ion at pH 3.0 is 0.048 and that K_{sp} values for barium oxalate and magnesium oxalate are 1.5×10^{-8} and 9.0×10^{-5}, respectively.

Solution

The fraction of oxalate (α_{ox}) in an oxalic acid solution at pH 3.0 could have been calculated from the acid dissociation constants and the pH, as in the last example of the last chapter. The concentration of Ba^{2+} can be found from the K_{sp} expression:

$$[Ba^{2+}][C_2O_4^{2-}] = [Ba^{2+}] * \alpha_{ox} C_t = 1.5 \times 10^{-8}$$

and

$$[Ba^{2+}] = 1.5 \times 10^{-8}/0.048 * 0.20 = 1.6 \times 10^{-6} M$$

The recovery factor for barium ion is

$$R = (0.010 - 1.6 \times 10^{-6})/0.010 = 0.9998 \text{ (quantitative)}$$

The concentration of magnesium ion can be found from its K_{sp} and our knowledge of the product $\alpha_{ox} C_t$:

$$[Mg^{2+}] = 9.0 \times 10^{-5}/0.048 * 0.20 = 0.0094 M$$

The recovery factor for magnesium ion is

$$R = (0.010 - 0.0094)/0.010 = 0.060 \text{ (not quantitative)}$$

(Notice that we assume that $C_t = 0.2$ even though precipitation of barium oxalate will reduce it to about 0.19. The approximation is close enough for our purposes.)

The separation factor is given by

$$\alpha = R_{Ba}/R_{Mg} = 0.9998/0.060 = 17$$

Notice that about 6% of the magnesium precipitates along with virtually all the barium. The magnesium ion left in solution has very little barium ion impurity; the barium has quite a bit of magnesium as an impurity.

For practice, try this calculation at pH 4 and pH 5.

The solubilities of metal hydroxide and metal sulfide precipitates vary over such a wide range that very good separations of metal ions can be made by adjusting the pH to values that will precipitate some but leave others in solution. Predictions can be made from simple calculations from K_{sp} values. Thus, it can be shown that Fe(III) will begin to precipitate as $Fe(OH)_3$ ($K_{sp} \sim 10^{-39}$) from a $0.01F$ solution at a pH of about 1.8, while Zn(II) will begin to precipitate as $Zn(OH)_2$ ($K_{sp} \sim 10^{-16}$) from a $0.01F$ solution at about pH 7. In principle, then, it is possible to separate Fe(III) and Zn(II) by carefully adjusting pH. Whether the precipitates isolated are pure depends on factors such as the extent of coprecipitation by adsorption and the rate at which precipitation occurs.

REFERENCES

1. H. H. Willard and N. K. Tang, *Industrial and Engineering Chemistry, Analytical Edition*, 9, 357 (1937).
2. L. Gordon, P. R. Ellefson, G. Wood, and O. E. Hileman, *Talanta*, 13, 551 (1966).

PROBLEMS

9.1. Describe the process of peptization. What effect might peptization have on the results of an analysis? How can peptization be avoided?

9.2. The addition of a small quantity of electrolyte to a solution containing a precipitate may prevent peptization. The addition of a large amount of electrolyte might result in the loss of more precipitate. Explain.

PROBLEMS

9.3. Describe the primary and counter-ion adsorbed layers for each precipitate in its corresponding electrolyte:
   (a) $BaSO_4$ in a solution of Na_2SO_4
   (b) $BaSO_4$ in a solution of $BaCl_2$
   (c) $AgCl$ in a solution of $AgNO_3$
   (d) $AgCl$ in a solution of Na_2SO_4

9.4. If silver chloride is precipitated from a solution containing excess silver ion, which of the following ions will probably be the principal counter-ion: NO_3^-, BrO_3^-, Na^+, or acetate ion? (Assume all ions are present at the same concentration.)

9.5. Distinguish between the following terms:
   (a) nucleation and growth
   (b) saturation and supersaturation
   (c) flocculation and peptization
   (d) isomorphism and occlusion
   (e) occlusion and adsorption
   (f) colloid and precipitate

9.6. Chemical analysis of a precipitate of $BaSO_4$ shows that it has the same concentration of Cs^+ ion as a contaminant as the solution from which the precipitate formed. What was the probable mechanism of contamination? How might the degree of contamination be reduced?

9.7. A solid material has a density of 2.25 g/cm^3. If a 1-g cube of the material is pulverized until the average particle size is 5×10^{-8} m, what is the approximate surface area of the sample? (Assume the final particles are cubes.)

9.8. According to von Weimarn, barium sulfate will precipitate almost immediately when equal volumes of $10^{-3}M$ $Ba(SCN)_2$ and $10^{-3}M$ $MnSO_4$ are mixed. However, when equal volumes of $10^{-4}M$ solutions of the same compounds are mixed, a precipitate does not appear for 1 week. Calculate the approximate concentration range over which Ba^{2+}, SO_4^{2-} solutions are "metastable." By what factors do the concentrations of ions in the solutions exceed the solubility of $BaSO_4$? (Assume the intrinsic solubility is about 2% of that predicted by a K_{sp} calculation, $K_{sp} = 1 \times 10^{-10}$ at 25°C.)

9.9. Trace quantities of the following pairs of radioactive ions may be present in the same solution. Find a carrier precipitate which could be used to separate the ions. A good carrier will not itself contain the ion being adsorbed.
   (a) CrO_4^{2-}, CH_3COO^-
   (b) SCN^-, CN^-
   (c) Pb^{2+}, Zn^{2+}
   (d) Ra^{2+}, Bi^{3+}

9.10. Barium ion contaminates $PbSO_4$ precipitates. How might you design a set of experiments to determine the mechanism of contamination (adsorption, occlusion, etc.)? You have at your disposal equipment to handle and study radioisotopes and a supply of $(Ba^{140})^{2+}$. This isotope has a half-life of about 12 days and decays with loss of a beta particle.

9.11. A 100-mg sample of barium nitrate contains about 1% silver nitrate as an impurity. Is it possible to precipitate the barium as barium sulfate without also precipitating silver sulfate? (Ignore adsorption as a source of contamination.) $K_{sp}(BaSO_4) = 1.0 \times 10^{-10}$ and $K_{sp}(Ag_2SO_4) = 1.2 \times 10^{-5}$.

9.12. Problem 9.11 is turned around: a 100-mg sample of silver nitrate contains about 1% barium nitrate as an impurity. Is it possible to precipitate the barium as barium sulfate without precipitating silver sulfate?

9.13. Copper(II) is to be precipitated as the hydroxide, $Cu(OH)_2$, $K_{sp} = 3 \times 10^{-20}$. At what pH will the recovery factor be 0.9999 if the original sample is $0.01F$ in Cu(II)?

9.14. Silver hydroxide has a K_{sp} of about 10^{-8}. In principle, is it possible to separate Cu(II) from Ag(I) as the hydroxides with a separation factor of 1000, assuming a mixture which is $0.01F$ in each metal ion?

9.15. Is it possible to have a recovery factor of 0.9999 for Ag(I) as the hydroxide if you add ammonium hydroxide to raise the pH of a $0.01F$ solution of silver nitrate? (Look for foreign ligand competition, as in the last chapter.)

9.16. Based on metal sulfide solubility product constants, predict which of the following separations are feasible with a recovery factor of 0.999 for one of the ions:
  (a) Fe(II), Cu(II)
  (b) Hg(II), Fe(II)
  (c) Zn(II), Sn(II)

Chapter 10

Gravimetric Methods

10.1 INTRODUCTION

Gravimetric methods of analysis rely on the isolation of a product of a reaction and the careful measurement of its weight. Gravimetric methods involve a number of chemical and physical phenomena. For example, the amount of water in a sample might be determined by measuring its weight before and after heating it in an oven. If conditions are carefully controlled, the amount of water volatilized from the sample can be calculated from the change in weight. As another example, the amount of carbon in a sample can be determined by heating it in a stream of CO_2-free air and trapping the CO_2 given off with an absorbant such as Ascarite (asbestos coated with NaOH). The change in weight of the Ascarite trap can then be used to calculate the weight of carbon in the original sample. Notice that both methods rely on the measurement of weight before and after a physical or chemical process.

In this chapter we are particularly interested in gravimetric methods which rely on precipitation to isolate the substance to be analyzed. We will look at gravimetry as a laboratory *process*, starting with sample treatment and ending with a calculation of results. The material in Chapters 8 and 9 has a direct bearing on gravimetric methods which rely on precipitation, and many of the ideas presented in those chapters will become more meaningful when put in the context of laboratory operations. The techniques and limitations of weighing were discussed in Chapter 1 and will not be discussed in more detail.

10.2 PROPERTIES OF SAMPLES AND PRECIPITATES

For a gravimetric method to be both accurate and precise the precipitated form of the material isolated must meet several basic criteria:

1. The precipitate must have low solubility. This was illustrated in Chapter 8. At least 99.9% of the analyte must be precipitated and recovered by filtration. Remember that a precipitate must also be insoluble in the solution used to wash it after filtration.
2. The precipitate should form large crystals. Large crystals are easy to filter and purify by washing.
3. The precipitate must be quite pure, or at least be easy to purify. Sources of contamination and possible remedies were discussed in Chapter 9.
4. The precipitate should have a high gram formula weight. Large samples are easier to weigh accurately than are small samples. The larger the gram formula weight of the precipitate the larger will be its weight for a given amount of sample.
5. The precipitate must have a known composition, or at least be convertible to a compound with a known composition. The composition cannot change while drying in an oven or during storage in a desiccator.

Very few precipitates meet all of these criteria. As a result special precautions must be taken in certain steps of every analysis to make sure that determinate errors do not arise. These special precautions, of course, are based on the results of careful research. For example, you will notice in the first experiment at the end of this chapter that calcium oxalate monohydrate must be dried in an oven at a temperature between 95 and 105°C. At higher temperatures too much water will be lost, and the composition of the precipitate will change. At lower temperatures the dihydrate will form, and again the composition of the precipitate will change. A large number of very careful experiments were run to find the temperature range over which the monohydrate is stable.

10.3 GRAVIMETRY AS A PROCESS

A successful gravimetric analysis involves a series of carefully executed steps. The general process described below is followed by the experiments at the end of the chapter and by most gravimetric methods.

1. *Obtain a sample.* Learn what you can about the sample, where it came from, how it was taken, how it was treated before you received it. You may be able to tell from the origin of the sample what kinds of compounds might be present which could interfere with the analysis you are to perform. You should also find out exactly what is needed in the analysis. An "estimation" will require less care than will an exacting assay.
2. *Dry the sample.* The purpose of drying a sample is to remove surface water. If the sample is unstable at oven temperatures, however, this step may lead to serious

errors. Determine experimentally (perhaps under a microscope) whether the sample decomposes at oven temperatures, and consider how drying might affect the outcome of the analysis.

3. *Weigh out portions of the sample.* Samples weighing more than 100 mg help reduce the importance of uncertainty in the weighing measurement (± 0.1 mg for most balances in undergraduate laboratories; see Chapter 1). Use what you know about the sample to approximate the amount of precipitate that will be isolated. Use enough sample to ensure that more than 100 mg of precipitate will be isolated, again to reduce weighing error.

Solid samples are usually weighed by difference, as explained in Chapter 1. Volatile liquids or solutions may be weighed by addition to a preweighed vessel.

4. *Dissolve the weighed portions.* Preliminary experiments should be done to see what will be needed to dissolve the sample. Many samples will dissolve in solutions of mineral acids such as HCl or HNO_3. Some samples may require the use of NaOH solutions. More unreactive samples may require "fusion": dissolving a sample in molten NaOH or KOH containing a strong oxidizer such as sodium peroxide. The details of such methods can be found in the older literature (see Recommended Reading list).

5. *Mask interferences.* Special care must be taken if the precipitate is likely to be contaminated by the coprecipitation of a compound thought to be present in the sample. There are no perfectly general remedies. You must know enough about the sample and consult the literature before trying to mask an interference.

6. *Precipitate the product.* As discussed in detail in Chapter 9, gradual precipitations produce the best results. A possible source of error may be impatience; make sure precipitation is finished before filtering the precipitate. Make sure that sufficient precipitating agent has been added to make the reaction quantitative.

7. *Digest the precipitate.* Digestion purifies the precipitate and encourages the growth of large, easy-to-filter crystals.

8. *Collect the precipitate by filtration.* A filtering device is cleaned and dried to constant weight (see Chapter 1) and then used to collect precipitate formed from an individual sample.

A number of filtering media are available, among them sintered glass filter crucibles, porcelain ("Gooch") crucibles, and filter paper. The cleaning and use of glass crucibles are described in detail in an experiment at the end of the chapter. Gooch crucibles require the use of a filtering mat prepared from either woven glass or a slurry of asbestos and are unlikely to be encountered in teaching laboratories. Once again, the literature should be consulted for the preparation of these filters (see Recommended Reading list). Filter paper may be used to collect some precipitates. "Ashless" (low-silica) filter paper is used to collect barium sulfate in the determination of sulfate. Filter paper containing $BaSO_4$ is carefully folded, placed in a porcelain crucible, and heated with a burner. The filter paper is carefully burned off, leaving the barium sulfate precipitate behind in the crucible.

The most important error in using a filter is loss of precipitate through the filtering medium. Glass filtering funnels can be purchased in three grades of coarseness:

course, medium, and fine, as discussed in Chapter 1. Fine filters are quite difficult to use; while they collect very finely divided precipitates, they filter very slowly. Adequate digestion makes it possible to use coarse or medium filters in most applications.

9. *Wash the precipitate.* Precipitates may be purified by washing with a dilute solution of electrolyte. This was discussed in Chapter 9. Peptization may result in loss of precipitate through the filter.

10. *Dry the precipitate.* The thermal stability of the precipitate must be known before trying to dry it in an oven. Surface water and certain electrolytes used to wash precipitates (some ammonium salts, for example) can be removed by drying for 1 to 2 hours at 120°C. If the weighing form of the precipitate is a hydrate, such as calcium oxalate monohydrate (see Experiment 1), oven temperature and drying time must be controlled very carefully to avoid the loss of either too little or too much water. It may be adequate to dry some precipitates in a stream of air, or simply in a desiccator, before weighing. Others will have to be heated in a furnace or with a burner to convert them to a stable weighing form.

There are many sources of error that can accompany the drying step. If the temperature of the oven is incorrect, part of the sample may be in an incorrect form for weighing. If a sample must be ignited, unburned filter paper may be left in the crucible. If a sample is heated too strongly during ignition, the sample may react with the crucible glaze and decompose. If the crucible tongs used to remove hot samples from the oven have been contaminated with materials spilled on the laboratory bench, the contaminants may be transferred to the crucible and cause high results. Your laboratory instructor will show you the proper techniques and provide you with information and anecdotes about many of these dangers.

11. *Weigh the precipitate.* The weight of the precipitate is determined by weighing the precipitate collected, washed, and dried in the filter, and subtracting the weight of the clean filter (found before step 8 above). The precipitate and filter should be brought to constant weight, as explained in Chapter 1.

Errors may arise in this step if the filter and precipitate are weighed before they have cooled to room temperature. Warm samples are buoyed up by convection currents in the balance chamber and weigh less than they should (see Chapter 1). Very hygroscopic samples may absorb water from the air while on the balance pan, producing high results.

12. *Calculate the results.* Section 10.5 is devoted to this subject.

10.4 GRAVIMETRIC METHODS

There are hundreds of gravimetric methods of analysis, far too many to discuss in any detail in an introductory text. Table 10.1 contains the most important methods for the most common elements. Also included are gram atomic weights and gram formula weights which might be useful in calculations.

10.5 CALCULATIONS FOR GRAVIMETRIC METHODS

TABLE 10.1 SELECTED GRAVIMETRIC METHODS

Element or species[a]	Precipitating agent	Product (GFW)[b]
Al	NH_3(aq)	Al_2O_3 (101.961)
	8-Hydroxyquinoline	$Al(C_9H_6ON)_3$ (459.43)
Ca	$C_2O_4^{2-}$	$CaC_2O_4 \cdot H_2O$ (146.10)
Carbonyl (aldehyde, R-CHO, or ketone)	2,4-Dinitrophenylhydrazine	$R-CH=NNHC_6H_3(NO_2)_2$ (209.14 + R)
Co	α-Nitroso-β-naphthol	Ignite to Co_3O_4 (240.977)
Fe	OH^-	Fe_2O_3 (159.692)
K, Cs, Rb	$PtCl_6^{2-}$	K_2PtCl_6 (486.00)
		Cs_2PtCl_6 (673.62)
K, NH_4^+, amines (RNH_3^+)	Sodium tetraphenylboron [$NaB(C_6H_5)_4$]	$KB(C_6H_5)_4$ (358.33)
Halides (Cl^-, Br^-, I^-) acid halides (RCOCl)	$AgNO_3$	AgCl (143.32), AgBr (187.78) AgI (234.77)
Ni	$C_4H_7O_2N_2$ (dimethylglyoxime)	$Ni(C_4H_7O_2N_2)_2$ (288.92)
Pb	Na_2CrO_4	$PbCrO_4$ (323.2)
PO_4^{3-}	Mg^{2+}, NH_4^+	$Mg(NH_4)PO_4$, ignite to $Mg_2P_2O_7$ (222.55)
Mg	NH_4^+, PO_4^{3-}	$Mg(NH_4)PO_4$, ignite to $Mg_2P_2O_7$ (222.55)
SO_4^{2-}	Ba^{2+}	$BaSO_4$ (233.39)
S^{2-}	Oxidize to sulfate	$BaSO_4$ (233.39)

[a] Atomic weights for the elements can be found on the inside cover of the text.
[b] Gram formula weights are calculated from 1975 International Union of Pure and Applied Chemistry (IUPAC) atomic weight values based on $^{12}C = 12.000$.

10.5 CALCULATIONS FOR GRAVIMETRIC METHODS

The calculations required for gravimetric analyses are much the same as those you performed in learning chemical stoichiometry in general chemistry. They become more complicated when it becomes necessary to track an analysis through a series of steps following a precipitation. These more difficult situations are much like the "road map" synthesis problems given in organic chemistry courses.

The following examples should provide enough background to help you solve even the most difficult gravimetric problems. Since it is common to work with samples weighing less than 1 g, we will express all weights in milligrams and use milligram formula weights (mGFW) and milligram atomic weights (mGAW) in calculations. This will let us avoid exponential notation. While it may take you a while to feel comfortable with milligram and millimole quantities, eventually it will save you trouble in calculations.

EXAMPLE 10.1

A 382.5-mg sample containing chloride is dissolved. The chloride is precipitated as AgCl with silver nitrate. The precipitate is collected, washed, dried, and found to weigh 523.5 mg. Calculate the percentage chloride in the sample.

Solution

The reaction is $Ag^+ + Cl^- = AgCl(s)$. From this we can see that 1 mmole of chloride in the sample will produce 1 mmole of silver chloride precipitate, that is,

$$\text{millimoles } Cl^- = \text{millimoles AgCl}$$

We must ultimately find the percentage of chloride in the sample. We will need the weight of chloride in the sample to calculate the percentage. Do not let this abstraction bother you. We cannot isolate chloride and weigh it, but we will treat it as if we can. We know that the millimoles of chloride is equal to the weight of chloride in milligrams divided by the milligram atomic weight of chlorine (35.45). The millimoles of silver chloride is similarly given by the weight of silver chloride in milligrams, divided by the milligram formula weight of silver chloride (143.33). The following relationship will hold because of the 1:1 stoichiometry of the reaction:

$$mg(Cl)/mGAW(Cl) = mg(AgCl)/mGFW(AgCl).$$

We can rearrange this expression and solve for mg(Cl):

$$mg(Cl) = [mGAW(Cl)/mGFW(AgCl)] * mg(AgCl) = (35.45/143.33)(523.5)$$
$$= 129.5 \text{ mg}$$

The percentage of chloride in the sample is found by dividing the weight of chloride by the weight of the original sample and multiplying by 100:

$$\%Cl^- = [mg(Cl)/mg(sample)](100) = 33.85\%$$

The ratio of the atomic weight of chlorine to the formula weight of AgCl is an example of a "gravimetric factor." This factor acts somewhat like a proportionality constant between the weight of sample and weight of precipitate. Its use is entirely optional. Some people prefer to cast gravimetric problems in a universal and memorizable form involving a factor, for example,

$$\text{weight material sought} = \left(\frac{\text{atomic weight material sought}}{\text{GFW precipitate}}\right) * (\text{weight of precipitate})$$

10.5 CALCULATIONS FOR GRAVIMETRIC METHODS

Tables of gravimetric factors are available in the literature. It is better for students to reason through these problems rather than to memorize (or mismemorize!) a pattern.

EXAMPLE 10.2

A 482.5-mg sample containing $MgCl_2$ and NaCl is dissolved in water and the magnesium is precipitated as $MgNH_4PO_4$ (magnesium ammonium phosphate). The precipitate is heated to 1050°C and converted to the stable weighing form, $Mg_2P_2O_7$ (magnesium pyrophosphate), of which 321.2 mg was obtained. Calculate the percentage of magnesium in the original sample.

Solution

This is a short "road map" problem. We can trace the problem as follows: 1 mmole of Mg produces 1 mmole of $MgNH_4PO_4$, and each millimole of $MgNH_4PO_4$ produces 0.5 mmole of $Mg_2P_2O_7$. We can ignore the intermediate compound magnesium ammonium phosphate in our calculations, and simply note that every millimole of magnesium pyrophosphate product corresponds to 2 mmole of magnesium in the original sample:

$$\text{mmole } Mg^{2+} = 2(\text{mmole } Mg_2P_2O_7)$$

Once again we know that $\text{mmole}(Mg^{2+}) = mg(Mg)/mGAW(Mg)$ and that $\text{mmole}(Mg_2P_2O_7) = mg(Mg_2P_2O_7)/mGFW(Mg_2P_2O_7)$. Substituting and rearranging gives an expression for milligrams of magnesium in the original sample:

$$mg(Mg) = [2*mGAW(Mg)/mGFW(Mg_2P_2O_7)] * mg(Mg_2P_2O_7)$$

The gravimetric factor is $2*mGAW(Mg)/mGFW(Mg_2P_2O_7)$. Notice that the factor of 2 appears in the numerator. A quick way to check the accuracy of the ratio is to see that magnesium appears twice in both the numerator and denominator.

Plugging in the numerical values, we find

$$mg(Mg) = [2(24.31)/(222.57)]*321.2 = 70.17 \text{ mg}$$

The percentage of magnesium in the original sample can be found by dividing the weight of magnesium by the weight of the original sample and multiplying the result by 100:

$$\%Mg = (\text{wt Mg/wt sample})*100 = 14.54\%$$

For practice, find the percentage of magnesium chloride in the original sample and the amount of magnesium ammonium phosphate recovered after the precipitation step.

EXAMPLE 10.3

A 731.8-mg sample containing phosphate (i.e., orthophosphate, PO_4^{3-}) is dissolved in dilute hydrochloric acid. Magnesium chloride and ammonium hydroxide are added, and magnesium ammonium phosphate precipitates. The precipitate is collected on ashless filter paper, then ignited at 1050°C to convert the magnesium ammonium phosphate to magnesium pyrophosphate, a more stable weighing form. The magnesium pyrophosphate is found to weigh 108.0 mg. Calculate the percentage of phosphate in the original sample.

Solution

Two millimoles of magnesium pyrophosphate are produced for every mole of phosphate in the original sample, according to the reaction

$$2MgNH_4PO_4 = Mg_2P_2O_7 + 2NH_3 + H_2O$$

The percentage phosphate is given by

$$\begin{aligned}
\%PO_4 &= \{2[mGFW(PO_4^{3-})]/mGFW(Mg_2P_2O_7)\}*mg(Mg_2P_2O_7)*100/mg\ sample \\
&= (2*94.97/222.56)*108.0*100/731.8 \\
&= 12.59\%
\end{aligned}$$

The stoichiometry for the reaction of phosphate to form pyrophosphate tells us that the factor of 2 must be included.

EXAMPLE 10.4

A 982.0-mg sample containing KCl, KI, and KNO_3 was dissolved in water and treated with excess silver nitrate. The precipitate, a mixture of AgCl and AgI, was collected, washed, dried, and found to weigh 804.8 mg. The mixed precipitate was then heated in a tube through which a stream of chlorine gas was passed, and the AgI in the sample was converted to AgCl. The final weight of the treated sample was 732.6 mg. Calculate the percentages of KCl and KI in the sample.

10.5 CALCULATIONS FOR GRAVIMETRIC METHODS

Solution

The overall process is outlined as follows:

$$KCl + KI + KNO_3 \ (982.0 \text{ mg}) \rightarrow AgCl + AgI \ (804.8 \text{ mg})$$
$$\downarrow Cl_2$$
$$AgCl \text{ only } (732.6 \text{ mg}) + I_2(g) \text{ lost}$$

Potassium nitrate does not react with silver nitrate. We begin by expressing the intermediate condition as an equation:

$$804.8 \text{ mg} = \text{mg(AgCl)} + \text{mg(AgI)}$$

We then express mg(AgCl) in terms of mg(KCl), and mg(AgI) in terms of mg(KI) in the original sample:

$$804.8 \text{ mg} = [\text{mGFW(AgCl)}/\text{mGFW(KCl)}]*\text{mg(KCl)}$$
$$+ [\text{mGFW(AgI)}/\text{mGFW(KI)}]*\text{mg(KI)}$$

$$804.8 = 1.923[\text{mg(KCl)}] + 1.415[\text{mg(KI)}]$$

This equation has two unknowns, and we must find a second equation to solve for both of them. We can make use of the last part of the chemical process, in which AgI in the sample is converted to AgCl:

$$732.6 \text{ mg} = \text{mg(AgCl) from KCl} + \text{mg(AgCl) from AgI}$$
or
$$732.6 = \text{mg(AgCl)} + [\text{mGFW(AgCl)}/\text{mGFW(AgI)}]*\text{mg(AgI)}$$
$$= \text{mg(AgCl)} + 0.6105[\text{mg(AgI)}]$$

We know how to express mg(AgCl) in terms of mg(KCl), and mg(AgI) in terms of mg(KI) from the first equation. We thus express the second equation as

$$732.6 = [\text{mGFW(AgCl)}/\text{mGFW(KCl)}]*\text{mg(KCl)}$$
$$+ 0.6105[\text{mGFW(AgI)}/\text{mGFW(KI)}]*\text{mg(KI)}$$

and substitute in values for the formula weights,

$$732.6 = 1.923[\text{mg(KCl)}] + 0.863[\text{mg(KI)}]$$

We now have two equations in two unknowns and can solve them to find mg(KI) = 130.8, mg(KCl) = 322.0. Dividing each by the weight of the original sample and multiplying the results by 100, we find

$$\%KI = 13.32\% \qquad \%KCl = 32.79\%$$

In trying to solve more complex gravimetric problems it is often a good strategy to look for a set of linear equations which will express the chemical "road map".

The problem set contains a variety of calculations of the type involved in gravimetric methods of analysis. As mentioned earlier, these problems are similar to stoichiometry problems assigned to students in general chemistry courses. They serve a dual purpose here: first, they should stimulate your thinking about processes involving more than one step, and second, they bring out some interesting descriptive chemistry not easily covered in an introductory course.

10.6 ADVANTAGES AND DISADVANTAGES OF GRAVIMETRIC METHODS

Gravimetric methods of analysis are highly accurate and precise in the hands of skilled analysts, particularly for samples containing more than 1% analyte. This is largely a result of the great accuracy with which we can weigh objects in the modern laboratory. With care, a 1-g sample can be weighed to an accuracy of ±0.1 mg, an uncertainty of 1 part in 10,000. This is far better than the accuracy of volumetric glassware, which is on the order of ±1 part in 1000 in routine (careful) work. Furthermore, gravimetric methods do not require the preparation of standard solutions. In practice, the accuracy of a full gravimetric method involving several steps can be on the order of 1 or 2 parts per thousand. Gravimetric methods can also be used with smaller samples. Microbalances make it possible to weigh samples of a milligram with an uncertainty of ±0.1 microgram. Considerably more care must be taken when using these delicate instruments, however. What usually limits the effectiveness of gravimetric methods for such small samples in the intrinsic solubility of the precipitate.

Despite all of these advantages, gravimetric methods have fallen into disfavor in the past 20 years. The major reason seems to be that it takes a great deal of time to do analyses gravimetrically, and the cost of skilled labor is very high. Most students have some trouble organizing three replicate gravimetric analyses. Imagine trying to raise your efficiency by working with a dozen samples simultaneously! Operations such as filtering are also very hard to automate. The result is that many laboratory managers forgo the great accuracy and precision of gravimetric methods and opt for quicker, less expensive methods.

REFERENCES

1. C. H. Hendrickson and P. R. Robinson, *Journal of Chemical Education,* 56, 341–342 (1979).
2. W. A. Hoffman and W. W. Brandt, *Analytical Chemistry,* 28, 1487–1489 (1956).

PROBLEMS

RECOMMENDED READING

As explained in the text, gravimetric methods are no longer the subject of much research. The older texts are generally good sources for the details of gravimetric methods. The following are some particularly thorough treatments. They hold many references to the older research literature.

Hillebrand, W. F., G. E. F. Lundell, H. A. Bright, and J. I. Hoffman, *Applied Inorganic Analysis,* 2nd ed., Wiley, New York, 1953.

Kolthoff, I. M., E. B. Sandell, E. J. Meehan, and S. Bruckenstein, *Quantitative Analysis,* 4th ed., Macmillan, New York, 1969.

Every text written for quantitative analysis at the introductory level contains a discussion of the calculations involved in gravimetric methods. Basic concepts of stoichiometry are discussed in all general chemistry texts.

PROBLEMS

10.1. Starting with exactly 1 g of material from column A, calculate the weight of substance in column B which can (in principle) be isolated by a gravimetric procedure.

A	B
$KBrO_3$	$AgBr$
KI	AgI
$MnCl_2$	$Mn_2P_2O_7$
P_2O_5	$Mg_2P_2O_7$
As_2O_3	As

10.2. What weight of the material in column A would you have to start with to isolate 0.8000 g of the corresponding material in column B?

A	B
$MgCl_2$	$Mg(C_9H_6ON)_2$
$NaCl$	$NaMg(UO_2)_3(C_2H_3O_2)_9 \cdot 6H_2O$
W	WO_3
$SbCl_3$	Sb_2O_4

10.3. The following list of sulfate salts is compiled from a manufacturer's catalog:
  (a) sodium sulfate, Na_2SO_4
  (b) nickel sulfate, $NiSO_4 \cdot 6H_2O$
  (c) aluminum potassium sulfate, $AlK(SO_4)_2 \cdot 10H_2O$

(d) aniline sulfate, $(C_6H_5NH_2)_2 \cdot H_2SO_4$
(e) ferrous ammonium sulfate, $Fe(NH_4)_2(SO_4)_2 \cdot 6H_2O$

When you dissolve 0.4000 g of each sample and add to each a 10% excess of barium chloride to precipitate $BaSO_4$ quantitatively, how many grams of $BaSO_4$ would you be able to isolate? Assume that each sample is 100% pure.

10.4. A 682.7-mg sample of a salt is treated with excess silver nitrate to precipitate AgCl. The precipitate weighs 237.6 mg. What is the purity of the salt if it is:
(a) $ZnCl_2$
(b) $BaCl_2 \cdot 2H_2O$
(c) $MgCl_2 \cdot 2H_2O$
(d) $NH_2OH \cdot HCl$

10.5. A 1.6543-g sample containing copper was dissolved in nitric acid and transferred quantitatively to a 100-mL volumetric flask. Three 25-mL portions were transferred to electrochemical cells and copper metal was plated out onto platinum electrodes. The electrodes were weighed before and after plating and were found to hold 0.0803, 0.0810, and 0.0806 g of copper metal. Calculate the percentage of copper in the original sample.

10.6. A 0.3050-g sample containing calcium was dissolved in acid and treated with an acidic solution of ammonium binoxalate. Several grams of urea were added, and hydrolysis consumed hydrogen ion, gradually raising the solution pH and precipitating calcium oxalate. The precipitate was collected, washed, and dried at 103°C for 2 hours to ensure a weighing form of $CaC_2O_4 \cdot H_2O$. If 0.2817 g of precipitate was isolated, what was the percentage of Ca in the original sample?

10.7. A sample of pure $CaC_2O_4 \cdot 2H_2O$ is decomposed in three distinct steps by raising its temperature.

Step 1: $CaC_2O_4 \cdot 2H_2O = CaC_2O_4 + 2H_2O(g)$
Step 2: $CaC_2O_4 = CaCO_3 + CO(g)$
Step 3: $CaCO_3 = CaO + CO_2(g)$

Calculate the weight lost in each step if the original sample of $CaC_2O_4 \cdot 2H_2O$ weighs 350.0 mg.

10.8. What is the volume of $0.1078 M$ H_2SO_4 required to precipitate quantitatively the metal ions from solutions of each of the following salts? (Assume that no excess is required.)
(a) 0.7900 g $BaCl_2 \cdot H_2O$
(b) 0.7667 g $Pb(OOCCH_3)_2 \cdot 3H_2O$ (lead acetate)
(c) 1.009 g $Sr(ClO_4)_2 \cdot 6H_2O$

10.9. Silica reacts with hydrofluoric acid to form volatile silicon tetrafluoride,

$$SiO_2 + 3H_2F_2 = H_2SiF_6 + 2H_2O$$
$$H_2SiF_6 + heat = SiF_4(g) + H_2F_2(g)$$

A 0.8497-g sample of mineral containing silica and metal oxides is treated with hydrofluoric acid and sulfuric acid. After strong heating with a burner to drive off SiF_4, the sample is found to weigh 0.4173 g. What is the percentage of silica in the sample? (Note: sulfuric acid prevents metal oxides from volatilizing as fluorides.)

10.10. Sodium tetraphenylboron, $Na^+[(C_6H_5)_4B^-]$, forms slightly soluble salts in cold acidic solution with potassium and ammonium ions.
  (a) The potassium in 100 mL of a 0.01075M solution of KCl is precipitated as the salt $K[B(C_6H_5)_4]$. What weight of precipitate can be isolated?
  (b) The solubility of potassium tetraphenylboron is about 7 mg/liter at 25°C. How large an error (absolute and relative) does this cause in the determination of potassium in part (a)?

10.11. A sample of brass weighing 1.3856 g is dissolved in hydrochloric acid. Concentrated nitric acid is then added to precipitate metastannic acid, H_2SnO_3. The precipitate is collected with the help of filter paper pulp in a quantitative filter paper, charred, and ignited in a crucible. The ignition product is stannic oxide, SnO_2. If 32.7 mg of stannic oxide is recovered, what is the percentage of tin in the brass sample?

10.12. Lead can be precipitated as lead chromate ($PbCrO_4$) from a solution in which chromate is gradually generated from Cr(III) by oxidation with bromate, BrO_3^-. If a 0.8650-g sample of brass contains 6.24% lead and it is necessary to generate a 100% molar excess of chromate, how many moles of potassium bromate must be added to a solution containing Cr(III)? The reactions are:

$$2Cr^{3+} + BrO_3^- + 5H_2O = 2CrO_4^{2-} + Br^- + 10H^+$$
$$Pb^{2+} + CrO_4^{2-} = PbCrO_4(s)$$

10.13. Lanthanum oxalate, $La_2(C_2O_4)_3 \cdot xH_2O$, is being investigated as a suitable precipitate for the gravimetric determination of lanthanum. If a sample of pure lanthanum oxide (La_2O_3) weighing 200.2 mg is dissolved and precipitated as lanthanum oxalate weighing 432.6 mg, how many waters of hydration are there in lanthanum oxalate? (What is the value of x in the formula?)

10.14. A sample containing only iron pyrites (FeS_2) and silica (SiO_2) weighing 0.8452 g was ignited in the air to convert the iron pyrites to ferric oxide (Fe_2O_3) and silica. The residue weighed 0.6629 g. What is the percentage of iron pyrites in the sample?

10.15. A mixture of ferric oxide (Fe_2O_3) and aluminum oxide (Al_2O_3) weighing 0.5542 g is heated in a stream of hydrogen gas and is found to lose 0.0554 g in weight. What is the composition of the original sample? (Al_2O_3 is not reduced by hydrogen.)

10.16. The phosphorus in a sample of detergent is converted to orthophosphate (PO_4^{3-}) by heating in acid for several hours. The orthophosphate is then precipitated as $Mg(NH_4)PO_4$, collected in a crucible, and ignited to the stable weighing form, $Mg_2P_2O_7$ (magnesium pyrophosphate).

(a) If the initial sample weighs 1.2103 g and 0.0454 g of magnesium pyrophosphate is collected, what is the percentage of phosphorus (as orthophosphate) in the sample?

(b) Using the information in part (a), calculate the weight of magnesium ammonium phosphate isolated in the intermediate step.

(c) If you are a detergent manufacturer trying to convince a purchaser that your detergent is "low in phosphorus," would you advertise the phosphorus content as percent phosphorus or percent phosphate? Explain, using gram formula weight comparisons.

10.17. A sample of chlorinated cyclohexane is taken from a section of a large fractionating column in a pesticide manufacturer's plant. A 0.4030-g sample is strongly heated with sodium metal in a sealed vessel. The chloride liberated is washed from the vessel and precipitated as silver chloride. If 1.061 g of AgCl is recovered from the original sample, what is the average number of chlorine atoms on the cyclohexane rings in the sample? (Cyclohexane is C_6H_{12}; dichlorocyclohexane is $C_6Cl_2H_{10}$.)

10.18. A 1.005-g sample of potassium iodate was reduced to potassium iodide, and the iodide ion was precipitated as silver iodide. The silver iodide was collected and subsequently heated in a stream of chlorine to convert it to silver chloride. The sample lost 0.4269 g during the conversion. Calculate the percentage purity of the potassium iodate.

10.19. Exactly 2 g of a mixture containing only silver chloride, silver bromide, and silver iodide lost 0.1000 g of weight when heated in bromine gas. The product of the bromine reaction lost 0.2000 g in weight when heated in chlorine gas. Calculate the percentage of each silver halide in the original mixture.

10.20. The phosphorus in 0.4587 g of a sample was converted to phosphate ion and subsequently precipitated as silver phosphate. The precipitate was then converted to silver chloride weighing 0.9250 g. Calculate the percentage of phosphorus in the original sample.

EXPERIMENT 10.1: Gravimetric Determination of Calcium as Calcium Oxalate Monohydrate

Introduction

Dried and weighed samples of an unknown containing calcium carbonate are dissolved with hydrochloric acid. Solutions are then filtered to remove silica. Solid urea and an acidified solution of ammonium oxalate are added to the solutions of sample, and the urea slowly hydrolyzes to generate oxalate ions homogeneously. Calcium oxalate precipitates gradually. When precipitation is complete the precipitate is collected by filtration with carefully cleaned and weighed sintered glass crucibles, washed with cold water and then acetone, and dried for 1 hour at 100 to 105°C. The pertinent reactions are

$$CaCO_3 + 2H^+ \rightarrow Ca^{2+} + CO_2 + H_2O$$
$$(NH_2)_2CO \text{ (urea)} + H_2O \rightarrow 2NH_3 + CO_2$$
$$2NH_3 + H_2C_2O_4 \rightarrow 2NH_4^+ + C_2O_4^{2-}$$
$$Ca^{2+} + C_2O_4^{2-} + H_2O \rightarrow CaC_2O_4 \cdot H_2O$$

The reaction of urea with water (hydrolysis) is slow at room temperature but can be accelerated by heating the solution. The rate at which ammonia forms determines the rate at which oxalate is produced from oxalic acid, and this determines the rate at which calcium oxalate precipitates.

CAUTION: Oxalate salts are poisonous.

Procedure

This experiment will require two laboratory periods for completion. During the first period you will clean and dry to constant weight three glass filter crucibles, dry your unknown sample, weigh out and dissolve three portions of the sample, and finally add urea and acidic ammonium oxalate to all three. You may then set your samples aside in your lab drawer (or in a designated safe place) and let the urea hydrolyze slowly at room temperature until the next lab period. During the second lab period you will check for completeness of precipitation, heat the solutions, filter, wash, dry, and weigh the precipitates.

1. Obtain a sample from your instructor, place it in a clean, dry weighing bottle, and dry it in a drying oven at 110°C for 1 hour to remove surface water.

2. Clean and label three sintered glass filter crucibles. Place the crucibles in a 400-mL beaker labeled with your name, cover the beaker with a watch glass, and place it in an oven at 110°C. After 1 hour remove the beaker and crucibles from the oven with beaker tongs. Using crucible tongs, transfer the glass filters to your desiccator. Allow them to cool for 30 minutes before weighing them. Record their weights, return them to the beaker and the oven, and repeat the process until the weights you measure agree within ±0.2 mg (constant weight).

3. While the filter crucibles are heating, weigh by difference 0.3- to 0.4-g samples of unknown into separate beakers. Record each sample weight to the nearest 0.1 mg.

4. Add about 75 mL of deionized water to each sample. Using a 10-mL graduated cylinder, add about 5 mL of $6M$ HCl, stirring the solution as you add acid. If you see wisps of white powder remaining after adding hydrochloric acid, heat your samples and filter them through number 42 filter paper. Be careful not to mix up samples at this stage. After filtering, wash each filter paper with five 5-mL portions of hot deionized water. Be sure that these washings remain with their proper filtered sample solutions.

5. Add five drops of methyl orange indicator to each solution and stir. The solutions should be pink at this point. If they are not, ask your instructor for help.

6. Add to the contents of each beaker about 20 mL of a saturated solution of ammonium oxalate in $0.3M$ hydrochloric acid (provided by instructor).

7. Dissolve about 15 g of urea in each solution. Even though the hydrolysis of urea is slow at room temperature, your solutions may start to get turbid in a few minutes. Cover each beaker with a watch glass and store the solutions in your lab desk until the next lab period. If at least 90 minutes remain in the first lab period, you may speed the hydrolysis by heating the covered solutions to near boiling until the indicator color changes from pink to yellow. This should take more than $\frac{1}{2}$ hour. Rinse any condensate on the watch glasses into the beakers.

8. After storage for several days in your desk the indicator should be yellow or a yellowish-pink (a pH of about 5), and the precipitation should be complete. Add dilute ammonium hydroxide dropwise until the indicator color is yellow or until you see no more calcium oxalate precipitating.

9. Heat the three solutions to near boiling while you set up an aspirator for your filter crucibles. Filter the hot solutions through the appropriate filter crucibles using aspirator suction. The excess oxalate ion in the hot solutions will suppress solubility of calcium oxalate in hot solutions. Use a rubber policeman and small volumes of ice-chilled deionized water to transfer the last remaining precipitate from the beaker. Deionized water will dissolve some calcium oxalate unless it is ice cold.

10. Wash the calcium oxalate crystals in each filter crucible with two 10-mL portions of chilled acetone. Draw air though the crystals for several minutes to remove the last traces of acetone.

11. Place all three crucibles in your labeled beaker, cover the beaker with a watch glass, and place it in an oven at 100 to 105°C for 1 hour. The temperature of the oven must not be allowed to go above 105°C.

12. Remove your crucibles from the oven and transfer them to your desiccator to cool for 30 minutes. Weigh the three crucibles to the nearest 0.1 mg. Return them to the 105°C oven, heat for another hour, and check for constant weight.

Report percent calcium in the sample.

Ions that form insoluble oxalate salts will interfere with the determination of calcium if they are at a sufficiently high concentration. The most important interferences are apt to be magnesium and iron, which are present in many limestone samples. Magnesium is known to postprecipitate on calcium oxalate if solutions are allowed to stand for long periods. In the author's experience postprecipitation is not a problem with limestone samples obtained from Thorn-Smith, Inc.

This experiment was adapted from Hendrickson and Robinson (Ref. 1).

EXPERIMENT 10.2: Gravimetric Determination of Chloride

Introduction

This is the standard, classical method for determining chloride in samples containing more than a few percent chloride. It involves the addition of a solution of silver nitrate to a dissolved portion of unknown sample and the precipitation of silver chloride. The precipitate is digested, collected in a glass filter crucible, washed, and dried in an oven. The method is both highly accurate and precise, and it has been used for many years as an introduction to gravimetric operations for undergraduate chemistry students.

The reaction is a simple 1:1 reaction of unknown chloride with excess silver ion to precipitate silver chloride,

$$Ag^+ \text{ (excess)} + Cl^- \text{ (unknown)} = AgCl \text{ (weighed)}$$

Procedure

1. Obtain a sample from your instructor. Transfer the sample to a clean, dry weighing bottle. Place the uncovered weighing bottle in a 150-mL beaker labeled with your name, cover the beaker with a watch glass resting on glass hooks, and dry in an oven at 110°C for 2 hours.

2. While the sample is drying, clean and dry three medium-porosity glass filter crucibles. Gently brush the sintered glass filters of the crucibles with dilute detergent solution to remove visible dirt. Assemble a water aspirator, filter flask, and rubber

filter ring of the correct size for the filter crucibles. Fill a crucible on the filtering apparatus about half full with $6M$ nitric acid, and draw the solution gently through the filter with aspirator suction. Rinse the crucible several times with deionized water. If you can still see old precipitate in the filter, disassemble the filtering apparatus and discard the nitric acid wash solution. Reassemble the filter apparatus and repeat the washing process with $1M$ ammonium hydroxide. Ammonium hydroxide should remove any silver chloride left by previous workers.

> CAUTION: Nitric acid is an oxidizing mineral acid and can burn your fingers if you are careless. Molar solutions of ammonia should not be left around the benchtop. Ammonia fumes are dangerous and quite disagreeable.

Rinse the crucibles liberally with deionized water and draw air through them for a few minutes. Using a pencil, label the glass crucibles #1, #2, and #3, and place them in a 400-mL beaker. Cover the beaker with a watch glass resting on glass hooks, and place the beaker in the oven at 110°C for 1 hour.

3. Remove the beaker containing your sample from the oven. Let the beaker cool for about 5 minutes before transferring the weighing bottle containing sample to your desiccator. Allow the crucibles to cool to room temperature in the desiccator (25 or 30 minutes). Using an analytical balance, weigh to the nearest 0.1 mg three portions of between 0.4 and 0.7 g of your unknown into clean 250-mL beakers. Be sure to number your beakers to correspond to the numbering of your filter crucibles. Be careful not to mix up your samples. To the sample in each beaker add 100 mL of deionized water and 1 mL of 1:1 nitric acid. Dissolve each sample with the aid of a stirring rod. Use a separate stirring rod for each sample; keep each stirring rod with its sample—do not change them. If you must remove a stirring rod from its solution, rinse it with deionized water: keep each sample in its own beaker.

4. Calculate the volume of $0.5M$ silver nitrate solution (provided) you will need to precipitate the chloride. Assume that the sample is pure sodium chloride, GFW = 58.5. For example, a 620-mg portion of your unknown sample, if it were pure NaCl, would contain 620/58.5 = 10.8 mmole of NaCl, and would require 10.8 mmole of silver ion to precipitate AgCl completely. If $0.5M$ silver nitrate solution is available you would use 10.8 mmole/0.5 mmole/mL, or 21.6 mL.

5. Add slowly (a few drops at a time) with vigorous stirring the amount of silver nitrate calculated for each unknown sample, and then add 10% extra. Heat the solutions containing precipitate almost to boiling to help the precipitate flocculate. Boiling must be avoided, as it will break up the larger clumps of crystals you are trying to form by heating. The precipitate will begin to darken slowly as a result of a reaction with light. In ordinary laboratory light on a 3-hour time scale, this darkening will not affect your results. If possible, avoid working in direct sunlight (see below).

After 20 or 30 minutes the precipitate will have fallen to the bottom of each

EXPERIMENT 10.2: GRAVIMETRIC DETERMINATION OF CHLORIDE

beaker, and the solutions will be quite clear. Add a drop of $0.5M$ silver nitrate to each beaker to test for completeness of precipitation. If a cloud of precipitate forms, add an additional 5% of silver nitrate solution, stir, and heat again.

6. Weigh the filter crucibles. While your AgCl precipitates are digesting, remove your filter crucibles from the oven and cool them in a desiccator for 30 minutes. Weigh each filter crucible to the nearest 0.1 mg on the analytical balance. Record each weight in your notebook. Return the crucibles to their beaker and place them in the oven at 110°C for an additional hour. After the hour is up, remove the crucibles from the oven and let them cool to room temperature in your desiccator. Weigh the crucibles again. If their weights agree within ±0.2 mg, they are at "constant weight" and can be used to collect precipitate. If there is insufficient time to filter the precipitates, store the solutions in your lab desk and the crucibles in your desiccator until the next lab period.

7. Collect the precipitates. Place weighed crucible #1 in the filter apparatus, and draw gently with the aspirator. Decant the clear solution from beaker #1, making sure that most of the precipitate stays in the beaker. When the clear solution is mostly decanted, add about 20 mL of wash solution ($0.01M$ nitric acid) to the precipitate and mix gently. Transfer portions of the precipitate and solution to the filter crucible. Hold the stirring bar across the lip and top of the beaker with one hand, and direct a stream of wash solution (in a plastic washing bottle) at precipitate which clings to the beaker. Use the rubber policeman to remove precipitate which still sticks to the beaker. Rinse every bit of precipitate into the filter crucible.

Wash the precipitate several times with 5-mL portions of the wash solution to remove silver ion which might adsorb on the precipitate. Catch the last washing by itself, and test for the presence of silver ion by adding a drop of $1M$ HCl. If you see a precipitate form, wash again with washing solution.

Draw the remaining liquid out of the filter crucible.

Repeat this process for crucibles and solutions #2 and #3.

8. Gently wipe the outside surfaces of the three filter crucibles to remove fingerprints. Place all three crucibles in the 400-mL beaker, cover with a watch glass resting on glass hooks, and dry in an oven at 110°C for 2 hours.

9. Cool, weigh, and repeat the process until constant weight has been achieved.

10. Calculate the percentage of chloride in the sample. The gram atomic weight of chlorine is 35.45, and the gram formula weight of AgCl is 143.33.

Notes and Special Precautions

1. Chloride is a common contaminant in reagents and water supplies. Test deionized water for the presence of chloride by adding a drop of $0.5M$ AgCl.

2. Silver nitrate is an expensive reagent. Take only what you need for precipitation. If you have any silver nitrate left, put it into a "silver wastes" bottle provided

by your instructor. You should also dispose of your AgCl precipitates in this bottle at the end of the experiment.

3. In Chapter 9 it was recommended that precipitation be carried out in hot solutions to avoid the formation of small crystals. In this particular experiment nitric acid is present in the solution containing chloride. Nitric acid may oxidize chloride to form chlorine and nitric oxide in hot solutions, and this would produce low results. Therefore we precipitate at room temperature.

4. Silver chloride is light-sensitive, and the precipitate should be kept out of strong light. The photoreaction produces silver metal on the surface of the precipitate:

$$2AgCl = 2Ag + Cl_2(g)$$

If the photoreaction occurs when there is excess silver ion present in the solution (during digestion), the chlorine liberated will react to form chloride in solution, and more AgCl will precipitate. This leads to high results. If the photoreaction occurs after excess silver has been removed (in the filter crucible, for example), the chlorine lost will cause low results. The loss or gain of weight is negligible for 3- or 4-hour exposures to laboratory light.

5. It is common practice to filter hot solutions; hot solutions move through filters more quickly than cold solutions (see the calcium oxalate experiment in this chapter). Unfortunately, silver chloride is too soluble at high temperatures for hot filtering to produce quantitative results, and you must filter room temperature solutions.

6. There are many potential interferences, notably thiocyanate, cyanide, sulfide, thiosulfate, ferro- and ferricyanide, bromide, and iodide. All of these species cause high results and must be removed or chemically altered before precipitation of AgCl. Cations such as mercuric, cadmium, Sn^{2+}, and others form soluble complexes with chloride and may produce low results. Reducing agents such as formaldehyde or SO_2 can reduce silver ion to silver metal and must be removed by oxidation before precipitation of AgCl (see Chapter 16).

EXPERIMENT 10.3: Gravimetric Determination of Lead in Brass

Introduction

In this experiment the lead in a sample of brass is precipitated as lead chromate. Rather than force a rapid precipitation by adding sodium chromate to the solution of lead, we generate chromate ion in solution (homogeneously) by reacting chromium(III) with the oxidizing agent, bromate:

$$2Cr^{3+} + BrO_3^- + 5H_2O \rightarrow 2CrO_4^{2-} + Br^- + 10H^+$$

EXPERIMENT 10.3: GRAVIMETRIC DETERMINATION OF LEAD IN BRASS

This reaction is quite slow at room temperature but can be accelerated by heating the solution nearly to boiling. As chromate ion forms it gradually precipitates lead chromate,

$$CrO_4^{2-} + Pb^{2+} = PbCrO_4(s)$$

The reaction of chromium(III) with bromate generates considerable hydronium ion, which has the unfortunate effect of increasing the solubility of lead chromate, and causing low results. In acidic solution chromate will form dichromate in a dimerization reaction,

$$2CrO_4^{2-} + 2H^+ = Cr_2O_7^{2-} + H_2O$$

Lead dichromate is more soluble than lead chromate, and the results will be low. Some of the extra hydronium ion is absorbed by a secondary reaction of bromide with the excess bromate,

$$5Br^- + BrO_3^- + 6H^+ \rightarrow 3Br_2(g) + 3H_2O$$

Gaseous bromine is lost from the solution during the reaction. Despite the help of this secondary redox reaction, we need to add an acetate buffer to absorb hydronium ion and achieve a quantitative precipitation.

Procedure

1. Clean three sintered glass filter crucibles. Set up an aspirator apparatus with a filter flask, rubber ring, hose connected to a liquid trap, and trap connected to an aspirator at the water tap. Rinse each filter crucible with several portions of dilute nitric acid to remove materials left from previous experiments. After the nitric acid, rinse each crucible with deionized water. Mark the crucibles #1, #2, and #3, and transfer them to a 400-mL beaker. Cover the beaker with a watch glass supported by glass hooks. Put the beaker in an oven at 120°C for 1 hour.

2. Obtain a brass sample (powder or chips) from your instructor. Transfer the sample to a weighing bottle. Rinse the sample with two 10-mL portions of acetone to remove oil that may be in the sample. Place the weighing bottle in a 150-mL beaker and put the beaker in the oven for $\frac{1}{2}$ hour to volatilize residual acetone. Remove the sample from the oven and let it cool in air. After it has cooled, weigh by difference three portions of the sample, each to the nearest 0.1 mg. Each portion should contain between 100 and 200 mg of lead. Ask your instructor for the approximate composition of your brass sample, and calculate how much sample you should weigh for each portion. Use labeled 150-mL beakers to hold the portions.

3. Dissolve the samples. Take your samples to the fume hood. Add 20 mL of 1:1 nitric acid to each beaker. Cover each beaker with a watch glass, and evaporate the solutions to less than 10 mL with heat from a Bunsen burner.

> CAUTION: Hot nitric acid is a powerful oxidizing agent and will cause serious burns. You must wear eye protection and work in a fume hood. Brown vapors (nitrogen dioxide) will be given off as the sample dissolves. Do not breathe this toxic gas.

Do not evaporate your samples to dryness. Avoid splattering sample onto the walls of the beakers. When the samples are less than 10 mL in volume, add about 30 mL of deionized water and continue to heat as before for 30 minutes. You may wish to remove your crucibles from the oven before the 30-minute heating.

4. As time permits, cool your crucibles for 30 minutes in your desiccator, and weigh them to the nearest 0.1 mg on an analytical balance. Return the crucibles to the oven for another hour. Again as time permits, remove them from the oven, let them cool to room temperature in your desiccator, and measure their weights. If the second weights agree with the first weights within ±0.2 mg, your crucibles are at constant weight. If they do not, repeat the heating, cooling, and reweighing process.

5. The dissolved samples will contain a white precipitate, an oxide of tin called metastannic acid, H_3SnO_3. This precipitate must be removed by filtering through quantitative filter paper (Whatman No. 42, for example). Collect the clear filtrate solutions in labeled 250-mL beakers. Wash the metastannic acid left behind in each filter paper with three 5-mL portions of warm 1:25 nitric acid. These washings will contain some lead and must be kept with their proper filtrate solutions.

6. Neutralize the filtrate solutions by adding $1M$ NaOH solution dropwise until a light precipitate forms and persists on swirling. Add 25 mL of acetate buffer (see below) to each solution. The buffer should dissolve the precipitate which formed. (If you must interrupt the experiment, this is a good place to do so.)

7. Add to each solution 20 mL of $0.1M$ chromium(III) nitrate and 20 mL of $0.2M$ potassium bromate (see note below). Heat all three solutions containing sample nearly to boiling over Bunsen burners. Within about 30 minutes the generation of chromate will be complete, precipitation of lead chromate will be finished, and the solution over the precipitates will be clear.

8. Collect the precipitates. The solutions containing the precipitate should be cooled to room temperature. Set up the aspirator apparatus and collect the precipitate from each sample in the filter crucible with the corresponding number. Filter as much clear liquid as possible before transferring precipitate crystals to a crucible. Use a rubber policeman and some cold deionized water to rinse all the precipitate out of each beaker. Wash the yellow crystalline precipitate with 10 mL of cold deionized water. Place the filter crucibles in the 450-mL beaker, place a watch glass on glass hooks over the beaker, and put the samples in the oven at 120°C for 1 hour. After an hour, cool the samples in your desiccator, then weigh them to the nearest 0.1 mg. Repeat the drying, cooling, and weighing process until the samples are at constant weight.

9. Calculate the percentage of lead in the sample.

Notes

The acetate buffer solution should be $0.6F$ in acetic acid and $2.0F$ in sodium acetate. Mix 16.4 g of sodium acetate with 3.6 mL of glacial acetic acid and dilute to 100 mL. You may wish to prepare enough buffer solution for some of your classmates to use.

CAUTION: Glacial acetic acid causes burns.

Your instructor will probably provide the solutions of chromium(III) nitrate and potassium bromate. If not, you can prepare the Cr(III) solution by dissolving 0.01 gram formula weight of $Cr(NO_3)_3 \cdot 9H_2O$ (GFW 400.15) in 100 mL of deionized water. You can prepare the potassium bromate solution by dissolving 0.02 gram formula weight of $KBrO_3$ (GFW 167.0) in 100 mL of deionized water. Your instructor should also provide a bottle for waste Cr(III) solution. It is a good idea not to flush chromium solutions down the sink and into your municipal water treatment system. Higher oxidation states of chromium are believed to cause cancer, at least in laboratory animals.

Interferences

Metal ions which form insoluble chromates may produce high results if they are present in high concentrations, or if the precipitation reaction is run in basic solution. You may wish to use the solubility product constants for chromates given at the end of the text to calculate the solubilities of silver, barium, and mercury(II) chromates at about pH 5.

The anions of common mineral acids will not interfere.

Acetate ion in the buffer will complex chromium(III) and will tend to slow the oxidation of Cr(III) to chromate [Cr(VI)]. Ammonia and ammonium salts should be avoided, as ammonium ion will compete with Cr(III) for oxidation by bromate. The product of the oxidation of ammonium ion is nitrogen gas.

Reference for this experiment: Hoffman and Brandt (Ref. 2).

Chapter 11

Precipitation Titrations

11.1 INTRODUCTION

In this chapter we will explore some titrimetric methods which are based on precipitation reactions. As you will see, the reaction of a Lewis acid with a Lewis base to form an insoluble product in a precipitation titration is analogous to the reaction of a Brønsted acid with a Brønsted base to form a weak electrolyte. The important difference is that in the precipitation titration a phase change accompanies the reaction. Precipitation titrations are a part of the "classical education" of analytical chemists. Like many other wet methods of analysis, they are becoming less and less important. In an introductory course in analysis, however, they can still be useful vehicles for reinforcing the concepts of solubility product, common ion effect, and adsorption.

We will begin the chapter by presenting a model system, the titration of chloride with standard silver nitrate solution, and discussing how a silver wire electrode can be used to monitor the titration and find the equivalence point. We will then compare the model with some real analytical data to see how closely theory agrees with practice. The second half of the chapter is devoted to applications: the titration of mixtures of halides, three important indicator methods for halides, and, finally, some more unusual direct and indirect methods.

11.2 THE MODEL SYSTEM: $Ag^+ + Cl^- = AgCl$

Imagine an experiment in which we take 50 mL of a $0.10F$ solution of NaCl and titrate it with a $0.10F$ solution of $AgNO_3$. We will monitor the progress of the titration by using a silver wire electrode, the potential of which depends on the activity of silver ion in the titration vessel (henceforth called a "cell"). We will measure the potential of the silver wire in comparison with a second electrode (called a reference electrode), whose potential is stable and well known. The measurement is very similar to that described in Chapter 6 for the measurement of hydronium ion activity with a glass electrode, a reference electrode, and a high-impedance voltmeter.† The Nernst equation, which will be discussed in detail in Chapter 14, relates the measured cell voltage to the activity of silver ion. The simplest representation of the Nernst equation for the response of the silver wire electrode is

$$E_{\text{wire}} = \text{constant} + 0.05916 \log a_{\text{Ag}^+}$$

where the term "constant" should be thought of as the measured value of the cell voltage when the activity of silver ion is unity, and the coefficient 0.05916 is a combination of natural constants and the absolute temperature.‡ It is often more convenient to describe this equation in terms of pAg, the negative logarithm of the silver ion activity,

$$E_{\text{wire}} = \text{constant} - 0.05916(\text{pAg})$$

pAg is totally analogous to pH, the unit of measure in acid-base titrations in Chapter 6.

It is not so important that you understand everything about this equation right now. It is important, though, to see that there is a linear relationship between the measurable cell voltage and the logarithm of the silver ion activity. Titration curves can be plotted with either cell voltages or pAg on the vertical axis and either volume or percent titrated on the horizontal axis.

Figure 11.1 shows a simulated curve for the model titration. There are three regions to consider to understand the titration curve: first, where there is more chloride than silver ion (0 to 99.9% titrated), second, where the amounts of chloride and silver ion are stoichiometrically equivalent (the equivalence point, 100% titrated), and third, where there is more silver than chloride ion (>100% titrated). In the first region silver ion activity is determined by K_{sp} and the activity of unreacted chloride and can be calculated by the approach taken in studying the common ion effect in Chapter 8. Remember that unreacted chloride suppresses the solubility of AgCl and therefore lowers the activity of silver ion. Dilution must be taken into account in this region

---

† Details of this kind of measurement appear in Chapter 15.

‡ "Constant" contains the standard potential of the silver ion/silver metal couple, the potential of the reference electrode, and the potentials which arise across electrode-solution junctions. The number 0.05916 is the value of the term $(2.303)*RT/nF$ at 298K for $n = 1$. Here R is the gas constant, T is the absolute temperature, F is the Faraday constant (96,487 coulombs per equivalent), and n is the number of electrons transferred in the reduction of a silver ion to the metal.

11.2 THE MODEL SYSTEM: Ag$^+$ + Cl$^-$ = AgCl

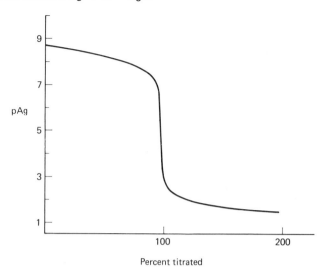

Figure 11.1 Simulated titration of $0.01F$ Cl$^-$ with $0.01F$ Ag$^+$.

because the activity of unreacted chloride depends on the total volume. To express this idea in the form of a general equation, let us say that we have a solution of sodium chloride ion at concentration C_L. We will then add V_M milliliters of silver nitrate solution of concentration C_M to an initial volume V_L mL of sodium chloride solution. The concentration of chloride is given by

$$[\text{Cl}^-] = (\text{mmole Cl}^- - \text{mmole Ag}^+)/(\text{total volume})$$
or
$$[\text{Cl}^-] = (C_L * V_L - C_M * V_M)/(V_M + V_L)$$

The activity of chloride is found by multiplying [Cl$^-$] by the activity coefficient of chloride in a solution of the ionic strength at the particular point in the titration. The ionic strength is calculated from the unreacted chloride, the original amount of sodium, and the amount of nitrate added with the silver ion titrant. The details are shown in Example 11.1. Once the activity of chloride has been calculated, the activity of silver ion can be determined from the thermodynamic K_{sp},

$$a_{\text{Ag}^+} = K^0_{sp}/a_{\text{Cl}^-}$$

At the equivalence point the only source of both silver ion and chloride ion is the dissociation of silver chloride, and their activities must be equal. The activity of silver ion is given by the square root of K^0_{sp}.

Once past the equivalence point, the activity of silver ion is determined only by the concentration of silver ion in excess and whatever silver ion is provided by dissociation of the silver chloride made in the titration. In most practical titrations (concentrations above $0.01F$), dissociation is not an important second source and can be ignored. Once again using C_M formal silver nitrate and C_L formal sodium chloride, when we add V'_M mL of silver nitrate after the equivalence point, we can calculate the silver ion concentration from

$$[Ag^+] = (\text{mmole Ag}^+ \text{ in excess})/(\text{total volume})$$
or
$$= (V'_M * C_M)/[V'_M + V_M(\text{eq. pt.}) + V_L]$$

where $V_M(\text{eq. pt.})$ represents the volume delivered to reach the equivalence point. Knowing the ionic strength, it is then possible to calculate the activity of silver ion.

For practice you might wish to use the equations presented to write a BASIC program to calculate a titration curve for 50 mL of $0.100F$ NaCl with $0.100F$ AgNO$_3$ and compare the results with those shown in Fig. 11.1.

EXAMPLE 11.1 Ag^+/Cl^- Titration Curve

Calculate pAg at the 50%, 100%, and 150% titrated points when 50 mL of a $0.100F$ solution of potassium chloride is titrated with $0.100F$ silver nitrate. $K^0_{sp} = 1.8 \times 10^{-10}$.

Solution

50% point: Having started with 5.00 mmole of chloride (0.1 mmole/mL * 50 mL), we have precipitated 2.5 mmole by the time the 50% point is reached. The solution volume is 75 mL and the ionic strength is $0.5 * [5 \text{ (from K}^+) + 2.5 \text{ (from Cl}^-) + 2.5 \text{ (from NO}_3^-)]/75$ mL = $0.067M$. The concentration of chloride is (2.5 mmole)/75 mL = $0.033M$. Its activity coefficient is about 0.83 in a solution of ionic strength 0.067 (Davies equation), and its activity is $0.83 * 0.033 = 0.028$. The activity of silver ion can then be calculated:

$$a_{Ag} = K^0_{sp}/a_{Cl^-} = 6.52 \times 10^{-9}$$
and
$$pAg = -\log(6.52 \times 10^{-9}) = 8.19$$

100% point: The activities of silver ion and chloride ion are given formally by the square root of K^0_{sp}, and are 1.34×10^{-5} (pAg = 4.87).

150% point: At this point 2.5 mmole of silver ion has been added in excess and the solution volume is 125 mL. The concentration of silver ion is 2.5/125 or $0.020M$. The ionic strength of the solution is $0.06M$ (5 mmole of K$^+$, 7.5 mmole of NO$_3^-$, and 2.5 mmole of Ag$^+$ in 125 mL) and the activity coefficient of silver ion is about 0.83 (Davies equation). The activity of Ag$^+$ is $0.83 * 0.020 = 0.0166$, and pAg = 1.78.

11.3 EFFECTS OF DILUTION

Figure 11.2 shows simulated curves for the titration of chloride with silver ion at the 0.1 and 0.01 F concentration levels. Both dilution and activity coefficient changes have been accounted for in each curve. The only point unaffected by dilution is the equivalence point. Notice that the equivalence point break is considerably smaller in the 0.01 F case than in the 0.1 F case. This will influence the way we use colored indicators to monitor endpoints, as we will see shortly.

11.4 REAL VERSUS IDEAL: SOME ACTUAL DATA

Idealized systems produce attractive plots for textbooks. But what kind of data can students expect to obtain in laboratory? Figure 11.3 shows experimental data obtained by students for the titration of 0.03981M KCl with 0.1069M AgNO$_3$ at 21°C, along with a computer-simulated titration curve for the same concentrations. The students measured the voltage of a clean silver wire dipped into the solution of chloride, as described in the first few paragraphs of this chapter. The reference electrode they used was a saturated calomel electrode (see Chapter 15). The experimental and simulated curves have been overlaid to show how similar they are. The simulation program corrects for dilution and temperature and uses the ionic strength at each point to calculate silver ion activity. It assumes that the K_{sp} of silver chloride is 1.27×10^{-10} at 21°C, that the standard potential of the silver ion/silver metal couple is 0.804 volt,

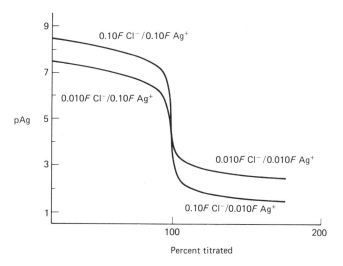

Figure 11.2 Effects of dilution on titration curves (simulated titrations of 0.10 and 0.010F Cl$^-$ with 0.10 and 0.010F Ag$^+$).

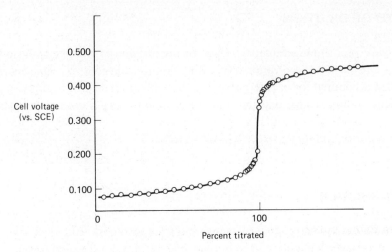

Figure 11.3 Real and ideal precipitation titrations. Real data (circled points): titration of $0.03981 M$ Cl^- with $0.1069 M$ Ag^+ at $21°C$. (Data of E. Peterson and C. Johnson, 1984.) See text for assumptions made in simulation.

and that the potential of the reference electrode is 0.244 volt. These last assumptions make the "constant" term in the Nernst equation equal to 0.559 volt.†

Adsorption, if it occurs to a significant extent, would be quite apparent when comparing a real curve with a simulation. The effect on a curve is to round the equivalence point break, making it less distinct. Before the equivalence point is reached, halide ions adsorb on the silver halide precipitate and halide activity in solution is lowered. This raises the activity of silver ion in solution, slightly raising the titration curve just before the equivalence point and rounding the break in the curve. After the equivalence point break, silver ion adsorbs on the precipitate, lowering the activity of silver ion and slightly lowering the experimental curve relative the simulated curve. We would expect there to be a point where silver ion replaces halide ion in the adsorbed layer, and it would be where the real and simulated curves cross. This behavior occurs at a system's "isoelectric point" and is accompanied in a real titration by a sudden flocculation of the precipitate (see Chapter 9). Although all halide ions adsorb on their silver halide precipitates, the extent of adsorption (and distortion of titration curves) is found to decrease in the order iodide, bromide, chloride and is most important when there is a great deal of precipitate (large surface area). The fact that at the $0.04F$ level an experimental titration curve can be so closely approximated by a simulation

---

† The values at $25°C$ are $K_{sp} = 1.8 \times 10^{-10}$, $E_{Ag^+/Ag} = 0.799$ volt, and $E_{ref} = 0.242$ volt. For temperature corrections see Owen (Ref. *1*). The simulation does not include the "liquid junction potential." An agar bridge was used to connect the reference electrode to the titration cell, and such connections cause an additional voltage drop of a few millivolts. This problem is discussed in Chapters 14 and 15. A correction of 5 millivolts produces better agreement.

11.5 TITRATION OF A MIXTURE OF HALIDES

that ignores adsorption indicates that adsorption is not an important problem at such concentrations.

11.5 TITRATION OF A MIXTURE OF HALIDES

Figure 11.4 shows both simulated and experimental curves for the titration of mixtures of potassium iodide and potassium chloride with silver nitrate at the $0.1F$ level. The simplest possible model treats the precipitations of AgI and AgCl as if they were totally independent, that is, as if adsorption and other secondary interactions did not exist.

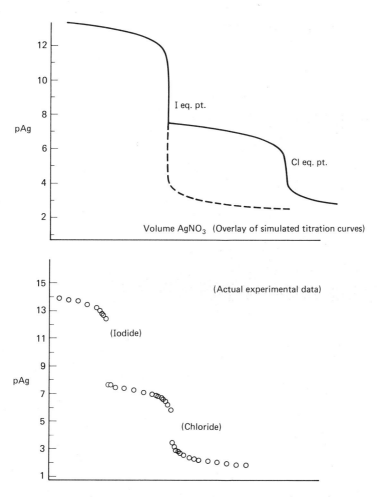

Figure 11.4 Titration of a mixture of halides. [Actual data for titration of 6.31 mmole of KI and 7.37 mmole of KCl in 50 mL of water at 21°C. From E. Peterson and C. Johnson, 1984.]

This amounts to drawing out individual titration curves and then simply cutting them out and pasting them together. The composite curve will show two clearly defined breaks. Does a mixed system really behave this way? Are the two equivalence point breaks at the correct positions of the two equivalence points?

EXAMPLE 11.2 Titration of Mixed Halides

A solution containing $0.1F$ KI and $0.1F$ KCl is titrated with $0.1F$ AgNO$_3$. Calculate pAg at the iodide equivalence point and at the point at which AgCl begins to precipitate. The thermodynamic solubility products of AgI and AgCl are 8.3×10^{-17} and 1.8×10^{-10}, respectively.

Solution

The equivalence point activity of silver ion in the titration of iodide is given by the square root of K_{sp}^0, that is, 9.1×10^{-9}; pAg = 8.04.

Silver chloride begins to precipitate when its solubility product is exceeded. The concentration of chloride is about $0.05F$ at this point (the titrant dilutes the sample), and the silver ion activity will be about $1.8 \times 10^{-10}/0.05 = 3.6 \times 10^{-9}$; pAg = 8.44. Chloride begins to precipitate before the iodide system reaches its equivalence point.

In the idealized titration in Example 11.2, silver chloride begins to precipitate before the equivalence point for silver iodide is reached. The inflection point of the first break anticipates the equivalence point; the true equivalence point lies just below the base of the first break. Extrapolating lines from the iodide break downward and from the chloride plateau backward gives a better estimate of the equivalence point volume, particularly when there is more chloride than iodide in the sample. If the early precipitation of chloride were the only complication, an unwary analyst might underestimate the amount of iodide and overestimate the amount of chloride by taking the equivalence point at the inflection point of the first break.

You should do some calculations to examine the idealized behavior of a solution containing $0.01F$ iodide and $0.1F$ chloride. Is the first equivalence point easier or harder to see than it is in Fig. 11.4? What about the case in which an equimolar mixture of bromide ($K_{sp}^0 = 4.9 \times 10^{-13}$) and chloride are titrated?

In many mixtures adsorption and coprecipitation are more important in determining the accuracy of a titration than is the obscuring of the first equivalence point. In the titration of iodide and chloride, for example, chloride adsorbs on the silver iodide precipitate and is coprecipitated. More silver is required to reach the iodide equivalence point, and the results are high. The chloride lost by coprecipitation is not titrated in the second part of the process, and thus the chloride results are low. The total amount of chloride and iodide can be determined quite accurately, but the iodide may be as much as 2% high and the chloride as much as 2% low. The formation of mixed crystals by a mechanism such as isomorphous replacement (see Chapter 9) can also be an important source of error. This is a particularly troublesome problem in mixtures of chloride and bromide; not only is the bromide endpoint reached after chloride begins to precipitate, but also the sharpness of the endpoint break is decreased by the formation of mixed crystals. Silver iodide-silver bromide and silver iodide-silver chloride mixed crystals are much less stable, and mixed crystal formation is less of a problem in these systems. Mixed halide systems received great attention in the 1930s, and more detailed discussions and original literature citations can be found in Laitinen's text (Ref. 2).

11.6 INDICATOR METHODS

11.6.1 The Mohr Method

Three classical indicator methods are used for the determination of halides by titration with silver ion, the *Mohr, Fajans,* and *Volhard* methods. The Mohr method is the oldest (1856) and simplest of the three and is still used extensively for chloride determinations and to a lesser extent for bromide. The indicator is chromate ion, present in solution at a concentration of about $0.005M$. In principle, chromate will form a brick red-colored precipitate with silver ion (Ag_2CrO_4, with $K_{sp}^0 = 2 \times 10^{-12}$) just at, or slightly after, the equivalence point of the silver-chloride titration. Figure 11.5 shows the theoretical value for pAg at which silver chromate forms if chromate activity

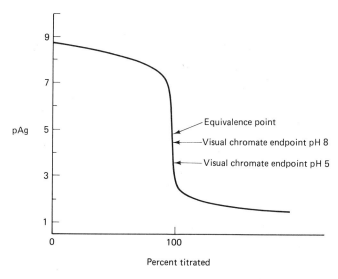

Figure 11.5 Mohr titration endpoints, pH 8 and 5.

is 0.005 and where this point occurs in the silver-chloride titration curve. A good visual endpoint actually requires a little more silver ion (enough red precipitate must form to be seen), and the necessary pAg is about 0.3 unit lower.

It should be apparent that the closer are the pAg at the equivalence point and the pAg necessary for a visible red, the more accurate will be the titration. If we assume that the silver ion activity needed for a red endpoint is 4×10^{-5}, we can approximate the relative error in determining chloride.

EXAMPLE 11.3 Mohr Titration

Calculate the error in the titration of chloride at the $0.1F$ and $0.01F$ levels, assuming that pAg 4.4 must be reached to see the silver chromate endpoint.

Solution

If we take the pK_{sp} of silver chloride as 9.74, we can calculate the pCl [i.e., $-\log(a_{Cl})$] to be 5.3 at the red endpoint (where pAg = 4.4). Assuming an activity coefficient of about 0.9 for a -1 ion at this ionic strength, the concentration of chloride is about $5.0 \times 10^{-6}/0.9 = 5.6 \times 10^{-6}M$. The original concentration of chloride, corrected for dilution, is $0.05M$. The error in the titration would be $5.6 \times 10^{-6}/0.05 = 0.01\%$. At the $0.01F$ chloride level, the error would be about 0.1%.

The accuracy of the Mohr method depends on the solution pH. Chromate ion is a weak base, $CrO_4^{2-} + H_2O = HCrO_4^- + OH^-$, $K_b = 2.8 \times 10^{-8}$ (K_a for the weak acid is 3.6×10^{-7}). Acidic solutions increase the solubility of Ag_2CrO_4 because protons compete with silver ion for the available chromate ion. Another way to say this is that the fraction of chromate is small in acidic solutions, and this raises the conditional solubility product of silver chromate (see Chapter 8) and increases its solubility. We would expect that as the solution became more acidic, the silver ion activity needed to see the red endpoint would *increase*. This could mean an appreciable positive titration error.

EXAMPLE 11.4 Mohr Chromate

Calculate pAg necessary to start the precipitation of silver chromate in a solution of $0.005F$ sodium chromate at pH 8. What is the effect of shifting pH to 5.0?

Solution

The expression for the fraction of chromate ion as a function of hydronium ion concentration is

11.6 INDICATOR METHODS

$$\alpha_{CrO_4^{2-}} = K_a/([H^+] + K_a)$$

The conditional solubility product is given by

$$K'_{sp} = K^0_{sp}/\alpha_{CrO_4^{2-}} = [Ag^+]^2 * C_{CrO_4}$$

pAg can be calculated from the second expression once the fraction of chromate is determined from the first. At pH 8 the fraction of chromate is 0.97, while at pH 5 the fraction is 0.036. At pH 8 the conditional solubility product is about 2.1×10^{-12}, while at pH 5 it is 5.6×10^{-11}. If we assume that the total concentration of chromate species is 0.005, the pAg value necessary to start precipitation at pH 8 is about 4.7, while that at pH 5 is 4.0. If the endpoint is to be visible, pAg will have to be about 4.4 at pH 8 and about 3.7 at pH 5.

It is important to see that a substantial overtitration occurs if one must add silver ion to lower pAg to 3.7. If the titrant is $0.1F$ silver ion and the solution volume is 100 mL, about 0.2 mL of titrant must be added beyond the equivalence point to lower pAg to 3.7.

The easiest way to control pH in the Mohr titration is to add sodium bicarbonate to the chloride solution before titrating. In Chapter 5 we found that the pH of a solution of the monoprotic form of a diprotic acid is roughly the average of the pK values, about 8.3 in the case of sodium bicarbonate. At pH > 10 hydroxide becomes an important competitor for silver ion, and overtitration becomes a serious problem once again. Ammonia forms stable complexes with silver ion and should not be used to raise solution pH.

11.6.2 The Fajans Method

The *Fajans* method for chloride uses an indicator in a novel way: the color change at the endpoint is produced when an indicator adsorbs on or desorbs from the precipitate. For the endpoint to be observed, the indicator must have a different color when adsorbed than it does in solution. A good example is the indicator dichlorofluorescein ("H_2DF"), which is yellow in solution and pink when adsorbed on silver chloride. The structure of this indicator is shown in Fig. 11.6. In the titration of a sample of potassium chloride with silver nitrate, prior to the equivalence point the particles of silver chloride adsorb untitrated chloride ions quite strongly as the primary adsorbed layer. Potassium ions make up the counterion layer around the negatively charged particles. Past the equivalence point silver ions become the primary adsorbed species, and nitrate ions become the secondary (counter) ions. When H_2DF is present its anion form DF^{2-} will displace the nitrate counterion layer and form an adsorbed layer of the silver salt of dichlorofluorescein around particles of the precipitate. The Ag-DF

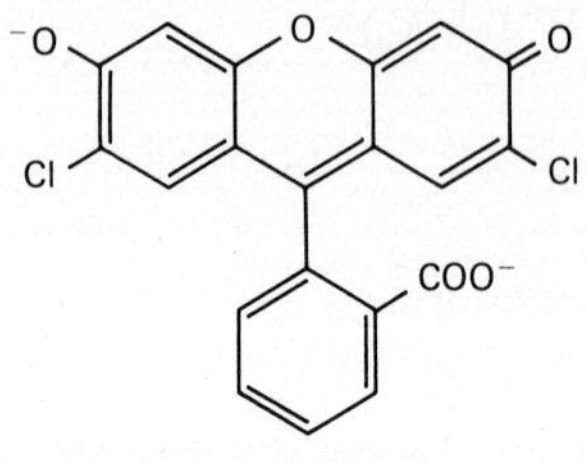

Figure 11.6 Dichlorofluorescein anion.

compound is not sufficiently insoluble to precipitate from solution, but requires that silver ions be stuck to a lattice.

$$AgCl:Cl^-:K^+ + Ag^+ = AgCl = AgCl:Ag^+:NO_3^-$$
(early)   (isoelectric)   (late)

$$AgCl:Ag^+:NO_3^- + DF^{2-} = AgCl:Ag^+:DF^{2-} + NO_3^-$$
(yellow)   (pink)

What are the complications? Solution pH can be a very important factor. The adsorbable form of the dye is the weak base anion form, and if the solution is too acidic adsorption will not occur. Dichlorofluorescein should not be used in solutions much more acidic than pH 4. Its relative, fluorescein, is a weaker acid and cannot be used below pH 6.5. The upper limit to pH is set by the formation of silver hydroxide species at about pH 10. Tetrabromofluorescein ("eosin") is a stronger Brønsted acid and can be used as an adsorption indicator at pH values as low as 2. Unfortunately, eosin has such an affinity for silver chloride that it displaces adsorbed chloride and gives a color change near the beginning of the titration rather than at the equivalence point. Other halide ions and thiocyanate are more strongly adsorbed on their silver precipitates and are not displaced by eosin before the equivalence point.

The visibility of the adsorption indicator color change depends on the surface area of the precipitate. Highly dispersed sols provide huge surface areas and the most visible color changes, while flocculated precipitates result in indistinct endpoints. Unfortunately, in halide solutions more concentrated than about $0.02F$ flocculation tends to occur about 1% before the equivalence point, and the sudden loss of surface area makes the endpoint less distinct. In such cases vigorous shaking of the solution can help maintain some sol and sharpen the endpoint.† A partially hydrolyzed starch called dextrin can be added to stabilize the silver chloride colloid (but not AgBr or AgI colloids) and improve the visibility of the color change. The optimum concentration range for Fajans method titrations is from about 0.005 to $0.02M$ halide.

---

† This is one of only a few methods based on a precipitation in which a colloidal precipitate is desirable.

11.6 INDICATOR METHODS

11.6.3 The Volhard Method

The *Volhard* method for chloride is based on a back titration of a carefully measured excess of silver ion (a solution of known concentration) with a standard solution of potassium thiocyanate (KSCN). Ferric ion is used as the indicator, and the endpoint is the formation of the red-colored species $FeSCN^{2+}$.

$$Cl^- + \text{excess } Ag^+ = AgCl + \text{excess } Ag^+ + \text{std. KSCN} = AgSCN(s)$$
[excess SCN^- produces red complex with Fe(III)]

EXAMPLE 11.5 Volhard Method

A 0.4260-g portion of a sample is to be analyzed for chloride by the Volhard method. The sample is dissolved in water and 20.00 mL of 0.1034M silver nitrate solution is added by pipet. Then 5 mL of Fe(III) indicator solution is added. Titration to the red endpoint requires 12.86 mL of 0.1194M potassium thiocyanate solution. Calculate the percent of chloride in the sample.

Solution

The number of millimoles of thiocyanate corresponds to the number of millimoles of silver ion left after the reaction with chloride in the sample. Thus the number of millimoles of chloride is given by

$$\text{mmole } Cl^- = \text{mmole } Ag^+ - \text{mmole } SCN^-$$
$$= 20.00 * 0.1034 - 12.86 * 0.1194 = 0.5325$$

The percentage of Cl^- in the sample is given by

$$\%Cl^- = (\text{mmole } Cl^- * GFW\ Cl) * 100/\text{mg sample}$$
$$= 0.5325 * 35.45 * 100/426.0 = 4.43\%\ Cl^-$$

There are a few factors that make the Volhard method somewhat less desirable than the Mohr and Fajans methods for the determination of halides. The first is that two standard solutions are required for the Volhard method, rather than one. This is a serious disadvantage if time or patience is in short supply. There are also some chemical problems to overcome, the principal one being the tendency of thiocyanate added in the second step to react with silver chloride precipitated in the first step. Silver thiocyanate is somewhat less soluble than silver chloride (K_{sp} approximately 10^{-12} vs. 10^{-10}), and to the extent that thiocyanate combines with AgCl, the results of the back titration are high. Two methods have been proposed to deal with this complication. The first is to flocculate the silver chloride precipitate to remove adsorbed silver ion, and then remove AgCl by filtration. The second, more novel, idea is to coat the AgCl precipitate by stirring in nitrobenzene. Thiocyanate cannot exchange with chloride through the nitrobenzene coating.

An additional problem with the Volhard method is that the formation constant for the $FeSCN^{2+}$ complex is rather small (~ 140), and a great deal of Fe(III) must be added to see the red color at the endpoint. A concentration of about $0.2F$ is required to insure against an indicator error larger than 0.1%, and this much Fe(III) gives a yellow color to the solution. If you are interested in this problem, you should calculate the equilibrium concentration of Fe^{3+} needed to produce a detectable red color. The concentration of $FeSCN^{2+}$ needed has been found to be about $6.4 \times 10^{-6} M$. What would be the concentration of SCN^- within about 0.1% of the Ag^+/SCN^- equivalence point?

The Volhard method holds some advantages over the Mohr and Fajans methods. Bromide, iodide, and thiocyanate are readily determined using the Volhard method. The Mohr method cannot be used for iodide and thiocyanate because chromate adsorbs strongly on AgI and AgSCN and gives a false endpoint.

11.7 APPLICATIONS

Titrations based on precipitation reactions have been used for the analysis of a wide variety of anions and cations. Some of the most important applications are listed in Table 11.1. Many of these are referred to in the problem set.

11.8 ENDNOTE: NORMALITY AND THE USE OF EQUIVALENT WEIGHTS

Equivalent weights and the normal concentration system, while traditionally of some importance in acid-base and oxidation-reduction titrations, are seldom used in precipitation and complexometric titrations. There is a good reason for this: it is quite difficult to use normality in an unambiguous way in titrations such as these. Nevertheless, students of analytical chemistry may eventually run across the use of normality in some application, and the system is therefore worth discussing.

In precipitation reactions *charge* is being neutralized, and so we define equivalent weights in terms of ionic charge. The equivalent weight of a cation being precipitated is defined as the gram formula weight divided by the cation charge,

$$\text{equivalent weight of cation} = \text{GFW/charge}$$

11.8 ENDNOTE: NORMALITY AND THE USE OF EQUIVALENT WEIGHTS

TABLE 11.1 PRECIPITATION TITRATIONS

Method	Species sought
Mohr	
Silver chromate endpoint	$\{Cl^-, Br^-\}$
Volhard	
Indirect determination with Ag^+; removal of precipitate	$\{Cl^-, PO_4^{3-}, CN^-, CO_3^{2-}, C_2O_4^{2-}, S^{2-}, CrO_4^{2-}\}$
Removal of precipitate unnecessary	$\{Br^-, I^-, SCN^-, AsO_4^{3-}\}$
Precipitate K^+ with known excess of tetraphenylboron (TPB^-); precipitate excess TPB^- with standard SCN^- solution	K^+
Fajans	
Direct determination with Ag^+; absorption indicators such as fluorescein, dichlorofluorescein, eosin	$Cl^-, Br^-, I^-, SCN^-, Fe(CN)_6^{4-}$
Titrate with potassium ferrocyanide; use diphenylamine plus a little ferricyanide for indicator	Zn^{2+}
Titrate with Cl^- to form Hg_2Cl_2; indicator bromphenol blue	Hg_2^{2+}
Titrate with Th^{4+} to precipitate ThF_4; indicator alizarin red S	F^-
Titrate with $Ba(OH)_2$ in 1:1 aqueous methanol; indicator alizarin red S	SO_4^{2-}

By definition, then, the equivalent weight of Fe^{2+} is 55.85/2 or 27.93, and that of Th^{4+} is 232.04/4 or 58.01.

The equivalent weight of an anion which reacts with the cation being precipitated is found by dividing the anion's gram formula weight by the number of cation equivalents (not moles) with which it reacts,

$$\text{equivalent weight of anion} = \text{GFW}/(\text{cation equivalents/mole of anion})$$

For example, consider the reaction

$$3Ag^+ + PO_4^{3-} = Ag_3PO_4(s)$$

The equivalent weight of silver in this reaction is the same as its gram formula weight, 107.87. The equivalent weight of phosphate is its GFW divided by 3 silver equivalents per mole of phosphate, or 31.66 g. One equivalent weight of silver (1 mole) will react with exactly one equivalent weight of phosphate ($\frac{1}{3}$ mole). A $0.1M$ solution of silver ion is also $0.1N$, but a $0.1M$ solution of phosphate is $0.3N$.

In the precipitation of barium sulfate,

$$Ba^{2+} + SO_4^{2-} = BaSO_4(s)$$

the equivalent weight of barium ion is its formula weight divided by 2, or 68.67. A $1M$ solution of barium ion is therefore $2N$. The equivalent weight of sulfate is its formula weight divided by 2, or 48.03. A $1M$ solution of sulfate is therefore $2N$ for this reaction. Notice that $\frac{1}{2}$ mole of barium ion (1 equivalent) will react with $\frac{1}{2}$ mole of sulfate ion (also 1 equivalent).

As a final example, consider the precipitation of ThF_4,

$$Th^{4+} + 4F^- = ThF_4(s)$$

The equivalent weight of thorium is 58.01 (see above), or $\frac{1}{4}$ mole. A $0.1M$ solution of Th^{4+} is therefore $0.4N$. The equivalent weight of fluoride is the same as its gram formula weight, 19.00 (i.e., 1 mole), and so a $0.1M$ solution of fluoride will also be $0.1N$. If you stretch your imagination a little, you will see that the positive charge in 10.0 mL of $0.1N$ Th^{4+} will be neutralized by the negative charge in 10.0 mL of $0.1N$ F^-, the definition of the equivalence point in a titration.

REFERENCES

1. B. B. Owen, *Journal of the American Chemical Society,* 60, 2229 (1938).
2. H. A. Laitinen, *Chemical Analysis,* McGraw-Hill, New York, 1960, chapter 12.

RECOMMENDED READING

Precipitation titrations are no longer a very active area of research, and the most important literature is more than 20 years old. The following works are classics in the area.

Kolthoff, I. M., and V. A. Stenger. *Volumetric Analysis,* vol. II, section C, 2nd ed., Interscience, New York, 1947.

Laitinen, H. A. *Chemical Analysis,* McGraw-Hill, New York, 1960. Chapter 12 covers precipitation titrations.

PROBLEMS

11.1. A 100-mL solution containing 10.50 mmole of NaCl and 9.95 mmole of NaI is titrated with $0.09972M$ silver nitrate. In the following problems assume that

the solubility product of silver chloride is 1.8×10^{-10} and that of silver iodide is 8.3×10^{-17}, and that activities and concentrations are equal.
(a) How many milliliters of silver nitrate solution are required to reach the equivalence point in the titration of iodide?
(b) Calculate the silver ion activity and pAg at the silver iodide equivalence point.
(c) Is pAg sufficiently small to start the precipitation of AgCl? What must it be?
(d) Calculate pAg when exactly half the iodide has been precipitated.
(e) Calculate pAg when exactly half the chloride has been precipitated.
(f) Calculate pAg at the silver chloride equivalence point.

11.2. Write a computer program to calculate points in a curve for the titration of the bromide in 50 mL of a $0.1000M$ solution of NaBr with $0.1000M$ silver nitrate. Be sure to account for dilution.

11.3. Could the formation of silver chromate be used to signal the equivalence point in the titration of bromide with silver ion? Consider titrating 25.00 mL of a $0.0500M$ solution of NaBr with $0.0500M$ AgNO$_3$. Use $K_{sp}(AgBr) = 4.9 \times 10^{-13}$ and $K_{sp}(Ag_2CrO_4) = 2.0 \times 10^{-12}$, and do not ignore dilution.

11.4. Explain as carefully as possible the role of solution pH in determining the pAg value at which the silver chromate endpoint will be seen.

11.5. The silver in a silver-copper coin is to be assayed by the Volhard method. If the coin weighs 2.218 g and contains about 60% silver:
(a) Calculate the approximate number of millimoles of silver in the coin.
(b) Describe how you would dilute a solution from a dissolved coin so that a 10.00-mL pipetted portion would contain about 0.5 mmole of silver.
(c) Describe how you would prepare a solution of KSCN which would supply about 0.5 mmole in about 40 mL.
(d) How would you standardize the solution in part (c)?
(e) Copper(II) is known to form complexes with thiocyanate ion. What might be the effect of copper on the results of the determination of silver by the Volhard method? How might you use a blank titration to estimate the effect of copper?

11.6. The success of the indirect Volhard method for chloride depends on the analyst's ability to isolate the silver chloride precipitate before back-titrating excess silver ion with thiocyanate.
(a) Why must AgCl be isolated before the back titration?
(b) Look at the solubility products of silver salts in Appendix II. Which silver compounds should not require isolation? Explain.

11.7. Thiocyanate might be determined by either direct or indirect Volhard methods. Which method is apt to be more accurate? Explain.

11.8. A 0.6272-g sample containing LiCl and NaCl is dissolved and diluted to 100.0 mL in a volumetric flask. A 25.00-mL portion is taken by pipet and titrated by adding 50.00 mL of $0.08791M$ silver nitrate to precipitate AgCl. Nitrobenzene is then added to coat the precipitate and the excess silver ion is back-titrated with 12.73 mL of $0.1020M$ KSCN solution. Calculate the percent chloride in the sample.

11.9. A 0.3728-g sample containing sodium fluoride is dissolved in deionized water and titrated with $0.0501M$ thorium(IV) nitrate solution. The reaction is

$$4F^- + Th^{4+} = ThF_4(s)$$

The endpoint is signaled at 28.48 mL by the color change in alizarin red S. Calculate the percent of fluoride in the sample.

11.10. A stockroom bottle containing either Na_2SO_4 or K_2SO_4 has lost its label. To establish its identity, you titrate the sulfate with barium hydroxide in methanol/water and watch for the alizarin red S endpoint. A 110-mg sample of the salt is dissolved in 25 mL of deionized water and is titrated with 8.35 mL of $0.0925M$ $Ba(OH)_2$. Is the salt sodium sulfate or potassium sulfate?

11.11. Iodoform, CHI_3, can be determined in acidified alcohol solution by adding an excess of standard silver nitrate solution and letting the reaction

$$CHI_3 + 3Ag^+ + 5H_2O = 3AgI(s) + HCOOH + 3H_3O^+$$

go to completion, and then back-titrating the unreacted silver ion with standard KSCN solution (Volhard).

A 10.00-mL portion of $0.0523M$ silver nitrate is added to a solution containing iodoform and nonreactive material. After 1 hour the silver iodide is filtered out and the excess silver ion is titrated with $0.0347M$ KSCN, 2.32 mL being required to reach the endpoint. How many milligrams of iodoform were present in the original sample?

11.12. Butyl mercaptan ($C_4H_{10}S$) in 100 mL of an alcohol solution is precipitated by adding 25.00 mL of $0.0274M$ $AgNO_3$. The unconsumed silver ion is then back-titrated with standard $0.01030M$ KSCN. What is the concentration of butyl mercaptan in the original solution if 5.72 mL of KSCN solution is required in the back-titration?

$$C_4H_{10}S + Ag^+ = C_4H_9SAg(s) + H^+$$

11.13. Barbiturate veronal (diethylbarbituric acid) can be determined by titration with silver nitrate in 5% borax buffer solution:

$$C_8H_{12}N_2O_3 + 2Ag^+ = C_8H_{10}Ag_2N_2O_3 + 2H^+$$

Silver chromate can be used for the endpoint indicator. If the veronal in a 100-mg tablet requires 9.63 mL of 0.01372M silver nitrate for titration, how many milligrams of veronal were present in the tablet?

11.14. Zinc is to be determined by titration with standard potassium ferrocyanide to form a complex double salt,

$$2Fe(CN)_6^{4-} + 3Zn^{2+} + 2K^+ = K_2Zn_3[Fe(CN)_6]_2(s)$$

A 874.6-mg sample containing zinc is dissolved in 50 mL of deionized water. To reach the endpoint, 23.12 mL of 0.02503M K$_4$Fe(CN)$_6$ solution is required. Calculate the percentage of Zn in the sample.

11.15. Interpret and reword each of the following statements in terms of equivalents and equivalent weights:
 (a) 4 mmole of NaF is required to precipitate 1 mmole of Th^{4+}.
 (b) 1 mmole of BaCl$_2$ is required to precipitate 2 mmole of Ag$^+$ or 1 mmole of Pb^{2+}.
 (c) $\frac{2}{3}$ mmole of Na$_3$PO$_4$ is required to precipitate 1 mmole of Ca^{2+}.

11.16. What are the equivalent weights of metal ion and anion in each of the following applications?
 (a) $Cu^+ + Cl^- = CuCl(s)$
 (b) $Ba^{2+} + 2F^- = BaF_2(s)$
 (c) $Ba^{2+} + C_2O_4^{2-} = BaC_2O_4(s)$
 (d) $Ba^{2+} + 2IO_3^- = Ba(IO_3)_2(s)$
 (e) $Al^{3+} + PO_4^{3-} = AlPO_4(s)$
 (f) $Ag^+ + CN^- = AgCN(s)$

11.17. How many milliliters of specified titrant will be needed to titrate the specified weight of reactant?
 (a) 0.1000N Th^{4+} and 0.1220 g NaF
 (b) 0.1058N Pb^{2+} and 0.1859 g Na$_2$SO$_4$
 (c) 0.1568N AlCl$_3$ and 0.3202 g NaH$_2$PO$_4$

11.18. How would you prepare each of the following solutions?
 (a) 0.100N sodium sulfate for the titration of BaCl$_2$.
 (b) 0.100N sodium fluoride for the titration of CaCl$_2$.
 (c) 0.200N calcium chloride for the titration of KIO$_3$.

11.19. What is the molarity of each of the following solutions used for precipitation titrations?
 (a) 0.0774N Pb^{2+}
 (b) 0.0923N Al^{3+}
 (c) 0.1072N Cl$^-$ for titration of Ag$^+$
 (d) 0.1072N Cl$^-$ for titration of Hg^{2+}
 (e) 0.0702N PO$_4^{3-}$ for titration of Ag$^+$

EXPERIMENT 11.1: Titrimetric Determination of Chloride by the Adsorption Indicator Method

A standard solution of silver nitrate is added to a solution containing chloride. The indicator, dichlorofluorescein, changes from a yellow color to a pink color at the equivalence point because it is adsorbed on colloidal silver chloride. The indicator mechanism is discussed in the text. You must prepare and standardize a solution of silver nitrate, using reagent grade potassium chloride, and then use the standardized solution to titrate triplicate samples of an unknown.

Procedure

1. Transfer about 3 g of reagent grade potassium chloride to a clean weighing bottle and dry it in a 110°C oven for 2 hours.

2. Obtain a sample from the instructor. Transfer the sample to a clean weighing bottle, and dry it in a 110°C oven for 2 hours.

3. Dissolve approximately 8.5 g of silver nitrate in about 500 mL of deionized water. Mix the solution well and transfer it to a 1-liter polyethylene bottle. Store the solution in the dark. Protect it from strong sunlight.

4. The silver nitrate solution will be about $0.1F$. Knowing that the gram formula weight of KCl is 74.46, calculate the amount of KCl to weigh to require about 40 mL of standard silver nitrate solution. Carefully weigh out three samples of appropriate weight into 250-mL Erlenmeyer flasks. Add 75 mL of deionized water, 1 mL of $0.5F$ acetic acid-$0.5F$ sodium acetate buffer, 5 mL of a 1% solution of dextrin, and six drops of dichlorofluorescein indicator solution (1 g per liter of ethanol-water solvent). Titrate the solution quickly with the silver nitrate stock, swirling vigorously, until the endpoint color change (yellow-green to pink) occurs. Calculate the concentration of the silver nitrate stock solution from the mean of the three trials.

5. Weigh as carefully as possible three 0.3- to 0.35-g samples of your unknown into clean Erlenmeyer flasks. Treat these samples just as you did the standard KCl samples, unless your instructor advises otherwise. Calculate and report percent chloride in the sample.

Notes

Solution pH must be between 4 and 10. The upper limit is set to avoid the formation of silver hydroxide complexes, while the lower limit avoids changes in the adsorption characteristics of the indicator. Keeping the solution at about pH 5 with buffer reduces the tendency of the indicator to adsorb on large particles of silver chloride.

There must be plenty of colloidal silver chloride present for the endpoint to be seen properly. Dextrin and agitation stabilize the colloid, but high ionic strength and high temperatures tend to force coagulation, reducing endpoint sharpness. Silver chloride is also sensitive to light. Reduction of silver ions to silver metal gives a gray cast to the solution and makes the endpoint hard to see. Doing the titration quickly and avoiding strong light will lessen the chances of interference from photoreduction.

The endpoint should be taken as the first persistent pink color.

A few metal ions interfere with this method. Both Fe^{3+} and Al^{3+} form hydroxide precipitates which adsorb chloride and cause low results. Ammonium ion forms ammonia at higher pH values and may slightly increase the solubility of AgCl. Anions which form insoluble silver salts will all cause results to be high. Examples are the halides, cyanide, thiocyanate, and sulfide.

EXPERIMENT 11.2: Titrimetric Determination of Chloride by the Volhard Method

The Volhard method for chloride is an indirect method in which an excess of standard silver nitrate solution is back-titrated with a standard solution of potassium thiocyanate:

$$Cl^- + \text{excess } Ag^+ \rightarrow AgCl(s) + Ag^+$$
$$Ag^+(\text{excess}) + SCN^- \rightarrow AgSCN(s)$$

The endpoint of the back titration is signaled by the reaction of excess thiocyanate with Fe^{3+} to form the red complex $Fe(SCN)^{2+}$. This method requires two standard solutions, as explained in the text.

Procedure

1. Transfer about 3 g of reagent grade potassium chloride to a clean weighing bottle and dry it in a 110°C oven for 2 hours.
2. Obtain a sample from the instructor. Transfer the sample to a clean weighing bottle and dry it in a 110°C oven for 2 hours.
3. Dissolve approximately 8.5 g of silver nitrate in about 500 mL of deionized water. Mix the solution well and transfer it to a 1-liter polyethylene bottle. Store the solution in the dark. Protect it from strong sunlight when you are using it. You may

standardize the solution with KCl by one of the methods described in these experiments.

Alternatively, if your instructor permits, you may use silver nitrate as a *primary standard*. Dry crystalline silver nitrate in an oven completely free of organic materials (including paper!) for 2 hours at 110°C. Examine the dried product carefully; if it is discolored it has probably undergone some reduction in the oven, and you should consider drying some more. If the salt is colorless or only slightly colored, weigh out 8.5 g to the nearest 0.1 mg, dissolve in a minimum volume of deionized water, and transfer quantitatively to a 500-mL volumetric flask. Dilute to the mark with deionized water. Calculate the concentration.

4. Primary standard grade KSCN is available. It should be dried for 2 hours at 110°C prior to use. Weigh to the nearest 0.1 mg sufficient KSCN (GFW 97.092) to prepare 100 mL of a $0.1F$ solution. Dissolve the salt in a small amount of water; then transfer it to a 100-mL volumetric flask and dilute to the mark with deionized water.

5. Weigh as carefully as possible three 0.3- to 0.35-g samples of your unknown into clean 250-mL Erlenmeyer flasks. Treat these samples just as you did the standard KCl samples unless your instructor advises otherwise. Dilute each sample with about 50 mL of deionized water, and add 5 mL of freshly boiled 1:1 nitric acid.

6. Titrate the first sample by adding a measured excess of standard silver nitrate solution from a buret. Swirl the contents of the flask continuously until the AgCl precipitate flocculates. Let the precipitate settle; then test for completeness of precipitation by adding another drop of silver nitrate solution. Record the volume of $AgNO_3$ you have added.

Next add 4 mL of nitrobenzene and swirl the contents vigorously to coat the AgCl precipitate. The solution over the precipitate should be clear. If it is not, swirl again.

Add 5 mL of $0.33M$ ferric nitrate solution. Back-titrate immediately with standard KSCN solution to the first faint orange color that persists for 30 seconds. Record the volume of standard KSCN solution.

7. Titrate the remaining samples by the same method. Calculate and report percent chloride in the sample.

Notes

Best results are obtained if less than 1 mL of standard silver nitrate is added in excess. Only a few milliliters of standard KSCN will be used in the back-titration step.

Use a drop of silver nitrate solution to test your deionized water for the presence of chloride.

Save all silver-containing residues and solutions.

For the calculation remember that

$$\text{mmole Cl}^- = \text{mmole Ag}^+ \text{ added} - \text{mmole KSCN}$$

EXPERIMENT 11.3: Titrimetric Determination of Chloride by the Mohr Method

The Mohr method for chloride is a direct titration in which a sample is titrated with standard silver nitrate solution. The endpoint is signaled by the formation of silver chromate:

$$2\text{Ag}^+ + \text{CrO}_4^{2-} = \text{Ag}_2\text{CrO}_4(s)$$

The endpoint error for this reaction can be rather large, as explained in the text. However, if the standardization of the silver nitrate solution and titration of the sample are both done with the method, the endpoint errors largely cancel. Details of the method and some of its limitations are discussed in the text.

Procedure

1. Obtain a sample from your instructor, transfer it to a clean dry weighing bottle, and dry it for 2 hours at 110°C.

2. Transfer about 3 g of reagent grade potassium chloride to a clean dry weighing bottle and dry it for 2 hours at 100°C.

3. Prepare stock $0.1M$ silver nitrate solution according to the directions given for the Volhard method (Experiment 11.2).

4. Standardize the silver nitrate titrant with triplicate titrations of standard KCl. Weigh to the nearest 0.1 mg three 0.30- to 0.40-g samples of dried KCl into 250-mL Erlenmeyer flasks. Dilute each sample with 50 mL of water from a graduated cylinder. To each sample add 2.0 mL of $0.10F$ K_2CrO_4 indicator solution. Titrate each sample with vigorous stirring until the red-pink indicator endpoint color persists for at least 20 seconds. Calculate the concentration of your stock solution. A relative standard deviation of 2 to 3 parts per thousand is desirable.

5. Ask your instructor for the approximate composition of your sample, and calculate the approximate weight you must take to use 30 to 40 mL of silver nitrate titrant. If your instructor directs, add 0.1 g of reagent grade sodium bicarbonate to each sample to adjust the pH to the range 6 to 10. Sodium bicarbonate may contain some chloride, so be prepared to run a blank titration.

6. From the endpoint volumes of your sample titrations calculate the moles of chloride and percent chloride in your unknown.

Notes

Vigorous swirling helps sharpen the endpoint in this experiment by helping to avoid occlusion of red Ag_2CrO_4 by AgCl. The pink cast of occluded silver chromate makes the endpoint indistinct.

Solutions should all be at about the same pH, temperature, and ionic strength before titration. All three factors influence the formation of silver chromate. To avoid determinate error, adjust the solution pH to between 6 and 10. Most samples prepared for instructional laboratories will not require pH adjustment, however.

The same compounds which interfere with the Volhard and Fajans methods also interfere with this method.

Chapter 12

Complexes and Complexometric Methods of Analysis

12.1 LEWIS ACIDS AND BASES

Chapters 5 and 6 described in detail the equilibria of protonic (Brønsted-Lowry) acid-base systems. A great deal of chemistry involves reactions which are totally analogous to Brønsted-Lowry acid-base reactions, but do not involve protons. For example, silver chloride, a covalent compound discussed in Chapter 8, is only a slightly stronger electrolyte than phenol:

$$AgCl(s) = Cl^-(aq) + Ag^+(aq) \qquad K = 1.8 \times 10^{-10}$$
$$C_6H_5OH(aq) = C_6H_5O^-(aq) + H^+(aq) \qquad K = 1.0 \times 10^{-10}$$

The fundamental difference between these two equilibria is that the dissociation of silver chloride involves both liquid and solid phases, while that of phenol involves only the liquid phase. The Lewis acid, silver ion, corresponds to the Brønsted acid, the proton, and the Lewis base, chloride ion, corresponds to the Brønsted base, phenolate ion. Just as we can use the response of a glass electrode to measure the pH of a solution of phenol and evaluate its dissociation constant, we can use an electrode made of silver wire to monitor the activity of silver ions in a solution of silver chloride and determine its dissociation constant.

The analogy can be pressed even further by noting that the polyprotic acids discussed in Chapter 5 (for example, phosphoric acid, H_3PO_4) correspond to the complex metal ion species discussed in Chapter 8 (for example, $AgCl_3^{2-}$). As you will see, the mathematical treatments of both kinds of equilibria are essentially identical. It is somewhat more convenient, though, to deal with formation constants in complex ion equilibria than with dissociation constants like those used in Chapter 5.

This chapter begins with some descriptive chemistry of metal ions (Lewis acids) and complexing agents (ligands, which are Lewis bases) which are important in analytical chemistry. Next, a detailed discussion of complexation equilibria is presented, along with algebraic and graphical methods for solving equilibrium problems. Finally, an extensive section on complexometric titrations is included to describe important applications.

12.2 METAL IONS IN SOLUTION

The complexation process involves the reaction of a simple (or "free") metal ion with one or more ligands to form a complex ion. The use of the term "free ion" needs some comment. The only ions which are at all free of interactions with other species are in the gas phase. Metal ions in solution are not really free at all, but are surrounded by solvent molecules. In the more inert solvents, for example, hexane, metal ion-solvent associations are quite weak, while in a highly polar amphiprotic solvent like water, associations are quite strong. Water is such a good acceptor and donor that it is intimately involved in the reactions of metal ions, and it cannot be neglected without the risk of drawing an incorrect conclusion about a chemical equilibrium. The oversimplified general reaction often seen in texts,

$$M + nL = ML_n$$

should always be understood to mean

$$M(H_2O)_n + nL = M(L)_n + nH_2O$$

to show that the addition of ligands to a metal ion in water is really the exchange of ligands for water.

A phenomenon which often accompanies solvation in water is *hydrolysis*. Small, highly charged metal ions associate so strongly with water that they are able to split off protons and form metal-hydroxide complexes,

$$M^+ + H_2O = M(OH) + H^+$$

Al (3+), Fe (3+), and Sn (4+) are particularly good hydrolyzing cations. Their chloride salts (among others) produce acidic solutions in water as a result of hydrolysis.

12.2.1 Metal Ion Coordination Number and Geometry of Complexes

The coordination number of a metal ion is defined as the number of covalent bonds it is able to form with ligands. It is an important factor in the formation of complexes because it dictates the stoichiometry and geometry of complexes. Coordination numbers range from 1 in simple ion pairs in the gas phase to 12 in some highly complex metal oxides. The extreme coordination numbers are not really important in analytical applications and will not be discussed here. By far the most important coordination numbers are 6 and 4. Structures based on these coordination numbers involve what is called cubic symmetry and are shown by the prototypes in Fig. 12.1. Six coordination can be viewed as a field of six ligands approaching the faces of a cube with a metal ion at its center. When all six ligands are identical [for example, the ammonia molecules in $Ni(NH_3)_6^{2+}$], a nearly perfect octahedral arrangement can be achieved.

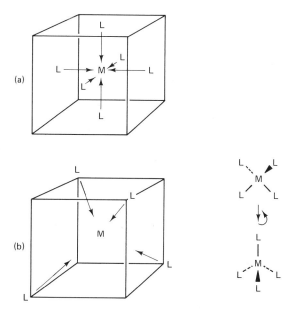

Figure 12.1 Six- and four-coordination to a metal ion. (a) Six-coordination: six ligands approach a metal ion from the faces of an imaginary cube. (b) Four-coordination: four ligands approach a metal ion from alternating corners of an imaginary cube. By rotating the cube up onto a corner it is possible to recognize the tetrahedral array. Another form of four-coordination involves four ligands approaching a metal ion from the four side faces of a cube. This configuration is called "square planar."

When more than one type of ligand bonds to a six-coordinate metal ion, geometric isomers (for example, *cis* and *trans* forms) are found. Whenever a field of ligands around a metal ion lacks a plane of symmetry, one should expect to find optical isomers (for example, *d* and *l* forms). Coordination compounds of Co(III) and Cr(III) are almost exclusively six-coordinate, and their chemistry is rich in isomers.

There are two important four-coordination geometries, the tetrahedral and square planar forms. The tetrahedral form can be viewed as a field of four ligands approaching alternate corners of a cube with a metal ion at its center. It is favored over six coordination when large ligands (Cl^-, Br^-, I^-, substituted amines, etc.) try to bond with relatively small metal ions, particularly those with a noble gas electron configuration or a full *d* shell [e.g., Zn^{2+} with a $4d(10)$ structure]. Geometric isomers do not exist in tetrahedral geometry, although optical isomers are possible when all four ligands are different.

The square planar form is like the six-coordinate form, but with two opposing ligands removed. This form crowds ligands more than does the tetrahedral form and is adopted by metal ions with eight *d* electrons, such as Ni^{2+}, Pd^{2+}, and Pt^{2+}, when they combine with very strong Lewis bases.

12.3 LIGANDS (LEWIS BASES)

While thousands of Lewis bases are known, perhaps only 20 or 30 are of great importance to analysts. What all Lewis bases have in common is, of course, electron pairs. Bases are conveniently classified according to the number of basic electron pairs (called the "denticity," a metaphorical term suggesting that bases have teeth which bite into a Lewis acid) and the identity of the atom to which they are attached. The simplest ligands have only one basic electron pair and are called "monodentate" ligands:

Halides: F^-, Cl^-, Br^-, I^-

Ammonia and amines: NH_3, CH_3NH_2, etc.

Carbon monoxide: CO

Cyanide: CN^-

These compounds have other electron pairs, but they are not involved in complexation.

There are also many "bidentate" ligands. Most of the important ones have nitrogen or oxygen atoms as bases, and many have conjugated rings. Several important examples are shown in Fig. 12.2. The bidentate nitrogen bases ethylenediamine, bipyridine, and 1,10-phenanthroline are strikingly similar in structure (note the —N—C—C—N— chain) and form five-membered rings with metal ions. The resulting compounds with rings are called "chelates," and the ligands used to form them are called "chelating agents." The chelating agent 2,4-pentanedione (or acetylacetone) can

12.3 LIGAND (LEWIS BASES)

lose the relatively acidic proton on the third carbon atom (pK_a = 8.2) and, by bonding through the oxygen atoms, can form a conjugated six-membered ring with a metal ion. This extra stabilization of the chelate ring enhances the stability of acetylacetone complexes. An example of a bidentate ligand with two different base atoms is 8-hydroxyquinoline (or "oxine"), which is also shown in Fig. 12.2. Like many other bidentates, oxine forms five-membered rings with metal ions.

There are a few tridentate and tetradentate ligands and even a pentadentate ligand, but only a few are of importance in analysis. Some of the examples shown in Fig. 12.3 are relatives of ethylenediamine and bond to the same kinds of metal ions. The tetradentate ligand nitrilotriacetic acid (NTA) (see Fig. 12.5) is a member of a class of compounds called aminocarboxylic acids and has been investigated extensively. It was used as a biodegradable phosphate substitute in detergents in the early 1970s,

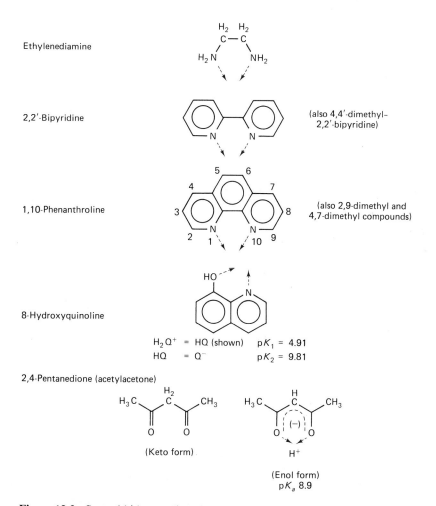

Figure 12.2 Several bidentate ligands.

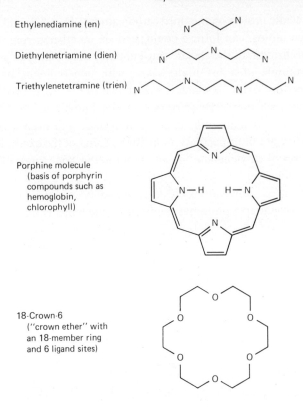

Figure 12.3 Polydentate ligands. For simplicity, carbon atoms are represented by intersecting lines and hydrogen atoms are omitted.

but was withdrawn from the market when studies with laboratory animals implicated it in birth defects (teratogenicity). New studies have shown it to be safe, and it is reappearing in detergent formulations. It helps detergents clean by complexing hard-water ions such as calcium and magnesium, which otherwise form soap scum deposits.

There are several hexadentate (six-toothed) ligands which are relatives of NTA. The best known of these aminocarboxylic acids is ethylenediaminetetraacetic acid (EDTA), which forms extraordinarily stable complexes with most metal ions. Since its synthesis in the 1930s and systematic study in the 1940s, EDTA has been used to determine directly or indirectly almost every element in the periodic table. The anion of EDTA has four terminal oxygen atoms and two branching nitrogen atoms and is able to use all six sites for bonding by wrapping around a cation in the manner of a claw. Figure 12.4 should give you an idea of how the EDTA anion must bend to do this.

Some other aminocarboxylate ion complexing agents are shown in Fig. 12.5. Even though they have greater or lesser numbers of bonding sites, they all behave

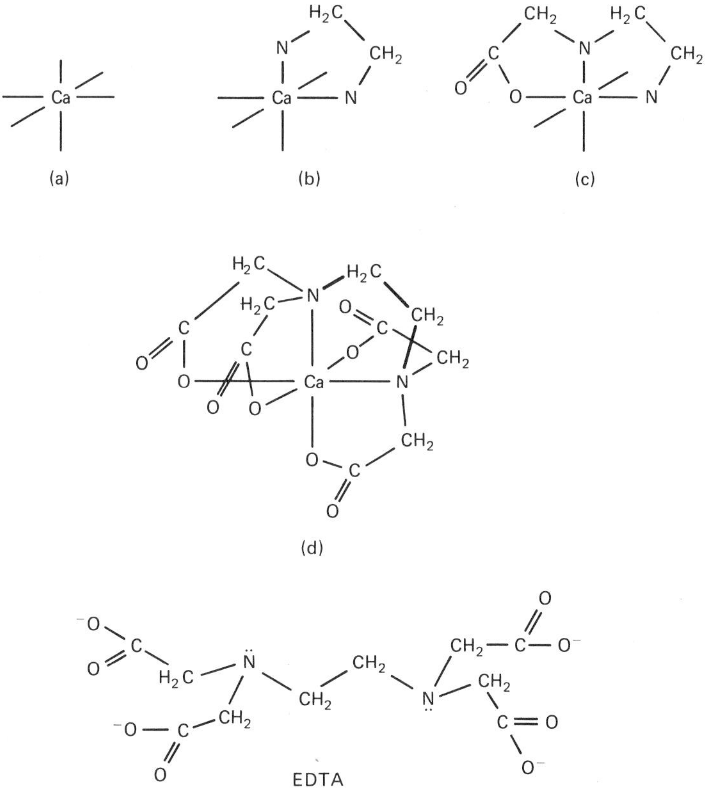

12.3 LIGAND (LEWIS BASES)

Figure 12.4 EDTA and its complex with calcium ion. (a) Six-coordinate calcium ion without ligands. (b) Coordination of nitrogens on EDTA spine showing the formation of a ring. (c) Coordination of one acetate group's oxygen, showing the formation of another five-membered ring. (d) Fully complexed calcium, showing the "claw" of EDTA around the calcium ion.

similarly to EDTA. Formation constants for their complexes with many metal ions are given in Appendix III. Part of that appendix is reproduced in Table 12.1.

Ligands are bases in reactions with protons as well as metal ions, and Brønsted acid-base reactions must be taken into account in complexation equilibria. EDTA has six base sites and can combine with as many as six protons. The most loosely held protons appear to be those on two of the acetate groups. There is strong evidence, in fact, that the neutral compound is really a "zwitterion," with two negative centers on the acetates and two positive centers on the protonated nitrogen atoms. These last protons are quite tightly held and seem to be the last ones removed when forming the tetra-anion, $EDTA^{4-}$. Figure 12.2 gives successive pK_a values for EDTA and several other aminocarboxylic acids.

The EDTA species of greatest interest to us is the tetra-anion, which is formed at high pH values. Rather than show the extremely complicated log distribution diagram

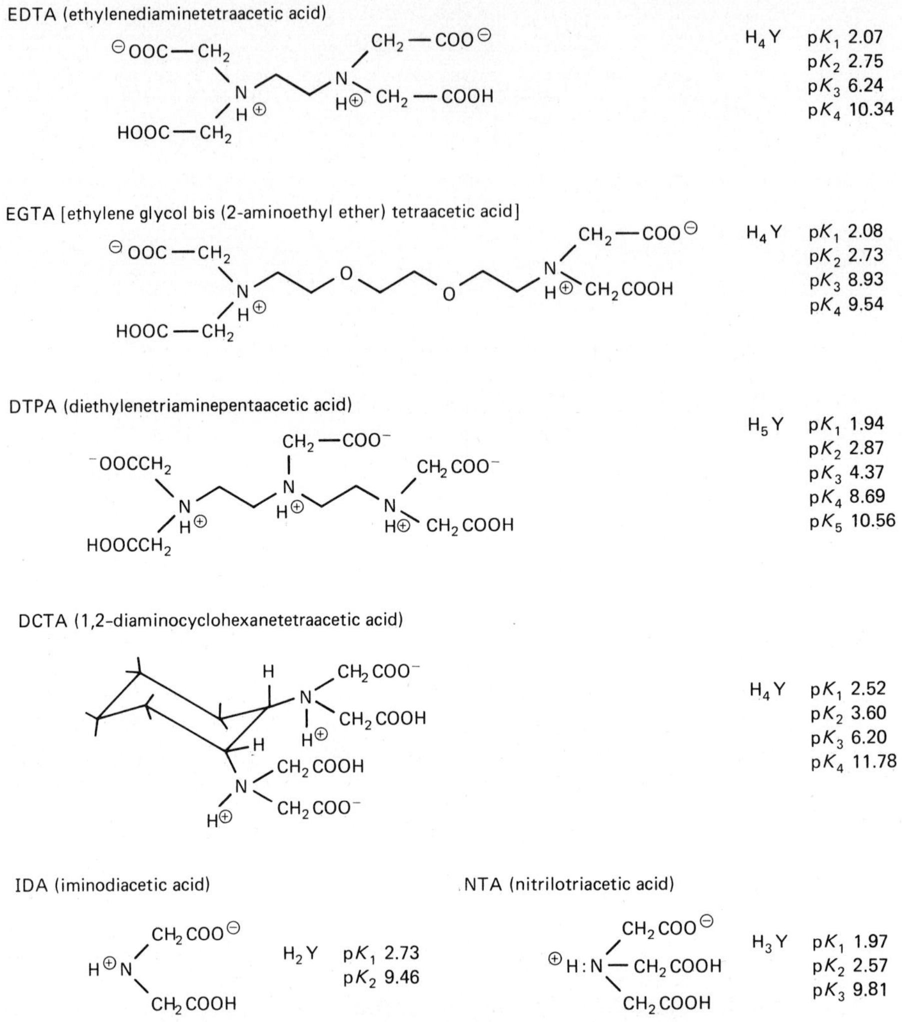

EDTA (ethylenediaminetetraacetic acid)

$^{\ominus}$OOC—CH$_2$\
　　　＼\
　　　　N—CH$_2$—CH$_2$—N\
　　／　H$^{\oplus}$　　　H$^{\oplus}$　＼\
HOOC—CH$_2$　　　　　　　CH$_2$—COO$^{\ominus}$\
　　　　　　　　　　　　　CH$_2$—COOH

H$_4$Y　　pK_1　2.07\
　　　　pK_2　2.75\
　　　　pK_3　6.24\
　　　　pK_4　10.34

EGTA [ethylene glycol bis (2-aminoethyl ether) tetraacetic acid]

$^{\ominus}$OOC—CH$_2$\
　　　＼\
　　　　N—CH$_2$—CH$_2$—O—CH$_2$—CH$_2$—O—CH$_2$—CH$_2$—N\
　　／　H$^{\oplus}$　　　　　　　　　　　　　　　　　　　　H$^{\oplus}$　＼\
HOOC—CH$_2$　　　　　　　　　　　　　　　　　　　　　　　　CH$_2$—COO$^{\ominus}$\
　　　　　　　　　　　　　　　　　　　　　　　　　　　　　CH$_2$COOH

H$_4$Y　　pK_1　2.08\
　　　　pK_2　2.73\
　　　　pK_3　8.93\
　　　　pK_4　9.54

DTPA (diethylenetriaminepentaacetic acid)

$^{-}$OOCCH$_2$　　　　　CH$_2$—COO^{-}　　　　CH$_2$COO^{-}\
　　　＼　　　　　｜　　　　　　　　／\
　　　　N—CH$_2$—CH$_2$—N—CH$_2$—CH$_2$—N\
　　／　H$^{\oplus}$　　　　H$^{\oplus}$　　　　H$^{\oplus}$＼\
HOOCCH$_2$　　　　　　　　　　　　　　　　CH$_2$COOH

H$_5$Y　　pK_1　1.94\
　　　　pK_2　2.87\
　　　　pK_3　4.37\
　　　　pK_4　8.69\
　　　　pK_5　10.56

DCTA (1,2-diaminocyclohexanetetraacetic acid)

(cyclohexane ring with two N substituents)

　　　H　　　CH$_2$COO^{-}\
　　　＼　／\
　　　　N\
　　　／｜＼\
　　　　｜　CH$_2$COOH\
　　　H$^{\oplus}$\
—H\
　　　　　　CH$_2$COOH\
　　　／\
　　N\
　／　＼\
H$^{\oplus}$　　CH$_2$COO^{-}

H$_4$Y　　pK_1　2.52\
　　　　pK_2　3.60\
　　　　pK_3　6.20\
　　　　pK_4　11.78

IDA (iminodiacetic acid)

　　　　CH$_2$COO$^{\ominus}$\
　　　／\
H$^{\oplus}$N\
　　　＼\
　　　　CH$_2$COOH

H$_2$Y　　pK_1　2.73\
　　　　pK_2　9.46

NTA (nitrilotriacetic acid)

　　　　　CH$_2$COO$^{\ominus}$\
　　　　／\
$^{\oplus}$H:N—CH$_2$COOH\
　　　　＼\
　　　　　CH$_2$COOH

H$_3$Y　　pK_1　1.97\
　　　　pK_2　2.57\
　　　　pK_3　9.81

Figure 12.5 Aminocarboxylate ligands (zwitterion forms). Carbons and hydrogens on chains and the ring in DCTA are unlabeled for clarity of the illustration.

for EDTA, it is simpler to tabulate values for the fraction of the tetra-anion as a function of pH (Table 12.2). These values will be used in Sections 12.5 and 12.6 dealing with EDTA equilibria and titrations.

12.4 STABILITY OF COMPLEXES

When we say that a complex is stable, we are really making a qualitative thermodynamic statement. A stable complex is formed by a process that has a negative standard free

12.4 STABILITY OF COMPLEXES

TABLE 12.1 FORMATION CONSTANTS FOR AMINOCARBOXYLATE LIGAND-METAL ION COMPLEXES ($\log K_f$)

Ligand[a]	Ba^{2+}	Ca^{2+}	Cd^{2+}	Cu^{2+}	Fe^{3+}	Hg^{2+}	Mg^{2+}	Ni^{2+}	Pb^{2+}	Zn^{2+}
EDTA	7.8	10.7	16.5	18.8	25.1	21.8	8.7	18.6	18.0	16.5
EGTA	8.4	11.0	15.6	17.0	20.5	23.2	5.2	12.0	13.0	12.8
DCTA	8.0	12.5	19.2	21.3	29.3	24.3	10.3	19.4	19.7	18.7
DTPA	8.8	10.6	19.0	20.5	27.5	27.0	9.3	20.0	18.9	18.0
IDA[b]										
(1:1)	1.7	2.6	5.3	10.5		10.8	2.9	8.3		7.3
(2:1)[c]			9.5	16.2				14.6		12.6
NTA[b]										
(1:1)	4.8	6.4	10.1	12.7	15.9	14.6	5.4	11.3	11.8	10.5
(2:1)[c]			14.5	16.3	24.3			15.8		13.5

[a] Full names and structures for these ligands are given in the text.
[b] IDA and NTA form complexes of the type ML_1 and ML_2, although not all values for ML_2 are available.
[c] Values in these rows are $\log \beta_2$, or $\log K_1K_2$.

energy change (ΔG^0). In more quantitative terms, the larger the standard free energy change, the larger is the formation constant (K_f^0) and the more stable is the complex, as shown by the relationship

TABLE 12.2 FRACTION OF $EDTA^{4-}$ AS A FUNCTION OF pH

pH	Fraction of $EDTA^{4-}$	pH	Fraction of $EDTA^{4-}$
0	3.95×10^{-22}	7.0	3.89×10^{-4}
0.5	3.88×10^{-20}	7.5	1.37×10^{-3}
1.0	3.66×10^{-18}	8.0	4.47×10^{-3}
1.5	3.10×10^{-16}	8.5	1.42×10^{-2}
2.0	2.00×10^{-14}	9.0	4.36×10^{-2}
2.5	7.65×10^{-13}	9.5	0.126
3.0	1.61×10^{-11}	10.0	0.314
3.5	2.22×10^{-10}	10.5	0.591
4.0	2.48×10^{-9}	11.0	0.820
4.5	2.54×10^{-8}	11.5	0.935
5.0	2.47×10^{-7}	12.0	0.979
5.5	2.22×10^{-6}	12.5	0.993
6.0	1.67×10^{-5}	13.0	0.998
6.5	9.33×10^{-5}	13.5	0.999

$$\Delta G^0 = -RT \ln K_f^0$$

Relationships such as this were developed in Chapter 3. Recall that R is the gas constant and T is the absolute temperature.

Thermodynamic stability is easily confused with inertness, which has to do with the rate of a chemical reaction rather than the extent to which it will ultimately go. Good examples are found in the reactions of cobalt(III) complexes, which are notoriously slow. Even though the reaction

$$Co(NH_3)_6^{3+} + 6H_3O^+ = Co(H_2O)_6^{3+} + 6NH_4^+$$

has an enormous equilibrium constant ($\sim 10^{25}$), $Co(NH_3)_6^{3+}$ will persist for days in acidic solution because it reacts so slowly. On the other hand, while the complex $Ni(CN)_4^{2-}$ is extremely stable,

$$Ni^{2+} + 4CN^- = Ni(CN)_4^{2-} \qquad K = 10^{22}$$

the replacement of CN^- with isotopically labeled cyanide ($^{14}CN^-$) is quite fast (half is replaced within 30 seconds of mixing). Complexes which react quickly are called *labile*, while those that react slowly are called *inert*.†

The stability of a complex is related to the strength of the acid and base that combine to produce it. While it is generally true that strong Lewis acids combine with strong Lewis bases to form stable complexes, there are many surprises. As indicated briefly in Chapter 4, a Lewis acid may appear to be strong in reactions with one base but weak in reactions with another. Rather than discuss in any detail the major contending theories of acid/base strength, it is probably more profitable to describe several factors which generally enhance acidity or basicity.

12.4.1 Size and Electronegativity

Small, highly charged cations are strong acids when they react with small, highly electronegative anions. Examples are Li^+ (ionic radius 90 picometers‡), Mg^{2+} (radius 86 pm), and other ions with inert-gas electronic structures from families IA and IIA of the periodic table, when reacting with fluoride, oxide, and hydroxide ion. These acids were called class A acids by Schwartzenbach (Ref. *1*), one of the first investigators to classify ions according to their reactivity. Later, Pearson¶ called these acids "hard

---

† A convenient criterion is completion within 1 minute at room temperature, a suggestion made by the inorganic chemist Henry Taube many years ago.

‡ 1 picometer (pm) = 10^{-12} meter or 0.01 angstrom. All radii in this section are experimentally determined six-coordinate radii tabulated by Shannon and Prewitt (Ref. *2*).

¶ See Pearson (Ref. *3*). This is neither the first nor last paper by Pearson on this subject, but is easy to read and available.

acids" in his hard-soft acid-base (HSAB) Theory. The compounds formed between these cations and anions are the traditional ionic compounds.

Larger metal ions which are more easily reduced (that is, take on electrons easily) are strong acids when they react with larger, less electronegative anions. Examples are Hg^{2+} (radius 116 pm) and Ag^+ (radius 129 pm) in reactions with bases such as sulfide (S^{2-}) and iodide (I^-). The *standard potential* (E^0) is the best measure of ease of reduction and will be discussed in detail in Chapter 14. Species which have large positive standard potentials, for example, mercuric ion,

$$Hg^{2+} + 2e^- = Hg \qquad E^0 = +0.8 \text{ volt}$$

are much easier to reduce than species which have negative standard potentials, for example, Zn^{2+},

$$Zn^{2+} + 2e^- = Zn \qquad E^0 = -0.76 \text{ volt}$$

The more easily reduced acids were called class B acids by Schwartzenbach, who noted that many are transition metal ions with large numbers of d-type electrons. Pearson calls these acids soft acids.

12.4.2 Effects of Substituents

Substituents can alter the acidity and basicity of species by both electron density effects and steric (size) effects. When a metal ion can react with several molecules of base, it becomes a weaker and weaker acid as successive base molecules attach themselves to it and electron density accumulates on the metal ion. This accumulation of charge is at least partly responsible for the decrease in the values of successive formation constants which is observed in almost all systems†:

$$Ag^+ + Cl^- = AgCl(aq) \qquad K_f = 10^{2.7}$$
$$AgCl(aq) + Cl^- = AgCl_2^- \qquad K_f = 10^{2.1}$$
$$AgCl_2^- + Cl^- = AgCl_3^{2-} \qquad K_f = 10^{0.7}$$
$$AgCl_3^{2-} + Cl^- = AgCl_4^{3-} \qquad K_f = 10^{0.5}$$

Steric interference among bulky ligands trying to crowd around small metal ions can also make bases appear to be weaker electron donors than they really should be. What might seem to be relatively minor structural differences between ligands can

---

† A statistical factor is also involved. There are more sites available on $Ag^+(H_2O)_4$ to attach a chloride ion than there are on the species $AgCl_3(H_2O)^{2-}$. Addition of a chloride ion to $AgCl_3^{2-}$ is thus a less probable event than addition of a chloride to aquated silver ion, and this lower probability is reflected in a smaller formation constant for the fourth step than for the first.

have surprisingly large effects on complex stability. For example, consider the base 1,10-phenanthroline (PHEN) and its close relative 2,9-dimethyl-1,10-phenanthroline (DMPHEN), which form covalent bonds with transition metals through their nitrogen atoms. Formation constants and structural formulas for the cadmium complexes of these two ligands appear below.†

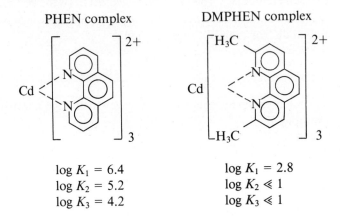

$\log K_1 = 6.4$ (PHEN)
$\log K_2 = 5.2$
$\log K_3 = 4.2$

$\log K_1 = 2.8$ (DMPHEN)
$\log K_2 \ll 1$
$\log K_3 \ll 1$

The first formation constant for the PHEN-Cd complex is a factor of 4000 larger than that for the DMPHEN-Cd complex, even though DMPHEN is a slightly stronger base toward protons ($pK_b = 7.85$) than PHEN ($pK_b = 9.1$). The difference in first formation constants is therefore interpreted as arising from the steric interaction of the two methyl groups with water molecules bound to cadmium ion at neighboring coordination sites. These groups also interfere with methyl groups on the second and third ligands as they approach in an attempt to bond, and as a result K_2 and K_3 are very small.

12.4.3 Nature of the Bond

Strong Lewis bases are almost always thought of as good sigma-bond donors. Ammonia and substituted amines and to a lesser extent oxygen and the halides are all sigma donors. But there are also Lewis bases which donate pi-type electrons to form complexes. An interesting example is the cyclopentadienyl anion (cp^-) in the sandwich compound ferrocene:

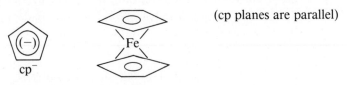

(cp planes are parallel)

† Data are from Yasuda et al. (Ref. 4).

12.4 STABILITY OF COMPLEXES

The bonds between the cyclopentadienyl anions and the iron(II) ion involve the pi-electron clouds of the ligand and vacant d-type metal orbitals pointed toward the ligand rings (rather than toward their centers). Pi-electron donors form stable bonds with transition metal ions in low oxidation states, that is, Pearson soft acids.

There are some interesting compounds which are sigma donors but also pi *acceptors* (acids). Like pi donors, pi acceptors form strong bonds with soft acids. Carbon monoxide, phosphorus, sulfur compounds, and to a lesser extent cyanide ion appear to act as pi acceptors. In the case of carbon monoxide there are *antibonding* molecular orbitals (pi*) of the proper symmetry and energy to overlap occupied metal d-type orbitals. The overlap of these orbitals allows electron density to be pulled away from the metal ion, decreasing its electron density and increasing its Lewis acidity. The sigma bond that forms between carbon and the metal is therefore strengthened by the enhanced acidity of the metal. At the same time, the addition of electron density to antibonding orbitals of the carbon monoxide ligand weakens the C≡O triple bond, an effect that can be observed in the infrared spectra of these complexes.

12.4.4 Formation of Rings

Many ligands with more than one base site are able to form chelate rings with metal ions. For example, one ethylenediamine molecule, with two nitrogen atoms, will react with a cadmium ion to form a species with a five-membered ring:

$$Cd^{2+} + H_2NCH_2CH_2NH_2 = Cd\begin{array}{c} NH_2-CH_2 \\ | \\ NH_2-CH_2 \end{array}$$

Ring formation appears to increase the stability of complexes, as shown by the overall formation constants for the following reactions:

$$Cd^{2+} + 4CH_3NH_2 = Cd(NH_2CH_3)_4^{2+} \qquad \log \beta_4 = 6.52$$
$$Cd^{2+} + 2H_2N(CH_2)_2NH_2 = Cd(NH_2(CH_2)_2NH_2)_2^{2+} \qquad \log \beta_2 = 10.62$$

The additional stability of the ethylenediamine complex, which most chemists call the "chelate effect," is really a result of the entropy change during formation. Recall that the logarithm of the overall formation constant is related to both the standard enthalpy and entropy of the reaction,

$$-RT \ln K^0 = \Delta G^0 = (\Delta H^0 - T \Delta S^0)$$

We would expect the standard *enthalpies* of these two reactions to be quite similar because CH_3NH_2 and $-CH_2NH_2$ should be equally strong bases. The observed difference must arise from the other term, the entropy change: each methylamine which bonds to a cadmium ion displaces one water molecule, while each ethylenediamine molecule displaces two water molecules. The ethylenediamine reaction thus creates more disorder in the solution than does the methylamine reaction, and it is accompanied by a positive ΔS^0.†

12.4.5 Effects of Solvent

Virtually all solvents have some electron donor or acceptor tendencies, and these will affect the measured stability of a complex. In solution all complexation reactions involve the displacement of solvent molecules,

$$M(\text{solvent})_6 + 6L = ML_6 + 6 \text{ solvent}$$

The stronger the solvent is as a base, the weaker the metal ion will appear to be as an acid in the reaction with the ligand. On the other hand, the solvent may also have electron acceptor properties, and the greater these are, the weaker the ligand appears to be as a base in reactions with a particular metal ion. It has been found, for example, that the complex $CoCl_4^{2-}$ is formed almost quantitatively when stoichiometric amounts of Co^{2+} perchlorate and chloride ion are mixed in nitromethane, a weakly basic solvent. When dimethyl sulfoxide, a much stronger Lewis base, is used, a large excess of chloride must be added to complex cobalt ion quantitatively.

We have surveyed the factors that influence the kinds of complexes that can form and the ways in which they form: metal ion acidity and coordination number, ligand basicity and denticity, and the effects of ring formation on complex stability. We are now ready to discuss the equilibria which exist in solutions of metal ions and ligands and then move on to analytical applications.

12.5 COMPLEXATION EQUILIBRIA

Complexation equilibria were introduced in Chapter 8 in the context of competing reactions. Examples used were the effects of excess chloride or ammonia on the solubility of silver chloride. Reactions such as

---

† Several other entropy factors should be considered in the analysis of the chelate effect. The entropy change is decreased, for example, by the loss of free rotation in the ethylenediamine molecules when they bond to the metal ion. See Chung (Ref. 5).

12.5 COMPLEXATION EQUILIBRIA

and
$$AgCl(s) + Cl^- = AgCl_2^-$$
$$AgCl(s) + NH_3 = Ag(NH_3)^+ + Cl^-$$

become important competitors for silver ion when large quantities of chloride ion or ammonia are present. Remember that the most important competitive reactions have the largest equilibrium constants (formation constants).

While the silver-chloride and silver-ammonia systems involved a great many species and at first looked extremely complicated, we found it possible to describe them fully by dealing with them in an orderly fashion. The methods we used involved conditional formation constants and material balance expressions. In the case of the formation of soluble silver chloride species, the material balance was written as

$$C_{Ag} = [Ag^+] + [AgCl(aq)] + [AgCl_2^-] + [AgCl_3^{2-}] + [AgCl_4^{3-}]$$

This was then used to create an expression for the fraction of uncomplexed ("free") silver ion in a series of steps you should review:

$$\alpha_{Ag^+} = \frac{1}{1 + K_1[Cl^-] + \beta_2[Cl^-]^2 + \beta_3[Cl^-]^3 + \beta_4[Cl^-]^4}$$

Remember that the beta notation was used to signify the product of several formation constants; for example, $\beta_2 = K_1 K_2$ and $\beta_4 = K_1 K_2 K_3 K_4$.

Once the fraction of free silver ion was found it was possible to calculate a conditional solubility product constant for AgCl,

$$K'_{sp} = C_{Ag} \times [Cl^-] = K^0_{sp}/\alpha_{Ag^+}$$

The conditional constant is a special kind of equilibrium constant that applies to a system under just one set of conditions. Remember that when the fraction of free silver ion is low (when a great deal of chloride is present), the conditional constant is large and the solubility is large. The K^0_{sp} is the thermodynamic solubility product constant and is independent of changing solution conditions.

Once the free silver ion concentration was known, the concentrations of all the other species could be calculated. Matters were made easier by writing the equations in such a way that there was a correspondence between the position of a complex species in the material balance expression and its position in the denominator of the fraction expression. For example, $AgCl_2^-$ was the third species from the left in the material balance and was found in the third term from the left in the denominator of the fraction. The fraction of $AgCl_2^-$ was therefore simply

$$\alpha_{AgCl_2^-} = \beta_2[Cl^-]^2/(1 + K_1[Cl^-] + \beta_2[Cl^-]^2 + \beta_3[Cl^-]^3 + \beta_4[Cl^-]^4)$$

As presented in Chapter 8, the next level of difficulty in complexation equilibria involved accounting for protonic acid-base competition at the same time as foreign ligand competition. The example used in that treatment was the calcium-oxalate-EDTA system, in which a "doubly conditional" formation constant was calculated. The doubly conditional constant, which we labeled K''_{sp}, depended on the fraction of free metal as well as the fraction of free ligand (oxalate). In the calcium-oxalate-EDTA system we lowered the fraction of free calcium ion by complexing it with EDTA, while at the same time lowering the fraction of free oxalate by working in acidic solutions, which favor the formation of binoxalate ion. The resulting low fractions made the doubly conditional solubility product constant much larger than the thermodynamic constant and dramatically increased the solubility of calcium oxalate.

We will use conditional formation constants and material balance expressions to study metal ion complex equilibria. For example, the formation of the calcium complex with EDTA can be represented by the reaction

$$Ca^{2+} + Y^{4-} = CaY^{2-}$$

The thermodynamic formation constant for this process is expressed as

$$K_f^0 = (a_{CaY})/(a_{Ca} * a_Y)$$

where charges are omitted for clarity. The concentration constant is related to K_f^0 by a ratio of activity coefficients,

$$K_f = K_f^0(f_{Ca} * f_Y/f_{CaY}) = [CaY]/[Ca][Y]$$

Most formation constants reported in the literature are concentration constants, determined in buffer solutions of high ionic strength. As a result, it is difficult to treat complexation equilibria with the same attention to detail as we gave Brønsted acid-base equilibria. In any given problem it may be necessary to combine formation constants determined at different ionic strengths. The result is that we must deal with more crude approximations in these calculations than we are accustomed to using in acid-base calculations.

Both metal ion and ligand can be involved in competing reactions. Our strategy will be to modify concentration formation constants for a complex by multiplying by fractions of free metal ion and free ligand. The doubly conditional formation constant for our calcium-EDTA complex example is then written as

$$K_f'' = K_f \alpha_{Ca} \alpha_Y = [CaY]/C_{Ca} * C_Y$$

The terms C_{Ca} and C_Y are the total concentrations of calcium and EDTA, respectively, which are not tied up in the Ca-EDTA complex. This last definition is important: C_{Ca} and C_Y do not include [CaY], the equilibrium concentration of the complex. The

12.5 COMPLEXATION EQUILIBRIA

fraction of Y^{4-} is calculated from the weak acid dissociation constants for EDTA and the pH of the buffer solution (as in Chapter 5). The fraction of free metal ion is calculated from the formation constants for complexes of the metal ion with other ligands that might be in solution (as in Chapter 8).

12.5.1 A Practical Question: Is the Reaction Complete?

Let us say that we must determine the amount of magnesium present in a sample by using EDTA as a titrant. We know that EDTA undergoes several Brønsted acid-base interactions and that the fraction of the tetra-anion, "Y^{4-}," can be calculated at any solution pH. By looking in the literature we find that magnesium ion will form a 1:1 complex with Y^{4-}, although the complex is not nearly as stable as most other metal-EDTA complexes:

$$Mg^{2+} + Y^{4-} = MgY^{2-} \qquad K_f = 5.0 \times 10^8$$

The literature also shows that magnesium forms a complex with hydroxide ion, although, once again, the formation constant is not very large:

$$Mg^{2+} + OH^- = MgOH^+ \qquad K_1 = 4.0 \times 10^2$$

If a solution containing aquated magnesium ion is made sufficiently basic, magnesium hydroxide will precipitate:

$$Mg^{2+} + 2OH^- = Mg(OH)_2(s) \qquad K_{sp} = 1.8 \times 10^{-11}$$

While it must be true that soluble magnesium hydroxide forms in basic solution, apparently no one has measured K for the reaction

$$MgOH^+ + OH^- = Mg(OH)_2(aq) \qquad K_2 = ?$$

The concentration of $Mg(OH)_2(aq)$ in solution at equilibrium with solid $Mg(OH)_2$ is the "intrinsic solubility" discussed in Chapter 8 and probably has a value of about 10^{-8} to $10^{-6} M$, like the hydroxides of nickel(II), cobalt(II), and other metals. By analogy to the treatment of the silver chloride system in Chapter 8 we know that

$$[Mg(OH)_2(aq)] = K_{sp} K_1 K_2$$

Taking $[Mg(OH)_2(aq)]$ equal to $10^{-7} M$, and knowing K_{sp} and K_1, we calculate K_2 to be $10^{1.1}$.

The question of whether a reaction is complete (meaning quantitative, 99.9% complete) was addressed in Chapter 3. Example 3.7 dealt with reactions with 1:1

stoichiometry (like EDTA-metal ion reactions) and showed that whether a particular equilibrium constant is large enough to make a reaction quantitative depends on the concentration of the reaction product. In the case of the formation of magnesium-EDTA complex, if the concentration of MgY is to be $0.1M$, the reaction will be quantitative if solution conditions are set to make the conditional formation constant larger than about 10^7. If, however, the concentration of MgY is to be $10^{-3}M$, the conditional constant must be about 10^9 for the reaction to be quantitative.

EXAMPLE 12.1 Completeness of the Mg-EDTA Reaction

Is the reaction of $0.01F$ Mg^{2+} with $0.01F$ EDTA quantitative at pH 8? If not, at what pH will it be quantitative?

Solution

We must calculate the conditional formation constant for the complex and then see if it is sufficiently large to make the reaction complete. We will calculate a fraction of the EDTA tetra-anion (Y^{4-}) and a fraction of uncomplexed magnesium ion at pH 8.

The fraction of Y^{4-} was calculated at several pH values and tabulated for convenience in Table 12.2. At pH 8 the fraction is 0.0045.

The fraction of free magnesium can be calculated from the material balance

$$C_{Mg} = [Mg^{2+}] + [Mg(OH)^+] + [Mg(OH)_2(aq)]$$

and is

$$\alpha_{Mg^{2+}} = 1/(1 + K_1[OH^-] + \beta_2[OH^-]^2)$$
$$= 1/[1 + 400(10^{-6}) + 5600(10^{-12})]$$

Matters are simplified because this value is not significantly different from unity. In other words, hydroxide ion is not an important competitor for magnesium in this solution. Notice that $[MgY^{2-}]$ is not included in this material balance.

The conditional constant is then calculated to be

$$K_f'' = K_f(\alpha_{Mg^{2+}})(\alpha_{Y^{4-}}) = 2.3 \times 10^6$$

Given this value for K_f'', is the reaction quantitative? Remember that when an MgY^{2-} complex unit dissociates, it creates one Mg^{2+} ion (either free or complexed by hydroxide) and one Y^{4-} ion (either free or complexed by hydronium ion). Therefore, at equilibrium $[MgY^{2-}] = (0.01 - C_{Mg})$ and $C_{Mg} = C_Y$. We can then use the expression for K_f'',

$$K_f'' = [MgY^{2-}]/C_{Mg}*C_Y$$

and substitute

$$K_f'' = (0.01 - C_{Mg})/C_{Mg}^2$$

12.5 COMPLEXATION EQUILIBRIA

We then solve this equation and find $C_{Mg} = 6.6 \times 10^{-5} M$. This means that $1.00 - [(6.6 \times 10^{-5})/0.01]$, or 99.3% of the magnesium is complexed with EDTA. This is not quantitative.

Since the magnesium-EDTA reaction is incomplete at the $0.01 M$ level at pH 8, it seems reasonable to ask under what conditions it would be complete. Given the facts that the fraction of Y^{4-} increases with increasing pH and the fraction of free magnesium decreases with increasing pH, it is necessary to find the optimum solution pH. Figure 12.6 is a plot of the logarithms of the fractions of Mg^{2+}, $MgOH^+$, and Y^{4-} and the logarithm of the conditional formation constant for the complex MgY^{2-} over the pH range 7 to 14. The effects of hydroxide ion competition for magnesium and proton competition for EDTA can be seen clearly in this figure; from pH 7 to 11 the conditional formation constant increases as the fraction of Y^{4-} increases but then decreases above pH 11 as the fraction of $MgOH^+$ increases. If the complexation of $0.01 F$ Mg^{2+} is to be complete, the conditional formation constant must be 10^8 or larger. Figure 12.6 shows that this is the case between pH 10.8 and 11.7. The pH for optimum formation of MgY^{2-} is 11.5. Note that the complexation of $10^{-3} F$ Mg^{2+} requires a conditional formation constant of 10^{10} for completeness and that such a constant cannot be achieved at any pH. This means that magnesium in millimolar solutions cannot be determined quantitatively by titration with EDTA.

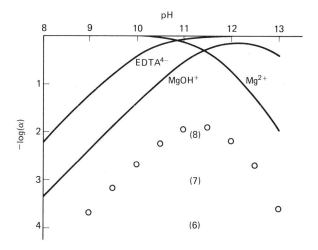

Figure 12.6 The Mg^{2+}-$MgOH^+$-$EDTA^{4-}$ system as a function of pH. This is a plot of the negative logarithm of the fraction of free magnesium, magnesium hydroxide, and the EDTA tetraanion. Small fractions have large numbers in this plot. Superimposed is a plot of $\log K_f''$ for the $Mg(EDTA)^{2-}$ complex as a function of pH. Notice that the maximum-value of $\log K_f''$ is 8.12 at pH 11.5.

12.6 COMPLEXOMETRIC TITRATIONS

Complexometric titrations are by far the most important application of complexation equilibria in analytical chemistry. In a manner analogous to acid-base and precipitation titrations, we react a complexing agent (usually the titrant) with a metal ion in a sample and watch for the equivalence point by using a colored indicator (visually or with the aid of a spectrophotometer) or a device to monitor the voltage of an electrode which responds to the activity of the metal ion. The general requirements of a complexometric titration were discussed in the preceding section; the stoichiometry of the reaction must be known, and the reaction must be quantitative (complete) at the equivalence point.

Complexometric titrations are best illustrated by plotting titration curves as $-\log(C_M)$ (i.e., pC_M) versus percent (or volume) titrated, as we did in the earlier chapter on precipitation titrations. Calculating a titration curve is rather easy; we need only know K_f'' and the initial formal concentration of metal ion (F_M). When K_f'' is large (greater than 10^8) the titration is totally analogous to a strong Brønsted acid-strong Brønsted base titration: pC_M is determined at any point up to the equivalence point by the concentration of metal ion uncomplexed with L, and it depends on dilution. For example, if 50 mL of $0.01F$ ligand (0.5 mmole) is added to 100 mL of $0.01F$ metal ion (1.0 mmole), 0.5 mmole of a 1:1 complex forms (concentration = $0.5/150$ = $0.00333M$) and 0.5 mmole of metal ion is left uncomplexed with L ($0.00333M$). If K_f'' is very large, there is little dissociation of the complex and $0.00333M$ can be taken as C_M. At the equivalence point we can calculate C_M if we recognize that it must be the same as the concentration of ligand in all its forms (except complexed with metal), C_L. Using the expression for the conditional formation constant, we find the general relationship as follows:

$$K_f'' = [ML]/C_M C_L = [ML]/C_M^2$$
$$C_M = ([ML]/K_f'')^{0.5}$$
$$pC_M = 0.5(\log K_f'' + pML)$$

If the complex is sufficiently stable, it simplifies matters to assume that the equilibrium concentration of ML is equal to the initial concentration of metal ion, corrected for dilution. The same approximation was used successfully in Chapter 5 to calculate the pH of a solution of acetic acid, a weak electrolyte which has a fairly large formation constant with protons,

$$K_f = [HOAc]/[H^+][OAc^-] = 1/K_a = 10^{4.76}$$

Past the equivalence point pC_M can be calculated from the expression for K_f'' by plugging in values for ligand concentration added in excess. The calculations are exactly

12.6 COMPLEXOMETRIC TITRATIONS

the same as those encountered in the suppression of solubility by the common ion effect in Chapter 8.

When the conditional formation constant of a complex is small, the titration of metal with ligand behaves like a weak acid-strong base or strong acid-weak base (Brønsted sense) titration. Under these circumstances dissociation of the weak complex decreases its concentration at equilibrium while augmenting the concentration of uncomplexed metal ion. Calculating a titration curve with this complication involves solving a quadratic equation, in much the same manner as shown in Chapter 5. In the case of a metal ion being titrated with a solution of ligand to form a 1:1 complex, we can assign the following variables:

C_M = total concentration of metal ion not complexed with L
V_M = initial volume of metal ion solution
C_L = total concentration of ligand not complexed with M
V_L = incremental volume of ligand solution

At any point in the titration curve we can write the following two material balances:

$$C_M + [ML] = [V_M/(V_M + V_L)]F_M$$
$$C_L + [ML] = [V_L/(V_M + V_L)]F_L$$

The coefficient $[V_M/(V_M + V_L)]$ corrects the initial (formal) concentration of metal ion for the effects of dilution; V_L mL of titrant solution is added to the V_M mL at the start of the titration. Subtracting the first expression from the second and rearranging, we find

$$C_L = C_M + [V_L/(V_M + V_L)]F_L - [V_M/(V_M + V_L)]F_M$$

We can then substitute this, along with the first material balance arranged to solve for C_M, into the expression for K_f''. In a series of steps we can then obtain a quadratic equation for [M],

$$C_M^2 + C_M\left(\frac{1}{K_f''} + \frac{V_L F_L}{V_L + V_M} - \frac{V_M F_M}{V_L + V_M}\right) - \frac{V_M F_M}{K_f''(V_L + V_M)} = 0$$

While a point-by-point calculation is quite time-consuming when done by hand, the following simple BASIC program can be used with a microcomputer:

```
100    DIM M(200): REM NUMBER OF POINTS CALCULATED
200    REM K=CONDITIONAL FORMATION CONSTANT
210    REM FM=FORMAL CONCENTRATION OF METAL ION SOLN.
220    REM VM=INITIAL VOLUME OF METAL ION SOLN.
```

```
230   REM FL=FORMAL CONCENTRATION OF LIGAND SOLN.
240   REM EQ=VOLUME REQUIRED TO REACH EQUIVALENCE PT.
900   READ K,FM,VM,FL
1000  EQ=FM*VM/FL
1100  L=1/K
1200  FOR VL= 1 TO 1.0*EQ STEP.01:REM 0 TO 100% TITRATED
1300  B= (L +VL*FL/(VM+VL) − (VM*FM)/(VM+VL))
1400  C= −VM*FM*L/(VM+VL)
1500  M(VL)= (−B+SQR(B*B−4*C))/2
1550  REM STORES METAL ION CONC AT EACH POINT
1600  NEXT VL
2000  **** (printed output or graphical output)****
3000  DATA
```

Examples of titration curves calculated with this algorithm are shown in Figs. 12.8 and 12.9 in the next section, where visual indicators and equivalence point sharpness are discussed. The program is useful for K_f'' up to about 10^{10}. For $K_f'' > 10^{11}$ computer single precision (seven decimals) cannot tell the difference between $(-B)$ and $SQR(B*B - 4*C)$ in line 1500, and returns a value $M(VL) = 0$. For such large K_f values a quadratic is unnecessary. Can you write a suitable approximation?

12.7 METALLOCHROMIC INDICATORS

As their name suggests, metallochromic indicators are molecules which change color when they form complexes with metal ions. In most applications the indicator is complexed with metal ion before the equivalence point. The complexing titrant, for example, EDTA, reacts with free metal ion and eventually metal ion tied up with metallochromic indicator. At the endpoint (one hopes close to the equivalence point), enough metal ion is taken from the indicator to change the color of the solution to that of the free indicator. To summarize:

initial solution:	M^+ + MIn (mixture, mostly M^+)
titration reaction:	M^+ + Y = M(Y)
just before endpoint:	MIn + Y = M(Y) + In(free)

[The colors of MIn and In(free) must be different. The endpoint is seen when the ratio In(free)/MIn > ~10.] As was the case with Brønsted-Lowry indicators, only very dilute solutions of indicator are necessary for the endpoint to be visible. A $10^{-5}M$ solution of most indicators is adequate and corresponds to 1 or 2 mg of indicator in 50 ml of solution.

For accuracy and precision, it is important that the metal-indicator complex be of about the same stability as the metal-titrant complex. If MIn is too stable, excess titrant will have to be added to see the endpoint color change. On the other hand, if MIn is much weaker than the metal-titrant complex, the endpoint will be premature.

12.7 METALLOCHROMIC INDICATORS

In either case the effect is to make the color change sluggish and indistinct, as will be discussed below.

Metallochromic indicators, being Lewis bases, have at least some affinity for protons, which must of course be considered in equilibrium calculations. One of the most popular indicators is Eriochrome Black T (or simply "Erio T"), which has the structure

[Structure of Eriochrome Black T with O$_2$S, *OH, *OH, N*=N, and NO$_2$ substituents]

Erio T is a triprotic acid which undergoes color changes during deprotonation:

$$H_3In(\text{red}) = H_2In^-(\text{red}) + H^+ \qquad pK_{a1} = 1.6$$
$$H_2In^-(\text{red}) = HIn^{2-}(\text{blue}) + H^+ \qquad pK_{a2} = 6.3$$
$$HIn^{2-}(\text{blue}) = In^{3-}(\text{orange}) + H^+ \qquad pK_{a3} = 11.6$$

A plot of pH vs. pC for Erio T, describing the acid-base behavior of the colored forms of the indicator, is shown in Fig. 12.7. The figure key tells how the figure can be used to calculate fractions of the various acid forms of the indicator at any specified pH. In the view of the analyst, the most interesting color change of Erio T is the blue/red color change which accompanies the exchange of one proton in the second step. A metal ion can react with the blue form (HIn^{2-}) to displace a proton and form a red metal-indicator complex,

$$HIn^{2-}(\text{blue}) + M^{2+} = MIn^-(\text{red}) + H^+$$

What appears to be critical to the color change is the formation of three donor-acceptor bonds and not so much the identity of the Lewis acid. Bonds can form with protons or metal ions, with little observable difference in color. Notice that MIn^- is equivalent to H_2In^-, with one metal ion supplying two acceptor sites. The location of donor sites on the indicator molecule is indicated with asterisks in the structural diagram above.

In a direct titration of metal ion with a complexing titrant like EDTA, the titrant competes with the indicator of available metal ion,

$$MIn^-(\text{red}) + Y^{4-} + H^+ = M(Y)^{2-} + HIn^{2-}(\text{blue})$$

The endpoint is taken as the point at which the last red color disappears from solution. Actually, the color change will occur over a range of metal ion concentrations. We

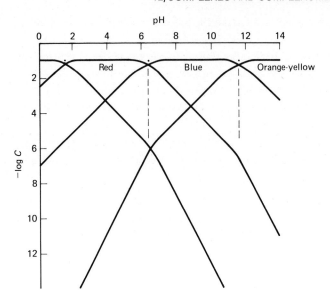

Figure 12.7 Plot of pH vs. $-\log C$ for Erio T indicator. This figure can be used to calculate approximate values for the fraction of the analytically important blue form, HIn^{2-}, at various pH values by simply subtracting 1 from the $-\log C$ value shown. Thus the fraction of HIn^{2-} is antilog $(1 - 1) = 1.0$ from about pH 7.4 to about pH 11.0. This is a graphical representation of the equilibria and is not meant to suggest that anyone would use Erio T as a $0.1F$ solution.

can calculate such a range from the conditional formation constant for the MIn^- complex, written as

$$K'_f = [MIn^-][H^+]/[M^{2+}]C_{In} = K_f \alpha_{HIn^{2-}}$$

This can be rearranged to give an expression for pM,

$$pM = \log K'_f + \log(C_{In}/[MIn^-]) + pH$$

In the pH range in which most EDTA titrations are performed with Erio T indicator (pH 9 to 11), the blue species, HIn^{2-}, predominates, and so it is a good approximation to set $[HIn^{2-}]$ equal to C_{In}, that is,

$$pM = \log K'_f + \log([HIn^{2-}]/[MIn^-]) + pH$$

If we can tell that the solution is distinctly red when the ratio $[HIn^{2-}]/[MIn^-]$ is 0.1 and distinctly blue when the ratio is 10, then the indicator color transition range is given by

12.7 METALLOCHROMIC INDICATORS

$$pM = \log K'_f \pm \log 10 = \log K'_f \pm 1$$

Notice that the transition range depends on the conditional formation constant of the MIn^- complex, which is a function of solution pH. The situation becomes more interesting when you remember that the size of the equivalence point break (mentioned in the last section) also depends on a pH-dependent conditional formation constant. The following example should help you sort out the complicated effects of pH.

EXAMPLE 12.2 Indicator Range

Calculate the indicator transition range for Erio T in the titration of magnesium in a solution at pH 10. The formation constant for $MgIn^-$ is 3×10^5. Acid dissociation constants for Erio-T: $pK_{a1} = 1.6$, $pK_{a2} = 6.3$, $pK_{a3} = 11.6$

Solution

First calculate the doubly conditional formation constant for the Mg-Erio T complex at pH 10, then calculate the range in pMg over which the color transition should occur. The conditional constant is given by

$$K''_f = K_f(\alpha_{HIn^{2-}})(\alpha_{Mg^{2+}})$$

Using an equation for $\alpha_{Mg^{2+}}$ from example 12.1 we find, $\alpha_{Mg^{2+}} = 0.96$ at pH 10.

Using the expression developed in Chapter 5 for the fraction of the monoprotic form of a triprotic acid, we find $\alpha_{HIn^{2-}} = 0.98$ at pH 10. The doubly conditional constant is thus $(3 \times 10^5)(0.98)(0.96) = 2.8 \times 10^5$.

The indicator range is given by $\log K''_f \pm 1 = 5.4 \pm 1$, or pMg = 4.4 to 6.4.

The three titration curves in Fig. 12.8 illustrate the effects of pH on the sharpness of the equivalence point break and the endpoint color change in the titration of magnesium with EDTA. Erio-T is the indicator. These curves are not experimental data, but have been calculated from literature values for formation constants as described in the last section. The formal concentrations of magnesium ion and EDTA were taken as 0.01, and dilution has been accounted for in the calculations. Notice that the curves are incomplete, spanning a range of 60 to 130% titrated (30 to 65 mL of $0.01F$ EDTA) rather than 0 to 150%.

The sharpness of the equivalence point break increases with pH up to about pH 10, then begins to decrease. This follows the pattern observed in the conditional formation constant with increasing pH shown in Fig. 12.6. An additional observation is that the endpoint color change (indicated by boxes drawn on the titration curves) is quite abrupt at pH 10, while at pH 8 and pH 11.5 it is quite gradual. Abrupt color changes occur when the conditional formation constants for the metal-indicator and

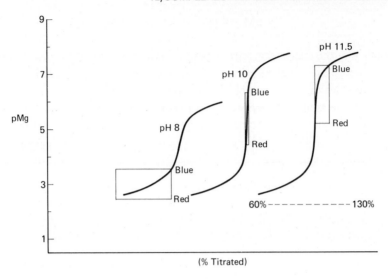

Figure 12.8 Titration of magnesium with EDTA using Erio T indicator.

metal-EDTA complexes are about equal. At pH 8 the magnesium-Erio T complex is so weak compared to the magnesium-EDTA complex that much of the indicator is freed before the 90% titrated point. At pH 11.5 the magnesium-Erio T complex is more stable than the magnesium-EDTA complex, and excess EDTA must be added to free the indicator. At pH 10 the color change starts at about 99% titrated and ends at about 101% titrated. With an initial solution of 50 mL of $0.01F\,Mg^{2+}$, this translates to a volume range of 49.5 to 50.5 mL of $0.01F$ EDTA. This is really too gradual a change for accurate and precise work, but might be acceptable for some semiquantitative studies.

Figure 12.9 illustrates a more difficult case, the titration of calcium with EDTA, using Erio T as an indicator. At all three pH values the color change is complete before the equivalence point is reached. The problem is that the calcium-Erio T complex cannot be made sufficiently stable by increasing the pH. The best results are obtained at pH 11.5, where the color change occurs within about 1% of the equivalence point. Higher pH values give unsatisfactory results because the principal form of the indicator above pH 11.5 is the fully deprotonated orange form. A red-to-orange endpoint is very difficult to see.

What are some alternative strategies for the determination of calcium? First, one should examine other indicators or mixtures of indicators. Most of the indicators related to Erio T and used at high pH values undergo red-to-blue or violet-to-blue color changes, and some people find these changes hard to see. Using an acid-base indicator to screen the red-to-blue color change seems to help. In an Experiment 12.2 presented later in this text, the acid-base indicator indigo carmine (yellow at pH > 12.5) is used along with the metallochromic indicator calcon (Eriochrome Blue Black R) to give a red-to-green color change in the titration of calcium. Calcon gives an

12.7 METALLOCHROMIC INDICATORS

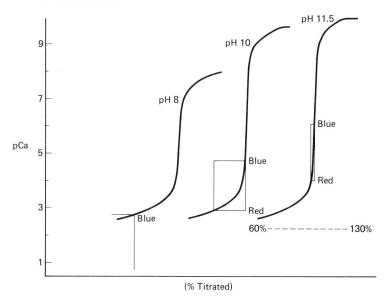

Figure 12.9 Titration of calcium with EDTA using Erio T indicator.

advantage in letting us work at higher pH values; its blue color zone extends a little beyond pH 13. Higher pH favors indicator-calcium complex stability, bringing the color transition range closer to the equivalence point. Some other metallochromic indicators, along with formation constants for complexes with several metal ions, are given in Table 12.3.

TABLE 12.3 POPULAR METALLOCHROMIC INDICATORS

Indicator	Known pK_a's	log K_f with metals[a]
Acid chrome dark blue	pK_4 = 7.6, pK_5 = 9.3, pK_6 = 12.4	Ca, 9.3; Ba, 6.2; Mg, 8.5
Calmagite	pK_2 = 8, pK_3 = 12.5	Ca, 6.1; Mg, 8.1; Zn, 12.5
Calcon (Eriochrome Blue Black R)	pK_1–pK_3: 1, 7.3, 13.5	Ca, 5.3; Cu, 21.5; Mg, 7.6; Zn, 12.5
Eriochrome Black T	pK_1–pK_3: 3.9, 6.4, 11.5	Ca, 5.4; Mg, 7.0; Mn, 9.6; Zn, 12.9
Murexide	pK_2 = 9.2, pK_3 = 10.9	Conditional constants, pH 11: Ca, 4.5; Ni, 10.2; Cu, 15.8
Xylenol orange	pK_1–pK_5: 2.6, 3.2, 6.4, 10.5, 12.3	

[a] These values are for reactions with deprotonated indicator anions, $M^{2+} + In^{3-} = MIn^-$. Log K_f values for reactions of monoprotonated forms are not well-known. In some systems they are 1.5 units smaller.

NOTE: While many thermodynamic constants are still unknown for these systems, a vast amount of experimental data is available describing color transition intervals for indicators in real solutions. Much of the early work is summarized in A. Ringbom, *Complexation in Analytical Chemistry*, vol. 16 of *Chemical Analysis*, Wiley-Interscience, New York, 1963.

Another alternative is to add a little magnesium ion ($10^{-4}F$) to the calcium sample in a pH 10 buffer (ammonia/ammonium chloride) before titrating with EDTA. The magnesium-Erio T complex is more stable than the calcium-Erio T buffer and is consumed after the calcium-EDTA equivalence point. This is an example of an "indirect" complexometric titration; the object is to determine calcium, but a second titration reaction (that of magnesium) signals the endpoint. It should be apparent that for this method to be accurate the same amount of magnesium must be present in calcium standards and unknown samples. Samples that contain their own magnesium (those, for example, derived from limestone or hard water) cannot be analyzed for calcium by this method. At pH 10 both calcium and magnesium are complexed by EDTA. Magnesium does not interfere in the determination with calcon at pH > 12.5 but precipitates as magnesium hydroxide.

Some Notes on "Hard Water" Ground water which has passed through limestone contains rather high concentrations of mineral ions such as calcium, magnesium, and iron. These ions can be a nuisance for two important reasons. First, when water containing them is boiled, they precipitate out as carbonate salts, coating and eventually plugging up water pipes and water heaters. Second, these ions react with soap, producing a precipitate of "soap scum" which sticks to washed clothing, limiting the ability of soap to clean. "Water hardness" is defined as the total concentration of calcium and magnesium, expressed as if it were all calcium carbonate. Hardness is most accurately determined by EDTA titration at pH 10 with Erio T indicator (see above and Experiment 12.1 at the end of the chapter). Commercial testing kits are available which use packaged buffer and a squeeze bottle of EDTA solution, the number of drops of which correspond to a specified number of milligrams of calcium carbonate. A quick titration can be done in a test tube.

Water hardness is commonly expressed in milligrams $CaCO_3$ per liter of water (or parts-per-million). An older set of units seems peculiar to current students of chemistry: grains per gallon. The following example illustrates the method of calculation.

EXAMPLE 12.3 Water Hardness
A 50-mL sample of water is buffered at pH 10. To titrate the magnesium and calcium to the Erio T endpoint, 15.65 mL of $0.01055M$ EDTA solution is required. Calculate the water hardness in parts per million and grains per gallon.

Solution

It is given that 15.65 mL of $0.01055M$ EDTA will consume $(15.65*0.01055) = 0.1651$ mmole of calcium and magnesium. We assume that the ions are all calcium. The 0.1651 mmole of calcium carbonate weighs $100.1(0.1651) = 16.53$ mg. To convert the concentration milligrams per liter, we divide by 0.050 liter, finding 330 mg/liter or ppm.

Since a grain is 64.8 mg and a gallon is 3.79 liters, we convert milligrams per liter to grains per gallon as follows:

$$(330 \text{ mg/liter})(3.79 \text{ liter/gal})/(64.8 \text{ mg/grain}) = 19.3 \text{ grain/gal}$$

This sample of water would definitely be called hard water, and most homeowners would want to use an ion exchange water softener to make the water more usable.

12.8 SPECIAL TECHNIQUES USING EDTA

EDTA reacts with so many metal ions under such a wide variety of conditions that a great deal of attention has been paid to making its reactions more specific. In addition, it has been necessary to find ways to speed up analyses involving slow reactions and drive thermodynamically unfavorable reactions to completion. In this section we will outline several important special techniques and then present a table listing methods for many metal ions.

12.8.1 Auxiliary Complexing Agents

If it is necessary to perform a titration with EDTA at high pH values, there is a risk that the metal ion of interest will precipitate as a hydroxide complex before any EDTA is added. The effects of a precipitation will be to slow the rate of equilibration between incremental additions of EDTA and to cause a fading endpoint. An auxiliary (or intermediate) complexing agent such as ammonia, triethanolamine, or tartrate can be added to prevent precipitation. The price that must be paid is that the equivalence point break becomes less dramatic; complexation of the metal ion raises the lower plateau of the plot of pM vs. volume of EDTA and makes the transition less sharp. In a NH_3/NH_4^+ buffer $0.1F$ ammonia is used as an auxiliary complexing agent in the titration of zinc with EDTA. Erio T is an excellent indicator for the equivalence point. In the absence of ammonia, zinc hydroxides precipitate.

12.8.2 Back Titrations

The technique of back-titrating was discussed in Chapters 6 and 11. Recall that back titrations are useful when a direct titration reaction is slow or incomplete or when a suitable indicator cannot be found. In the present context a carefully measured amount (more than required for a complete reaction) of a standard solution of EDTA is added to a sample containing the metal ion which we will call the primary ion. The excess EDTA is then titrated with a standard solution of a second metal ion to a metal-indicator color change (e.g., blue to red with Erio T). The endpoint color change is the opposite of that seen in a direct titration of the second metal ion with EDTA. An important restriction is that the second metal ion not form a more stable complex with EDTA than the primary metal ion, otherwise the second metal ion will pull EDTA away from the primary metal ion and the results of the analysis will be low. In acidic solutions zinc is often used as the second ion, along with xylenol orange as an indicator. In basic solutions magnesium ion and Erio T are often used.

Back titrations can be used to avoid precipitation of primary metal ion hydroxides in basic solutions. EDTA is added to an acidic solution of the primary ion, which is made basic just before the back-titration step. Example 12.4 describes the method used for the titration of aluminum, which precipitates in solutions more basic than about pH 5.

Another use for the back-titration technique is to overcome what is called "indicator blocking." When a primary metal ion-indicator complex is either stronger or much more inert than the primary ion-EDTA complex, no color change can be seen at the equivalence point, and the indicator is said to be blocked by the primary ion. By using a second ion-indicator pair the primary ion-indicator pair problem can be avoided altogether.

EXAMPLE 12.4 Back Titration

Exactly 10 mL of Al^{3+} stock solution was transferred to a titration flask, and 25.00 mL of 0.01532F EDTA was added by pipet. The pH of the solution was adjusted to 5.0 with a hexamethylenediamine hydrochloride buffer (HMDH$_2^{2+}$, pK_{a1} = 3.1, pK_{a2} = 4.2). The sample was back-titrated with 18.32 mL of 0.01020F Pb^{2+} solution to the xylenol orange endpoint. What is the concentration of the Al^{3+} stock solution?

Solution

The number of millimoles of EDTA tied up with aluminum is given by

$$\text{mmole as Al(Y)} = \text{mmole Y added} - \text{mmole Pb}^{2+}$$
$$= 25.00(0.01532) - 18.32(0.01020)$$
$$= 0.1961 \text{ mmole}$$

The concentration of the aluminum stock solution is

$$0.1961 \text{ mmole}/10.00 \text{ mL} = 0.01961 M$$

12.8.3 Indirect Titrations

EDTA titrations can be extended to include *anions* by using indirect techniques. An anion may be precipitated by adding an excess of standard metal ion solution. The precipitate is then collected and washed, and the filtrate solution is titrated with standard EDTA. Sulfate, sulfide, carbonate, and oxalate can all be determined in this way.

An alternative is to collect the precipitate, add excess EDTA solution to it, and boil until it dissolves. The excess EDTA can then be back-titrated with a standard metal ion solution, for example, Mg^{2+}.

12.8.4 Displacement Titrations

Metal ions which form extremely stable EDTA complexes may not have suitable indicators, because pM at the equivalence point in an EDTA titration may lie several units above any indicator color range. If an excess of some less stable second metal ion-EDTA complex is added to the solution of the primary ion, the second metal will be displaced and can then be titrated directly with EDTA. A good example of a primary metal is mercury, which forms an extraordinarily stable EDTA complex (log K_f = 25.1). The displacement reaction with magnesium-EDTA is

$$Hg^{2+} + \text{excess } MgY^{2-} = HgY^{2-} + Mg^{2+}$$

The titration reaction is then

$$Mg^{2+} + Y^{4-} = MgY^{2-} \quad \text{(Erio T endpoint, pH 10)}$$

12.8.5 Masking and Demasking

Masking is a widely used technique for controlling interferences and titrating mixtures of ions. Cyanide, fluoride, and thiourea are among the simplest and most popular masking agents. To determine calcium in a sample containing iron(II), which also forms a stable EDTA complex, one adds sodium cyanide to a basic solution of the sample (never add cyanide to acidic samples—poisonous HCN gas forms). Cyanide makes a complex with the iron(II) and masks it but does not tie up calcium. The calcium can then be titrated with standard EDTA solution. Table 12.4 lists several masking agents, along with ions that they effectively mask. All masking agents are sensitive to solution pH, and the lists of masked ions apply only to the specified pH ranges. Most of the information in this table could be predicted from the principles of HSAB theory described in Section 12.4.1.

A demasking agent causes the release of a metal from a masking complex. A

TABLE 12.4 COMMON MASKING AGENTS AND THE IONS THEY WILL MASK[a]

Species	Ag^+	Al^{3+}	Ba^{2+}	Bi^{3+}	Ca^{2+}	Cd^{2+}	Co^{2+}	Cu^{2+}	Fe^{3+}	Hg^{2+}	Mg^{2+}	Mn^{2+}	Ni^{2+}	Pb^{2+}	Sb^{3+}	Sn^{2+}	Zn^{2+}
Inorganic																	
Ammonia	+					+	+	+					+				+
Cyanide	+					+	+	+		+			+				+
Fluoride		+			+				+		+					+	+
Hydroxide	+	+				+		+	+	+	+	+	+	+		+	+
Phosphate				+	+				+					+			
Thiocyanate	+			+				+	+	+			+				
Organic																	
Thiourea				+					+								
2,3-Dimercapto-1-propanol	+			+		+	+	+	+	+		+	+	+	+	+	+
Triethanolamine	+			+				+	+	+	+	+			+	+	+
EDTA	+	+	+	+	+	+	+	+	+	+	+	+	+	+	+	+	+

[a] These masking effects are general guidelines and are based on available K_f data. A plus sign means that the potential exists for adequate masking. A blank space means that generally complex formation is inadequate for masking. This table can be used as a first step in designing titration procedures.

metal ion masked with cyanide is easily demasked by a reaction with formaldehyde in a buffered solution:

$$4CH_2O + Zn(CN)_4^{2-} + 4H_2O = Zn^{2+} + 4H_2C(CN)OH + 4OH^-$$

The product is a much weaker base than cyanide ion and will not tie up zinc.

Metal ions can be demasked from fluoride ion by adding boric acid to form the tetrafluoroborate ion, BF_4^-. Metal complexes with thiourea, $(NH_2)_2C=S$, can be destroyed by boiling with hydrogen peroxide.

A somewhat more subtle method involves adding another metal ion to pull away a masking ligand. For example, if Ni^{2+} has been masked by cyanide, it can be demasked by silver ion:

$$Ni(CN)_4^{2-} + 2Ag^+ = Ni^{2+} + 2Ag(CN)_2^-$$

The released nickel can then be titrated with EDTA for analysis.

12.9 ENDNOTE: WHY EDTA IS SO POPULAR

Complexometric methods of analysis, particularly titrations, would be a far less important part of analytical chemistry today if EDTA had not been prepared and investigated in the 1940s. Before that time the only really important complexing agent was cyanide ion, which reacts with several metal ions, notably silver and gold. The ultimate limitation of cyanide is that it is unidentate and must form complexes in a series of steps. Even though the formation constant for each step may be fairly large ($>10^5$ is common), the fact that there are several complex species competing for the available cyanide ion causes a titration curve to be drawn out, with a poorly defined equivalence point break. For example, cadmium reacts with cyanide in a series of steps,

$$Cd^{2+} + CN^- = CdCN^+ \qquad \log K_1 = 5.5$$
$$CdCN^+ + CN^- = Cd(CN)_2 \qquad \log K_2 = 5.1$$
$$Cd(CN)_2 + CN^- = Cd(CN)_3^- \qquad \log K_3 = 4.6$$
$$Cd(CN)_3^- + CN^- = Cd(CN)_4^{2-} \qquad \log K_4 = 3.6$$

The overall formation constant is quite large ($\log \beta_4 = 18.8$)—larger, in fact, than the formation constant for cadmium-EDTA complex ($\log K = 16.5$). In terms of thermodynamics (i.e., the large β_4), cyanide should be an excellent titrant for cadmium. Figure 12.10 shows that the stepwise equilibria so complicate the situation that the titration curve is unacceptable. The great virtue of EDTA is that it has six base sites and accomplishes all its work in one step.

The calculations used to create Fig. 12.10 involve solutions to third- and fourth-

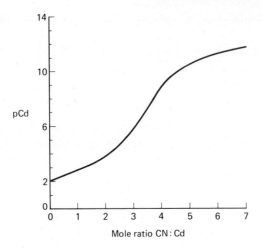

Figure 12.10 Titration of Cd^{2+} with cyanide.

order equations. The method used to solve these equations and the computer program used to create the figure appear in Appendix 6.

REFERENCES

1. G. Schwartzenbach, *Advances in Inorganic Chemistry and Radiochemistry,* 3, 257 (1961).
2. R. D. Shannon and C. T. Prewitt, *Acta Crystallographica,* B25, 925 (1969).
3. R. G. Pearson, *Journal of Chemical Education,* 45, 643 (1968).
4. M. Yasuda, K. Sone, and K. Yamasaka, *Journal of Physical Chemistry,* 60, 1667 (1956).
5. C.-S. Chung, *Inorganic Chemistry,* 18, 1321 (1978).

RECOMMENDED READING

Basolo, F., and R. C. Pearson. *Coordination Chemistry,* Benjamin, New York, 1964. This short monograph is a good introduction to metal ion compounds and their geometry. Although it is rather old, it is still available in libraries and can be read by students who have studied only general chemistry.

Perrin, D. D. *Masking and Demasking of Chemical Reactions,* vol. 33 of the series *Chemical Analysis,* I. M. Kolthoff and P. J. Elving, eds., Interscience, New York, 1970. A thorough treatment of the subject, difficult for beginners.

Pribil, R. *Applied Complexometry,* vol. 5 of the Pergamon Series in Analytical Chemistry, R. Belcher, D. Betteridge, and L. Meites, eds., Pergamon, New York, 1982. An excellent survey and bibliography of the more recent literature.

Ringbom, A. *Complexation in Analytical Chemistry,* vol. 17 of the series *Chemical Analysis,* I. M. Kolthoff and P. J. Elving, eds., Interscience, New York, 1963. A classical treatment of the theory and practice of complexometry, although quite challenging for beginning students.

PROBLEMS

12.1. A solution of EDTA is standardized with high-purity copper. Three successive 25-mL portions of the 0.02697M standard copper solution require 24.23, 24.12, and 24.30 mL of EDTA solution. What is the concentration of the EDTA solution, based on the mean of these results?

12.2. A 50-mL sample of copper requires 31.25 mL of EDTA for titration to the equivalence point. If the EDTA solution is 0.00725M, what is the concentration of copper expressed in milligrams per liter?

12.3. A 100-mL sample of tap water containing both calcium and magnesium is buffered at pH 10 and titrated with 0.01562M EDTA. To reach the Erio T endpoint 11.58 mL is required.
  (a) Both calcium and magnesium react with EDTA at this pH. What is the sum of the calcium and magnesium concentrations in the water sample?
  (b) How would you alter conditions to mask magnesium and determine calcium alone?

12.4. A sample of swimming-pool water is to be analyzed for calcium ion. A 100-mL portion is titrated with 0.01020M EDTA at pH 12.5 using Calcon indicator, 15.44 mL being required to reach the endpoint.
  (a) What is the concentration of calcium ion in milligrams per liter (parts per million)?
  (b) If the swimming pool has a volume of 3.75×10^5 gallons, what weight of calcium oxalate monohydrate might be recovered in a quantitative precipitation of calcium?

12.5. Barium ion forms rather weak complexes with the complexing agent DTPA (see Fig. 12.5) and hydroxide ion:

$$Ba^{2+} + DTPA^{5-} = Ba(DTPA)^{3-} \quad K_f = 10^{8.8}$$
$$Ba^{2+} + OH^- = BaOH^+ \quad K_f = 10^{0.6}$$

  (a) Using pK_a values from Fig. 12.5, calculate the fraction of DTPA^{5-} at pH 10.6.
  (b) Ignoring for the moment the competing reaction of hydroxide with barium ion, calculate the conditional constant for the formation of Ba(DTPA)$^{3-}$ at pH 10.6.
  (c) Repeat (b) taking into account the formation of BaOH$^+$.
  (d) Imagine mixing equal volumes of 0.01F BaCl$_2$ and 0.01F DTPA buffered

at pH 10.6. Calculate the sum of the concentrations of Ba^{2+} and $BaOH^+$ in the resulting solution. Is the sum of the concentrations less than 0.001 times the equilibrium concentration of $Ba(DTPA)^{3-}$? (That is, is the formation quantitative?)

12.6. Salicylic acid ("HSal") forms stable complexes with Cu^{2+} and with protons:

$$Cu^{2+} + Sal^- = CuSal^+ \qquad K_1 = 10^{10.6}$$
$$CuSal^+ + Sal^- = Cu(Sal)_2 \qquad K_2 = 10^{7.9}$$
$$HSal = H^+ + Sal^- \qquad K_a = 10^{-2.9}$$

25 mL of $0.020F$ Cu^{2+} is added to 25 mL of $0.20F$ salicylic acid, and the pH is adjusted to 3.0.
(a) Calculate the fraction of free salicylate ion (Sal^-) at pH 3.0 and the conditional constants for the formation of the copper-salicylate complexes.
(b) Calculate the concentrations of all the copper-containing species at equilibrium. [Hint: assume that almost all of the copper is complexed as $Cu(Sal)_2^-$; write a material balance for Sal species.]

12.7. Copper(II) forms complexes with EDTA ($K_f = 10^{18.8}$) and DCTA ($K_f = 10^{21.3}$). The pK_a values for these ligands are shown in Fig. 12.5. The thermodynamic formation constants suggest that the Cu(DCTA) complex is more stable than the Cu(EDTA) complex. Which is the more stable complex in a solution buffered at pH 5? [Ignore competition from copper(II) hydroxide species.]

12.8. Nitrilotriacetic acid (NTA) forms complexes with cadmium(II):

$$Cd^{2+} + NTA^{3-} = Cd(NTA)^- \qquad K_{f1} = 10^{10.1}$$
$$Cd(NTA)^- + NTA^{3-} = Cd(NTA)_2^{4-} \qquad K_{f2} = 10^{4.4}$$

(a) Account for the fact that K_{f1} is much larger than K_{f2}.
(b) Speculate on the bonding and structure of the two complexes. Remember that water molecules are Lewis bases.

12.9. NTA and IDA are structurally similar ligands (see Fig. 12.5) with similar weak acid dissociation constants. NTA is a tridentate ligand, while IDA is a bidentate ligand in reactions with cadmium(II):

$$Cd^{2+} + IDA^{2-} = Cd(IDA) \qquad K_{f1} = 10^{5.3} \qquad K_{f2} = 10^{4.1}$$
$$Cd^{2+} + NTA^{3-} = Cd(NTA)^- \qquad K_{f1} = 10^{10.1} \qquad K_{f2} = 10^{4.4}$$

(a) Offer an explanation for the difference in first formation constants. The Cd-ligand bond energies (enthalpies) should be similar for the two complexes, yet the formation constants are quite different.
(b) Using pK_a values from Fig. 12.5, calculate the fraction of $Cd(NTA)^-$ in a

solution which is $0.01F$ in Cd^{2+} and $0.1F$ in NTA and buffered at pH 5.0. Is the Cd^{2+} complexed quantitatively by NTA in this solution?

(c) Do the calculations in part (b) using IDA rather than NTA.

12.10. IDA is structurally very similar to half an EDTA molecule (see Fig. 12.5), the difference being the ethylene connecting group in EDTA. What other factor, besides the ethylene group, might be responsible for the difference observed in the following formation constants?

$$Ni^{2+} + 2IDA^{2-} = Ni(IDA)_2^{2-} \quad \beta_2 = 10^{14.6}$$
$$Ni^{2+} + EDTA^{4-} = Ni(EDTA)^{2-} \quad K_f = 10^{18.6}$$

12.11. The reaction of Fe^{3+} with EGTA has a very large equilibrium constant:

$$Fe^{3+} + EGTA^{4-} = Fe(EGTA)^- \quad \log K_f = 20.5$$

(a) Using pK_a values for EGTA from Fig. 12.5, calculate the *minimum* pH for the quantitative complexation of $0.01F\ Fe^{3+}$ with $0.01F$ EGTA, ignoring competition from the formation of $Fe(OH)^{2+}$.

(b) If the first two formation constants for Fe(III) hydroxides are $10^{11.1}$ and $10^{10.7}$, *estimate* the pH for the quantitative complexation of $0.01F\ Fe^{3+}$ with $0.01F$ EGTA.

12.12. One of the few complexometric titrations with a monodentate ligand that produces good analytical results is the classic determination of chloride with mercuric ion,

$$2Cl^- + Hg^{2+} = HgCl_2$$

Diphenylcarbazone forms a violet colored complex with the first excess of mercuric ion.

A sample weighing 0.2632 g is dissolved in water and titrated with $0.05732M$ mercuric nitrate. If 17.86 mL of titrant is required to reach the endpoint, what is the percentage of chloride in the original sample?

12.13. Zinc forms two complexes with ethylenediamine (en):

$$Zn^{2+} + en = Zn(en)^{2+} \quad K_1 = 10^{5.9}$$
$$Zn(en)^{2+} + en = Zn(en)_2^{2+} \quad K_2 = 10^{5.2}$$

A solution is prepared to be just like the equivalence point solution in the titration of Zn^{2+} with en, that is, $0.1F\ Zn^{2+}$ and $0.2F$ en. Assume also that the solution is sufficiently basic that competing Brønsted acid-base reactions are unimportant. Calculate the concentrations of Zn^{2+}, $Zn(en)^{2+}$, and $Zn(en)_2^{2+}$.

From the relative concentrations of these species, can you conclude anything about the suitability of en as a titrant for the determination of zinc at the 0.1F level? (Hint: Assume [en] $\approx 10^{-4}M$ to start.)

12.14. The thorium [Th(IV)] and lanthanum [La(III)] in a solution are to be determined by titration with EDTA. A 1.140-g sample containing both elements and inert material is dissolved in nitric acid, and the pH of the resulting solution is adjusted to 3.0 with ammonia. The thorium is titrated with 0.0532M EDTA; 19.02 mL is required to reach the xylenol orange endpoint. An additional 25.00 mL of 0.0532M EDTA is then added by pipet, and the pH is raised to 5.5 to complex the lanthanum quantitatively. Unreacted EDTA is then back-titrated with 8.23 mL of 0.04875M lead nitrate solution (to the point where the xylenol orange color change reverses). Calculate the percent thorium and percent lanthanum in the sample.

12.15. A 1.343-g sample of an alloy containing only copper, zinc, and tin is dissolved in nitric acid. Tin precipitates as metastannic acid and is removed by filtration. The filtrate solution containing Cu(II) and Zn(II) is carefully collected, transferred to a 250-mL volumetric flask, and diluted to volume with water.

A 25-mL portion of the solution is taken for determination of the sum of the copper and zinc concentrations by titration with EDTA at pH 5. For the titration 32.32 mL of 0.04360M EDTA is required.

A second 25-mL portion is treated with thiourea to mask the Cu(II). Titration of the Zn(II) requires 20.28 mL of 0.04360M EDTA.

Calculate the percentages of Cu, Zn, and Sn in the sample.

EXPERIMENT 12.1: Titrimetric Determination of Calcium (Samples Containing No Magnesium)

Calcium can be determined *indirectly* with the help of magnesium-Erio T indicator complex in solutions buffered at pH 10. Direct titration of calcium would require that the calcium-Erio T complex be sufficiently stable to give a sharp and accurate endpoint:

$$Ca(In)^- + H^+ + EDTA^{4-} = Ca(EDTA)^{2-} + HIn^{2-}$$
$$\text{(red)} \qquad\qquad\qquad\qquad\qquad\qquad \text{(blue)}$$

Unfortunately the calcium-Erio T complex is not very stable. Magnesium ion forms a more stable complex with Erio T than does calcium and can be used to signal EDTA added in excess of the amount of calcium in the sample:

$$\text{reaction:} \quad Ca^{2+} + EDTA^{4-} = Ca(EDTA)^{2-}$$
$$\text{excess EDTA:} \quad EDTA^{4-} + Mg(In)^- + H^+ = Mg(EDTA)^{2-} + HIn^{2-}$$
$$\qquad\qquad\qquad\qquad \text{(red)} \qquad\qquad\qquad\qquad \text{(blue)}$$

In this way calcium is determined indirectly; the endpoint is signaled by a change in complexation of magnesium. Magnesium can be added to the buffer solution to ensure a supply of the ion for the indicator reaction. Since the same amount of buffer is used in both the standardization and unknown titrations, the volume of EDTA used to titrate the magnesium will be compensated. If samples contain their own magnesium, it will also be titrated along with magnesium added and will produce high results.

Titrant Solution

Prepare the $0.02F$ solution of $Na_2EDTA*2H_2O$ as described in the direct titrimetric determination of calcium.

Buffer Solution

Work in the fume hood. Prepare a solution of pH 10 buffer by dissolving 32 g of ammonium chloride in 100 mL of deionized water. Add 285 mL of concentrated ammonia and dilute the resulting solution to 500 mL. Add about 0.1 g of magnesium chloride to the solution and mix thoroughly.

CAUTION: Ammonia fumes are toxic irritants.

Standardization

Prepare 100 mL of a stock solution of primary standard grade calcium carbonate as described in Experiment 12.2 on the direct titration of calcium.

Pipet a 10-mL portion of the standard solution into a clean 250-mL Erlenmeyer flask. Add 25 mL of buffer solution and a spatula tip of Erio T indicator (may be mixed with KCl), and titrate with the EDTA solution to the red-to-blue endpoint. The true endpoint is the first blue which persists for at least 20 seconds.

The Unknown

Follow the directions given in the direct titrimetric method. Add cyanide to complex heavy metals *only after adding pH 10 buffer*. Once again, 0.1 g of KCN will suffice.

Titrate to the red-to-blue endpoint.

Calculate percent Ca in the unknown sample.

> CAUTION: Potassium cyanide is extremely poisonous. Wash your hands thoroughly with soap after working with KCN. Dispose of cyanide solutions only in approved vessels.

EXPERIMENT 12.2: Titrimetric Determination of Calcium (Samples Containing Magnesium)

Calcium may be determined in the presence of magnesium by titration with standard EDTA solution at pH 12.5. A mixed acid-base and metallochromic indicator is used to signal the endpoint. The acid-base indicator indigo carmine turns yellow when the pH is high enough for hydroxide to mask magnesium. The metallochromic indicator calcon is red when complexing calcium (before the endpoint) and blue when the calcium is complexed by EDTA (after the endpoint). The combination of yellow and red, and yellow and blue, produces a distinct endpoint color change of rose-red to green if the indicators are mixed in the proper proportions.

Titrant Solution

Prepare a $0.02F$ solution of EDTA by dissolving the appropriate weight of disodium EDTA ($Na_2EDTA \cdot 2H_2O$, GFW 372.24) in 1 liter of deionized water. A few drops of 50% NaOH will help the salt dissolve quickly. Since EDTA can complex ions in glass, its concentration might change over a period of weeks if stored in a glass bottle. Store EDTA solutions in plastic bottles.

EXPERIMENT 12.2: TITRIMETRIC DETERMINATION OF CALCIUM (SAMPLES CONTAINING MAGNESIUM)

Standardization

Standardize the EDTA solution with primary standard grade calcium carbonate (GFW 100.09). Dry about 2 g of $CaCO_3$ in a weighing bottle at 110°C for at least 1 hour. After cooling in a desiccator, weigh out as accurately as possible 0.8 g of $CaCO_3$ into a clean 150-mL beaker. Add about 10 mL of deionized water to the weighed calcium carbonate, followed by several drops of $6M$ hydrochloric acid. Swirl the solution and add hydrochloric acid until the sample dissolves. Heat the dissolved sample with a Bunsen burner to drive off carbon dioxide, but be careful not to boil or spatter the sample. Let the solution cool before transferring it quantitatively to a 100-mL volumetric flask. Rinse the beaker several times with deionized water, transferring each rinse to the volumetric flask. Dilute the contents of the flask to the mark with deionized water, and mix the solution thoroughly by inverting the flask.

Pipet a 10-mL portion of the calcium stock solution into a clean 250-mL Erlenmeyer flask containing about 20 mL of deionized water. Add two or three drops of indigo carmine indicator (0.5% wt/vol in water); then add $1M$ NaOH dropwise until the solution just turns yellow (pH 12.5). Add two or three drops of calcon indicator (0.5% wt/vol in methanol) and immediately titrate with EDTA to the rose-to-green color change. Titrate rapidly to avoid decomposition of the indicator.

If carbonate ion is present, calcium carbonate may precipitate when NaOH is added. Calcium carbonate will dissolve in excess EDTA, but the slowness of dissolution causes the endpoint to fade. The true endpoint will be a green color which persists for more than about 20 seconds.

The Unknown

Dry your unknown for 2 hours at 110°C. You can assume that your unknown is an impure sample of calcium carbonate and will contain no more than 40% calcium. At your discretion you may weigh out one large sample, dissolve it in acid, boil to remove carbon dioxide, transfer to a 100- or 250-mL volumetric flask, dilute, and take appropriate portions by pipet. Alternatively, you may weigh out individual samples into Erlenmeyer flasks, dissolve them, boil, dilute, and titrate. Information in Chapter 1 about weighing errors and volumetric errors should help you decide which technique to use.

Some unknown samples are powdered limestone and will contain some silica, which is not soluble in hydrochloric acid. Gentle heating of the solution while you add hydrochloric acid will drive off CO_2 and help you decide if there is silica residue. Silica will not interfere and need not be filtered out. It will adsorb some indicator during the titration, however, and may make the endpoint somewhat more difficult to judge (you will see specks of darker color).

The dissolved unknown sample is prepared in much the same way as the standards. However, since the unknown may also contain iron and other metals, it is necessary to use cyanide as a masking agent. Once the $1M$ NaOH has been added and the pH has been adjusted to 12.5 with the help of indigo carmine, add about 0.1 g of potassium cyanide. Swirl the solution to dissolve the KCN, add calcon indicator, and titrate immediately with EDTA solution. Report the percentage of Ca in the unknown sample.

> CAUTION: Potassium cyanide is a deadly poison. Ingestion of 200 mg can be lethal. Cyanide can be absorbed through the skin. When added to acidic solutions, cyanide will produce HCN, which is volatile and extremely toxic. Use extreme caution when working with KCN. Wash your hands frequently and thoroughly. Do not dispose of cyanide solutions in the sink. Your instructor will provide a specially marked bottle for cyanide wastes.

Chapter 13

UV-Visible Absorption Spectrometry and Colorimetric Methods of Analysis

13.1 INTRODUCTION

Absorption spectrometry is one of the most interesting and active areas of analytical chemistry. The ability of many molecules and ions to absorb light in the ultraviolet (UV) and visible regions of the spectrum provides an excellent means to study their chemical reactions and detect their presence in samples. Even in the most basic introduction to analytical methods based on the absorption of light, a tremendous amount of information must be given. This chapter is divided into three major parts. Section 13.2 deals with the properties of light, the spectrum, and the absorption phenomenon. Section 13.3 through 13.5 cover Beer's law and its most important applications in analysis and the characterization of complexes. Section 13.6 is devoted to a description of the instrumentation needed to make absorption measurements. What should emerge from this long chapter is a clear view of why spectrometric measurements are useful and important and how they are made.

The 14th edition of *Standard Methods for the Examination of Water and Wastewater* describes a rapid and simple method for the determination of ferrous iron [Fe(II)] with the ligand 1,10-phenanthroline, described in the last chapter. The complexation reaction forms an orange-red species,

$$Fe^{2+} + 3(Ophen) = Fe(Ophen)_3^{2+}$$

where "Ophen" is

A measure of the intensity of the orange-red color, called the *absorbance,* is found by experiment to be proportional to the concentration of the complex over a certain concentration range. The key to the analysis is that the absorbance is also proportional to the original concentration of ferrous iron when an excess of *o*-phenanthroline is present in solution. Here is an excerpt from the standard method (Ref. *1*):

> Procedure a. Preparation of calibration curves:
> 1. Range 0 to 100 μg (micrograms) Fe/100 mL final solution—Pipet 2.0, 4.0, 6.0, 8.0, and 10.0 mL standard iron solution (1.00 μg Fe/mL) into 100 mL volumetric flasks. Add 1.0 mL of hydroxylamine hydrochloride solution and 1 mL sodium acetate solution (20% solution by weight) to each flask. Dilute each to about 75 mL with distilled water, add 10 mL phenanthroline solution (1 mg/mL, saturated), dilute to volume, mix thoroughly, and let stand for 10 minutes. Measure the absorbance of each solution in a 5-cm cell at 508 nm against a reference blank prepared by treating distilled water with the specified amounts of all reagents except the standard iron solution. . . . From the data obtained, construct a calibration curve for absorbance against milligrams of iron.
> 2. Range 50 to 500 μg Fe/100 mL final solution—Follow the procedure specified in the preceding paragraph, but use 10.0 . . . 50.0 mL of standard iron solution and measure the absorbance values in 1 cm cells.

A table is provided which helps the analyst choose the cell path length appropriate for particular Fe concentration ranges. For example, it is claimed that 50 to 200 μg Fe in a final volume of 50 mL can be determined with a cell with a 1-cm light path; 10 to 40 μg Fe in a final volume of 100 mL can be determined in a cell with a 10-cm light path.

Several experiments must have been done by the chemists who wrote the "cookbook" instructions just quoted. What kind of experiment showed that color intensity could best be measured at a wavelength of 508 nanometers? Could measurements be made at other wavelengths? What experiment showed that a 5-cm path length was needed for the 2.0 μg Fe/100 mL solution? Is there some more concise way of expressing the relationship that seems to exist between concentration and required path length, other than presenting a table? What assumptions underlie the use of the table? We will need a brief review of the properties of light and the electromagnetic spectrum before answering these and other questions.

13.2 ELECTROMAGNETIC RADIATION—LIGHT

13.2.1 Electromagnetic Radiation as Waves

Electromagnetic radiation (EMR) is known in several familiar forms. The most familiar is the radiation we can see, visible light. Radio waves, microwaves, heat, and x-rays

13.2 ELECTROMAGNETIC RADIATION—LIGHT

are other forms. The motion of EMR through space involves oscillating electrical and magnetic fields and is described in terms of waves—wavelength, frequency, velocity, and amplitude. Figure 13.1 represents the electrical and magnetic field vectors of EMR, which oscillate at right angles to each other and to the direction in which the wave travels. The vectors are represented by arrows whose lengths are proportional to the force of the fields, and called *amplitude*. The oscillations are regular and periodic and are described mathematically by a sine or cosine function. The rate at which the vectors change size is described by a *frequency* (v), that is, a number of oscillations per second (units: sec^{-1} or hertz). The distance between amplitude peaks along the line of travel is called the *wavelength* (λ). Traditionally, wavelengths in the ultraviolet and visible regions of the EMR spectrum (see below) have been expressed in angstrom units (1 Å = 1 × 10^{-10} m) and in millimicrons (1 mμ = 1 × 10^{-9} m). In the SI system of units the proper unit is the nanometer (1 nm = 1 × 10^{-9} m). The velocity of an EMR wave is given by the product of its wavelength and its frequency,

$$v = \lambda * v \quad \text{(i.e., distance} * \text{time}^{-1})$$

The velocity at which an EMR wave travels depends on the medium through which it moves. In a vacuum all EMR travels ("propagates") at the same immense speed, the speed of light (c), 2.997925 × 10^8 m/sec. This is the greatest speed at which EMR can travel. In air and denser media EMR travels at lower speeds because the electrical field vectors of the EMR interact with those of the media and slow the wave. Since a source of EMR fixes the frequency of its emission, as the velocity of a light wave decreases in a medium its wavelength must also decrease.

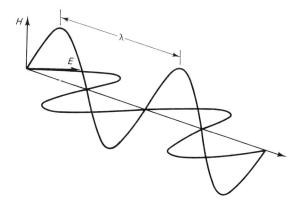

Figure 13.1 Representation of a propagating wave of electromagnetic radiation. H and E represent the magnetic and electric field vectors, respectively. A wavelength is shown as λ. The frequency of the wave is the number of complete vector oscillations per second. The wave number is the number of oscillations per centimeter on the line of propagation.

EXAMPLE 13.1 Electromagnetic Units and Concepts

Compare the frequency of an ultraviolet light wave of wavelength 250 nm with that of a blue-green light wave of wavelength 500 nm.

Solution

The equation in an earlier paragraph tells us that frequency is given by velocity divided by wavelength. Assuming that both waves are traveling at the same speed, when the wavelength doubles, the frequency is halved. If the speed of light is assumed to be c (2.998×10^8 m/sec), the frequency of 250-nm light is 8.339×10^{16} sec^{-1}:

$$\nu = c/\lambda = (2.998 \times 10^8 \text{ m/sec})/(250 \times 10^{-9} \text{ m})$$
$$= 8.339 \times 10^{16} \text{ sec}^{-1}$$

The frequency of 500-nm light is half of this value, 4.17×10^{16} sec^{-1}.

EXAMPLE 13.2 Velocity of Light

The speed of yellow light (590 nm) in glass is about two-thirds of its speed in a vacuum. What is its wavelength in glass?

Solution

Remember that the light source fixes the frequency of the emission. Therefore it must be true that

$$\nu = \text{a constant} = c/\lambda = v/\lambda$$

where v is the speed in glass. If light in glass travels at $\frac{2}{3}$ times its speed in a vacuum, its wavelength in glass must be $\frac{2}{3}$ times its wavelength in a vacuum, or $590 * \frac{2}{3} = 393$ nm.

(NOTE: light of 393 nm is barely visible violet. When we look at a yellow lamp through a piece of glass it is yellow, not violet. Can you explain?)

The ratio of the speed of light in a vacuum to that in some medium is called the *index of refraction* of the medium (n, a dimensionless quantity),

$$n = c/v$$

In many materials refractive index will change with wavelength. The change in refractive index with wavelength ($dn/d\lambda$) is called *dispersion*. Media which have this property are called "dispersive media" and can be very useful in making, for example, prisms for instruments which must measure intensities of light at different wavelengths.

13.2.2 Electromagnetic Radiation as Particles

While the wave picture of light is very useful in describing light's interactions with electrical fields, it is not adequate to describe its behavior when it is absorbed or emitted by atoms or molecules. In these situations light behaves like energetic particles, called *photons*. The fact that in some experiments light behaves as waves, while in others it behaves as particles, makes the nature of light hard to understand and fascinating. Fortunately, light never behaves as both waves and particles in the same experiment.

The energy of a single photon of light is proportional to its frequency,

$$E = h*\nu$$

The proportionality constant is called *Planck's constant* (6.6262×10^{-27} erg sec or 6.6262×10^{-34} joule sec) after Max Planck, the physicist who conceived of the bundles of energy called "quanta." Since the product of frequency and wavelength is constant in a particular medium, it follows that the energy of a photon is inversely proportional to its wavelength (when it acts like a wave),

$$E = h*v/\lambda \quad \text{(or } h*c/\lambda \text{ in vacuum)}$$

Some spectroscopists prefer to use *wave number* (ω, omega) rather than wavelength to describe light, and to write the equation for energy as

$$E = h*v*\omega$$

Wave number has the dimension of inverse length (m^{-1} or more often cm^{-1}). Notice that energy and wave number are directly proportional.

EXAMPLE 13.3 Energies and Wave Numbers

Calculate the energy and wave number of a photon which exhibits a wavelength of 500 nm in a vacuum.

Solution

The energy of the photon is given by $E = hc/\lambda$,

$$E = 6.6262 \times 10^{-34} \text{ J sec} * (2.998 \times 10^8 \text{ m/sec})/500 \times 10^{-9} \text{ m}$$
$$= 3.97 \times 10^{-19} \text{ J (per photon)}$$

An Avogadro's number of photons has an energy of 2.39×10^5 J or 239 kJ. This is the equivalent of 57 kcal (per mole) and is equal to the energy of many chemical reactions. Visible light has sufficient energy to promote valence electron transfers and cause chemical reactions to occur in many systems.

A wavelength of 500 nm corresponds to a wave number of 2.00×10^{-3} nm^{-1}, or 2.00×10^3 cm^{-1}. You may remember from general chemistry that the Rydberg constant is about five times larger than this (109,800 cm^{-1}) and corresponds to the ionization enthalpy for the hydrogen atom. You may also recall from organic chemistry that the carbonyl stretching frequency occurs in the infrared spectrum at about 1750 cm^{-1}. This is considerably less energy than 20,000 cm^{-1}.

13.2.3 The Electromagnetic Spectrum

The full range of energies and wavelengths of EMR is called the *electromagnetic spectrum* (EMS) and is shown in Fig. 13.2. The limits of each region are defined arbitrarily, and exact boundaries have no physical significance.

The part of the EMS we are most interested in is a tiny section containing part of the ultraviolet (UV), the visible (Vis), and part of the near infrared (NIR). These regions cover a wavelength range of 160 to 1000 nm and are shown in the expanded part of Fig. 13.2. Colors seen by the human eye are indicated both above and below the spectrum. Colors above the line are the colors of light in wavelength regions. Colors below the line are *complements* of the colors above the line. For example, light

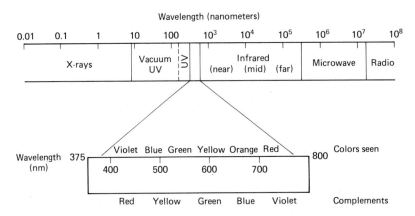

Figure 13.2 The electromagnetic spectrum. If a sample appears to be orange, it absorbs light in the blue-green region of the visible spectrum. The fact that the eye is more sensitive to yellow-green light than to blue or red light complicates this simple complementary relationship. It can be used, however, for a first guess about absorption characteristics.

13.2 ELECTROMAGNETIC RADIATION—LIGHT

of 500-nm wavelength appears blue-green to human eyes. An object which absorbs 500-nm light but transmits (or reflects) all other wavelengths will appear red-orange to human eyes. Permanganate solutions are purple, meaning that they transmit purple light. They absorb green and blue-green wavelengths. Copper(II) solutions transmit blue light and thus absorb yellow light. The Fe(II)–*o*-phenanthroline complex described at the beginning of the chapter absorbs the greatest amount of light at 508 nm (blue-green) and transmits red-orange light. Remember that a source which emits all visible wavelengths will appear white. A sample which transmits all wavelengths from a white source will appear to be colorless. A sample which absorbs all visible wavelengths will appear black.

13.2.4 Emission and Absorption of Electromagnetic Radiation; Spectra

You will recall from earlier chemistry courses that atoms and molecules have a number of discrete, quantized energy levels. In atoms these energy states correspond to various configurations of electrons in atomic orbitals. In molecules the situation is much more complicated. In addition to electrons in atomic orbitals, there are electrons in orbitals which belong to the molecule as a whole, the so-called molecular orbitals. The energy of an atomic orbital is set primarily by the number of protons in the nucleus and the number of electrons in neighboring atomic orbitals. The energy of a molecular orbital is set by the number and nature of the atoms in the molecule *and* the distances which separate them. Since molecular vibrations and rotations change the distances between atoms in molecules, we should expect molecular and atomic spectra to be quite different.

At room temperature there is so little thermal agitation that all atoms and most molecules exist in their lowest electronic energy state, called the *ground state*. It is possible to excite atoms and molecules to higher electronic energy states by several processes, including exposing them to sufficiently energetic light, high temperatures, or high-voltage discharges (high-energy collisions do the exciting) or bombarding them with high-energy particles such as electrons. In this chapter we are interested in how molecules are excited when they absorb light:

$$\text{molecule} + \text{photon} \rightarrow \text{molecule}^\ddagger$$
$$\text{(ground state)} \qquad\qquad \text{(excited state)}$$

Excited states are stable for only short times, typically 10^{-6} to 10^{-9} second. The excited atom or molecule will then relax to the ground state or a lower excited state with the release (emission) of energy in the form of light or heat (collisions), or both.

$$\text{molecule}^\ddagger \rightarrow \text{molecule} + \text{photon (or kinetic energy)}$$
$$\text{(excited state)} \qquad \text{(ground state)}$$

Molecules can also be excited to higher vibrational and rotational states, but with considerably less energy than is required in electronic excitation. While it may take 30,000 cm^{-1} (light at a wavelength of 333 nm, in the UV) to excite an electronic transition, it may take only 3000 cm^{-1} to excite a vibration (3.33 micrometers, about the wavelength of a C—H stretch in the infrared) and 300 cm^{-1} to excite a rotation (wavelength 33.3 μm).

The absorption and emission of light by individual atoms and molecules in low-pressure gases appears in very sharp spectral lines. Atoms in the gas phase at low pressure act largely independently of each other, and their energy states are quite well defined. An example is shown in the *emission spectrum* of Fig. 13.3. If we increase the pressure and temperature of the gas sufficiently, collisions distort the energy levels and broaden spectral lines to the point where emission and absorption appear to be virtually continuous over a large range of wavelengths. Both sharp line and continuous emission light sources are important in chemical spectroscopy.

When molecules are in the liquid state or dissolved in solvent, their absorption and emission bands broaden. Examine the *absorbance spectra* of benzene in Fig. 13.4. In the gas phase benzene molecules absorb light in a series of very intense narrow bands near 250 nm (in the UV). In ethanol solvent most of the sharp spectral bands are lost as a result of solvent interactions and collisions in the liquid phase. The benzene absorbance lines centered at 250 nm arise from an electronic excitation between benzene molecular orbitals and have an energy that corresponds to about 40,000

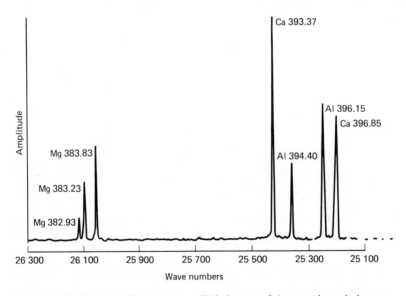

Figure 13.3 Emission line spectrum. This is part of the atomic emission spectrum from a magnesium-aluminum-calcium light source for an atomic absorption spectrometer. The wavelength range is 380.2 to 398.4 nm, expressed in wave numbers. [From W. K. Yuen and G. Horlick, *Analytical Chemistry*, 49(9), 1446 (1977); reproduced with permission of the American Chemical Society.]

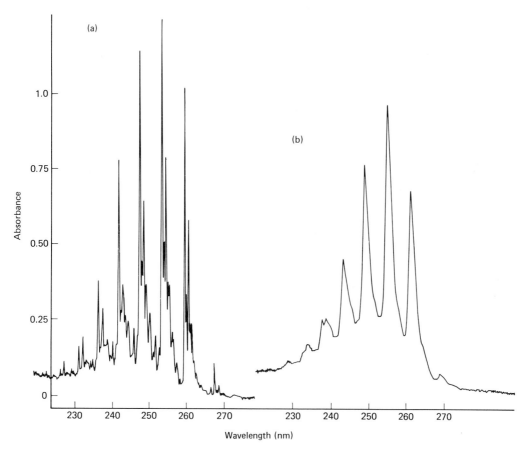

Figure 13.4 Absorbance spectra of benzene. (a) Benzene vapor in air, 1 cm path length, 0.2 nm bandpass. (b) $4.5 \times 10^{-4} M$ benzene in ethanol, 1 cm path length, 0.2 nm bandpass. In the gas phase vibrations of benzene molecules split the electronic transition absorptions. Solvent interactions destroy most of the vibrational detail.

cm^{-1}. The sharp spikes are vibrational bands superimposed on the main electronic transition. Rotational bands, in turn, split the vibrational bands, although these are so close together that the instrument cannot resolve them. As an exercise, see if you can locate lines in the spectrum which differ by a few thousand or a few hundred wave numbers from the main electronic band and which correspond to vibrational excitations.

13.3 ABSORPTION SPECTROSCOPY

"Absorption spectroscopy" is a general term for experiments performed to study the way molecules and atoms absorb light. In this section we will investigate the basic relationships of absorption spectroscopy. Methods based on the emission of light are important in their own right but are subjects for more advanced courses.

An absorbance spectrum of a $8.9 \times 10^{-5} M$ solution of the complex tris(*o*-phenanthroline) iron(II) is shown in Fig. 13.5. An absorbance spectrum is a plot of *absorbance* vs. wavelength and is obtained by changing the wavelength of light passed through a sample and measuring how much is absorbed.

When a light beam of radiant power P_0 passes through a completely uniform sample of absorbing species with concentration C in a sample cell with a light path of length b, the power of the light is lowered to a level we will call P. The *transmittance* of the sample is defined as the ratio of the radiant power levels,

$$T = P/P_0$$

and is often expressed as a percentage, $\%T$. Radiant power has the units of energy per unit area (joules per square centimeter for example), and transmittance is dimensionless.

The *absorbance* of the sample is defined as the negative $\log_{10}$ of the transmittance,

$$A = -\log_{10}(T) = \log(P_0/P)$$

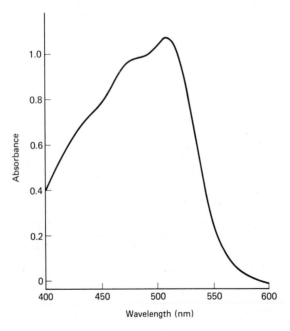

Figure 13.5 Absorbance spectrum of $8.9 \times 10^{-5} M$ tris(1,10-phenanthroline)Fe(II).

13.3 ABSORPTION SPECTROSCOPY

As the radiant power of the light passed by a sample decreases, *A increases* logarithmically. Absorbance has no units, although many people give it its own dimension, the "absorbance unit."

13.3.1 Lambert's Law

It has been found that without exception the absorbance of light by a uniform absorbing material increases directly with the thickness of the absorbing material, that is,

$$A = kb$$

where b is the path length of light through the material (its breadth) and k is a proportionality constant.

EXAMPLE 13.4 Lambert's Law
A 0.50-mm thickness of green glass has an absorbance of 0.025 at a wavelength of 710 nm. What are the absorbance and transmittance of a 0.50-cm thickness of the same glass?

Solution

The transmittance of the 0.50-mm glass is $10^{-A} = 0.94$. Lambert's law says that since the 0.50-cm thickness of glass is 10 times the thickness of the 0.50-mm piece of glass, its absorbance should be $10(0.025) = 0.25$. This corresponds to a transmittance of $T = 10^{-A} = 0.56$. It is important to note that transmittance *does not* decrease linearly with increasing path length.

13.3.2 Beer's Law

Within certain limitations, to be discussed soon, there *may* be a direct proportionality between absorbance and concentration, expressed as

$$A = abC$$

In this equation C is the concentration, b is again the path length, and a is a proportionality constant called the *absorptivity*. Since A is a dimensionless quantity, the units of absorptivity must be inverse concentration, inverse length. For example, when C is expressed in mg/dL (mg/100 cm³) and b is in cm, a will have the units dL/mg cm. When C is expressed in moles per liter and b is in cm, a is given the special symbol ϵ (epsilon) and the special name *molar absorptivity*. The absorptivity and, of course, molar absorptivity are *intensive* quantities and do not depend on the amount of material present (see Chapter 3). Molar absorptivities for many compounds are tabulated in the literature. Ideally, these values are independent of the instruments with which they were obtained, but it pays to be skeptical. As you will see, there are several sources of instrument error that can affect results.

EXAMPLE 13.5 Molar Absorptivity; Absorptivity

A $4.12 \times 10^{-5} M$ solution of the complex $Fe(Ophen)_3^{2+}$ has a measured absorbance of 0.48 at $\lambda = 508$ nm in a sample cell with path length 1.00 cm. Calculate the molar absorptivity, then the absorptivity in units of milligrams of Fe per liter. (A 0.04 mM solution of the complex is also 0.04 mM in iron, and the gram atomic weight of Fe is 55.85).

Solution

Using Beer's law, we calculate the molar absorptivity:

$$\epsilon = A/bC = 0.48/(1 \text{ cm})(4.12 \times 10^{-5} \text{ mole/liter})$$
$$= 1.17 \times 10^4 \text{ liter/mole cm}$$

To express the concentration as milligrams of Fe per liter, we multiply 4.12×10^{-5} by 55.85 g Fe/mole, and by 1000 mg/g, to obtain 2.30 mg/liter. The absorptivity is then

$$0.48/(1 \text{ cm})(2.30 \text{ mg/liter}) = 0.209 \text{ liter/mg cm}$$

Molar absorptivity can serve as a rough measure of a method's *detection limit,* that is, the smallest signal that can be read above the background noise of the instrument. With inexpensive equipment, for example, it is possible to measure absorbances as small as 0.01 to 0.02 reliably. Compounds with large molar absorptivities can be detected at very low concentrations. On the basis of theoretical arguments which are beyond the scope of this text, the maximum molar absorptivity we could expect to find would be about 150,000. The minimum concentration observable for a solution of such a compound in a 1-cm cell would be about $7 \times 10^{-8} M$. Compounds which absorb in the visible region with molar absorptivities greater than about 2×10^4 are rather rare. Quite a few have molar absorptivities in the range 1000 to 10,000, and

13.3 ABSORPTION SPECTROSCOPY

spectrophotometric studies in the concentration range 10^{-6} to $10^{-3} M$ are common in the analytical literature.

EXAMPLE 13.6 Detection Limit

Assuming that the molar absorptivity of the $Fe(Ophen)_3^{2+}$ complex is 1.17×10^4 liter/mole cm at 508 nm, calculate the detection limit for the analysis of Fe if the smallest reliable photometric measurement is 0.01 absorbance unit. Compare the detection limits with a 1-cm and a 5-cm cell.

Solution

In a 1-cm cell:

$$C = A/\epsilon b = 0.01/1.17 \times 10^4(1)$$
$$= 8.5 \times 10^{-7} M$$

This concentration corresponds to $8.5 \times 10^{-7}(55.85) = 4.8 \times 10^{-5}$ g/liter. This can also be expressed as 48 parts per billion (wt/vol). In a 5-cm cell the detection limit will be one-fifth this value, or slightly less than 1 μg/100 ml (or 10 ppb).

EXAMPLE 13.7 Concentration Ranges

If the absorbance range is to be 0.01 to 1.0 in a determination, what concentration ranges for Fe can be managed in 1- and 5-cm cells?

Solution

In a 1-cm cell the minimum concentration is 4.8 μg/100 mL (see last example). The maximum concentration will be 480 μg/100 mL. In a 5-cm cell the minimum concentration will be 0.96 μg/100 mL and the maximum will be 96 μg/100 mL. Go back to the first section of this chapter and read the directions for the analysis for iron. Do the ranges make sense?

The procedure quoted at the beginning of the chapter called for the preparation of a *calibration curve,* a plot of absorbance vs. concentration. An example is a student-generated calibration curve for the analysis of iron, shown in Fig. 13.6. Also called a *working curve,* the plot is actually a plot of Beer's law for the $Fe(Ophen)_3^{2+}$ complex. The slope of the line is observed to be constant, and is equal to ϵb. Since the cell had a 1-cm path length, the slope is the molar absorptivity, 11,760 liter/mole cm.† The absorbance of a sample containing an unknown amount of iron can be located on the working curve, and a line can be dropped to the concentration axis to find its concentration. In this example, the unknown had an absorbance of 0.58 and a corresponding concentration of 270 μg/100 mL.

Alternatively, the equation for the line that best fits the data can be obtained by linear regression, and the equation can be solved for concentration, given the experimental absorbance. Regression analysis of the data in Fig. 13.6 gives the equation

$$A = 1.176 \times 10^4 C + 7.619 \times 10^{-3}$$

where 7.619×10^{-3} is the absorbance at $0.0M$ concentration. Beer's law predicts that this value should be zero. Remember that *all* the data points are used to calculate the line which has this intercept and that each point has some experimental error (random, we hope). The regression calculation produces a variance of 5.74×10^{-4}, which we can interpret as a measure of the "goodness of fit." A variance of 5×10^{-4} is acceptable for beginners. Experienced workers should be able to obtain variances one-tenth this size. Remember that the best straight line minimizes the deviations of the experimental absorbances and values on the line ($A_{\text{fit}} - A_{\text{meas}}$). The sum of the squares of the deviations divided by $(n - 1)$ is the variance (see Chapter 2). A small variance means that the points all lie close to the fitted line.

13.4 APPLICATIONS OF BEER'S LAW

13.4.1 The Analysis of Mixtures

Beer's law may be applied to the determination of species in a mixture. In the simplest case the absorbance of two species at some wavelength can be expressed as a sum,

$$A_{\text{total}} = A_1 + A_2$$

where A_1 and A_2 are the absorbances of species 1 and 2, respectively. If both species obey Beer's law and they do not react with each other to form a third or fourth species, then the following expression is valid:

$$A_{\text{total}} = \epsilon_1 b C_1 + \epsilon_2 b C_2$$

---

† This result is obtained from five data points and is more reliable than a single value taken from a single absorbance reading. The line is straight only if Beer's law is strictly obeyed. Some experimental scatter should be expected, especially with a great many analysis steps and inexpensive equipment.

13.4 APPLICATIONS OF BEER'S LAW

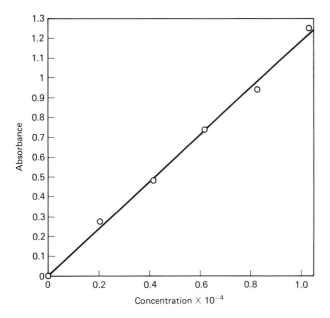

Figure 13.6 Beer's law plot for the determination of Fe(II) with 1,10-phenanthroline. $Y = 7.619 \times 10^{-3} + (1.176)X$; variance = 5.737×10^{-4}. (Data of T. Weber.)

In principle, if we measure the absorbance (A_{total}) at two separate wavelengths and we know ϵ_1 and ϵ_2 at both wavelengths, then we have enough information to perform an analysis of the mixture.

EXAMPLE 13.8 Analysis of a Two-Component Mixture

A stainless steel razor blade is to be analyzed for chromium and manganese. After dissolving the 0.5770-g blade in acid and using persulfate to oxidize chromium to dichromate ($Cr_2O_7^{2-}$) and manganese to permanganate (MnO_4^-), the solution is transferred to a 1-liter volumetric flask and diluted to the mark with $3M$ sulfuric acid. Portions of the 1-liter solution are transferred to sample cells. Absorbances are measured at 440 and 545 nm, wavelengths established by previous experiments to be correct for the analysis of permanganate and dichromate. The absorbance at 440 nm (mostly due to dichromate) is found to be 0.348. The absorbance at 545 nm (mostly due to permanganate) is found to be 0.504. Calculate %Cr and %Mn in the razor blade.

Molar absorptivities of the species at the analytical wavelengths are tabulated,

Wavelength	A_{total}	ϵ ($Cr_2O_7^{2-}$)	ϵ (MnO_4^-)
440	0.348	369	95
545	0.504	11	2350

[Molar absorptivities are from Lingane and Collat (Ref. 2). The analyst would normally determine these from Beer's law plots for separate solutions of pure dichromate and pure permanganate at the two wavelengths.]

Solution

Begin by writing an equation for the absorbance at each wavelength:

$$A_{440} = \epsilon_{Cr}C_{Cr} + \epsilon_{Mn}C_{Mn}$$
$$= 369C_{Cr} + 95C_{Mn}$$
$$A_{545} = \epsilon_{Cr}C_{Cr} + \epsilon_{Mn}C_{Mn}$$
$$= 11C_{Cr} + 2350C_{Mn}$$

This pair of equations has two unknowns, C_{Cr} and C_{Mn},

$$0.348 = 369C_{Cr} + 95C_{Mn}$$
$$0.504 = 11C_{Cr} + 2350C_{Mn}$$

and can be solved simultaneously to give $C_{Cr} = 8.88 \times 10^{-4}M$ and $C_{Mn} = 2.10 \times 10^{-4}M$. Notice that C_{Cr} at this point is the concentration of dichromate. The concentration of chromium is *twice* this value (there are two Cr in $Cr_2O_7^{2-}$), or $1.78 \times 10^{-3}M$.

Solving for %Cr and %Mn:

$$\%Cr = 2*C_{Cr}*1 \text{ liter}*51.996 \text{ g/mole}*100/0.5770 \text{ g sample}$$
$$= 16.0\%$$
$$\%Mn = C_{Mn}*1 \text{ liter}*54.938 \text{ g/mole}*100/0.5770 \text{ g sample}$$
$$= 2.00\%$$

(Note: The analysis of a real sample is more complicated than is suggested by this example. Other species present in the sample may absorb at the wavelengths used in the analysis, and correction factors must be applied. See the source of the molar absorptivities cited above for complications in steel analysis.)

The determination of the weak-acid dissociation constant of the indicator bromcresol green is presented as laboratory Experiment 13.3 at the end of this chapter. It is based on the analysis of mixtures of the protonated (yellow) and deprotonated (blue) forms of the indicator at different pH values in the vicinity of pK_a. Molar absorptivities are calculated at two wavelengths for two indicator forms in very acidic and very basic solutions. These experimental values are then used to calculate concentrations in buffered solutions of the indicator. Activity corrections are applied, and a value for the dissociation constant is calculated.

13.4.2 Photometric Titrations

The proportionality that can exist between absorbance and concentration can allow us to monitor a titration that produces or consumes a species which absorbs light. An excellent example is the determination of copper ion by titration with EDTA. Figure 13.7 shows some experimental data for the determination of about 10 micromoles of Cu(II) in 10 mL of buffer solution held in a cell with a 5-cm path length and 14-mL volume. Each point in the plot marks the addition of 50- or 100-μL (0.05 or 0.1 mL) portions of 0.00445M EDTA to the buffered Cu(II) solution. After thorough mixing, absorbances are read at 320 nm (in the UV). At 320 nm the Cu(EDTA) complex has a molar absorptivity of about 350 liter/mole cm. Uncomplexed Cu(II) and EDTA have negligible molar absorptivities at this wavelength. Notice in the titration curve that the absorbance increases as the titration proceeds, then levels out after the equivalence point. The equivalence point can be found by extrapolating the rising line and plateau lines back to an intersection point. Detailed instructions for this titration are given at the end of the chapter in Experiment 13.4.

In some systems it may be necessary to monitor the concentration of the titrant, rather than the reaction product, during the titration. Figure 13.8 shows the curve for the titration of 5.00 mL of a 2.29 mM solution of Fe(II) with a 6.11 mM solution of potassium permanganate, monitored at a wavelength of 545 nm, where permanganate absorbs (ϵ = 2350 liter/mole cm). The chemical reaction is

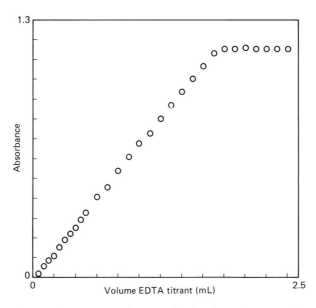

Figure 13.7 Spectrophotometric titration of Cu(II) with EDTA. Experimental absorbance data have been corrected for dilution. See text for experimental details. (Data of T. Weber and C. May.)

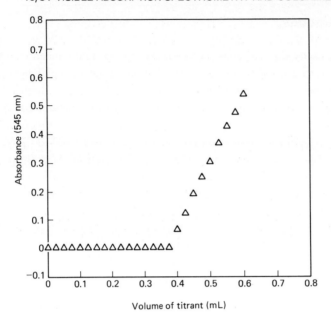

Figure 13.8 Spectrophotometric titration of Fe(II). Spectrophotometric titration curve for the reaction of 5.0 mL of 2.29 mM Fe(II) with 6.11 mM potassium permanganate. (Data of C. Priewe.)

$$5Fe(II) + MnO_4^- + 8H^+ = 5Fe(III) + Mn(II) + 4H_2O$$

Once the equivalence point is reached the permanganate concentration rises, and the absorbance at the permanganate peak wavelength increases. Some people call an endpoint like this a "dead start" endpoint. The equivalence point volume is determined by extrapolating the rising line back to the horizontal axis.

13.4.3 Determining Ligand:Metal Combining Ratios

UV-visible spectrometry can be a very useful tool for studying the composition of metal-ligand complexes and determining their formation constants. We can learn the ratio of ligand to metal in a complex ML_n by finding the absorbance of the complex in solutions in which there are different ratios of ligand and metal concentrations. Three important techniques merit some discussion: the mole ratio, continuous variations, and slope ratio methods.

Mole Ratio Method (Yoe-Jones Method) In this method the analyst prepares a series of solutions in which the formal concentration of one of the species (usually the metal ion) is held constant, while that of the other species is varied. A plot of absorbance of

13.4 APPLICATIONS OF BEER'S LAW

the complex against the ratio of moles in the set of solutions is then prepared. If the formation constant for the complex is sufficiently large, the plot will consist of two lines which can be extrapolated to intersect at the mole ratio in the complex. Figure 13.9 shows a mole ratio plot for the Fe(II)–1,10-phenanthroline complex. There can be little doubt as to where the lines intersect and as to the composition of the complex. The large value of β_3 for the complex (about 10^{21}) causes there to be very little curvature near the intersection point. If β_3 were smaller, there would be more curvature, and a decision as to the composition of the complex would be more difficult.

Method of Continuous Variations In this method a series of solutions containing both metal ion and ligand are prepared. Throughout the series the concentrations of both metal ion and ligand are varied. For example, a series of solutions could be prepared in which ratios of metal ion and ligand concentrations are 1:9, 2:8, 3:7, ..., 9:1. These ratios cover the full range of common metal-ligand combining ratios. The experimentalist then measures the absorbance of each solution at the wavelength where the complex absorbs, and plots absorbance versus the concentration ratio of metal ion to ligand. A plot for a hypothetical 2:1 complex with overall formation constant 10^{10} is shown in Fig. 13.10. A maximum absorbance occurs at a mole ratio corresponding to the combining ratio in the complex. In the figure the M/L ratio is 3.3:6.7, or 1:2. The formation constant for the complex can be calculated from the difference between the absorbance at the intersection ("infinite formation constant" case) and the experimental absorbance of the solution of the stoichiometric ratio mixture. We will discuss the calculation later in section 13.5.3 on deviations from Beer's law.

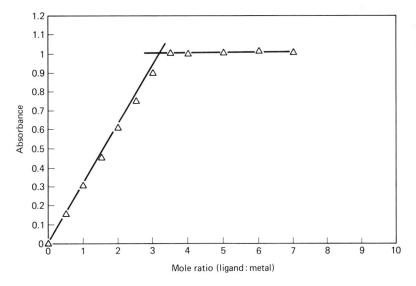

Figure 13.9 Mole ratio plot for Fe(II)-phenanthroline complex. Absorbance of the complex was measured at 508 nm. (Data of C. Priewe, 1985.)

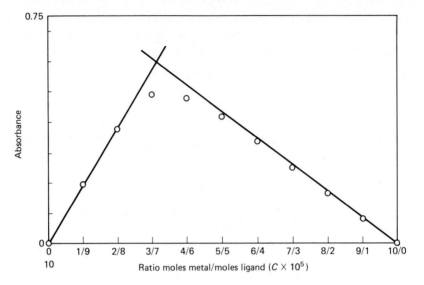

Figure 13.10 Continuous variations plot for ML_2 complex. Simulated data for a continuous variations study of a 2:1 complex with $K_f = 10^{10}$, molar absorptivity 2.00×10^4 at the analytical wavelength.

An experimental difficulty encountered with both the mole ratio and continuous variations methods is that metal ion complexes with small formation constants produce vague results. Incomplete reactions cause great curvature in these plots; extrapolated straight lines, as in Fig. 13.10, might then as easily be drawn to intersect at a ratio of 2.5:7.5 as 3.3:6.7. The slope ratio method may be better in such cases.

Slope Ratio Method This procedure is quite useful for determining combining ratios for weak complexes. For the method to be useful, K_f for the complex must be large enough that the complex is less than 0.1% dissociated in the presence of an excess of either metal ion or ligand. It is also necessary that Beer's law be followed in excess metal ion or ligand (often high ionic strength solutions) and that only one complex be formed.†

Consider the general reaction,

$$m\text{M} + n\text{L} = \text{M}_m\text{L}_n$$

When ligand L is present in large excess the concentration of complex is $(1/m)$ times the initial concentration of metal ion, assuming negligible dissociation; that is,

---

† Actually all three methods require that only one complex form. In the case of Fe-Ophen complexes this criterion is met. Many systems cannot satisfy this requirement, and studies using these methods are of limited value.

$$[M_m L_n] = (1/m)C_M$$

If Beer's law is obeyed at the wavelength where the complex forms, then

$$A_m = \epsilon b[M_m L_n] = \epsilon b(1/m)C_M$$

and a plot of absorbance (called A_m here) versus C_M will be linear, with slope $\epsilon b/m$.

When metal ion M is present in large excess the concentration of complex is $(1/n)$ times the initial concentration of ligand, again assuming negligible dissociation of the complex. A similar relationship exists between absorbance and initial concentration of the ligand, C_L,

$$A_l = \epsilon b[M_m L_n] = \epsilon b(1/n)C_L$$

A plot of absorbance (A_l here) versus C_L will be linear with slope $\epsilon b/n$.

The slopes of the two plots (i.e., A_m/C_M and A_l/C_L) have a ratio which is the combining ratio of ligand to metal ion in the complex:

$$\frac{A_m/C_M}{A_l/C_L} = \frac{\epsilon b/m}{\epsilon b/n} = n/m$$

13.5 LIMITATIONS TO THE USE OF BEER'S LAW

Here is some advice worth remembering: *prove* that Beer's law holds for a system, never simply assume that it does. In contrast to Lambert's law, to which no exceptions have ever been observed, there are systems for which Beer's law does not strictly apply. Deviations from Beer's law arise from either physical or chemical effects. In this section we will discuss some of those effects.

13.5.1 Physical Limitations

Beer's law strictly applies to ideal solutions, that is, dilute solutions. Most solutions behave ideally at concentrations at or below $10^{-3}M$. At higher concentrations absorbing species may be influenced by charge distributions in neighboring solute species, and changes in molar absorptivities and peak wavelengths may be seen. Ionic strength may also affect molar absorptivity if the electronic transitions being stimulated by absorption of light are sensitive to electrostatic interactions with ions surrounding the molecules. These effects are treated in advanced courses.

The refractive index of a solution is known to affect molar absorptivity. A correction can be applied using the equation

$$\epsilon' = \epsilon * n/(n^2 + 2)^2$$

where ϵ' is the corrected molar absorptivity, ϵ is the observed molar absorptivity, and n is the refractive index. For example, the refractive index of pure water is 1.33, while that of $2M$ HCl is 1.35. This difference changes the correction factor from 0.0936 (water) to 0.0924 ($2M$ HCl), a decrease of about 1%. The effect is seldom significant at concentrations below $0.01M$, but it should be considered in more concentrated solutions. When very weakly absorbing species must be studied (for example, in infrared spectroscopy), concentrated solutions are needed and this effect can limit accuracy and precision.

13.5.2 Chemical Effects

Deviations from Beer's law are found when absorbing species react chemically with each other or with the solvent. Dimerizations and dissociations provide excellent examples of this kind of behavior. The deviation is usually more apparent than real and occurs because absorbance is plotted against an incorrectly evaluated concentration. Figure 13.11 shows hypothetical visible spectra of a weak acid, HA, and its conjugate base, A^-. The weak acid dissociation constant is assumed to be 10^{-4}. At 500 nm (blue-green absorbed, red-yellow transmitted) the undissociated weak acid absorbs with molar absorptivity 1050 liter/mole cm. At 700 nm (red-yellow absorbed, blue-green transmitted) the anion has a molar absorptivity of 830. At 600 nm the weak acid and anion have the same molar absorptivity, 480 liter/mole cm.

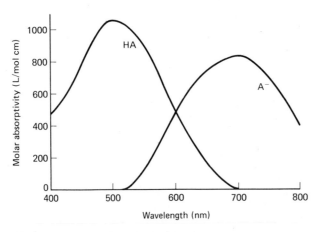

Figure 13.11 Absorbance spectra of hypothetical weak acid and conjugate base. Notice that molar absorptivity, not absorbance, is plotted. The spectra are overlaid as if recorded separately and put together only for demonstration purposes.

13.5 LIMITATIONS TO THE USE OF BEER'S LAW

Figure 13.12 shows Beer's law plots that could be obtained from solutions of the weak acid in pure water at three fixed wavelengths, 500, 600, and 700 nm, in a cell of 1 cm path length. Notice that at 500 nm the Beer's law plot curves upward; the molar absorptivity of HA seems to increase as its concentration increases. At 700 nm the plot bends to the right; the molar absorptivity of A^- seems to decrease with concentration. The plot for absorbances measured at 600 nm stays perfectly straight; Beer's law seems to hold. The cause of the curvatures is dissociation of HA. It is a property of all weak electrolytes that the fraction dissociated increases as the formal concentration decreases. Thus, for a dissociation constant of 10^{-4}, when the formal concentration of HA is $1 \times 10^{-3} M$, [HA] is really $7.3 \times 10^{-4} M$ (27% dissociated). When the formal concentration of HA is $5 \times 10^{-4} M$, [HA] is really $3.2 \times 10^{-4} M$ (36% dissociated). In a study of the absorbance of solutions of NaA in pure water, the same phenomenon *lowers* the concentration of A^- relative to the formal concentration as the formal concentration is increased. For practice, do a calculation and verify this statement.

Notice that the Beer's law plot at 600 nm is quite straight. At this wavelength the molar absorptivities of HA and A^- are the same, and whatever HA dissociates to form A^- will absorb just as strongly. The dissociation process still goes on, but its effects are hidden. This kind of spectral point is called an *isosbestic point* (isosbestic means having the same absorbance). Such points, when they exist, are useful for determining total (or analytical) concentrations.

Under what conditions will a Beer's law plot for a weak acid be a straight line? If we perform Beer's law dilutions in solutions buffered at a particular pH, the fraction dissociated will be fixed (see Chapter 5) and the Beer's law line will be straight. If we

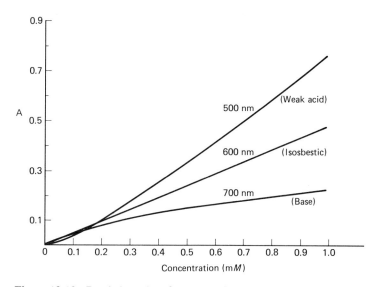

Figure 13.12 Beer's law plots for weak acid indicator ($pK_a = 4.0$).

wish to determine accurately the molar absorptivity of the weak acid, we must do the Beer's law studies in solutions of pH 1.5 to 2 units lower than pK_a ("zero" percent dissociated). Think about the conditions needed to determine accurately the molar absorptivity of the weak base anion.

13.5.3 A Diversion: Taking Advantage of a Limitation

What appears at first to be a limitation of Beer's law turns out to be a valuable tool for determining a dissociation constant, as shown in the following example.

EXAMPLE 13.9 Determining K_a

A series of solutions of a colored indicator is prepared in $0.1F$ HCl, and absorbance measurements show that Beer's law is obeyed over a concentration range of 1×10^{-5} to $5 \times 10^{-4} M$ at the wavelength at which the acid form of the indicator absorbs. The $1 \times 10^{-4} F$ solution in the series has an absorbance of 0.37, measured in a 1-cm cell. A second $1 \times 10^{-4} F$ solution of the indicator is prepared in pure water. The absorbance at the peak wavelength for the acid form is 0.23. Calculate the fraction dissociated in pure water and the pK_a of the indicator.

Solution

Solving the Beer's law expression for ϵ, using the data from the $1 \times 10^{-4} F$ solution of acid, we find

$$\epsilon = A/bC = 0.37/(1)(1 \times 10^{-4}) = 3.7 \times 10^3 \text{ liter/mole cm}$$

In the pure water solution of indicator, we calculate the concentration of acid form to be

$$C = A/b\epsilon = 0.23/(1)(3.7 \times 10^3) = 6.2 \times 10^{-5} M$$

The fraction dissociated is given by

$$\alpha = (1 \times 10^{-4} - 6.2 \times 10^{-5})/1 \times 10^{-4} = 0.38$$

The dissociation constant can be calculated if we recognize that the concentrations of hydronium ion and base form of the indicator are equal (using a charge balance, ignoring hydroxide):

13.5 LIMITATIONS TO THE USE OF BEER'S LAW

$$K_a = [H^+][A^-]/[HA] = (3.8 \times 10^{-5})^2/(6.2 \times 10^{-5}) = 2.3 \times 10^{-5}$$

or $pK_a = 4.64$.

Alternatively, we can let C be the total concentration of indicator species. The concentrations of H^+ and A^- can then be expressed as fractions of C,

$$[H^+] = [A^-] = C*\alpha \quad \text{and} \quad [HA] = C*(1 - \alpha)$$

Therefore,

$$K_a = C^2\alpha^2/C(1 - \alpha) = C\alpha^2/(1 - \alpha) = 1 \times 10^{-4}(0.38)^2/(0.62)$$

and $K_a = 2.3 \times 10^{-5}$.

13.5.4 Deviations Due to Instrumentation

Beer's law applies only when the light striking a sample is of "one wavelength," or *monochromatic*. As you will see when we discuss the design of spectrometers, "one wavelength" is an annoyingly vague requirement. All instruments pass light in a range of wavelengths centered on a nominal value. If we set the instrument dial to 500 nm and it is correctly calibrated, the instrument will transmit light of 500 nm at maximum power. At lower power, light of 501 and 499 nm will also be passed. At lower power still, we might find light of 502 and 498 nm. The range of wavelengths passed is called the *bandwidth*. The range of wavelengths passed at half (or greater) peak power is called the *effective bandwidth*. The size of the bandwidth depends on the ability of the instrument to spread out light (its dispersion) and the width of the entrance and exit ports, called the slits. Slit widths must be large enough to let enough light through to the detector so that it can respond. A simple instrument with low dispersion and an insensitive detector will pass very wide range of wavelengths to a sample, perhaps 10 nm on either side of the nominal wavelength. Sophisticated instruments with high dispersion and very sensitive detectors may allow only 0.01 nm to pass on either side of the nominal wavelength.

A wide range of wavelengths can cause a loss of spectral detail and deviations from Beer's law if there are large changes in absorbance at wavelengths in the range. Figure 13.13 shows two spectra obtained from benzene vapor in the UV, one with an effective bandwidth of 0.1 nm, the other with an effective bandwidth of 1 nm. Sharp spectral details evident in the 0.1-nm spectrum are almost completely lost in the 1-nm spectrum. The detector averages the light power passed within the bandwidth, and narrow absorbance lines can simply be missed when the bandwidth is large.

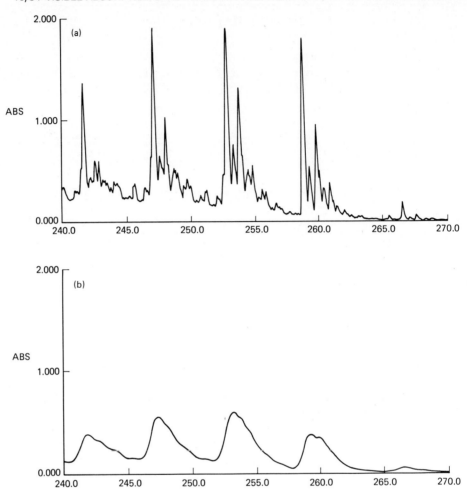

Figure 13.13 Effect of bandpass on the resolution of sharp peaks (benzene vapor absorbance spectra). (a) 0.10 nm bandwidth, benzene vapor in air, 1 cm path length, scan rate 4 nm/min; (b) 1.0 nm bandwidth, same sample as (a), scan rate 4 nm/min.

Eight absorbance peaks for a hypothetical compound assumed to obey Beer's law over a range of concentrations are drawn in Fig. 13.14. Two identical bandwidths are drawn, centered at two wavelengths, one on the peak and the other on the shoulder of the peak. Simulated Beer's law plots obtained at both wavelengths are drawn next to the spectra. The plot with absorbance measured at the peak (marked P) obeys Beer's law quite well, as does the plot from the shoulder (marked S) measured with a very narrow bandwidth. When wider bandwidths are used, Beer's law plots from shoulder absorbance data curve dramatically, as suggested by the inset sketch. The principle is this: when there is a large change in absorbance within the bandwidth, deviation from

13.5 LIMITATIONS TO THE USE OF BEER'S LAW

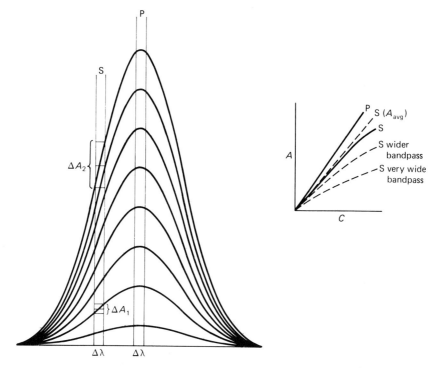

Figure 13.14 Deviation from Beer's law resulting from measurement of absorbance on peak shoulder. Effect of bandwidth, $\Delta\lambda$, on measured absorbances at peak wavelength (P) and at wavelength on shoulder (S). A set of eight simulated spectral peaks are overlaid to help illustrate the effect. In the Beer's law plot sketched in the inset, P marks the line obtained at wavelength P with bandwidth $\Delta\lambda$. $S(A_{avg})$ marks the line one could obtain with a very narrow bandwidth. The other lines marked S might be obtained with larger and larger bandwidths; the larger the bandwidth, the more negative the deviation from Beer's law.

Beer's law behavior can be expected. Notice in the band on the shoulder that the absorbances at the extremes of the bandwidth become more and more different as the height of the peak (and concentration of species) increases. The detector responds to light power transmitted over the bandwidth. Since absorbance decreases logarithmically as power increases, the absorbance calculated from the power over the whole bandwidth is smaller than the absorbance observed at the center of the band. The greater the difference in absorbance at the edges of the band, the greater will be the departure from the actual absorbance at the band center. If one must measure absorbances on the shoulder of a peak, it is therefore essential to use a very narrow bandwidth to minimize the differences in absorbance. The fixed wide bandwidths (20 nm or more) of inexpensive instruments makes them almost useless for studies of spectra with narrow absorbance bands.

13.6 INSTRUMENTATION

In this section we will consider the design of instruments which are used to perform experiments like those described earlier in this chapter. The treatment will deal only with instruments used to study the absorption of UV-visible light. Instruments used to study the phenomena of emission, fluorescence, phosphorescence, and scattering are also employed in important analytical techniques, but are best left to a course in instrumental methods of analysis.

13.6.1 Instrument Components

"Spectrophotometers" are instruments used to study the absorption of light. They are made up of at least six components:

1. A stable *source* of electromagnetic radiation.
2. A device to select a restricted range of wavelengths to pass on to a sample. We will call this a *monochromator*.
3. A *sample container*, which is itself transparent to EMR from the source.
4. A *detector* which converts an EMR signal to an electrical signal and may amplify it.
5. A *signal processor* which receives the detector signal and may amplify and/or perform mathematical operations on it. It may also convert a continuous wave (analog) signal to a digital signal.
6. An *output device* which gives an interpretable signal to the chemist. There may be a visual or hard copy record of the experiment.

Figure 13.15 shows a basic block diagram for the six components. The arrangement of blocks in a straight line is indeed meant to suggest that absorption spectrophotometers make in-line measurements and can be built on a single optical bench. All instruments made to measure EMR use the same basic components, although their arrangements may differ. For example, in emission spectroscopy the sample serves as a source, and the arrangement is source–monochromator–detector–processor. In atomic absorption spectroscopy the monochromator receives light *after* it passes through the sample. The source sends very narrow "discrete" wavelengths to the sample. In fluorescence spectroscopy there are two monochromators, one before and one after the sample. The sample emits light, and the emission is measured at right angles to

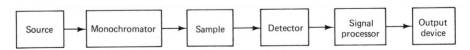

Figure 13.15 Block diagram of a UV/visible spectrometer.

13.6 INSTRUMENTATION

the incident beam against a black background. Again, the basic components are all there, but arranged as required by the particular experiment.

Let us look at the components individually and then describe some simple and inexpensive instruments used in the UV-visible region of the spectrum.

Sources Two basic types of sources are used in the UV-visible region, *continuous* sources and *discrete* sources. Continuous sources emit light at all wavelengths over a wide range of wavelengths, while discrete sources emit only certain very narrow lines. There are two important continuous sources in the UV-visible, the hydrogen (or deuterium) gas discharge lamp and the tungsten filament lamp:

H or D Lamp Both high- and low-voltage lamps are available. Both types produce continuous EMR in the range of 180 to 380 nm. High-voltage lamps require about 2000 V to sustain a gas discharge between aluminum electrodes, and the intense heat released usually requires special water cooling for the lamp. In a low-voltage lamp a heated metal oxide filament provides a stream of electrons to a metal anode (positive charge) when only about 50 V is applied. The electrons strike and excite H_2 or D_2 molecules, which dissociate and emit light. Since ordinary glass absorbs UV light, H and D lamps have quartz windows, which pass light with wavelengths as short as 180 nm. In the visible region the continuous radiation of these sources has superimposed on it high-power discrete lines, making them less useful.

Tungsten Filament Lamp The most commonly used source for studies in the visible and near-infrared regions (350 to 2500 nm) is a lamp with a tungsten metal filament. In most instruments sufficient voltage is applied to run the lamp at temperatures of about 2900 K. Higher intensities can be obtained by raising the temperature of the filament. Unfortunately, at higher temperatures tungsten becomes more volatile and the life of the filament is shortened. In "quartz-halogen" lamps a tungsten filament is sealed in a quartz bulb containing some iodine vapor. At high temperatures volatilized tungsten will react with iodine on the hot quartz surface to form WI_2. This material diffuses back to the filament, where tungsten is redeposited. This chemical cycling prolongs the life of tungsten filaments at high operating temperatures.

Regardless of the design of a source, its output must be both intense and stable. Even though modern detectors are quite sensitive, remember that only a small portion of the source's output is passed to the sample. This is especially true when very narrow bandpasses are needed for high-resolution work. Since the power of a source varies exponentially with applied voltage, it is important to use carefully regulated power supplies to ensure light signal stability.

Monochromators (Wavelength Selectors) In many UV-visible spectral studies only a very narrow range of wavelengths can be passed from the source to the sample. A monochromator is a device that selects wavelengths to send to the sample. A beam of light from a source passes into a monochromator through a narrow aperture called an entrance slit. In the monochromator the beam is collimated ("linearized") by lenses or a mirror and then passed on to a filter or a dispersing device such as a grating or a prism. A filter will absorb all but a desired set of wavelengths, while a dispersing device will cause all but a narrow range of wavelengths to deviate from the optical path. A focusing lens or mirror then focuses the beam of light (actually an image of the entrance slit) on the plane of an exit slit, where a portion will pass on to the sample. The simplest monochromators rely on filters; they are inexpensive, lightweight, and portable and can be designed for specific analyses in the field. Grating and prism instruments are more complex, more delicate, more selective, and far more expensive than filter in-

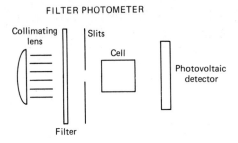

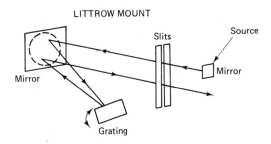

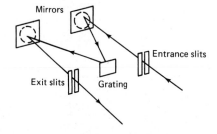

Figure 13.16 Basic monochromators.

struments. They can also be made to scan spectra automatically and are preferred in methods development and in very careful analytical work.

Figure 13.16 shows a filter photometer and two very popular monochromator designs, the Littrow and Czerny-Turner designs. Light beams composed of only one discrete wavelength are drawn to show how these devices absorb or disperse light. The measure of how well a dispersing device does its job is *linear dispersion, $d\lambda/dx$,* at the exit slit. A high-quality monochromator might have a linear dispersion of 1 nm per millimeter of exit slit width. If the exit slit were opened to 0.1 mm, a bandwidth of about 0.1 nm would pass on to the sample.

Let us examine the properties of filters, gratings, and prisms more closely.

Filters *Absorption filters* absorb certain parts of the spectrum and are made of colored glass or dyed gelatin sandwiched between two glass plates. Absorption filters have very large bandwidths, generally greater than 40 nm. *Cutoff filters* are absorption filters which have sharply defined wavelength boundaries. By combining two cutoff filters it is possible to isolate a narrow region of the spectrum, as shown in Fig. 13.17.

Interference (Fabry-Perot) filters rely on optical interference to isolate a narrow band of wavelengths. An interference filter (IF) is made by sandwiching a very thin layer of optically transparent material (calcium fluoride, for example) between two

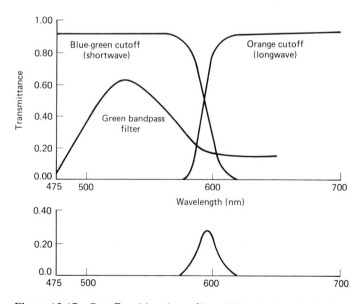

Figure 13.17 Cutoff and bandpass filters. Note that vertical axis is transmittance, *not absorbance.* The lower transmittance spectrum would be obtained if the blue-green cutoff and orange cutoff were used together. The resulting bandwidth is about 25 nm, centered at 595 nm. The green bandpass filter is only one piece of filtering material and has a much larger bandwidth than the combined cutoff filters.

semitransparent metallic films (usually silver), coated on the inside surfaces of glass plates. The thickness of the inner layer determines the wavelengths of light passed with maximum power, according to the equation

$$m\lambda = 2dn$$

where m is an integer (1, 2, 3, etc.), d is the thickness of the layer, and n is the refractive index of the inner layer (at λ). When a beam of parallel ("collimated") light waves strikes the IF perpendicularly, a portion of the beam passes through the semitransparent silver layer, while the rest is reflected back toward the source. The portion passed strikes the second silver layer, and part of it is reflected back into the inner layer. If the reflected portion is of the wavelength (or an integral number of wavelengths) satisfying the equation, it is reflected off the inner surface of the first silver layer *in phase* with incoming light of the same wavelength. This particular wavelength is thus selectively reinforced, while light of other wavelengths is decreased in power by destructive interference. Figure 13.18 shows the Fabry-Perot effect with nonperpendicular light, simply to show the reflection process. Perpendicular incidence is needed for interference to occur. Interference filters pass bandwidths of about 10 nm, a smaller range than provided by absorption filters. Since an IF reflects so much light back at the source, it will transmit only about 10% of the incident light.

Gratings Two types of diffraction gratings are made, transmission and reflection gratings. While the theory behind the operation of a transmission grating is easier to

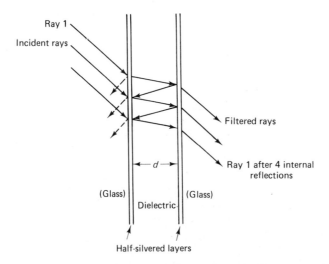

Figure 13.18 An interference filter. Reflections within the interference filter are shown by arrows. Angular incidence is indicated only to help show how reflection can be reinforced by incident rays. In an instrument the incident rays are perpendicular to the filter.

understand, reflection gratings are much more commonly used, and we will discuss them. A master grating is made by carefully machining or etching closely spaced grooves on a highly polished rectangular plate, then coating it with a highly reflective layer of metal. The distance between the grooves is about the same size as the wavelength of light the grating is to diffract. For a UV-visible monochromator (180 to 800 nm), a grating with 1000 to 3000 lines per millimeter is needed. Gratings used in instruments are replicas of master gratings, made by a process somewhat like that used to take a plaster cast of a footprint. Figure 13.19 shows an edge-on view of an *echellette* grating, one with grooves cut at an angle (ϕ), called the *blaze angle*, relative to the base of the grating. The line drawn perpendicular to the base of the grating is called the "grating normal." Two parallel light rays of the same wavelength are drawn separated from each other by the distance d, the width of a groove. They strike the grating at an angle α to the grating normal and are diffracted at an angle β. Because the rays strike at an angle, ray 2 would have to travel farther than ray 1 if the rays were to stay together throughout diffraction. If the extra travel of ray 2 corresponds to one wavelength (or an integral number, m, of wavelengths), then rays 1 and 2 can be reflected *in phase* and will reinforce each other. The distance is given by $(x - y)$ in Fig. 13.19. If you examine the figure, you will see that $x = d \sin \alpha$ and that $y = d \sin \beta$. The condition for reinforcement is

$$m\lambda = d(\sin \alpha - \sin \beta)$$

When the incoming and outgoing rays are on opposite sides of the grating normal (as shown), a negative sign is used in the equation. When they are on the same side of the grating normal the extra travel distance is $(x + y)$ and a positive sign is used in

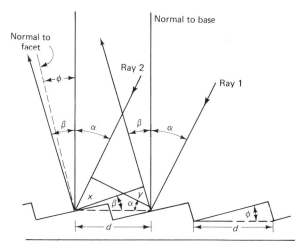

Figure 13.19 Diffraction at a reflection grating. The view is from the edge of the reflection grating. Light rays strike at angle α and are diffracted at angle β.

the equation. The number m in the equation is an integer called the *order* number. Light of $1 \times \lambda$, $\lambda/2$, $\lambda/3$, $\lambda/4$, ..., will pass at the same pair of angles α and β. As order number increases, power decreases quite rapidly.

When incoming light strikes the grating along the grating normal (i.e., $\alpha = 0$), light of a particular wavelength called the *blaze wavelength* is diffracted and reflected at the blaze angle. Reflection and diffraction combine to produce the maximum light power at this wavelength. The equation relating blaze wavelength to blaze angle is

$$m\lambda = d[\sin(0) + \sin(2\beta)]$$

or

$$m\lambda = 2d \sin \phi$$

for small blaze angles.

EXAMPLE 13.10 Diffraction Grating

A reflection grating is ruled with 1000 lines/mm and a blaze angle of 15°. Calculate the wavelengths passed at the blaze angle.

Solution

Solve the blaze angle equation for λ, $\lambda = 2d \sin \phi/m$. A ruling of 1000 lines/mm gives a d value of $(1/1000) = 0.001$ mm, or 1000 nm. In the first order $\lambda = 2(1000 \text{ nm}) \sin(15) = 517.6$ nm. In the second order light of $517.6/2 = 258.8$ nm is passed, but at much lower intensity than first-order light. Third-order light of 172.5 nm also passes at 15°, but at very low intensity.

In the example just given it was shown that at least three wavelengths of light are reinforced and emerge at the same angle from a grating. The only way to isolate one wavelength is to filter out shorter multiple-order wavelengths before the beam strikes the grating. Filters that do this are called "order sorters."

EXAMPLE 13.11 Diffraction Grating

Light strikes the reflection grating used in the last example at an angle of 30°. Calculate the wavelengths passed at 5, 10, and 15°, on the opposite side of the grating normal.

Solution

The equation used is $m\lambda = d(\sin \alpha - \sin \beta)$. The negative sign is used because the incident and diffracted rays are on opposite sides of the grating normal. In the first order, $\lambda = 412.8$ nm at 5°, 326.3 nm at 10°, and 241.2 nm at 15°.

13.6 INSTRUMENTATION

Notice in the last example that on changing the angle α from 5 to 10° we observe a change in wavelength of 86.5 nm. Changing α from 10 to 15° changes the wavelength by 85.1 nm, almost exactly the same amount. To a first approximation, a grating disperses light equally well over a very wide range of wavelengths, resulting in a linear wavelength scale and simpler instrumentation.

Prisms Instruments with prism monochromators are no longer very common. Although prisms were at one time less expensive than gratings of comparable quality, they are now more expensive. The major difference between prisms and gratings involves linearity of dispersion, that is, the wavelength scale. While gratings have fairly constant dispersion over the UV-visible wavelength scale (see last example), quartz prisms show much greater dispersion in the UV than in the visible, a less desirable situation. Prisms, on the other hand, do not suffer higher-order wavelength overlap and therefore transmit very little unwanted (stray) light.

One of the most widely used prism monochromator designs is the Littrow configuration, shown in Fig. 13.16. It was used in the first large-scale production spectrometer, the Beckman DU, produced in the 1940s and still found in some laboratories. The Littrow mount is very economical of both space and optics. One mirror element can be used to both focus and collimate light. Light passes twice through the Littrow prism, located where a grating is shown in the figure.

Sample Containers Sample containers, called cells or cuvettes, must be transparent to light in the spectral region being studied. In the UV-visible-NIR region quartz and fused silica cells are useful over the range 200 to 3300 nm, while optical glass is restricted to a range of 350 to 2200 nm. Colorless disposable plastic cells are available for use in the visible region.

Cells are made in a variety of shapes and path lengths. The most common are rectangular cells with 1 cm path length. Cells with 2, 5, and 10 cm path length are glass cylinders with end windows made of glass, quartz, or fused silica. Cells with path lengths as small as 0.05 mm are commercially available. Some inexpensive instruments use cylindrical test tube cells. Reflections from curved surfaces and variations in path length make these cells unsuitable for careful work.

In some studies it may be necessary to use a pair of cells whose transmission characteristics are nearly identical. "Matched" pairs of cells can be purchased which are made from the same batch of glass, quartz, etc. and whose path lengths are the same within ±0.01 mm. Cells must be kept spotlessly clean. Even if carefully matched cells are used, fingerprints, grease, and film deposits on the cell faces can ruin otherwise carefully planned experiments.

Detectors Many detectors are available for work in the UV-visible region. Table 13.1 is a list of common detectors, their best wavelength ranges, and their signal outputs.

TABLE 13.1 PHOTODETECTORS FOR THE UV-VIS-NEAR IR

Detector	Sensor element	Range (nm)	Output
Photovoltaic cell	Semiconductor on metal	400–800	Current or voltage
Phototube	Group I, V oxide	200–1000	Current
Photomultiplier tube	Group I, V oxide	160–1000	Current
Photodiode array	Silicon photodiode	200–1100	Current

Photovoltaic Cells Also called barrier layer cells, photovoltaic cells are simple and rugged devices that generate their own current and voltage when illuminated and require no external power supply. A schematic drawing of a photovoltaic cell is shown in Fig. 13.20. A plate of conductive metal such as iron is used as one electrode. A thin film of a semiconducting metal such as selenium is deposited on its surface. A very thin layer of silver is then deposited on top of the selenium surface to act as a second electrode, able to collect electrons. When light strikes the Se-Ag interface, electrons are transferred from the Se to the Ag. The resulting positive charge centers in the Se can be filled with electrons from the iron layer only with great difficulty, and a voltage develops across the Fe-Se boundary. Migration of the centers in Se causes a small but measurable current of perhaps 10 to 100 microamperes in most applications.

The spectral response of a selenium photovoltaic cell is close to that of the human eye, with maximum sensitivity at about 550 nm but with some sensitivity in the UV and near IR. These detectors are lightweight and inexpensive and are very popular for portable equipment. Unfortunately, they show response fatigue (signal decreases over time), which may limit the accuracy of instruments in which they are used.

Phototubes A phototube is made from a light-sensitive cathode (negative electrode) shaped as a half-cylinder, partially surrounding a wire anode (positive electrode),

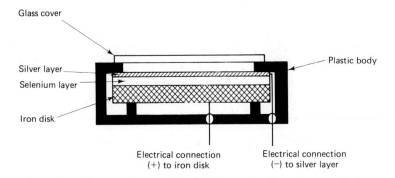

Figure 13.20 Sketch of a selenium photovoltaic cell.

both sealed in an evacuated glass envelope in the manner of an old-fashioned radio tube (see Fig. 13.21). An external power supply holds the voltage of the cathode at about −90 volt versus the anode. When light strikes the cathode, loosely held electrons are ejected from the cathode and drawn to the anode. This flow of electrons constitutes a current, which can subsequently be amplified by a signal processor. The current flowing in the device is proportional to the radiant power striking the photocathode.

The spectral response of a phototube is determined by the metals used to make the cathode and whether the tube envelope is glass or quartz. At least a dozen cathode formulations are popular, among them Sb-Cs, with a peak sensitivity at 350 nm and useful response as far in the red as 700 nm. Ga-As cathodes extend the response to about 950 nm and provide very uniform sensitivity down into the UV. Cs-Te formulations operate only in the UV, from 120 to 350 nm.

Phototubes produce small currents even when in the dark. These "dark currents" arise principally from thermal excitation of electrons from the cathode at room temperature. Dark currents can be reduced by cooling phototubes in liquid nitrogen.

Photomultiplier Tubes A photomultiplier tube (PMT) can be thought of as a phototube with a built-in amplifier system. Like a phototube, a PMT has a photocathode. Instead of having one anode, however, a PMT has a series of anodes called dynodes, which are coated with photoemissive material. The dynodes are held at increasingly positive voltages with respect to the cathode; dynode 1 might be held at +100 V, dynode 2 at +200 V, dynode 3 at +300 V, and so forth. A schematic view of a PMT with nine dynodes is shown in Fig. 13.22. Light enters the PMT and strikes the photocathode. Ejected electrons are drawn to the first dynode, where they eject several more electrons. These electrons are drawn to the second dynode, where collisions eject even more electrons, and so on until by the ninth dynode an avalanche of perhaps 10^6 or 10^7 electrons reaches the anode. The initial one-photon impact is therefore magnified a million or more times by the dynodes.

Photodiodes A photodiode (PD) is a semiconducting device which will allow current to flow only when it is struck by light. Current passed by PDs is proportional to radiant power over a wide range of intensities. Furthermore, PDs are much more

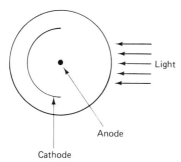

Figure 13.21 Schematic drawing of a vacuum phototube.

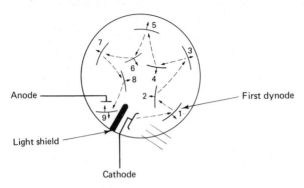

Figure 13.22 Schematic drawing of a photomultiplier tube.

sensitive than ordinary phototubes (but not PMTs). A real virtue of PDs is that they can be microminiaturized, and thousands can be created on a single computer chip in a large regular array of rows and columns. The conducting status of each diode can be checked by a computer every few milliseconds. When a dispersed light signal from a monochromator is shone on a large diode array, light intensities at nanometer (or shorter) intervals can be read from the diode array. An entire visible spectrum can be measured in a matter of a few seconds, with a resolution comparable to that of PMT instruments requiring scan times of several minutes. Photodiode array spectrometers equipped with microcomputers are still more expensive than PMT spectrometers used for common chemical applications, but costs should become comparable in a few years.

Signal Processors A signal processor is an electronic device which takes the voltage or current signal from a detector and converts it to a form which is intelligible to the experimentalist. In many "double-beam" instruments (see below) the detector signal is modulated, containing signals from a sample cell alternating with signals from a cell containing only solvent. The important data are the differences between the alternating signals, and these can be obtained electronically. Simple electronic circuitry can be used to convert light power ratios (transmittance) to absorbances and to filter out noise from weak signals.

If digital signals can be obtained from a signal processor a variety of interesting and important data manipulations can be performed quite easily. A computer can be used to add spectra (to compare with a mixture of absorbing species) or subtract spectra (compensating for a nonflat baseline or an impurity in a solvent), integrate spectral peaks, or take their derivatives. Derivative spectroscopy can be particularly effective in the analysis of complex spectra, where small spectral peaks are superimposed on larger peaks (as shoulders, for example). Figure 13.23 shows simple normal and derivative spectra, simulated by a computer program which adds normal distribution

13.6 INSTRUMENTATION

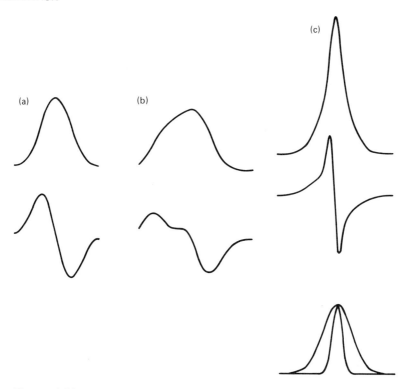

Figure 13.23 Normal and first-derivative spectra. (a) A single spectral peak simulated by a normal distribution curve. The derivative of the normal curve is plotted below, and shows an inflection at the peak wavelength. (b) A close pair of peaks. The peak on the left is of lower intensity than the main peak. The derivative reveals where the shoulder peak has maximum intensity. (c) The top peak appears to be a single sharp spectral peak. It was calculated by summing the amplitudes of two normal peaks, one narrower than the other, shown at the bottom. The derivative plot shows that the sharp peak is not Gaussian. The derivative is quite sensitive to the change in slope, which would probably be missed by the experimentalist.

curves. The first derivative of a spectral trace is really a plot of the slope of the spectrum as a function of wavelength. Moving up an absorbance band, the slope increases until it reaches a maximum, then decreases again as the peak is approached. At the peak the slope is zero. As the absorbance falls, the slope increases but with opposite sign, until it passes a maximum and again falls to zero. Figure 13.23a shows the derivative plot of a single peak. Figure 13.23b shows the derivative plot of a peak with a shoulder. A normal scan is not sufficiently selective to allow the shoulder peak to be characterized, but the peak is much more clearly evident in the derivative plot. Figure 13.23c shows that a derivative plot can reveal the presence of a second peak when a normal scan gives no hint of its presence. A general discussion of derivative spectroscopy can be found in Refs. 5 and 6.

Readout Devices   Inexpensive devices use meters to display absorbance or transmittance, and permanent records must be made manually. On more expensive instruments a spectrum can be recorded on paper with a strip chart recorder. With computerized instruments the signal is displayed on a video monitor and, after data manipulation and labeling, is sent to a printer or plotter.

13.6.2 Spectrophotometric Instruments

Now that we have discussed the individual components of UV-visible instruments, it is time to assemble some complete instruments. We will begin with simple colorimeters and filter photometers, then describe the single-beam Bausch & Lomb Spectronic 20 and some more sophisticated double-beam designs.

Visual Colorimeters   These are the simplest and least expensive devices. They rely on the human eye as a detector, daylight as the light source, and have no monochromator. An unknown is prepared in a sample cuvette and the intensity of its color is compared visually with either a series of standard solutions or a set of colored plastic disks. The method is not as crude as one might imagine. Precision on the order of 10% (relative standard deviation) is possible, and that is often adequate for field work. A major drawback to visual comparison is that mixtures of absorbing species are impossible to work with; mixtures require the use of a monochromator. Colorimeters are available from several sources. The Hach Company of Loveland, Colorado, sells colorimeters and chemical kits for on-site analysis of water pollutants. Visual colorimeters are also used for the routine monitoring of chlorine and pH in swimming pools.

Filter Photometers   The simplest devices use a tungsten lamp as a source, a variable aperture to control light intensity, a filter wheel to select among a set of fixed wavelengths, a photovoltaic cell detector, and a meter for output. To use a filter photometer, one follows a series of simple steps. First, a filter is chosen for the wavelength range in which the sample absorbs. A cuvette containing solvent (water, for example) is then placed in the light path, and the aperture is opened or closed to produce an absorbance reading of 0.0 on the output meter. A sample is then placed in the light path and its absorbance is read. A Beer's law plot allows the connection to be made to concentration.

The Hach model DR/1A colorimeter has seven filters for the visible region, with spectral bandpasses between 26 and 37 nm. A tungsten source is powered by four D-cells in a battery pack. Either 1-inch or 1-cm cuvettes can be used in the sample housing. The detector is a silicon photovoltaic cell. The instrument can be used to test more than 45 water quality parameters and costs less than $400.

13.6 INSTRUMENTATION

Single-Beam Spectrophotometer In contrast to filter photometers, which allow only a restricted set of fixed wavelengths, a spectrophotometer allows wavelengths to be selected continuously over the entire visible or UV-visible spectrum. Depending on the design and quality of components, spectrophotometers are available for a few hundred dollars to more than $30,000. We will look carefully at the Bausch & Lomb Spectronic 20, a single-beam grating instrument commonly found in undergraduate laboratories. The instrument has a range of 340 to 625 nm, extendable to 950 nm with a special red-sensitive phototube and filter, and a preset bandpass of 20 nm. Figure 13.24 shows the light path for a Spectronic 20. Light from a tungsten lamp enters the monochromator housing at the entrance slit, is collimated with the "objective lens," and passes to a grating with 600 lines/mm. Dispersed light is passed to the exit slit via the "light control," a V-shaped slot which is moved in and out of the light path to partially obscure the light beam. The "occluder" is a piece of metal which blocks the light path completely until a cuvette is placed in the sample compartment. When the occluder is out of the light path, light passes to the sample and then to the photomultiplier tube detector. The position of the grating is set by a dial on the top of the instrument which drives the cam shown in the diagram. The cam is machined to compensate for the slight nonlinearity of dispersion of the grating and ensure a linear wavelength scale. There are two phototubes in the instrument. One measures light passing through the sample, while the other measures the output of the source directly. Fluctuations in the intensity of the tungsten source are compensated for by taking the difference in the signals of the two phototubes. The output of the instrument is presented on a meter on the instrument face. Recorder output is optional.

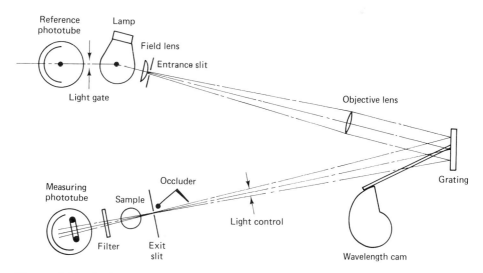

Figure 13.24 Optical system schematic for Bausch & Lomb Spectronic 20. (Reproduced with permission of Bausch & Lomb, Rochester, New York.)

To use the instrument, one sets the monochromator to the wavelength at which the sample absorbs. The instrument is then set to 0.0% transmittance with nothing in the sample compartment. The light beam is occluded when no sample is present, and the meter is adjusted to read only the "dark current" signal from the photomultiplier tube. Next a reference cuvette containing pure solvent (or a suitable blank, as in Experiment 13.1 at the end of this chapter) is placed in the sample compartment and the meter is set to 100%T by turning a dial which moves the light control in and out of the optical path. The dial does not open or close the slits; they are fixed on a Spectronic 20. A cuvette containing the sample is then placed in the sample compartment and the sample's absorbance (or transmittance) is read from the meter. Remember that three adjustments are made prior to a sample reading: wavelength is set, the meter is set to zero with an empty sample compartment, and the meter is then set to 100%T (0.0 absorbance) with solvent or blank. This process is necessary because the Spectronic 20 is a single-beam instrument.

Double-Beam Spectrophotometer More expensive and complicated instruments use a double-beam design, in which sample and reference cuvettes are always present in the instrument. The light path is switched between sample and reference by a "chopper," a segmented mirror which reflects and transmits the light path in two directions, as shown in Fig. 13.25. The detector sees signals from the two paths as an alternating signal at a frequency of the chopper, perhaps 10 cycles per second. The signal processor takes the difference between the signals and sends it to the output device. You can think of the double-beam instrument as performing the single-beam experiment just described 10 times every second. Double-beam instruments also scan wavelengths automatically.

13.7 SUMMARY

This chapter began with a quoted description of a colorimetric procedure for the determination of iron in drinking water. The goal was to learn enough about UV-visible spectrophotometry to be able to *understand* the procedure, rather than find out how to follow the directions as if it were a cooking recipe. We began with a discussion of the properties of light and the behavior of material as it interacts with light. We discussed absorbance spectra and how they are obtained and saw the absorbance spectrum of the complex used in the analysis of iron. We then discussed the Beer's law working curve that is the foundation of a spectrophotometric method of analysis and used that to gain some insight into the directions for determining iron given at the beginning of the chapter. We expanded the Beer's law treatment to include photometric titrations, very useful quantitative assay techniques for many compounds. We then moved on to a discussion of the mole ratio, continuous variations, and slope ratio methods for determining the ligand-to-metal ratio in a complex. Experimental data showed that the Fe–*o*-phenanthroline complex is a 3:1 complex. Beer's law meth-

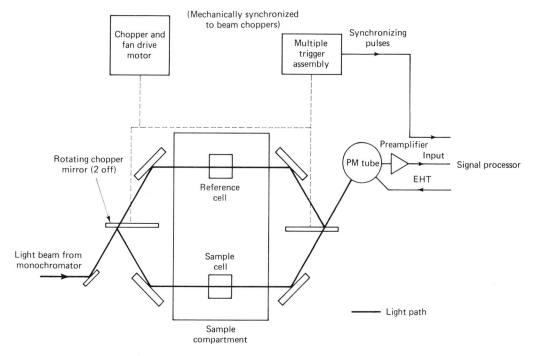

Figure 13.25 Double-beam cell compartment and beam-switching apparatus. Two synchronized chopper mirrors are used to alternate the light signal between sample and reference beams. When the left-hand mirror is in the path, light travels through the sample cell. The right-hand mirror will be out of the path and light will travel to the detector (PM tube). When the left-hand mirror is out of the path, the right-hand mirror will be in the path and light will pass through the reference cell. (From Varian Techtron Model 635 instruction manual.)

ods were then described for the determination of the formation constant of a metal ion complex—making the best of a deviation from ideal Beer's law behavior.

The second half of the chapter was devoted to the instrumentation used to obtain results like those cited in the directions for the analysis of iron. We discussed in some detail the design and construction of simple UV-visible instruments and the ways in which their components work to produce meaningful data.

All of this information has served only as an introduction to one of the most powerful and useful methods of acquiring analytical and structural information. If you have mastered the text material and have had some hands-on experience in the laboratory, you are well on the way toward learning about chemical compounds.

REFERENCES

1. *Standard Methods for the Examination of Water and Wastewater,* 14th ed., American Public Health Association, American Water Works Association, and Water Pollution Control Federation, 1975.

2. J. Lingane and J. Collat, *Analytical Chemistry,* 22, 166 (1950).
3. H. V. Malmstadt and E. G. Gohrbrandt, *Analytical Chemistry,* 26, 442 (1954); and references therein.
4. P. B. Sweetser and C. E. Bricker, *Analytical Chemistry,* 25, 253 (1953).
5. E. J. Meehan, "Fundamentals of Spectrophotometry" in *Treatise on Analytical Chemistry,* 2nd ed. Part 1 vol. 7, chapter 2, Wiley, New York, 1981.
6. A. J. Giese and C. S. French, *Applied Spectroscopy,* 9, 98 (1955).

RECOMMENDED READING

Ewing, G. W. *Instrumental Methods of Chemical Analysis,* 5th ed., McGraw-Hill, New York, 1985. A well-illustrated clear explanation of basic spectrophotometry.

Olsen, E. D. *Modern Optical Methods of Analysis,* McGraw-Hill, New York, 1975. Chapters 1 and 2 deal with and extend the material presented here. The treatment is logical and easy to read.

Rossotti, F. J. C., and H. Rossotti, *The Determination of Stability Constants,* McGraw-Hill, New York, 1961. Chapter 3 gives an advanced treatment of the theory behind the methods of Yoe and Jones and the slope ratio method. Beginners will have to work hard to deal with this treatment, but it is careful and detailed.

Standard Methods for the Evaluation of Water and Wastewater, 14th ed., American Public Health Association et al., 1975. This is a compendium of standard methods, many of which use spectrophotometry.

PROBLEMS

13.1. Light of wavelength 510 nm is blue-green in color.
  (a) Express 510 nm in angstrom units and in micrometers.
  (b) Calculate the frequency of this light.
  (c) What is the velocity of this light in a vacuum? In water ($n = 1.33$)?

13.2. The refractive index of toluene is 1.498 at 590 nm (yellow light). Calculate the speed of yellow light in toluene.

13.3. Permanganate ion absorbs light in a band centered at about 545 nm. Express this wavelength in angstrom units and in micrometers. What is the energy of light of this wavelength, expressed in kilojoules per mole?

13.4. A sample absorbs light only in the red region of the spectrum. What color will it appear to the human eye?

13.5. A sample is yellow. In what wavelength region should it absorb light?

13.6. The Rydberg constant, 109,800 cm^{-1}, is the ionization energy of the hydrogen atom expressed in wave number units. To what wavelength of light (in a vac-

PROBLEMS

uum) does this correspond? In what region of the spectrum can light of this wavelength be found?

13.7. A convenient rule of thumb is that electronic transitions in molecules require about 10 times the energy of molecular vibrations, which in turn require about 10 times the energy of molecular rotations. If a certain molecular electronic transition absorbs light at 500 nm (20,000 cm^{-1}), at what wavelengths might you look to see vibrational and rotational absorbances?

13.8. Convert the following %T values to absorbances:
- (a) 10%T
- (b) 99%T
- (c) 99.9%T
- (d) 1%T
- (e) 20%T
- (f) 50%T

13.9. Convert the following absorbances to %T:
- (a) 1.0
- (b) 0.1
- (c) 0.01
- (d) 0.434
- (e) 2.0
- (f) 0.301

13.10. A green complex absorbs at 720 nm. If a $10^{-3}M$ solution of the complex has a transmittance of 0.400 in a 1.0-cm cell, what should the transmittance of the same solution be in a 5.00-cm cell?

13.11. A sample has a transmittance of 0.300 in a 1.00-cm cell. By what factor must one increase cell path length to decrease the transmittance of the sample to 0.100?

13.12. A 10.0-mg sample of a natural product of unknown molecular weight is dissolved in 1 liter of diethyl ether. The resulting solution has an absorbance of 0.152 in a 1.0-cm cell at 320 nm. Calculate the absorptivity of the compound.

13.13. A 55.0-mg sample of a pure compound of molecular weight 310 is dissolved in 100 mL of acetone. A portion of the resulting solution has an absorbance of 0.230 measured at 430 nm in a 1.0-cm cell. Calculate the absorptivity of the compound in units dL/mg cm and the molar absorptivity.

13.14. Chromate ion has a molar absorptivity of about 300 liter/mole cm at 456 nm. What is the concentration of chromate in a sample which has an absorbance of 0.324 at 456 nm in a 1-cm cell?

13.15. Nitrite ion can be determined at trace levels by reaction with sulfanilic acid and α-naphthylamine to form a red azo dye in acidic solutions. The molar absorptivity of the red dye is about 40,000 at 520 nm. What is the lowest concentration of nitrite you could determine as the azo dye in a 1.0-cm cell? Assume that 1 mmole of nitrite produces 1 mmole of red dye and that the lowest reliable absorbance reading of an instrument is 0.01.

13.16. Lead(II) forms a very stable complex with the ligand dithizone. The complex is formed in basic aqueous solution and extracted into chloroform. The complex has a molar absorptivity of 6.5×10^4 at 520 nm in chloroform. An analyst collects and melts about 500 g of snow. A 100-g portion of the resulting water is made basic with ammonia, treated with cyanide and citrate to mask other metal ions, and then extracted with 10 mL of a solution of dithizone in chloroform. The chloroform solution is transferred to a 5-cm cell, and an absorbance of 0.04 is measured at 520 nm. Calculate the concentration of lead in the snow (in micrograms per liter or parts per billion), remembering that lead is concentrated in the extraction process. The atomic weight of lead is 207.2.

13.17. A Beer's law working curve for a colorimetric analysis is found to be linear. Regression analysis of the data gives the following equation for the "best line" through the points:

$$y = (1.728 \times 10^4)x + 1.02 \times 10^{-3}$$

with a variance of 4.2×10^{-5}.

(a) An unknown sample has an absorbance of 0.289 when treated in the same way as the working curve standards. What is the concentration of the unknown?

(b) What is the significance of the number 1.02×10^{-3}?

(c) Explain the importance of the variance. What would the data be like which gave a variance of 4.2×10^{-3} instead of 4.2×10^{-5}? Sketch a working curve to illustrate your answer.

13.18. A weak acid indicator has the following spectral characteristics, measured in a 1-cm cell:

Solution	A (450 nm)	A (500 nm)	A (550 nm)
0.1 mM HA in 0.1F NaOH	0.623	0.099	0.10
0.1 mM HA 0.1F HCl	0.051	0.099	0.818

A $1 \times 10^{-4} F$ solution of HA is buffered at pH 5.0. A portion is transferred to a 1-cm cell. The absorbances at 450 and 550 nm are found to be 0.270 and 0.374, respectively. Calculate pK_a for the indicator, remembering that K_a = $[H^+][A^-]/[HA]$ and that $[H^+]$ is fixed by the buffer.

13.19. A chemist wants to measure the formation constant for a weak 1:1 metal ion complex, ML. A mixture of $1.0 \times 10^{-4} F$ M and $0.01 F$ L has an absorbance of 0.525 at 625 nm in a 1-cm cell. M does not absorb at this wavelength, but L is known to have a molar absorptivity of 20. A solution which is $1.0 \times 10^{-4} F$ in M and $1.0 \times 10^{-4} F$ in L has an absorbance of 0.298 at 625 nm in a 1-cm cell. Calculate K_f for the complex, ML. (Hint: At equilibrium the second solution has the same composition as a $1.0 \times 10^{-4} F$ solution of ML.)

PROBLEMS

13.20. Aspirin (acetylsalicylic acid) gradually hydrolyzes in a moist environment to form salicylic acid and acetic acid. This reaction limits the shelf life of aspirin. Salicylic acid absorbs in the UV at 296 nm with molar absorptivity 3500. Aspirin absorbs with molar absorptivity 80 at 296 nm. At 266 nm both compounds have a molar absorptivity of 570.

A 0.325-g aspirin tablet is ground up and dissolved in 1 liter of chilled pH 4 buffer. The absorbances measured at 296 and 266 nm in a 1-cm cell are 0.486 and 1.026, respectively. What is the percentage of salicylic acid in the tablet? (GFW aspirin = 180.16, GFW salicylic acid = 138.12.)

13.21. The acid form of an acid-base indicator has a molar absorptivity of 3050 at 510 nm. An isosbestic point for the weak acid-conjugate base pair occurs at 620 nm, with molar absorptivity 1020. In a spectrum of a solution of the indicator in pure water the absorbance at 510 nm is 0.494, while that at 620 nm is 0.210. Calculate the fraction dissociated and pK_a of the weak acid.

13.22. Iron(III) and salicylate ion form a complex which absorbs at 525 nm.
  (a) Sketch a photometric titration curve for the determination of Fe(III) sulfate with sodium salicylate, monitored at 525 nm.
  (b) EDTA forms a much stronger complex with Fe(III) than does salicylic acid. Sketch a photometric titration curve for the titration of Fe-salicylate complex with EDTA, monitored at 525 nm.

13.23. Mixtures of Fe(III) and Cu(II) have been resolved by titration with EDTA in solutions at pH 2. The titration is monitored at 725 nm, where Cu (EDTA) absorbs but Fe(EDTA) does not. Knowing that Fe(EDTA) is much more stable than Cu(EDTA), sketch the photometric titration curve you would expect for the determination of 1.25 mmole of Fe and 2.05 mmole of Cu with $0.005M$ EDTA.

13.24. Recall that calcium can be determined by titration with EDTA at pH 12 in the presence of calcon indicator. The endpoint color change is red → blue, as EDTA displaces calcon from the red calcium-calcon complex. Design a photometric method for titrating Ca^{2+} and predict the shape of a photometric titration curve.

13.25. p-Nitrophenolate ion has a molar absorptivity of 2.0×10^3 at 400 nm. The weak acid form does not absorb at 400 nm. 10 mL of 0.10 mM p-nitrophenol is titrated directly in a 5-cm path length cell with $0.01M$ NaOH. Calculate the absorbance of the solution after the addition of 0.05, 0.075, 0.100, 0.125, and 0.150 mL of base.

13.26. Iron reacts with sulfosalicylic acid (SSA) to form a complex which absorbs at 430 nm, with molar absorptivity of about 6000 liter/mole cm in chloroacetic acid buffer (pH 3).
  (a) What concentration of iron complex will give an absorbance of 1.0 in a 1-cm cell?

(b) You are given a set of ten 50-mL volumetric flasks, ferric sulfate nonahydrate (GFW 562.02), sulfosalicylic acid (GFW 218.19), and 1 liter of buffer solution. Describe the solutions you would prepare to determine the ligand-to-metal ratio for the iron-SSA complex, using the mole ratio method.

(c) With the material in part (b), describe the solutions you would prepare for a continuous variations study.

(d) The Fe(SSA)$_3$ complex is quite stable. Sketch the expected experimental results from parts (b) and (c).

13.27. An interference filter is to be used in an instrument to isolate light of wavelength 500 nm. How thick must a layer of dielectric material be if its refractive index is 1.46 and the filter must pass light of 500 nm in first order?

13.28. A reflection grating is ruled with 1250 lines/mm and a blaze angle of 14°. Assume that white light is incident at an angle of 10° (α).

(a) What first-order wavelengths will be diffracted at 10 and 20° on the same side of the grating normal? Sketch the grating and light paths.

(b) Define "blaze wavelength." Calculate the blaze wavelength for this grating.

(c) What other wavelengths are passed at the angles in part (a)? How can these wavelengths be removed?

13.29. Two sharp spectral emission lines of wavelengths 500.10 and 501.50 nm strike a reflection grating at the blaze angle, 15°. The grating is ruled with 800 lines/mm.

(a) Calculate the difference in the angles at which the two wavelengths are diffracted.

(b) Using basic trigonometry, calculate the separation of the spectral lines on a projection screen held 1 m away from the grating. Calculate the dispersion as "linear dispersion," with units nanometers per millimeter.

13.30. A reflection grating must be turned to sweep a spectrum across the exit slit of a monochromator. A grating with 1250 lines/mm and a blaze angle of 20° is aligned so that light is incident at an angle of 14°. Calculate the wavelength diffracted at 25° (same side of normal). The grating is turned so that the angle of incidence is 15°. The light reaching the exit slit is now diffracted at 26°. How much of the spectrum has been swept in turning the grating by 1°?

13.31. What are the advantages and disadvantages of using gratings as opposed to filters to select wavelengths of light?

EXPERIMENT 13.1: Colorimetric Determination of Manganese in Steel

Manganese can be determined accurately and precisely by oxidizing manganese(II) to permanganate ion (MnO_4^-) and measuring the amount of permanganate colorimetrically. A steel sample containing about 1% manganese is dissolved in nitric acid. The resulting solution is then treated with ammonium peroxodisulfate and then potassium periodate to oxidize manganese quantitatively to the +7 oxidation state. The sample is then diluted appropriately, a portion is poured into a spectrophotometer cuvette, and the optical absorbance of the sample is measured at a wavelength where permanganate absorbs light. The absorbance of the solution is then compared with the absorbance of a solution prepared from a standard steel sample, and the percentage of manganese in the unknown is calculated.

A steel sample containing manganese metal is dissolved in nitric acid according to the following reaction:

$$3Mn^0 + 2NO_3^- + 8H^+ \rightarrow 3Mn^{2+} + 2NO(g) + 4H_2O$$

The nitric oxide produced in this step may interfere with the subsequent oxidation by periodate and must be removed by boiling and by the addition of ammonium peroxodisulfate:

$$2NO + 3S_2O_8^{2-} + 4H_2O \rightarrow 2NO_3^- + 6SO_4^{2-} + 8H^+$$

Other oxides of nitrogen and carbon are also oxidized by peroxodisulfate. Unreacted peroxodisulfate is reduced by water at boiling temperatures:

$$2S_2O_8^{2-} + 2H_2O \rightarrow 4SO_4^{2-} + O_2(g) + 4H^+$$

In theory, peroxodisulfate should be able to oxidize manganese to the +7 state. Unfortunately, the oxidation step is too slow to be practical, and it is necessary to use potassium periodate as the final oxidant:

$$2Mn^{2+} + 5IO_4^- + 3H_2O \rightarrow 2MnO_4^- + 6H^+ + 5IO_3^-$$

The presence of excess iodate keeps the permanganate from decomposing. The only potentially serious interfering metals are chromium and iron. Phosphoric acid is added to the oxidized steel sample to complex and mask the iron. Steel samples containing more than a few tenths of a percent chromium should not be analyzed by this method.

Procedure

1. Weigh by difference into a labeled 250-mL beaker a 0.6- to 0.8-g sample of unknown steel to the nearest 0.1 mg. Your instructor may tell you to weigh out a larger or smaller sample depending on the amount of manganese in the steel (1% Mn assumed here).

2. Repeat step 1 using a sample of steel whose percentage of Mn is known. You will save time and trouble if the weight you take of this standard is within about 10 mg of the weight of your unknown steel.

3. Bring the beakers containing sample and standard to the fume hood.

> CAUTION: You must wear eye protection when working with acids and in the hood.

Add 50 to 60 mL of 1:3 nitric acid to each beaker, then cover them with watch glasses. Heat the samples until they are dissolved. Boil both solutions for a few minutes to help remove NO gas. You may see black specks of carbon remaining after the metal dissolves. If a brown precipitate forms (manganese dioxide), it may be dissolved by adding about 0.1 g of sodium bisulfite ($NaHSO_3$) and heating for a few minutes.

> CAUTION: Be very careful when you handle these solutions. They can burn your skin very badly.

4. Remove the burners and let the solutions cool for 5 minutes. Remove the watch glasses and carefully sprinkle 1 g of ammonium peroxodisulfate into each beaker. Return the watch glasses to the beakers and boil the solutions for 10 or 15 minutes. The carbon flecks should disappear and the excess peroxodisulfate will decompose.

5. After the solutions have cooled for a few minutes, add 15 mL of 85% phosphoric acid and 0.5 g of potassium periodate to each. Stir with glass rods (not metal spatulas!). Be careful not to cross-contaminate the samples. Boil both solutions gently for 3 or 4 minutes. Let them cool for a few minutes before adding an additional 0.2 g of potassium periodate. Boil the solutions for another 3 to 4 minutes to complete the oxidation.

6. Transfer both solutions to labeled 500-mL volumetric flasks, chill them to room temperature in an ice-water bath, and dilute to volume with deionized water.

7. Using deionized water as a blank, measure the absorbance of your standard and unknown solutions at a wavelength of 545 nm (see Notes below).

8. If your unknown sample and standard weights were within a few milligrams of each other, you may calculate the percentage of Mn in the unknown steel by multiplying the percentage in the standard by the ratio of the absorbance of the unknown to that of the known. This is a simple application of Beer's law,

$$A_x = k_x * C_x \qquad A_{std} = k_{std} * C_{std}$$

If $k_{std} = k_x$, then

$$C_x = C_{std} * (A_x/A_{std})$$

You may use %Mn for C in these equations. If your unknown sample and standard were not close in weight, you will have to calculate the molar absorptivity of permanganate from the absorbance in the standard solution and use that number to calculate the concentration in the unknown.

Notes

Ask your laboratory instructor for detailed operating instructions for the spectrophotometer you use. Simple table-top single beam instruments have similar controls, and their operation is summarized below:

1. Turn on the instrument and let it warm up for about 15 minutes.
2. Set the wavelength control to the analytical wavelength, 545 nm for this experiment.
3. With the cuvette holder empty and cover closed, set the meter to 0.0% transmittance (infinite absorbance) with the dark-current control.
4. Place a cuvette containing blank solution (deionized water here) in the cuvette holder, and close the cover. Turn the light control knob to set 100% transmittance (0.00 absorbance) on the scale. Now both ends of the scale are calibrated.
5. Place a cuvette containing a sample in the cuvette holder and close the cover. Read the absorbance from the scale and record it. You may wish to read transmittance from the scale and later convert it to absorbance.
6. Recheck zero and 100%T and recheck the absorbance of your solution.

EXPERIMENT 13.2: Colorimetric Determination of Iron(II) (1,10-Phenanthroline Method)

Iron in drinking water and wastewater is quickly and easily determined by reaction with the ligand 1,10-phenanthroline ("Ophen") to form the orange-red tris-ligand complex, $Fe(Ophen)_3^{2+}$. Solutions of this complex are quite stable, and Beer's law is obeyed over a wide pH range (pH 3 to 9). In this method solutions are buffered at about pH 3.5 to avoid the possible precipitation of iron salts with potentially interfering anions and to ensure rapid formation of the complex (rapid color development).

Reagents

1. Hydroxylamine hydrochloride solution ($NH_2OH \cdot HCl$). This is a 10% wt/vol solution in deionized water, which has had its pH adjusted to 4.5 (glass pH electrode) with sodium acetate. Hydroxylamine is a reducing agent and is added to iron samples to convert Fe(III) in the sample to Fe(II). Your instructor may provide this solution.

2. Sodium acetate solution. Prepare this solution by dissolving about 100 g of sodium acetate in deionized water to make 1 liter of solution. Your instructor may also provide this solution.

3. o-Phenanthroline solution. This solution is about 0.3% wt/vol in deionized water and is essentially a saturated solution. This material is rather expensive and your instructor will most likely provide it.

4. Standard iron solution (you must prepare this). The solution will be 0.050 mg of Fe per milliliter of solution. There are two options available:

   (a) Weigh out as carefully as possible 0.35 g of analytical reagent grade ferrous ammonium sulfate, $FeSO_4 \cdot (NH_4)_2SO_4 \cdot 6H_2O$. Dissolve in about 50 mL of deionized water containing 1 mL of concentrated sulfuric acid (DO NOT PIPET BY MOUTH!). Transfer the solution to a 1-liter volumetric flask and dilute to the mark with deionized water.

   (b) Alternatively, weigh 0.05 g of freshly cleaned reagent grade iron wire as carefully as possible, and dissolve in 10 mL of 1:1 HCl. Do not use nitric acid or bromine water to oxidize the iron. Transfer the resulting solution quantitatively to a 1-liter volumetric flask and dilute to the mark with deionized water.

The Working Curve

Using a pipet, transfer a 5.00-mL portion of the standard iron solution to a small beaker and add a drop of bromphenol blue indicator. Titrate the acid in this iron solution by adding sodium acetate solution from a small graduated pipet until the indicator turns green (the first blue color is acceptable). Note the volume of acetate solution you used to titrate the acid, and discard the solution containing the 5 mL of iron stock solution.

Transfer a second 5.00-mL portion of the iron stock to a 50-mL volumetric flask and add by pipet 1 mL of hydroxylamine solution and 5 mL of o-phenanthroline solution. Next add the same quantity of sodium acetate solution as was required in the titration step. Do not dilute to volume yet; set the flask aside for at least 10 minutes to let the complex form. When diluted to the mark, the solution in this flask should be about 5 μg Fe/mL, that is, 5 ppm (wt/vol). Calculate the exact concentration, using the concentration of your stock solution.

EXPERIMENT 13.2: COLORIMETRIC DETERMINATION OF IRON(II)

Because sodium acetate may contain some iron, you must prepare a blank solution. Prepare the solution using everything called for in the previous paragraph *except* the stock iron solution.

Set up the spectrophotometer as directed by your instructor. Pour a few milliliters of the 5 ppm Fe solution you have just prepared into a clean sample cell, and measure the absorbance of the solution at wavelengths in the range 450 to 550 nm, using the blank solution to set 0.0 absorbance at each wavelength. Check the absorbance at 10-nm intervals. The absorbance maximum will be at about 508 nm. The spectrum is presented as Fig. 13.5 in the text. You will measure absorbances for your working curve at the peak wavelength.

Prepare at least four other standards as described above, covering an absorbance range of 0.1 to 1.0. Use Beer's law to calculate the concentrations of the solutions you will prepare.

Plot your data as absorbance versus concentration. Is Beer's law obeyed?

Analysis of Your Unknown Sample

Your instructor will provide instructions for pretreatment of your sample. If your sample is prepared from ferrous ammonium sulfate, do not dry the sample.

For samples containing 5 to 14% iron, weigh a 150- to 200-mg portion and dissolve in 50 mL of deionized water containing 1 mL of concentrated sulfuric acid. Transfer the solution to a 500-mL volumetric flask and dilute to the mark with deionized water. Transfer a 5-mL portion of the resulting solution to a small beaker and repeat the bromphenol blue test. Using another 5-mL portion of your sample, follow the instructions given for the 5 μg Fe/mL standard. Let the prepared solution stand for 10 minutes before diluting to 50 mL. Measure the absorbance at the analytical wavelength. Repeat the analysis on a second portion of your sample as time permits.

For Water Samples Water samples should be stabilized at the time of collection by adding 2 mL of concentrated hydrochloric acid per 100 mL of sample. At the time of analysis transfer a 50-mL portion of the water sample to a clean 100-mL volumetric flask. Titrate with sodium acetate solution to the bromphenol blue endpoint. Discard this solution and prepare another, without the indicator. Add the volume of sodium acetate solution you just determined, then add 2 mL of hydroxylamine hydrochloride solution and 10 mL of *o*-phenanthroline solution. Set the solution aside for 10 minutes, then dilute to 100 mL with deionized water. Measure the absorbance of the solution at the analytical wavelength.

Use linear regression to calculate an absorptivity from the slope of your working curve (see Chapter 2 for the linear regression method). Use the absorptivity to calculate the concentration of your unknown sample. Alternatively, locate the absorbance of your unknown sample on the working curve line, and read the corresponding value for concentration. Convert milligrams of Fe per liter to milligrams of Fe in your sample, and finally percent Fe. If you are working with a water sample, report iron as parts per million.

Interferences

Many metal ions and anions interfere in the determination of iron. The following is a brief list:

Cyanide, phosphate, fluoride, and nitrite: species which complex iron

Chromium, zinc (>10 times Fe concentration), copper, and cobalt (concentration > 5 ppm): species which produce colored complexes with Ophen or will tie up Ophen

Bismuth, cadmium, mercury, silver: precipitate Ophen

Strong oxidizers: may overwhelm reducing power of hydroxylamine, leave some Fe(III)

If cyanide, phosphates, or nitrates are present, boil the solution after adding acid. Phosphates may still be a problem and may require extraction (see below).

If concentrations of other metal ions are not too high, additional o-phenanthroline may be added to compensate for their presence.

Extraction Iron may be extracted from hydrochloric acid solutions ($6M$ HCl) by shaking in a separatory funnel for 1 minute with an equal volume of diisopropyl ether. Iron transfers to the ether phase as $HFeCl_4$ and can be stripped back out with subsequent equilibration with deionized water.

> CAUTION: Diisopropyl ether may develop peroxides and become explosive. Shake some potassium iodide solution with a sample of diisopropyl ether. If iodine forms, discard the batch of ether.

EXPERIMENT 13.3: Determination of the pK_a^0 of Bromcresol Green

UV-visible spectrometry can be used to determine the dissociation constant of a weak acid when it and its conjugate base absorb light. In this experiment you will determine the pK_a of the indicator bromcresol green (HBG), by studying visible spectra of its solutions in a series of acetate buffers.

EXPERIMENT 13.3: DETERMINATION OF THE pK_a^0 OF BROMCRESOL GREEN

The weak acid dissociation equilibrium can be represented by

$$HBG^- = H^+ + BG^{2-} \quad (H^+ \text{ represents } H_3O^+)$$

with the expression for K_a

$$K_a = \frac{[H^+][BG^{2-}]}{[HBG^-]} \quad \text{(brackets denote } concentrations\text{)}$$

Taking negative logs and rearranging gives

$$pK_a = pH + \log\frac{[HBG^-]}{[BG^{2-}]}$$

In this experiment we will fix pH with buffers and measure $[HBG^-]$ and $[BG^{2-}]$ using Beer's law and the intensity of their absorbances in the visible region of the spectrum. Bromcresol green has two distinct colors, *blue* in basic solution, and *yellow* in acidic solution. Let λ_b be the wavelength of maximum absorbance in the blue solution and λ_y be the wavelength of maximum absorbance in the yellow solution. Let A_{HBG^-} and $A_{BG^{2-}}$ represent the absorbances of HBG and BG$^-$, respectively.

Since absorbances are additive, we can say that

$$A^{\lambda_b} = A^{\lambda_b}_{HBG^-} + A^{\lambda_b}_{BG^{2-}}$$
$$A^{\lambda_y} = A^{\lambda_y}_{HBG^-} + A^{\lambda_y}_{BG^{2-}}$$

Substituting in Beer's law ($A = \epsilon bC$, for $b = 1.00$ cm) gives

$$A^b = \epsilon^{\lambda_b}_{HBG^-}[HBG^-] + \epsilon^{\lambda_b}_{BG^{2-}}[BG^{2-}]$$
$$A^y = \epsilon^{\lambda_y}_{HBG^-}[HBG^-] + \epsilon^{\lambda_y}_{BG^{2-}}[BG^{2-}]$$

The simplest of all possible situations would be that $\epsilon^{\lambda_b}_{HBG}[HBG^-]$ would be negligible in the first equation and $\epsilon^{\lambda_y}_{BG}[BG^{2-}]$ negligible in the second. But nature is seldom so kind. What we must do is measure $\epsilon^{\lambda_b}_{HBG^-}$ and $\epsilon^{\lambda_y}_{HBG^-}$ in an acidic solution (when $[HBG^-] \gg [BG^{2-}]$) and $\epsilon^{\lambda_b}_{BG^{2-}}$ and $\epsilon^{\lambda_y}_{BG^{2-}}$ in a basic solution (when $[BG^{2-}] \gg [HBG^-]$). The results can then be used to calculate values for $[HBG^-]$ and $[BG^{2-}]$ in buffered solutions where pH $\approx pK_a$, and so calculate pK_a.

Measuring *concentrations* allows us to calculate an approximate value of the constant K_a. The glass electrode measurement is an *activity* measurement (the pH 4.00 buffer defines a_{H^+} according to the NBS). If we had values for the activity coefficients of HBG$^-$ and BG^{2-}, our measured K_a would approach the thermodynamic value. Using the earlier expression and knowing, for example, that $a_{BG^{2-}} = f_{BG^{2-}}[BG^{2-}]$, we see that

$$pK_a^0 = pH + \log\frac{[HBG^-]}{[BG^{2-}]} + \log\frac{f_{HBG^-}}{f_{BG^{2-}}}$$

The activity coefficients can be approximated with the Davies equation

$$-\log f_i = \frac{0.511 z_i^2 \mu^{1/2}}{1 + 1.5\mu^{1/2}} - 0.2\mu z_i^2$$

where μ is the ionic strength and z_i is the ionic charge.

Procedure

A 0.100% (weight to volume) solution of bromcresol green (GFW 698.04) in water is available. Acetic acid, perchloric acid, and sodium acetate are available. Prepare all solutions to have an ionic strength close to 0.05M. The pH 4.00 standard buffer for the glass electrode is also a 0.05M solution.

1. Find λ^b and λ^y and get approximate values for $\epsilon_{HBG^-}^b$, $\epsilon_{BG^{2-}}^b$, $\epsilon_{HBG^-}^y$ and $\epsilon_{BG^{2-}}^y$.

Transfer 200 μL of 0.100% HBG stock to two 25-mL volumetric flasks. Dilute one to the mark with 0.05M perchloric acid and the other with 0.05M sodium acetate.

Calculate the approximate pH of both solutions. Then calculate the approximate concentrations of HBG$^-$ and BG^{2-} in both solutions assuming $pK_a \sim 5$. Is [HBG$^-$] $\gg$ [BG^{2-}] in the acid solution? Is [BG^{2-}] $\gg$ [HBG$^-$] in the basic solution? What are the factors?

Record spectra of both solutions between 800 nm and 350 nm *on the same section of chart paper*. What are λ^b and λ^y? Calculate approximate molar absorptivities at these wavelengths.

2. Using the molar absorptivities just obtained, prepare sets of standard solutions of HBG$^-$ in 0.05M HClO$_4$ and in 0.05M sodium acetate which will have absorbances in the range of 0.2 to 0.8 absorbance units. Four or five standards per set are adequate. Plot A vs. concentration for both species at λ^b and λ^y, and calculate refined values for the molar absorptivities.

3. Prepare four acetic acid/sodium acetate buffers over the pH range of 3.8 to 5.4. Adjust ionic strengths with sodium perchlorate if necessary. Remember that acetic acid is a weak acid and doesn't contribute to ionic strength.

Calculate the amount of HBG$^-$ to add to each of four 25-mL volumetric flasks. When you have transferred the right amount of HBG$^-$, dilute each volumetric flask to the mark with a different buffer.

Pour the contents of each volumetric flask into a small beaker. Measure the pH of each solution before running a full visible spectrum. Record all spectra on one section of chart paper. Overlaying the spectra will allow you to see an isosbestic point if one exists.

4. Using your absorbance data, solve four pairs of simultaneous equations and generate four values for pK_a (formal constant).

Using the Davies equation calculate activity coefficients for HBG^- and BG^{2-}. Calculate four values for pK_a^0 (thermodynamic constant) and take an average. Do you see trends in the data? How do the buffer solutions differ in their composition?

Bromcresol green (HBG^-)
(tetrabromo-m-cresol-sulfon-phthalein)
gfw 698.04

Dissociable proton $pK_a \sim 5$

EXPERIMENT 13.4: Spectrophotometric Titration of Cu(II) with EDTA

The use of a spectrophotometer to monitor the progress of a titration allows an analyst to work at much lower concentrations than would otherwise be possible. For example, the lower limit of concentration for titrations monitored with electrodes is about $10^{-3} M$. The use of long-path spectrophotometer cells and microburets makes it possible to work at the $10^{-6} M$ level, or even lower. In very careful work precision of better than 0.5% (relative standard deviation) can be achieved using a spectrophotometer.

In this experiment you will titrate Cu(II) in a sample using the complexing agent ethylenediaminetetraacetic acid (EDTA) in solutions buffered at pH 4.2. Stock solutions of Cu(II) and EDTA will be provided. The Cu(II) solution will be a standard, and its concentration will appear on the label. You will standardize the EDTA solution with the Cu(II) solution. You will then use the standardized EDTA solution to titrate an unknown copper sample. Unless your instructor tells you otherwise, your sample will contain between 0.2 and 1.3 mg of copper.

Procedure

1. You must prepare 0.5 liter of pH 4.2 buffer. Use 5 mL of glacial acetic acid and 4.5 g of sodium acetate.

> CAUTION: Glacial acetic acid causes burns. Protect your eyes and work in a hood.

Check the pH of the buffer with a glass electrode and meter, and adjust the pH to 4.2 with either acetic acid or sodium acetate.

2. Run a survey spectrum of the Cu(II)-EDTA complex.

    A. Prepare 25 mL of a millimolar solution of Cu(II) in pH 4.2 buffer. Pour some of the solution in a 1-cm cuvette and record the spectrum over a wavelength range of 350–250 nm. Use pH 4.2 buffer in a reference cuvette.

    B. Prepare 25 mL of a millimolar solution of EDTA in pH 4.2 buffer. Record the spectrum in a 1-cm cuvette over the same wavelength range, again using buffer in the reference cuvette.

    C. Prepare 25 mL of millimolar Cu-EDTA complex in pH 4.2 buffer and record the spectrum as in parts A and B.

Use the spectra you have recorded to find a wavelength at which the molar absorptivity of Cu(EDTA) is about 1000. What are the molar absorptivities of Cu(II) and EDTA at this wavelength?

3. Standardize the EDTA stock solution (triplicates). Pipet exactly 1 mL of the Cu(II) standard into a 100-mL volumetric flask and dilute to volume with buffer. Calculate the concentration of Cu(II) in this solution. Transfer 10.00 mL of this solution by pipet to a clean 5.00-cm pathlength cuvette. Use 10 mL of pH 4.2 buffer in a 5.00-cm reference cuvette. Set the instrument to the wavelength you chose in the last step. Add EDTA stock solution to the sample cuvette in 20 or 25 µL portions using an adjustable micropipet. After each addition of EDTA, cap and gently shake the cuvette before measuring absorbance. Record and plot absorbance after each addition of titrant. Locate the equivalence point volume on the plot, and use it to calculate the molar concentration of EDTA. Average the results of three titrations.

4. Titrate the unknown (triplicates). You will be given a small volume of solution containing Cu(II). Transfer the sample to a 100 mL volumetric flask and dilute to the mark with pH 4.2 buffer. Pipet exactly 10.00 mL of this solution into a clean 5.00-cm cuvette, and titrate with standard EDTA solution, just as you did in part 3. Record and plot data, locate equivalence points, and calculate milligrams Cu in the sample. Remember that you are titrating one-tenth of the copper in the sample in each titration.

Questions

    (a) Why is a pH of 4.2 better for this analysis than a pH of 7 or a pH of 2? (consider doubly conditional K_f)

EXPERIMENT 13.4: SPECTROPHOTOMETRIC TITRATION OF Cu(II) WITH EDTA

(b) Sweetser and Bricker monitored this titration at 745 nm, where the molar absorptivities of Cu(II) and the Cu(EDTA) complex are about 80 and 20 L/mole cm, respectively. What limitations are placed on the method by working at 745 nm?

(c) If the volume change which accompanies the addition of titrant were not negligible, how would the titration curve be affected?

Chapter 14

Oxidation-Reduction Equilibria and Electrochemical Cells

14.1 INTRODUCTION

This chapter is the first in a series of three chapters which deal with fundamental aspects of electroanalytical chemistry. The discussion in this chapter focuses on electrochemical cells and oxidation-reduction equilibria. The relationships between the free energy change in a redox reaction and the electromotive force (*emf*) of a cell are explored, along with the concept of reversibility, liquid junction potential, and some electrochemical cells of practical importance. Chapter 15 is concerned with the measurement of electrochemical cell *emf* ("potentiometry") and the evaluation of the activities of hydronium ions and other ions using ion-selective electrodes. Chapter 16 contains a survey of titrimetric methods of analysis based on redox reactions.

14.2 OXIDATION-REDUCTION REACTIONS

When a few pieces of copper metal are dropped into a solution of silver nitrate a spontaneous chemical reaction occurs. Silver ions are reduced to form silver metal on the surface of the copper,

$$Ag^+ + e = Ag$$
(reduction→)

At the same time and at the same rate, copper metal is *oxidized* to form copper ions,

$$Cu = 2e + Cu^{2+}$$
$$(\text{oxidation} \rightarrow)$$

The two reactions represented by these equations are called *half-reactions*. They can be added or subtracted to give a balanced *redox* reaction,

$$2Ag^+ + Cu = 2Ag + Cu^{2+}$$

Notice that while electrons appear in half-reactions, they subtract out of balanced redox reactions. Since we have observed that copper metal reduces silver ion, we know that the reaction proceeds to the right under the conditions of the experiment. Another way to describe the process is by saying that Ag^+ is a sufficiently strong oxidizing agent to oxidize copper metal under the conditions of the experiment.

14.3 ELECTROCHEMICAL CELLS

It is possible to get some useful work from this spontaneous chemical reaction if we isolate the two half-reactions in an electrochemical cell and connect them by an interruptible circuit, like that shown in Fig. 14.1. When the electrical connections are made the voltage drop across the cell is called the "galvanic electromotive force," or more simply the *emf* of the cell. As you will see, the emf of a cell depends on the activities of ions, the partial pressures of gases (if they are present), and temperature. Cell emf is related to the free energy change in the chemical reaction, that is, the driving force of the reaction.

In Fig. 14.1 the left-hand cell compartment holds a copper metal electrode in a solution of copper nitrate. The right-hand compartment contains a solution of silver

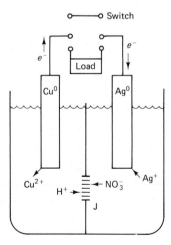

Figure 14.1 Silver-copper cell.

14.3 ELECTROCHEMICAL CELLS

nitrate and a silver metal electrode. The two compartments are connected at two points. Wires lead to a switch which can be used either to connect the half-cells directly or to add an electrical load (a light bulb, for example) to the circuit. The full circuit is completed by a junction port in the cell wall (J) which permits the flow of current (ions can move) but not the mixing of bulk quantities of solution. Such a junction might be made of fritted glass, asbestos or cotton fibers, or unglazed porcelain. When the switch is set to connect the half-cells, the reaction proceeds just as if there were no barrier and the solutions could mix. With the switch in the other position, the flow of current in the wires can be used to light the bulb. The voltage under load will not be the same as that calculated from equations we will use later. The behavior of cell voltages when current is drawn is an important part of practical electrochemistry and will be discussed at the end of this chapter.

The cell electrode at which reduction occurs is called the *cathode*. In the silver-copper cell described here the cathode is the silver electrode. As positively charged ions move from the solution to the electrode surface, they carry a positive charge, which tends to accumulate on the electrode. The positive charge draws electrons from the external circuit.

Oxidation always occurs at the *anode* of a cell. In the silver-copper cell the anode is the copper electrode. As copper is oxidized it moves from the electrode to the solution and leaves electrons behind, giving a negative charge to the anode. Electrons are released by the anode to the external circuit and flow to the silver cathode, where they are used in the reduction of silver ions. Positive ions move through the solution from anode compartment to cathode compartment across the junction to complete the circuit.

14.3.1 Cell Diagrams

Rather than draw out complicated pictures of beakers and electrodes or use long verbal descriptions of cells, we rely instead on a systematic method of representing cells with lines and symbols. The silver-copper cell can be represented by

$$(-)Cu \,/\, Cu^{2+}\, (a = 1) \,//\, Ag^+\, (a = 1) \,/\, Ag(+)$$

Slashes or diagonal lines are used to indicate where voltage drops could develop in a cell. A single slash represents a phase change (solid/solution) or a junction (solution/solution, a "liquid junction"). A double slash represents a "salt bridge," a special junction device containing two solution/solution interfaces, which will be discussed later. It is also common practice to indicate activities (a), formal concentrations (F), or gas pressures (P) in these representations. The charges on the metals are their actual electrostatic charges, concerning which further comment is needed.

14.3.2 Conventional Wisdom

In 1953 the International Union of Pure and Applied Chemistry (IUPAC) established conventions for the signs of electrodes and cell emf's. By convention, a schematic such as

$$\text{Cu} / \text{Cu}^{2+} \, (a = 1) \, // \, \text{Ag}^+ \, (a = 1) \, / \, \text{Ag}$$

is always written with the reduction half-reaction on the right side and the oxidation half-reaction on the left side. This tells the reader that the spontaneous reaction is

$$2\text{Ag}^+ + \text{Cu} \rightarrow 2\text{Ag} + \text{Cu}^{2+}$$

By convention, the emf of the cell is *assigned* the same sign as the electrostatic sign of the right-hand electrode, that is,

$$E_{cell} > 0 \quad \text{when} \quad (E_{right} - E_{left}) > 0$$
and
$$E_{cell} < 0 \quad \text{when} \quad (E_{right} - E_{left}) < 0$$

It is important to point out that there is no necessary connection between the sign of the electrode and the sign of the cell emf; they are simply made to be the same by international agreement.

The IUPAC convention for the sign of a cell emf is consistent with the thermodynamic requirement that the free energy change of a spontaneous reaction be negative,

$$\Delta G_{reaction} < 0 \quad \text{(spontaneous reaction)}$$

The free energy change for a cell reaction is related to the emf of the cell by

$$\Delta G_{cell} = -nFE_{cell}$$

which will be discussed in the next section. For the time being, notice that when E_{cell} is positive, G_{cell} is negative, and the reaction can proceed spontaneously as written.

Now consider what would happen if we were to write the silver-copper cell as

$$\text{Ag} / \text{Ag}^+ \, (a = 1) \, // \, \text{Cu}^{2+} \, (a = 1) \, / \, \text{Cu}$$

A reader knowing the IUPAC convention would *assume* that the copper electrode had a positive electrostatic charge and was the cathode and that the silver electrode was negatively charged and was the anode. However, in the laboratory a voltmeter would show that the silver electrode was positively charged, indicating that the cell diagram was written contrary to convention.

14.3 ELECTROCHEMICAL CELLS

When we speak of the emf of a half-cell, say that of the Ag^+/Ag couple, we actually mean the emf of the *full cell*

$$Pt, H_2(g)(P = 1.00 \text{ atm}) / H_3O^+ (a = 1.00) // Ag^+(a = 1.00) / Ag$$

In the laboratory it is found that the silver electrode has a positive charge, meaning that the reaction

$$2Ag^+ + H_2(g) \rightarrow 2Ag^0 + 2H^+$$

proceeds spontaneously. The cell emf (0.80 V) is given a positive sign and the special name standard potential (E^0). The left-hand half-cell serves as a universal reference, against which all other half-cell emf's can (in principle) be measured, and is called the standard hydrogen electrode (SHE). The SHE is assigned a voltage of 0.000 V.†
A table of standard potentials appears in Appendix IV.

The practice recommended by the IUPAC and followed in this text is to present standard potentials *written only as reductions*. Some texts in inorganic and physical chemistry, most notably Latimer's *Oxidation Potentials* (Ref. *1*), tabulate emf data for oxidation half-reactions. For example, Latimer wrote the "standard oxidation potential" of the bromide/bromine couple as

$$2Br^- = Br_2 + 2e^- \qquad E^0 = -1.087 \text{ V}$$

To make this notation conform to IUPAC rules, we simply reverse the reaction and change the sign:

$$Br_2 + 2e^- = 2Br^- \qquad E^0 = +1.087 \text{ V}$$

When using the older literature it is important to be sure of the sign convention used by the author.

14.3.3 Combining Half-Cell Reactions

We know from thermodynamics that the maximum work that can be obtained from a chemical system is given by the change in free energy of the system in the course of

---

† Some half-cell emf's cannot be measured experimentally, and thus some standard potentials are calculated from free energy data. Recall that the standard free energies of $H_2O(l)$ and H_3O^+ are the same (-237.18 kJ/mole) and that of $H_2(g)$ is zero. Since ΔG^0 for the electrochemical half-reaction

$$2H_3O^+ + 2e^- = H_2(g) + H_2O$$

is zero, it is logical to make it the zero point of the standard potential scale.

a reaction.† In an electrochemical cell, the standard free energy change is proportional to the standard cell potential (a function of standard potentials of the half-reactions, as you will see) according to the expression

$$\Delta G^0 = -nFE^0_{cell}$$

The standard state is one in which all reactants and products are present in their most stable form (solid, liquid, or gas) at 25°C and at unit activity. The proportionality factor n is the number of faradays generated in the cell reaction, and F is the faraday constant, the charge of 1 mole of electrons (96,486.6 coulombs).

As stated before, free energy change in a cell reaction is a measure of the maximum available work and depends on the quantity of material reacting; that is, it is an extensive property. In contrast, the standard potential is a measure of the driving force of a reaction and is an intensive property. Extensive properties can be summed, but intensive properties cannot.‡ Thus we can sum the quantities nFE^0 for a set of half-reactions (or nE^0 in volt-equivalents, since F is a constant), but not simply the standard potentials by themselves.

Consider once again the cell reaction

$$2Ag^+ + Cu = 2Ag + Cu^{2+}$$

The emf of the cell can be found by combining the standard potentials of the two half-reactions in the following way:

$2(Ag^+ + e = Ag)$	$E^0 = +0.800$ V
$-(Cu^{2+} + 2e = Cu)$	$-E^0 = -(0.337)$ V
Net: $2Ag^+ + Cu = 2Ag + Cu^{2+}$	$+0.463$ V

Even though the silver half-reaction equation must be multiplied by 2 to balance the full cell reaction, it appears that the standard potential is simply added to that of the silver couple without multiplying. Obscured by this very simple treatment is the fact that free energies (nFE^0) must be combined, rather than just standard potentials. The following equations show why the outcome is nevertheless the same:

$2(Ag^+ + e = Ag)$	0.800	$2[1F(0.800)]$
$-Cu^{2+} + 2e = Cu$	$-(0.337)$	$-[2F(0.337)]$
$2Ag^+ + Cu = 2Ag + Cu^{2+}$	$+0.463$ V	$+F(0.926)$
	(E^0's added directly)	(Free energies added)

† This reaction is a hypothetical, thermodynamically reversible reaction. For all real reactions the work is less than the free energy change.

‡ The distinction is similar to that between the properties heat (extensive) and temperature (intensive). If we take two pails each containing 1 liter of water at 20°C and pour their contents into a larger pail, the result is 2 liters of water at 20°C with twice the heat content of an individual liter of water. Thus heat is additive but temperature is not.

14.3 ELECTROCHEMICAL CELLS

We can get the value of E^0 for the full cell by dividing the full-cell free energy change [$F(0.926)$] by nF for the balanced full reaction, that is, $2F$. The result is 0.463 V, the same as the value found by simply subtracting the standard potentials.

When we set out to calculate the standard potential of a new half-reaction from two other half-reactions, we must keep in mind the fact that we are no longer dealing with a balanced full reaction. For example, the standard potential of the half reaction

$$Fe^{3+} + 3e = Fe$$

which cannot be measured in the laboratory, can be calculated from two other half-reactions:

$Fe^{3+} + e = Fe^{2+}$	0.771 V	$1F(0.771)$
$Fe^{2+} + 2e = Fe$	-0.440 V	$2F(-0.440)$
$Fe^{3+} + 3e = Fe$	0.331 V	$F(-0.110)$

If we divide $F(-0.110)$ by nF for the new half-cell reaction, that is, $3F$, we find $E^0 = -0.037$ V. By simply adding the standard potentials of the original half-reactions we do not adequately account for the free energy change in the third reaction. This can be done only by combining values of nFE^0.

14.3.4 Cell Spontaneity

The free energy change occurring in a full cell reaction allows us to judge the spontaneity of the reaction. The standard potential of the full cell allows us to calculate the thermodynamic equilibrium constant, knowing that

$$\Delta G^0 = -RT \ln K^0 = -nFE^0_{cell}$$

where R is the gas constant and T is the absolute temperature. At 298 K, this expression becomes

$$\log K^0 = \frac{nE^0_{cell}}{0.05916}$$

When E^0_{cell} is positive, $\log K^0$ is positive, K^0 is greater than one, and the cell reaction is spontaneous (ΔG^0 is negative).

Spontaneous cells are commonly referred to as galvanic cells or voltaic cells (after Galvani and Volta, the "grandfathers" of electrochemistry). Designation of electrode signs and the names anode and cathode in galvanic cells have been discussed above. Later you will be introduced to *electrolysis* cells, cells that are forced to operate in a nonspontaneous direction by the application of an external voltage. In such cells the sign conventions are just the reverse of those in galvanic cells: a cathode (still the site of reduction) is negative, and an anode (site of oxidation) is positive. Cell currents flow in opposite directions from those in galvanic cells.

EXAMPLE 14.1

Calculate E^0_{cell}, ΔG^0_{cell}, and K^0 for the spontaneous reaction between zinc and silver chloride. Diagram the galvanic cell and designate the anode and cathode and their polarities.

Solution

The following half-reactions can be used:

$$Zn^{2+} + 2e^- \rightarrow Zn \qquad E^0 = -0.763 \text{ V}$$
$$AgCl + e^- \rightarrow Ag + Cl^- \qquad E^0 = +0.222 \text{ V}$$

For a cell reaction to run spontaneously, the algebraic sum of the standard potentials must produce a positive E^0_{cell}, that is, a negative free energy change. To get such a free energy change, we subtract the zinc half-reaction from that of silver chloride and balance the electrons:

$$2(AgCl + e^- \rightarrow Ag + Cl^-) \qquad 2E = 2(+0.222) = +0.444 \text{ V-eq}$$
$$\underline{Zn \rightarrow Zn^{2+} + 2e^-} \qquad \underline{2E = 2(+0.763) = +1.526 \text{ V-eq}}$$
$$Zn + 2AgCl \rightarrow Zn^{2+} + 2Ag + 2Cl^- \qquad \qquad \qquad 1.970 \text{ V-eq}$$

We can get the standard free energy change quite easily by multiplying 1.970 volt-equivalents (V-eq) by -96486.6 coulombs per equivalent (the faraday),

$$\Delta G = -nFE^0 = -190.1 \text{ kJ (per mole)}$$

or

$$\frac{-190.1 \text{ kJ/mole}}{4.184 \text{ kJ/kcal}} = -45.43 \text{ kcal/mole}$$

The equilibrium constant is given by

$$K^0 = 10^{nE^0_{cell}/0.059}$$
$$= 10^{33.4} = 2.4 \times 10^{33}$$

The standard potential of the cell is $\Delta G^0/2F = 0.985$ V, since two electrons are exchanged in the reaction. Knowing from K^0 that the reaction

$$Zn + 2AgCl \rightarrow Zn^{2+} + 2Ag + 2Cl^-$$

14.3 ELECTROCHEMICAL CELLS

is spontaneous, we draw the cell schematically with the silver electrode on the right-hand side

$$\text{anode } (-)\text{Zn} / \text{Zn}^{2+} / \text{AgCl}, \text{Cl}^-, \text{Ag}(+) \text{ cathode}$$

Recall that the cathode is always the site of reduction.

14.3.5 Cell Reversibility

Much to the confusion of students, electrochemists speak of reversibility in two ways, as chemical reversibility and electrochemical (thermodynamic) reversibility. A cell is chemically reversible if, when the direction of cell current is reversed, the same half-reactions occur but in the reverse direction. To illustrate these ideas consider the cell

$$\text{Pt}, \text{H}_2 / \text{H}^+ // \text{Cu}^{2+} / \text{Cu}$$

which implies the reaction

$$\text{H}_2 + \text{Cu}^{2+} \rightarrow \text{Cu} + 2\text{H}^+$$

If we apply a voltage from an outside source which opposes the galvanic emf and reverses the direction of current flow, we observe the reaction

$$\text{Cu} + 2\text{H}^+ \rightarrow \text{Cu}^{2+} + \text{H}_2$$

This is just the reverse of the galvanic reaction. The cell is *chemically* reversible.

However, if we take this cell and substitute a platinum metal electrode for the copper cathode, we create a chemically irreversible cell. The spontaneous reaction is the same as before, but the reverse reaction at the platinum electrode on the right-hand side is

$$2\text{H}_2\text{O} \rightarrow \text{O}_2 + 4\text{H}^+ + 4e^-$$

since there is no copper metal. The net cell reaction is simply

$$2\text{H}_2\text{O} = 2\text{H}_2 + \text{O}_2$$

The platinum cathode is an irreversible electrode unless it is coated with copper metal, in which case the forward and reverse reactions are again the same.

Thermodynamically reversible cells are those which are always in a state of equilibrium. In such cells the rate of electron transfer among species and between species and electrodes is so fast that the system can respond immediately to very small changes in cell emf. Imagine a galvanic cell such as that in Fig. 14.2 which is connected to an external emf that is of exactly the same magnitude but opposite polarity. If the cell is thermodynamically reversible, only a tiny ("infinitesimal") difference in the external emf will make the direction of the cell reaction reverse.

A chemically irreversible cell is never in a state of equilibrium and can never be operated reversibly in the thermodynamic sense. A chemically reversible cell can approach thermodynamic reversibility only when very little current flows through the cell. A rapidly discharging cell will not behave reversibly and a measurement of its emf will have little meaning. Experiments requiring careful cell emf measurements rely on the use of measuring devices which draw very little current. This will be discussed in Chapter 15.

Reversible and Irreversible Electrodes The simplest electrodes consist of a metal in contact with a solution of its own ions and are sometimes called "class I electrodes." Examples are

$$// \ Zn^{2+} \ / \ Zn \tag{a}$$
$$// \ Ag^{+} \ / \ Ag \tag{b}$$
$$// \ H^{+} \ / \ H_2, \ Pt \tag{c}$$

and are all reversible electrodes. The hydrogen electrode (c) in contact with hydrogen ions is usually included in this group, since it is equivalent to electrodes (a) and (b) except for the presence of platinum metal to provide an electrical path for the nonconductive hydrogen gas.

Electrodes of the "second kind," or class II, involve a metal, a slightly soluble salt of the metal, and a solution of some other salt containing the anion of the metal salt. The most important example is the saturated calomel electrode (SCE), the most common reference electrode in electrochemistry,

$$// \ KCl \ (saturated) \ / \ Hg_2Cl_2 \ / \ Hg \qquad E^0 = +0.2444 \ V$$

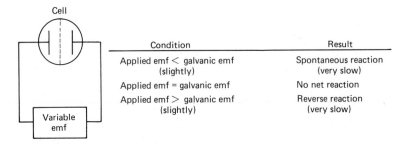

Figure 14.2 Thermodynamic reversibility in an idealized cell.

14.3 ELECTROCHEMICAL CELLS

Electrodes such as the SCE behave reversibly. Reactions involve oxidation of the metal or reduction of the salt, both chemically reversible with respect to the anion. In the case of the SCE the metal oxidation and complexation with chloride may be viewed as two steps which sum to give a net oxidation half-reaction:

$$2Hg \rightarrow Hg_2^{2+} + 2e^-$$
$$\underline{Hg_2^{2+} + 2Cl^- \rightarrow Hg_2Cl_2(s)}$$
$$2Hg + 2Cl^- \rightarrow Hg_2Cl_2(s) + 2e^-$$

The reduction reaction is exactly the reverse. The matter of complexation and its effects on cell potential will be taken up shortly.

"Class III" electrodes are considerably more rare but are of some interest in that they can be used to measure the activities of otherwise inscrutable ions. Electrodes in this class consist of a metal in equilibrium with two slightly soluble salts with a common anion. A good example is the calcium oxalate / silver oxalate / silver half-cell,

$$// \ CaCl_2 \text{ solution } / \ CaC_2O_4(s), \ Ag_2C_2O_4(s), \ Ag$$

This half-cell involves the set of reactions

$$2e^- + 2Ag^+ \rightarrow 2Ag$$
$$Ag_2C_2O_4(s) = 2Ag^+ + C_2O_4^{2-}$$
$$Ca^{2+} + C_2O_4^{2-} = CaC_2O_4(s)$$

The electrode is reversible with respect to calcium ion. It is necessary in electrodes such as this that the second salt (CaC_2O_4 here) be more soluble than the first salt ($Ag_2C_2O_4$ here). In principle, a class III electrode is a class II electrode known to be reversible, which is modified to respond to the activity of a third ion.

A fourth type of reversible electrode consists of two soluble species in equilibrium at an inert electrode, for example

$$// \ Fe^{3+}, \ Fe^{2+} \ / \ Pt$$
$$// \ Ce^{4+}, \ Ce^{3+} \ / \ Pt$$

These are legitimately class I electrodes, but function without transfer of metal ions between the electrode and solution interface. Platinum does not take part in the half-cell reaction, but serves simply to make a chemically inert electrical connection to the soluble redox species. When one member of the soluble redox pair is absent, the half-cell becomes irreversible for reasons discussed previously.

The design, construction, and use of electrodes from these classes will be taken up in Chapter 15.

14.3.6 Effects of Species Activities on Cell EMF: The Nernst Equation

Although we have not discussed the matter, you have probably anticipated that the standard free energy change is not the only factor governing the emf of an electrochemical cell. It turns out, of course, that the activities of the electroactive species in the cell influence the emf of the cell, in a manner first explained by Nernst.

The free energy change involved in the general reversible cell reaction

$$aA + bB = cC + dD$$

is given by the difference in free energy between products and reactants:

$$\Delta G_{cell} = (cG_C + dG_D) - (aG_A + bG_B)$$

Each free energy term in this expression is a function of species activity. For example in the case of species C,

$$G_C = G_C^0 + RT \ln a_C^c$$

in which all the terms have their usual significance; G_C^0 is, of course, the free energy of species C when $a_C^c = 1.0$.

Putting together this equation and the set of equations like it for all species, generates the expression for the free energy change for the cell,

$$\Delta G_{cell} = \Delta G_{cell}^0 + RT \ln \frac{a_C^c a_D^d}{a_A^a a_B^b}$$

Knowing that $\Delta G_{cell} = -nFE_{cell}$, we can write this equation as

$$E_{cell} = E_{cell}^0 - \frac{RT}{nF} \ln \frac{a_C^c a_D^d}{a_A^a a_B^b}$$

or (at 25°C)

$$= E_{cell}^0 - \frac{0.05916}{n} \log \frac{a_C^c a_D^d}{a_A^a a_B^b}$$

The last expression is the Nernst equation in its most familiar form. Notice that the negative sign accompanies the ratio (products/reactants) in the log term. A positive sign can be used with the ratio (reactants/products) in the log term. The log term contains the equilibrium constant for the redox reaction, when E_{cell} is 0 V. As is the case with the thermodynamic constants you studied in Chapter 3, pure solid phases involved in the reaction are assigned unit activity. Similarly, in dilute solutions water

14.3 ELECTROCHEMICAL CELLS

is assumed to have unit activity. In systems with gaseous components, the pressure of a gas (in atmospheres) is freely substituted for activity.

The direct application of the Nernst equation in the form of the last equation to actual laboratory work may be difficult, since we usually know concentrations of species rather than their activities. Recall from Chapter 3 that the activity of a species is equal to the product of its activity coefficient (f) and its concentration (in brackets). Therefore we can express the Nernst equation as

$$E = \underbrace{E^0 - \frac{0.059}{n} \log \frac{f_C^c f_D^d}{f_A^a f_B^b}}_{E^{0\prime}} - \frac{0.059}{n} \log \frac{[C]^c[D]^d}{[A]^a[B]^b}$$

$E^{0\prime}$ is a standard potential associated with concentration measurements and is analogous to the concentration equilibrium constant, K, cousin of the thermodynamic constant, K^0.

It is rather common for one of the reactants or products of a cell reaction to react with water or some other component of the solution and be present in more than one form. For example, in dilute HCl solutions, ferric iron can be expected to be involved in reactions such as

$$Fe^{3+} + Cl^- = FeCl^{2+}$$
$$FeCl^{2+} + Cl^- = FeCl_2^+$$
$$Fe^{3+} + H_2O = FeOH^{2+} + H^+, \text{ etc.}$$

Instead of trying to calculate the equilibrium concentration of Fe^{3+} from equilibrium constants, it is more practical to use a form of the Nernst equation with analytical (formal) concentrations, denoted $C_{Fe^{3+}}$, for example. With formal concentrations the Nernst equation becomes

$$E = E^f - \frac{0.059}{n} \log \frac{C_C^c C_D^d}{C_A^a C_B^b}$$

where E^f is the *formal potential.*

By definition, the formal potential is the potential of a half-cell for which the logarithmic concentration term of the Nernst equation is zero. Either all concentrations are $1F$, or the ratio of the formal concentrations is unity. Formal potentials are ordinarily measured in solutions which contain a substantial excess of added electrolyte ("supporting electrolyte") relative to the concentrations of members of the reduction couple. Members of the couple whose concentrations are fixed by the electrolyte (H^+, OH^-, Cl^-, etc.) are not written into a Nernst equation involving formal potentials. The formal potential is related to a standard potential in the way that a conditional equilibrium constant is related to a thermodynamic constant (see Chapter 8).

EXAMPLE 14.2
Given the cell SHE // Fe^{3+}, Fe^{2+} (1M HCl) / Pt, calculate the potential of the cell at $C_{Fe^{2+}} = 0.1M$, $0.001M$, $10^{-5}M$ at constant $C_{Fe^{3+}} = 10^{-2}M$. $E^0(Fe^{3+}/Fe^{2+}) = 0.771$ V, $E^f(Fe^{3+}/Fe^{2+}) = 0.70$ V.

Solution

Use the expression for the full cell potential

$$E_{cell} = E_{right} - E_{left}$$
$$= E^f(Fe^{3+}/Fe^{2+}) - 0.059 \log \frac{C_{Fe^{2+}}}{C_{Fe^{3+}}} - E_{SHE}$$
$$= 0.70 - 0.059 \log \frac{0.1}{0.01} - 0.000 \text{ V}$$
$$= 0.70 - 0.059 \text{ V} = \underline{0.64 \text{ V}}$$

For
$C_{Fe^{3+}} = 10^{-2}M$, $C_{Fe^{2+}} = 10^{-3}M$, $E = 0.76$ V
$C_{Fe^{3+}} = 10^{-2}M$, $C_{Fe^{2+}} = 10^{-5}M$, $E = 0.88$ V

Notice that an increase in the ratio of concentrations of oxidized form to reduced form makes the half-cell potential more positive. This means that the half-cell becomes more strongly oxidizing. Furthermore, in a cell such as this where both members of the couple are soluble and there is no dilution-dependent dimerization process, the potential of the cell does not depend on dilution if ionic strength is kept constant. So long as the *ratio* of the concentrations of oxidized and reduced forms and the ionic strength is constant, cell potential will be constant.

EXAMPLE 14.3
Consider the cell

$$\text{SHE} // Cr_2O_7^{2-}, H^+(0.1M \text{ HCl}), Cr^{3+} / Pt,$$

with half-reaction

$$Cr_2O_7^{2-} + 14H^+ + 6e = 2Cr^{3+} + 7H_2O$$

Calculate the potential of the cell when the concentration of dichromate is $0.1F$ and that of Cr^{3+} ion is $0.01F$. $E^f = 0.93$ V; $E^0 = 1.33$ V.

14.3 ELECTROCHEMICAL CELLS

Solution

The equivalent forms of the Nernst equation are

$$E = E^0(Cr_2O_7^{2-}/Cr^{3+}) - \frac{0.059}{6} \log \frac{a_{Cr^{3+}}^2}{a_{Cr_2O_7}a_{H^+}^{14}}$$

$$= E^{0\prime}(Cr_2O_7^{2-}/Cr^{3+}) - \frac{0.059}{6} \log \frac{[Cr^{3+}]^2}{[Cr_2O_7^{2-}][H^+]^{14}}$$

$$= E^f(Cr_2O_7^{2-}/Cr^{3+}) - \frac{0.059}{6} \log \frac{C_{Cr^{3+}}^2}{C_{Cr_2O_7^{2-}}}$$

Substitution gives $E = 0.93 - \frac{0.059}{6} \log \frac{(0.01)^2}{0.1} = 0.96$ V. The fact that the formal potential of the couple in 0.1M HCl is lower than the standard potential is the anticipated result: decreased solution acidity should decrease the oxidizing power of dichromate. Notice that the Nernst equation for the formal potential does not include a concentration term for hydrogen ion; [H$^+$] is held constant by the electrolyte. Formal potentials are always given with a specified electrolyte. Formal potentials for a variety of couples are included in the table of standard potentials (Appendix IV).

14.3.7 Effects of Complexation on Cell EMF

Formation of an Insoluble Complex The oxidizing or reducing powers of a metal ion can be changed dramatically by complexing it with an anion. For example, aquated silver ion, Ag$^+$(aq), is a rather strong oxidizing agent ($E^0 = 0.799$ V). When complexed with chloride to form insoluble AgCl, however, its oxidizing power is reduced considerably ($E^0 = 0.222$ V). This effect can be understood in terms of standard free energies. Figure 14.3 shows standard free energies for Ag$^+$(aq), Ag^0, AgCl, and AgI, as well as for the mixtures Ag^0 + Cl$^-$ and Ag^0 + I$^-$. Notice that the difference in standard free energies of reactants and products is quite large for the reduction of silver ion,

$$Ag^+(aq) + e \rightarrow Ag^0 \qquad \Delta G^0 = -77.1 \text{ kJ/mole} \leftrightarrow 0.799 \text{ V}$$

and considerably smaller for the reductions of silver chloride and silver iodide

$$AgCl + e \rightarrow Ag^0 + Cl^- \qquad \Delta G^0 = -21.8 \text{ kJ/mole} \leftrightarrow E^0 = 0.222 \text{ V}$$
$$AgI + e \rightarrow Ag^0 + I^- \qquad \Delta G^0 = 14.6 \text{ kJ/mole} \leftrightarrow E^0 = -0.151 \text{ V}$$

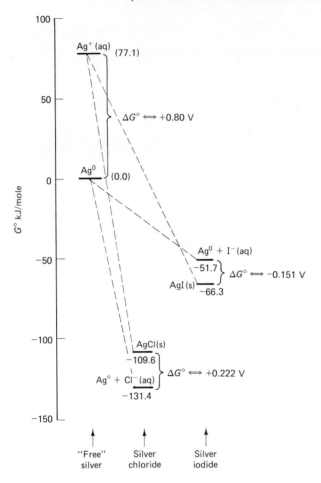

Figure 14.3 Free energy changes for silver species.

Figure 14.3 shows that the presence of the halogen anions lowers the standard free energies of both reactants (AgX) and products (Ag + X$^-$), but lowers the reactants *more*, thus decreasing $\Delta G°$ for the reduction. Notice that the standard free energy of AgI is *lower* than that of Ag^0 + I$^-$, and so $\Delta G°$ is positive for the reduction of AgI ($E°$ is negative).

The relationship between the formation constant for a complex and its standard potential can be developed by using the AgCl/Ag couple. Consider an electrochemical cell in which the half-cell reactions involve the couples Ag$^+$ + e = Ag and AgCl + e = Ag + Cl$^-$:

$$\text{Ag} \,/\, \text{AgCl} \,/\, \text{Cl}^- \,//\, \text{Ag}^+ \,/\, \text{Ag}$$

14.3 ELECTROCHEMICAL CELLS

When the activities of Cl^- and Ag^+ are unity,

$$E_{cell} = E^0_{cell} = E^0_{Ag^+/Ag} - E^0_{AgCl/Ag} = 0.799 - 0.222 = 0.577 \text{ V}$$

The sign of E_{cell} tells us that when the anode and cathode are connected, Ag^+ is reduced to Ag^0 on the right and Ag^0 is oxidized to AgCl on the left. When the system reaches equilibrium $E_{cell} = 0$ and so

$$E^0_{Ag^+/Ag} - 0.05916 \log \frac{1}{a_{Ag^+}} = E^0_{AgCl/Ag} - 0.05916 \log a_{Cl^-}$$

Rearranging gives

$$E^0_{AgCl/Ag} = E^0_{Ag^+/Ag} - 0.05916 \log \frac{1}{a_{Ag^+} a_{Cl^-}}$$

Since K_f for AgCl is given by $1/(a_{Ag^+} a_{Cl^-})$,

$$E^0_{AgCl/Ag} = E^0_{Ag^+/Ag} - 0.05916 \log K_f$$

The solubility product constant is given by $1/K_f$, and

$$E^0_{AgCl/Ag} = E^0_{Ag^+/Ag} + 0.05916 \log K_{sp}$$

is perhaps an equally useful expression.

Two Soluble Complexes The way in which the complexation of both species of a redox couple influences the reduction potential of the couple can be seen by examining the citric acid complexes ($C_6H_8O_7$ or "H_4Cit") of ferric and ferrous iron. The relative formation constants tell us that the ferric citrate species is more stable than the ferrous citrate species:

$$Fe^{3+} + H_2Cit^{2-} = FeH_2Cit^+ \qquad K_{III} = \frac{[FeH_2Cit^+]}{[Fe^{3+}][H_2Cit^{2-}]} = 10^{6.3}$$

$$Fe^{2+} + H_2Cit^{2-} = FeH_2Cit \qquad K_{II} = \frac{[FeH_2Cit]}{[Fe^{2+}][H_2Cit^{2-}]} = 10^{2.12}$$

The Nernst equation for the reaction $Fe^{3+} + e^- \rightarrow Fe^{2+}$ is

$$E = E^0(Fe^{3+}/Fe^{2+}) - 0.059 \log \frac{[Fe^{2+}]}{[Fe^{3+}]}$$

If we take the ratio of equilibrium constants above and rearrange them to solve for the ratio of ferrous to ferric iron concentrations, we find

$$\frac{[Fe^{2+}]}{[Fe^{3+}]} = \frac{K_{III}}{K_{II}} \frac{[FeH_2Cit]}{[FeH_2Cit^+]}$$

When this is substituted into the Nernst equation, we find

$$E = E^{0\prime}(Fe^{3+}/Fe^{2+}) - 0.059 \log \frac{K_{III}}{K_{II}} - 0.059 \log \frac{[FeH_2Cit]}{[FeH_2Cit^+]}$$

Part of this expression can be recognized as the Nernst equation for the half-reaction $FeH_2Cit^+ + e \rightarrow FeH_2Cit$, that is,

$$E = E^{0\prime}(FeH_2Cit^+/FeH_2Cit) - 0.059 \log \frac{[FeH_2Cit]}{[FeH_2Cit^+]}$$

At equilibrium the last two equations equal zero volts. When we set them equal to each other, we see the effect of complexation quantitatively

$$E^{0\prime}(FeH_2Cit^+/FeH_2Cit) = E^{0\prime}(Fe^{3+}/Fe^{2+}) + 0.059 \log \frac{K_{II}}{K_{III}}$$

In $1M$ NaClO$_4$, $E^{0\prime}(Fe^{3+}/Fe^{2+}) = 0.74$ V, and so

$$E^{0\prime}(FeH_2Cit^+/FeH_2Cit) = 0.74 + 0.059 \log 10^{-4.2}$$
$$= \underline{0.49 \text{ V}}$$

From the relative potentials we can tell that ferric iron has become a weaker oxidizing agent when complexed with citric acid. The +3 oxidation state is stabilized relative to the +2 oxidation state. On the other hand, for the 1,10-phenanthroline complexes of iron, we find

$$Fe(Ophen)_3^{3+} + e \rightarrow Fe(Ophen)_3^{2+} \qquad E^{0\prime} = 1.06 \text{ V } (1M \text{ H}_2\text{SO}_4)$$

The fact that $E^{0\prime}$ for this couple is higher than that for the uncomplexed species in the same electrolyte (+0.68 V) means that complexed iron(III) is a stronger oxidizer: $Fe(Ophen)_3^{2+}$ is a more stable complex than $Fe(Ophen)_3^{3+}$. At this point you should be able to calculate the ratio of stability constants from the $E^{0\prime}$ values for complexed and uncomplexed species.

Effects of pH A great many half-reactions involve hydrogen ions or hydroxide ions. Such reactions invariably involve oxidized and reduced species which contain different

14.3 ELECTROCHEMICAL CELLS

numbers of oxygen atoms. For example, consider the vanadium(V)/(IV) reduction couple,

$$VO_2^+ + 2H^+ + e^- \rightarrow VO^{2+} + H_2O$$

The Nernst equations

$$E = E^{0\prime}(VO_2^+/VO^{2+}) - 0.059 \log \frac{[VO^{2+}]}{[VO_2^+][H^+]^2}$$

and

$$E = E^f(VO_2^+/VO^{2+}) - 0.059 \log \frac{C_{VO^{2+}}}{C_{VO_2^+}}$$

apply for this couple. If $[VO^{2+}] = C_{VO^{2+}}$ and $[VO_2^+] = C_{VO_2^+}$, then

$$E^f(VO_2^+/VO^{2+}) = E^{0\prime}(VO_2^+/VO^{2+}) + 0.059 \log [H^+]^2$$

Since pH is approximately the negative logarithm of $[H^+]$, this expression becomes

$$E^f(VO_2^+/VO^{2+}) = E^{0\prime}(VO_2^+/VO^{2+}) - 0.059(2)\text{pH}$$

$E^{0\prime}$ for the couple is about 1.0 V. At pH = 1, the formal potential for the couple is about 2(0.059 V) = 0.12 V lower than at pH = 0.

The oxidizing power of species such as UO_2^{2+}, MnO_4^-, IO_3^-, $Cr_2O_7^{2-}$, and AsO_4^{3-} is strongly dependent on solution acidity and will be discussed in Chapter 16.

14.3.8 An Additional Problem: Liquid Junction Potential

It was mentioned earlier that an electrical potential arises when the solutions in two separate half-cell compartments are joined. This so-called liquid junction potential occurs when ions not present in both systems diffuse across the region of the junction from an area of higher concentration to one of lower concentration. The force of diffusion depends on the concentrations of the ions, while the rate depends on the ease with which the ions move through solution (mobility). Differences in the mobilities of ions tends to produce a charge separation at the junction and thus a potential drop.

There are three liquid junction situations which can develop, as shown in Figure 14.4a–c. In situation (a) (type I liquid junction), there are two solutions of the same electrolyte at different concentrations. There is diffusion of both H^+ and Cl^- from right to left, but H^+ is the more mobile ion and outruns Cl^-, producing a net positive charge on the left side of the junction. Using the convention $E_{\text{right}} - E_{\text{left}}$, the liquid junction potential is negative in this case. In some experimental situations $E_{\text{l.j.}}$ can be

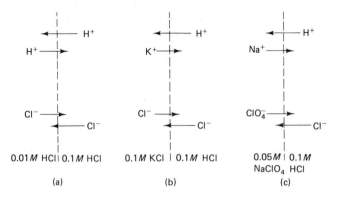

Figure 14.4 Liquid junctions.

approximated by calculations, as discussed by Lingane (Ref. 2) and authors he refers to.

In situation (b) (type II junction) there are two solutions of the same concentration with one ion in common. In this example, Cl^- does not diffuse because there is no appreciable concentration gradient at the junction. Hydrogen ions are more mobile than potassium ions, however, and the left-hand side builds a net positive charge. In this case $E_{l.j.}$ is negative.

Situation (c) is a type III liquid junction and is a very complicated situation. Here the solutions differ in both concentration and species, $0.1M$ HCl and $0.05M$ $NaClO_4$. All ions diffuse across the junction, but the sign and (particularly) the magnitude of $E_{l.j.}$ are hard to predict. Unfortunately, type III junctions are fairly common.

It is possible to minimize (but not eliminate) liquid junction potential from cells by using a "salt bridge" as a junction. Such a bridge is made by pouring a hot solution of salt containing agar into a U-shaped glass tube and allowing it to cool and gel. Two popular recipes are 30% KCl/3% agar/67% water and 10% KNO_3/3% agar/87% water. The inverted U tube can make a connection between two beakers serving as half-cell compartments. If used more than once, these bridges must be stored in solutions of the appropriate electrolyte and inspected regularly for gel tube separation or growth of bacteria. The huge concentration of salt in bridges such as these lowers liquid junction potential to 1 or 2 mV.

There is only one type of cell which truly has no liquid junction, the amalgam concentration cell. An example is the zinc amalgam cell,

$$\text{Zn(Hg)} \ / \ \text{Zn}^{2+} \ / \ \text{Zn(Hg)}$$
$$C_1 \quad C_3 \quad C_2$$

The half-reactions and full cell reaction are

14.3 ELECTROCHEMICAL CELLS

$$\begin{array}{rl} & Zn(Hg)(C_1) \rightarrow Zn^{2+}(C_3) + 2e^- \\ & \underline{Zn^{2+}(C_3) + 2e^- \rightarrow Zn(Hg)(C_2)} \\ \text{Net:} & Zn(Hg)(C_1) \rightarrow Zn(Hg)(C_2) \end{array}$$

The half-reactions are indeed legitimate chemical reactions, but the net reaction is simply a transfer of zinc from one amalgam pool to the other. The full cell emf is independent of concentration of zinc ions and, as is the case in all true "concentration cells," the equilibrium constant for the process is unity ($E^0_{\text{right}} - E^0_{\text{left}} = 0$). The cell emf is given by

$$E_{\text{cell}} = \frac{0.059}{2} \log \frac{C_1}{C_2}$$

Table 14.1 shows liquid junction potentials obtained for particular junctions. Most measurements like these are made when a cell without a liquid junction (such as the zinc amalgam cell) has a liquid junction added specifically to study the potential drop at the junction.

14.3.9 Practical Cells

The efforts of a great many electrochemists, particularly in the industrial sector, are devoted to the efficient production and storage of electrical power in cells. This section contains a description of a few classical and commercially important electrochemical cells. The discussion is by no means definitive or exhaustive, but it may help bring together some of the concepts introduced in the chapter.

TABLE 14.1 LIQUID JUNCTION POTENTIALS

Junction	$E_{\text{l.j.}}$ (mV)
0.1F NaCl/0.1F KCl	−6.4
0.1F HCl/0.1F KCl	+27
0.1F NaCl/3.5 F KCl[a]	−0.2
0.05F H$_2$SO$_4$/3.5F KCl[a]	+4
0.1F NaOH/3.5F KCl[a]	−2.1

[a] Corresponds to liquid junction potential at salt bridge-electrolyte interface.

The following terms are often encountered in descriptions of electrochemical cells:

Battery: collection of two or more cells.

Primary cell: electrochemical energy converter with internal storage of chemicals, which cannot be recharged. A "one shot" power cell.

Fuel cell: primary cell with external storage of reactants.

Secondary cell: electricity storage device which takes in and dispenses only electricity.

Rechargeable power cell: cell whose discharge reactions are reversible.

Storage density: measure of the maximum amount of electricity (coulombs) that can be drawn per unit weight of chemical reactant.

$$\text{S.D.} = \text{coulombs/gram} = \frac{96486.6(n)}{\text{grams substance}}$$

(Elements with high atomic weights have low storage densities.)

Energy density: measure of the energy that can be drawn from a unit weight of substance; the product of storage density and cell voltage. A cell with a large storage density can have a small energy density if cell voltage is small. (Units: coulombs volts/gram = watts seconds/gram.)

Power density: rate at which a device can release the energy of each cell; the product of cell voltage and current. [A cell with large energy density can have a small power density if cell resistance is large.]

Primary Cells The earliest chemical/electrical energy converting device was created by Alessandro Volta in 1800 during experiments performed to test the theory that electric charge could come from the junction of two metal disks. The "voltaic pile" was made from zinc and silver disks in pairs separated by brine-soaked sheets of pasteboard. While we now realize that the electrical current developed in the brine electrolyte rather than at the zinc-silver contact, Volta's work showed that electricity could be gotten from the reaction of chemical elements. A few years afterward (1807), Sir Humphrey Davy built a voltaic pile of over 250 metal plates and was able to reduce the elements sodium and potassium electrochemically. Davy went on to isolate barium, magnesium, calcium, and strontium. Electrochemistry became an important method of isolating elements and a method vital to chemistry.

The first really practical long-term generator of electrical energy was that designed by Daniell in 1836. The cell consisted of a zinc anode dipped in a sulfuric acid solution separated from a copper solution/copper metal electrode (cathode) by a porous clay vessel. A modern representation is shown in Fig. 14.5. The spontaneous reaction should be quite familiar by now,

14.3 ELECTROCHEMICAL CELLS

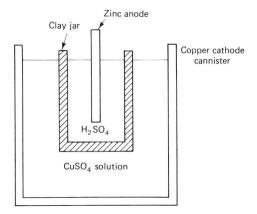

Figure 14.5 Daniell cell.

$$Cu^{2+} + 2e \rightarrow Cu^0$$
$$\underline{Zn^0 \rightarrow Zn^{2+} + 2e}$$
$$Cu^{2+} + Zn^0 \rightarrow Cu^0 + Zn^{2+}$$

Cell voltages of about 1.1 V can be obtained from this cell.

The Leclanché dry cell was an important development in the portability of primary cells. In a dry cell the electrolyte is immobilized in the form of an electrically conductive paste. A modern version of the Leclanché cell is shown in Fig. 14.6. The anode reaction is $Zn \rightarrow Zn^{2+} + 2e$, while the cathode reaction is

$$2MnO_2 + 2H^+ + 2e \rightarrow Mn_2O_3 + H_2O$$

The consumption of protons in the paste means that hydroxyl ions are generated. These are involved in side reactions such as

$$Zn^{2+} + 2OH^- \rightarrow ZnO + H_2O$$

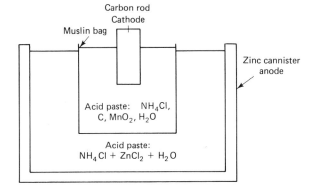

Figure 14.6 Leclanché cell.

which render the cell only partially rechargeable and, in practice, a primary cell. The relatively large energy density of about 500 W sec/g under low current drain have made these cells popular for use in flashlights, portable radios, and toys.

A more modern version of the Leclanché cell, the manganese alkaline dry cell, can be made by substituting KOH for NH_4Cl in the past electrolyte. In modern designs the zinc electrode is placed in the center of the cell array. Cell voltage is about 1.6 V and energy density about 550 W sec/g at higher current levels than the dry cell can provide.

Alkaline mercury cells are fundamentally the same as alkaline manganese cells, except that a mercuric oxide-covered graphite electrode is used as a cathode with KOH in the electrolyte. These cells are comparable to their manganese counterparts in energy density (390 W sec/g) but can sustain a high current draw without a substantial decrease in voltage (1.35 V). Their long life, both in service and on the shelf, has made them popular for use in photographic equipment, hearing aids, and digital watches.

Energy Storage Devices (Secondary Cells) The lead-acid storage battery (Planté, 1860) is one of the oldest and most lastingly popular rechargeable devices made. The reliability of these batteries and their simplicity, low cost, and high current output account for their almost universal use in automobiles and trucks.

A single cell ($E = 2.05$ V) is made of two sheets of lead metal with a porous sheet spacer between them soaked in sulfuric acid (saturated with lead sulfate). During a charging cycle Pb^{2+} is deposited on the Pb cathode,

$$Pb^{2+} + 2e \rightarrow Pb \text{ (cathode)}$$

On the second electrode, Pb^{2+} from solution is oxidized to Pb^{4+}, which subsequently hydrolyzes and deposits on the plate,

$$Pb^{2+} + 6H_2O \rightarrow PbO_2 + 4H_3O^+ + 2e \text{ (anode)}$$

During a discharge the reverse reactions occur with production of large currents. Three or six cells are combined in electrical series to produce a 6- or 12-V battery.

A limitation of the lead-acid battery results from a process called *sulfation,* which is a spontaneous discharge reaction occurring even when the battery is not under load. At the lead metal electrode the pair of reactions

$$Pb + SO_4^{2-} \rightarrow PbSO_4 + 2e$$
$$2H^+ + 2e \rightarrow H_2$$

take place, while at the lead oxide surface

$$PbO_2 + 2H_2SO_4 + 2e \rightarrow PbSO_4 + 2H_2O + SO_4^{2-}$$

and

$$Pb + SO_4^{2-} \rightarrow PbSO_4 + 2e$$

14.3 ELECTROCHEMICAL CELLS

both occur. The result of both pairs of reactions is the coating of the battery electrode surfaces with lead sulfate, lowering the efficiency of the desired reactions and shortening the life of the battery.

A second problem with lead-acid batteries is their low energy density (70 W sec/g). Anyone who has tried to lift a dead battery out of a car on a January morning knows that a massive battery is required to generate enough power to start an automobile engine.

The nickel-cadmium (NiCad) battery is probably the second most popular commercial battery at present. Diagrams of a NiCad cell during discharge and recharge are shown in Fig. 14.7. NiCad batteries have enjoyed increasing popularity in recent years in solid-state devices such as hand-held calculators. They have a relatively low cost and can be operated through hundreds of charge/discharge cycles. The average energy density for NiCad cells is about 150 W sec/g, twice that of the lead-acid battery.

A relative newcomer (1960) to the field of energy storage devices is the silver-zinc cell. This high-energy cell (400 W sec/g) uses silver oxide (Ag_2O) and zinc electrodes. During discharge the cathode reaction is

$$Ag_2O + H_2O + 2e^- \rightarrow 2Ag + 2OH^-$$

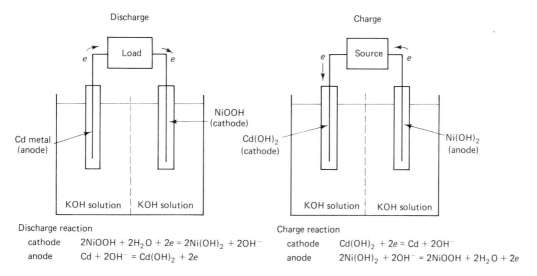

Figure 14.7 Nickel-cadmium cell. (All electrodes have metal cores.)

Discharge reaction:

$$2NiOOH + 2H_2O + 2e = 2Ni(OH)_2 + 2OH^- \quad \text{(cathode)}$$

$$Cd + 2OH^- = Cd(OH)_2 + 2e \quad \text{(anode)}$$

Charge reaction:

$$Cd(OH)_2 + 2e = Cd + 2OH^- \quad \text{(cathode)}$$

$$2Ni(OH)_2 + 2OH^- = 2NiOOH + 2H_2O + 2e \quad \text{(anode)}$$

while the anode reaction is

$$Zn + 4OH^- \rightarrow ZnO_2^{2-} + 2H_2O + 2e$$

The charging cycle reactions are just the opposite of these. Although the presence of slow side reactions limits these cells to less than about 100 charging cycles, their high energy density makes them highly desirable secondary cells.

Lithium Primary Cells Several recently invented primary cells are based on lithium chemical systems. Lithium metal is an excellent anode material because it is the most powerful reducer known ($E^0 = -3.05$ V) and has a very low density. Thus cells built with lithium anodes are both powerful and lightweight; that is, they have high energy densities. When hermetically sealed, these devices are excellent implantable power sources for cardiac pacemakers. Schematics for two lithium-based devices are shown along with full cell reactions in Figs. 14.8 and 14.9. The lithium-iodine cell uses a cathode made of a mixture of iodine (I_2) and polyvinyl pyridine (PVP). The Wilson Greatbatch model 752 Li-I_2 cell is a hermetically sealed canister of volume 9.5 cm^3, weighing only 27 g, with a nominal voltage of 2.8 V and an energy density of almost 400 W sec/g (compare with 70 W sec/g for a lead-acid cell). The lifetime of a model 752 cell is about six years under pacemaker load conditions.

Another interesting lithium cell is the lithium-silver chromate cell. Many of these are produced as button-type cells for electronic watches and pacemakers. The anode is again lithium metal, while the cathode is a mixture of silver chromate powder and graphite pressed into a pellet. The electrolyte is typically a solution of lithium perchlorate in propylene carbonate solvent. A sketch of the cross section of an SAFT model Li210 cell is shown in Fig. 14.9. The cell has a high operating voltage (3.2 V) and an extremely high energy density ($\sim$470 W sec/g).

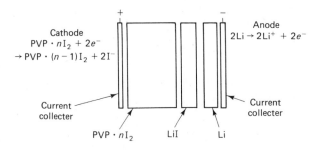

Cell reaction: PVP · nI_2 + 2Li → PVP · $(n-1)I_2$ + 2LiI + 2 faradays

Figure 14.8 Lithium-iodine primary cell. (Courtesy of Cardiac Pacemakers, Inc., St. Paul, Minnesota.)

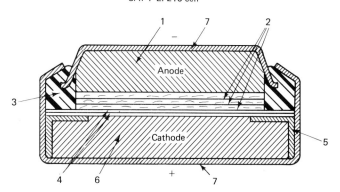

Figure 14.9 Lithium-silver chromate primary cell. 1, Metallic lithium anode; 2, polypropylene separator; 3, propylene grommet; 4, barrier; 5, retaining ring; 6, silver chromate cathode; 7, stainless steel case. (Developed by the Societé des Accumulateurs Fixes et de Traction (SAFT), France. Courtesy of Cardiac Pacemakers, Inc., St. Paul, Minnesota.)

The exploration and creation of new electrochemical energy devices is one of the most interesting areas of chemistry. With recent developments in energy use and the need for improved management of energy, the search for highly efficient, lightweight, low-cost means of storing energy is certain to intensify.

REFERENCES

1. W. M. Latimer, *Oxidation Potentials,* 2nd ed., Prentice-Hall, Englewood Cliffs, N.J., 1952.
2. J. J. Lingane, *Electroanalytical Chemistry,* 2nd ed., Interscience, New York, 1958.
3. G. F. Smith and F. P. Richter, *Analytical Chemistry,* 16, 380 (1944).

PROBLEMS

14.1. Under what conditions is a cell said to be reversible? Carefully distinguish between chemical and thermodynamic reversibility.

14.2. Calculate the emf of each of the following cells, using standard potentials. Identify the anode and cathode in each cell. SHE and SCE are reference electrodes discussed in the text.
 (a) SHE // $Tl^+(10^{-3}M)$ / Tl
 (b) SHE // $Zn^{2+}(10^{-3}M)$ / Zn

(c) Tl / Tl$^+$(10^{-3}M) // Zn^{2+}(10^{-3}M) / Zn
(d) SHE // pH 4.00, P_{H_2} = 1.00 atm / Pt
(e) SCE // Ag$^+$(0.01M) / Ag
(f) SCE // Cu^{2+}(0.02M), Cu$^+$(0.001M) / Pt
(g) Pt / Cu^{2+}(0.02M), Cu$^+$(0.001M) // Ag$^+$(0.01M) / Ag

14.3. Give the shorthand representations of the cells whose net reactions are given. Calculate E^0 for these cells from standard electrode potentials.
(a) $\frac{1}{2}H_2 + Ag^+ = H^+ + Ag$
(b) $Zn + 2Tl^+ = Zn^{2+} + 2Tl$
(c) $2AgCl + H_2 = 2Ag + 2Cl^- + 2H^+$
(d) $S_2O_8^{2-} + 2I^- = I_2 + 2SO_4^{2-}$

14.4. Knowing the standard potentials for the two half-reactions

$$Ag^{2+} + e = Ag^+ \quad \sim 1.95 \text{ V}$$
$$Ag^+ + e = Ag^0 \quad +0.7995 \text{ V}$$

calculate the standard potential for the half-reaction

$$Ag^{2+} + 2e = Ag^0$$

14.5. Using data from Appendix IV, calculate the standard potential for the half-reaction

$$Ti^{3+} + 3e = Ti^0$$

14.6. The silver-zinc cell described in the chapter relies on the half-cell reactions

$$Ag_2O + H_2O + 2e = 2Ag + 2OH^- \quad E = +0.342 \text{ V}$$
$$ZnO_2^{2-} + 2H_2O + 2e = Zn + 4OH^- \quad E = 1.216 \text{ V}$$

Calculate the equilibrium constant for the full cell reaction and the potential of the full cell. Which electrode is the cathode and which is the anode in the cell?

14.7. The silver-zinc cell in Problem 14.6 uses an alkaline paste. Describe the cell reaction and calculate the theoretical voltage (unit activities) for a silver-zinc cell using an *acidic paste*. (Use silver metal and zinc metal electrodes.)

14.8. The lithium-silver chromate cell described in the text makes use of the half-reaction

$$Ag_2CrO_4 + 2e = 2Ag^0 + CrO_4^{2-} \quad E = +0.446 \text{ V}$$

PROBLEMS

If you were setting out to increase the voltage of a lithium-based cell, would there be any reason to explore the possibility of substituting silver ferrocyanide for silver chromate?

$$Ag_4Fe(CN)_6 + 4e = 4Ag + Fe(CN)_6^{4-} \qquad E = +0.19 \text{ V}$$

14.9. Given the equilibrium constant, $K = 15$, for the reaction

$$Br_2(aq) + Br^- = Br_3^-$$

and $E^0(Br_2/2Br^-) = 1.085$ V, calculate $E^0(Br_3^-/3Br^-)$.

14.10. Smith and Richter (Ref. 3) measured reduction potentials for systems of the type

$$Fe(Phen)_3^{3+} + e = Fe(Phen)_3^{2+}$$

where "Phen" represents various substituted 1,10-phenanthrolines (ferroins). Calculate the ratio of stability constants of the ferrous and ferric complexes given that in $1M$ sulfuric acid, $E^f(Fe^{3+}/Fe^{2+}) = 0.68$ V and the formal potentials of the Phen couples are

nitro ferroin	1.25 V
chloro ferroin	1.11 V
2,2'-dipyridyl ferroin	0.97 V

Which ferroin species forms the most stable complexes with Fe^{2+}?

14.11. Calculate the solubility product of lead sulfate, given

$$E^0(Pb^{2+}/Pb) = -0.126 \text{ V}$$
$$E^0(PbSO_4/Pb) = -0.3563 \text{ V}$$

Be sure to derive a Nernst relationship governing K_{sp}.

14.12. The cell

$$\text{Ag, } Ag_2SO_4(s) \left/ \begin{array}{c} Hg_2SO_4(aq) \text{ saturated} \\ Ag_2SO_4(aq) \text{ saturated} \end{array} \right/ Hg_2SO_4(s), \text{ Hg}$$

has an emf of 0.140 V. Write a spontaneous chemical reaction for the cell and calculate the free energy change for the reaction.

14.13. In the lead-acid storage battery, lead oxide is reduced at the cathode to Pb^{2+} according to the reaction:

$$PbO_2 + 4H_3O^+ + 2e = Pb^{2+} + 6H_2O \quad E^0 = 1.455 \text{ V},$$

while at the anode, lead is oxidized to Pb^{2+},

$$Pb = Pb^{2+} + 2e^-$$

for which

$$Pb^{2+} + 2e^- = Pb \quad E^0 = -0.126 \text{ V}$$

Using this information, calculate the standard potential of the half-reaction

$$PbO_2 + 4H_3O^+ + 4e^- = Pb + 6H_2O$$

14.14. Given the information

$$\begin{aligned} Cu^{2+} + 2e^- &= Cu & E^0 &= 0.337 \text{ V} \\ Cu^+ + e &= Cu & E^0 &= 0.521 \text{ V} \\ Cu^{2+} + e &= Cu^+ & E^0 &= 0.153 \text{ V} \end{aligned}$$

calculate K and ΔG (kJ/mole) for the reaction

$$2Cu^+ = Cu + Cu^{2+}$$

14.15. What is the reduction potential of the Sn(IV)/Sn(II) couple in $1M$ HCl when the ratio of the concentrations of Sn(IV) to Sn(II) is 0.01, 0.1, 1.0, 10.0, 100? The formal potential of the couple is 0.14 V vs. SHE.

14.16. The solubility product of mercurous sulfate is 6.2×10^{-7} at 25°C, and the cell

$$Pt \:/\: H_2(735 \text{ mm}) \:/\: H_2SO_4(10^{-3}M) \:/\: Hg_2SO_4(s), Hg$$

has an emf of 0.863 V at 25°C. Calculate the standard potential of the half-reaction

$$Hg_2^{2+} + 2e = 2Hg$$

14.17. Consider the cell $Tl \:/\: Tl^+ \:/\!/\: Np^{4+}, Np^{3+} \:/\: Pt$.

$$E^0(Tl^+/Tl) = -0.3363 \text{ V, and } E^0(Np^{4+}/Np^{3+}) = +0.147 \text{ V}.$$

(a) Which of the electrodes will be positive if all ions are present at unit activity?
(b) What is the equilibrium constant for the cell reaction?

(c) If the activity of Tl^+ is $1.00 M$ and that of Np^{4+} is $10^{-6} M$, what would the activity of Np^{3+} have to be for the cell emf to be 0.100 V?

14.18. Interpret the observation that the standard potential of the Fe(III)/Fe(II) couple is 0.771 V, while the formal potential of the couple in $1M$ HCl is +0.70 V.

14.19. Consider a power cell whose representation is

$$Ag^0, AgCl(s) \,/\, Cl^- \,/\, Hg_2Cl_2(s), Hg^0$$
$$E^0(Hg_2Cl_2/Hg) = +0.268 \text{ V}, \quad \text{and} \quad E^0(AgCl/Ag) = +0.222 \text{ V}.$$

(a) For unit activity of chloride ion, how many cells would have to be connected in series to generate a voltage of 1.0 V?

(b) Using Nernst expressions, show why you would expect the voltage to drop suddenly when the cell goes "dead." (Hint: Examine the concentration dependence of the full cell reaction.)

(c) What will cause failure of the cell?

Chapter 15

Potentiometry

15.1 INTRODUCTION

The subject of this chapter is the measurement of ionic activities and concentrations via the measurement of electrochemical cell potential. The basic ideas should be familiar by now. In Chapter 6 it was necessary to introduce the use of a glass membrane electrode to describe the measurement of solution pH during an acid-base titration. In Chapter 11 an experiment was described in which a silver wire was used to monitor the activity of silver ion during an argentimetric titration of chloride. These experiments have many things in common. First, they both involve electrochemical cells. One of the electrodes of the cell responds to the activity of a species in solution and is called an *indicator electrode;* it "indicates" the activity of an ion by its potential. The other electrode of each cell provides a stable reference potential against which we can measure the potential of the indicator electrode. The potential of the *reference electrode* should not be influenced by the activities of species in the test solution. What kinds of solution species can we monitor potentiometrically? How do we build electrodes to respond to these species? How do we treat experimental data obtained in potentiometric studies?

15.2 INDICATOR ELECTRODES

15.2.1 A Quick Review

The simplest indicator electrode we can build consists of a metal wire dipped in a solution of its ions. The example in Chapter 11 was a silver wire in a solution of silver nitrate. The wire indicates the tendency of the half reaction

$$Ag^+ + e = Ag^0$$

to occur. From Chapter 14 we know that this is a reversible system. We also know that the Nernst equation describing its behavior is

$$E(\text{wire}) = E^0(Ag^+/Ag) + 0.05916 \log a_{Ag^+} \qquad (25°C)$$

where E^0 is the standard potential and a_{Ag^+} is the activity of silver ion. We called this a class I electrode in Chapter 14.

Another very simple indicator electrode is a platinum or gold wire dipped in a solution containing metal ions. An electrochemical couple such as Fe^{3+}/Fe^{2+}, where both species are soluble, can be studied with this kind of indicator electrode. The Nernst equation showing the response of a platinum wire to iron activities is

$$E(\text{wire}) = E^0(Fe^{3+}/Fe^{2+}) - 0.05916 \log [a(Fe^{2+})/a(Fe^{3+})] \qquad (\text{at } 25°C)$$

When both oxidized and reduced forms of iron are present the system acts reversibly. In practice, we can use a platinum electrode to monitor the progress of titrations involving even irreversible half-reactions. See Chapter 16 for details. Notice that the material in the wire electrode itself (i.e., Pt) does not appear in the Nernst equation. It indicates the potential of the half-reaction.

15.2.2 Ion-Selective Electrodes

Ion-selective electrodes (ISEs) are indicator electrodes whose active elements are membranes which respond to the activity of ions or molecules in solution. The term "ion-specific electrode" is also applied to these electrodes, but usually by real optimists. The best possible analytical device would respond to the activity of only one ion in a mixture of several ions with similar chemical properties at similar concentrations. True specificity is quite hard to obtain in real experiments.

The most general definition of a membrane is that it is a barrier between two solutions across which an electrical potential can develop. Some membranes are permeable to ions (like a biological membrane), while others are impermeable (like a glass or silver sulfide membrane). Some membranes conduct electrical charge by the motion of ions and/or electrons, while others have extremely high electrical resistance and thus conduct very poorly. We will examine more carefully the behavior of three types of membranes: glass, inorganic salt disks and single crystals, and inert membranes used to hold ion exchange materials.

Glass Membrane Electrodes A glass membrane electrode is made by fusing a thin spherical glass membrane to the end of a glass tube. The membrane holds a buffer solution which connects in turn to an internal reference electrode by way of a flowing

15.2 INDICATOR ELECTRODES

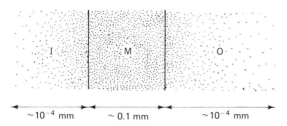

Figure 15.1 Glass membrane zones (dimensions not drawn to scale).

junction. When a fresh, dry glass membrane electrode is dipped into an aqueous solution, its surfaces soak up water (they "hydrate"). Within a few hours a slice of the glass membrane would resemble the sketch in Fig. 15.1. Three zones are shown in Fig. 15.1: an inner zone (I) at the membrane-internal buffer interface, a middle zone (M) consisting of the glass silicate lattice, and an outer zone (O) at the membrane-test solution interface. The I and O zones are swollen by water. All three zones contain the kinds of cations found in glasses, typically sodium and calcium and, in some formulations, lithium. Some of the cations in the I and O zones are rather free to exchange with cations in the surrounding solutions. Hydronium ions are very mobile and will diffuse into the swollen zones and exchange at cation sites. The result is an imbalance in the ions in exchangeable sites across the membrane which causes an electrical potential drop. Apparently there is very little motion of ions or electrons across the middle zone: the electrical resistance of even very thin glass membranes (100 μm, for example) is several million ohms.†

The most important experimental problem is that of measuring accurately the potential drop across the high-resistance glass membrane. Unfortunately, this potential is only one of several that arise in the electrode system. Figure 15.2 shows that the glass membrane electrode is really a half-cell electrode in a full electrochemical cell. A potential will arise at every phase boundary drawn in the figure: (1) the internal reference electrode potential, (2) the potential drop between the reference electrode solution and the membrane, (3) the potential drop between the external (test) solution

† Buck (Refs. 1 and 2). Vesely et al. (Ref. 3) discuss mechanisms for the development of potential at a glass membrane.

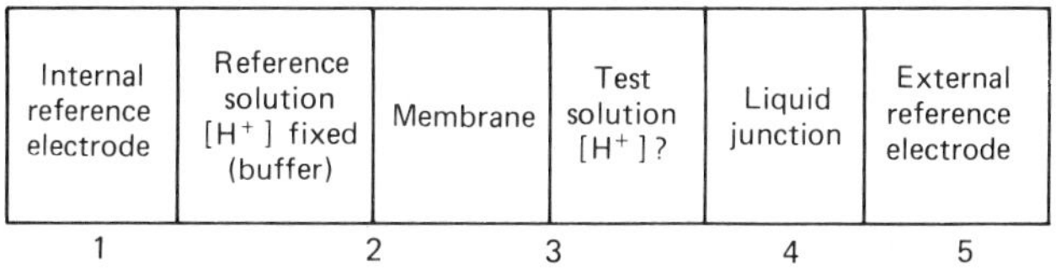

Internal reference electrode	Reference solution [H$^+$] fixed (buffer)	Membrane	Test solution [H$^+$]?	Liquid junction	External reference electrode
1	2	3	4	5	

Figure 15.2 Diagram of electrochemical cell with glass electrode.

and the membrane, (4) the potential across the flowing junction, and (5) the potential of the external reference electrode. In an actual experiment the composition of the internal reference electrode solution never changes, and potentials 1 and 2 are constant. If built properly, the external reference electrode will also have a constant potential. Potentials 3 and 4, however, will change whenever the test solution is changed. It may be possible to reduce potential 4 (see Section 14.3.8), but it cannot be eliminated totally. Thus potential 3, the membrane potential, can be closely approximated by the measured potential, but will not be exactly equal to it.

The measurement problem is confounded by the appearance of a small (i.e., a few millivolts) "asymmetry potential" of the membrane, arising from differences in ion exchange abilities of the inner and outer swollen zones. These differences can arise from adsorbed films, a history of severe dehydration, or abrasion. To keep asymmetry potentials small you should treat glass membranes very carefully. Do not allow them to dehydrate, and never touch them with your fingers.

From all of this you should conclude that no single measurement of hydronium ion activity will be possible with a single potential reading from a glass electrode. The best practical compromise is to say instead that a glass membrane electrode responds to hydronium ion activity according to the following expression:

$$E(\text{electrode}) = \text{constant} + 0.05916 \log a_{H^+} \quad (25°C)$$

The "constant" in this expression contains potentials 1, 2, 4, and 5 from Fig. 15.2. It will vary from electrode to electrode and from experiment to experiment with the same electrode due to changes in liquid junction and asymmetry potentials. In the laboratory we calibrate the response of the electrode and the meter we use to measure its potential with one or more standard pH solutions. We then measure the potential of the system with the test solution relative to the standard and *assume* that the "constant" in the Nernst equation is the same for both solutions.

The National Bureau of Standards has defined pH values to within ±0.005 pH unit for standard buffers spanning a large part of the pH range (see Table 15.1). These buffers are commercially available and can be used to calibrate a glass membrane electrode and a pH meter. Normal calibration procedure has been discussed in Chapter 6.

Ideally, a glass membrane electrode should respond only to hydronium ion activity. In practice, however, other cations can affect the membrane potential. A Nernst equation can be written taking into account other cations:

$$E(\text{electrode}) = \text{constant} + 0.05916 \log [a_1^{1/m} + (K_{12} a_2)^{1/m}]$$

Let us identify hydronium ion as ion number 1 and sodium ion as ion number 2. The terms a_1 and a_2 are then the activities of hydronium ion and sodium ion, respectively. The superscript $1/m$ is usually very close to unity for a cation with a +1 charge. The term K_{12} is called a *selectivity coefficient:* the larger K_{12}, the greater will be the inter-

15.2 INDICATOR ELECTRODES

TABLE 15.1 NBS STANDARD BUFFERS

	pH at			
Formulation	10°C	20°C	25°C	30°C
Potassium tetraoxalate, 0.05M	1.670	1.675	1.679	1.683
Potassium bitartrate, saturated at 25°C	—	—	3.557	3.552
Potassium biphthalate, 0.05M	3.998	4.002	4.006	4.015
KH_2PO_4, 0.025M + Na_2HPO_4, 0.025M	6.923	6.881	6.863	6.853
Sodium tetraborate, 0.01M	9.332	9.225	9.180	9.139

Source: From R. G. Bates, *Determination of pH*, 2nd ed., Wiley, New York, 1973.

ference from sodium ion in the determination of hydronium ion. The equation tells us that when $K_{12}a_2$ is very small (very low sodium activity and very low coefficient), the electrode will respond only to hydronium ion. We might expect, however, that

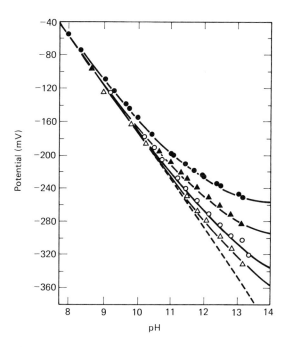

Figure 15.3 Response curve of Beckman general-purpose electrode. From R. P. Buck, J. H. Boles, R. D. Porter, and J. A. Margolis, *Analytical Chemistry,* 46, 255 (1974). Reproduced by permission of the American Chemical Society.

TABLE 15.2 GLASS COMPOSITION (MOLE PERCENT) AND ION RESPONSE

	Mole percent			Response
Al_2O_3	Na_2O	SiO_2	Other	
1	21.4	72.2	6 CaO	H^+
4	27	69		NH_4^+ in H^+
5	27	68		K^+ in Na^+, $K_{NaK} = 20$
18	11	71		$K_{NaK} = 2800$
19.1	28.8	52.1		Ag^+; $K_{HAg} = 10^5$
25	—	60	15 Li_2O	Li^+
—	—	80	10 Li_2O 10 CaO	"Low alkaline error" H^+

when hydronium ion activity is very low and sodium ion activity high (say $0.1 F$ NaOH in water), the electrode will be influenced substantially by sodium ion. The response of a Beckman general-purpose glass membrane electrode with sodium as interfering ion is shown in Fig. 15.3. From this set of curves Buck et al. determined the selectivity coefficient $K_{H^+Na^+}$ to be 10^{-12}. The Beckman type E-2 electrode was found to be less sensitive to sodium ion activity, with $K_{H^+Na^+} = 10^{-16}$.

The departure from linearity of a plot like that in Fig. 15.3 at high pH is called the *alkaline error*. While the alkaline error makes it difficult to make accurate pH measurements at high pH, it has been used to advantage in measuring the activity of sodium ion. Eisenman (Ref. 4) extended the sodium error range for a glass membrane to *lower* pH levels by increasing the amount of aluminum oxide (Al_2O_3) in the formula of H^+-responsive glass. The result was a sodium-responsive glass electrode. Table 15.2 shows how dependent the response of glass membranes is on glass composition.

EXAMPLE 15.1 Alkaline Error

If the selectivity coefficient for the Beckman general-purpose electrode is 10^{-12} for Na^+ in the presence of H^+, calculate the approximate error in millivolts when $a_{Na^+} = 1.0$ and $a_{H^+} = 10^{-12}$.

Solution

Use the Nernst equation which includes the selectivity factor, and assume $1/m = 1.0$,

$$E = \text{constant} + 0.05916 \log [10^{-12} + 10^{-12}(1.0)]$$
$$= \text{constant} - 0.05916(11.70)$$

15.2 INDICATOR ELECTRODES

The error in the measured potential should be given by

$$E(\text{error}) = 0.05916(12.0 - 11.7) = 0.018 \text{ V, or } 18 \text{ mV}$$

This result means that an experimentalist ignorant of the alkaline error would conclude that the pH of the solution is 11.7 rather than 12.0 if the pH meter were calibrated to indicate 1 pH unit change for every 0.05916-V change in potential.

Popular Glass Electrode Designs   There are two basic designs for glass electrodes: an electrode which is used with a separate reference electrode (usually an SCE) or the far more popular combination electrode, which contains a glass electrode and reference electrode in one electrode body. Both designs are shown in Fig. 15.4. Notice the position of the liquid junction on the combination electrode.

In theory, it is possible to adapt a glass membrane electrode to measure the activity of any species which reacts with hydronium ion in an equilibrium. A design which has served as a prototype for many other applications is the Severinghaus CO_2-responsive electrode (Ref. 5), shown in Fig. 15.5. In this design the glass membrane tip of the pH electrode is covered with a cap of Teflon holding a solution of sodium bicarbonate. Carbon dioxide passes through Teflon rather easily, dissolves in the bicarbonate solution, and alters the equilibria,

$$CO_2 + H_2O = H_2CO_3$$
$$H_2CO_3 = H^+ + HCO_3^-$$

A change in CO_2 in a sample will therefore cause a change to occur in the pH of the solution in the Teflon cap, and the potential of the glass membrane electrode will change.

Guilbault (Ref. 6) and others (Refs. 3, 7, 8) have described electrodes which respond to biologically active compounds such as urea, glucose, and amino acids. In perhaps the most novel design (Ref. 8) an ammonia-responsive electrode is made from a pH electrode whose tip is covered with a membrane, much like the CO_2 electrode described above. The membrane in this device holds a millimolar solution of ammonium chloride. The ammonia-responsive electrode is then attached to an airtight cell which contains pH 8.5 buffer (Tris/TrisH$^+$; see Chapter 6) and a Teflon stir bar coated with the enzyme urease. The enzyme reacts with urea in a sample of, say, blood and releases ammonia:

$$OC(NH_2)_2 + 2H_2O + H^+ = HCO_3^- + 2NH_4^+$$
(urea)

At pH 8.5 the ratio of free ammonia to ammonium ion is about 0.18. Ammonia gas is released to the air space in the chamber and diffuses to the membrane-covered pH

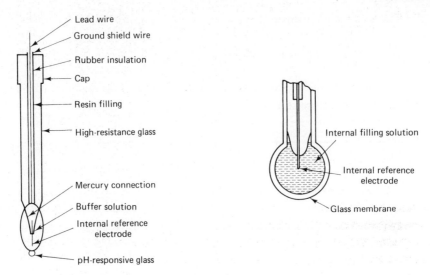

Figure 15.4 Single and combination glass electrodes. (Single electrode from H. Willard, L. Merritt, J. Dean, and F. Settle, *Instrumental Methods of Analysis,* 6th ed., Van Nostrand, New York, Figure 22.1. Combination electrode sketch courtesy of Sargent–Welch Scientific Co., Skokie, Illinois.)

electrode, where it is detected. Equilibrium in the cell is achieved within 2 or 3 minutes, and the method is accurate and precise. One great virtue of the cell is that sodium and potassium ions, present at much higher concentrations in blood than is urea, cannot leave the solution along with ammonia and therefore do not interfere. Similar enzymatic methods have been developed for several important amino acids (Ref. *3*).

Solid-State Electrodes Solid-state electrodes use as their membranes either a pressed disk of an inorganic salt (e.g., silver sulfide), a mixture of salts (a "heterogeneous" electrode), or a single crystal (e.g., lanthanum fluoride, LaF_3). Solid-state membranes generally must be able to conduct electricity to a slight extent and be no more soluble than about $10^{-6}M$. It is generally thought that the solubility of the membrane determines the lower limit of its response to ion activity.

The silver sulfide electrode (Fig. 15.6) has a membrane made of pressed polycrystalline Ag_2S and responds to the activities of sulfide ion and silver ion. The motion of silver ions in the membrane may account for the conductivity of the membrane. By adding either CuS, PbS, or CdS to the silver sulfide before pressing the membrane, sensitivity to Cu^{2+}, Pb^{2+}, or Cd^{2+} can be obtained. When crystals of silver halides (except fluoride) or silver thiocyanate are pressed into a disk with Ag_2S, the resulting AgX/Ag_2S electrode will respond to halide ion or thiocyanate activity. A sulfate-responsive electrode has been made by pressing a disk of 32 mole% Ag_2S, 31 mole% PbS, 32 mole% $PbSO_4$, and 5 mole% CuS (Ref. *9*).

Single-crystal electrodes can be highly selective and sensitive. For example, a

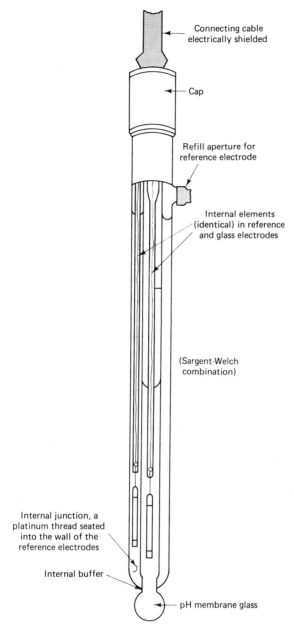

Figure 15.4 (*continued*)

single crystal of lanthanum chloride containing some europium(III) (a "doping agent") can be mounted and sealed at the end of an electrode body. The potential drop across the crystal will respond to fluoride ion activities as low as 10^{-6}. See Chapter 4 in Vesely (Ref. *3*).

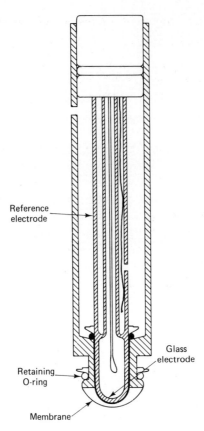

Figure 15.5 CO$_2$-responsive electrode. (Sketch courtesy of Instrumentation Laboratories, Inc., Andover, Massachusetts.)

Another type of solid-state membrane, a "Pungor electrode," named for its inventor (Ref. *10*), can be made by mixing a slightly soluble salt with silicone rubber and curing the mixture until it solidifies. The resulting rubber disks can be cemented

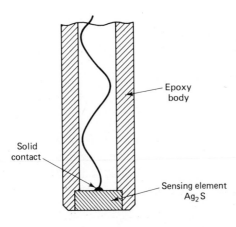

Figure 15.6 Silver sulfide solid-state electrode. (Sketch courtesy of Orien Research, Inc., Boston, Massachusetts.)

15.2 INDICATOR ELECTRODES

with conductive epoxy to a lead wire and sealed on the end of a glass tube. As you might imagine, these electrodes are physically quite rugged.

Liquid Ion Exchange Membrane Electrodes (LIEMs) The responsive element in an LIEM electrode is a chemically inert porous membrane which covers a reservoir of liquid ion exchange resin (see Fig. 15.7). When the electrode is dipped into a solution containing exchangeable ions, a potential develops at both the ion exchange liquid/membrane and the membrane/test solution boundaries. The potential is a function of the activity of the exchangeable ions. The electrode shown in Fig. 15.7 is an Orion calcium-responsive LIEM electrode, designed with a double concentric tube and a self-contained AgCl/Ag reference electrode. The ion exchange material is an alcohol solution of the calcium salt of bis(2-ethylhexyl)phosphoric acid ("d2EHP"), held in the outer concentric tube. The ion exchanger can diffuse through the membrane and exchange Ca^{2+} at the membrane/test solution interface without dissolving in the test solution.

Table 15.3 contains a list of ion exchange compounds used in LIEMs and the ions they can be used to study. The last four compounds listed are used in a subclass of LIEMs called "neutral carrier" electrodes. Valinomycin, nonactin, and monactin are antibiotics. Figure 15.8 shows how valinomycin chelates potassium ion. Several crown polyethers (called "cryptates") are shown in Fig. 15.9. It is easy to see how

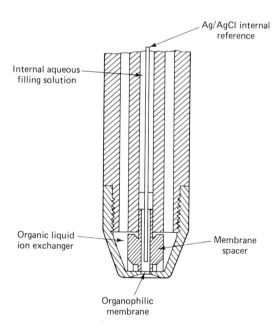

Figure 15.7 Liquid ion exchange membrane electrode. (Sketch courtesy of Orien Research, Inc., Boston, Massachusetts.)

TABLE 15.3 LIQUID ION EXCHANGE MATERIALS

Material	Responsive to ions
Tetraphenylarsonium/phosphonium salts	I^-, NO_3^-, SCN^-
Tetraalkylammonium salts	Br^-, NO_3^-
Bathophenanthroline Fe, Ni complexes	NO_3^-, fluoroborate
Bis(O,O'-diisobutyldithiophosphato)M^{2+} salts	M^{2+} = Ni^{2+}, Cd^{2+}, Pb^{2+}
Cyclic polyethers (cryptates)	K^+, Na^+
Dioctylphthalate	R_4N^+
Valinomycin, nonactin, monactin	K^+ in presence of Na^+, Ca^{2+}

these cyclic polyethers might tie up an ion or a cation-anion pair. The mechanism of neutral carrier electrodes is an interesting subject for research. It may be quite similar to the mechanism by which living membranes differentiate ions.

15.3 REFERENCE ELECTRODES

We need to measure the potential of an indicator electrode against a reliable second half-cell potential. We will call this second half-cell a reference half-cell or simply a reference electrode. We can use an electrode called the "Standard hydrogen electrode" (SHE), an example of which is shown in Fig. 15.10, to measure indicator electrode potentials against the zero point of the thermodynamic free energy scale. The half-reaction for the SHE is

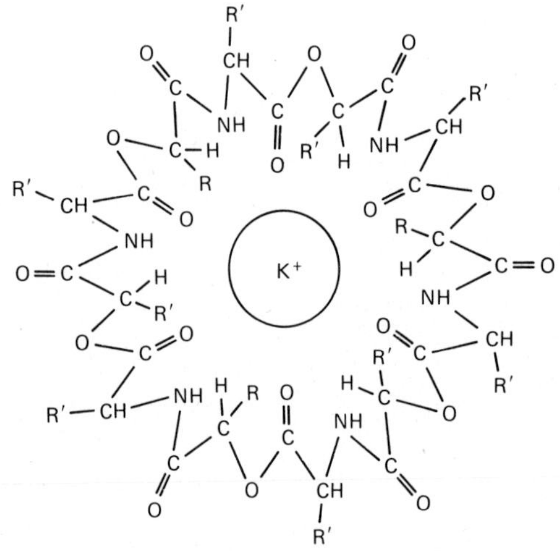

Figure 15.8 Valinomycin.

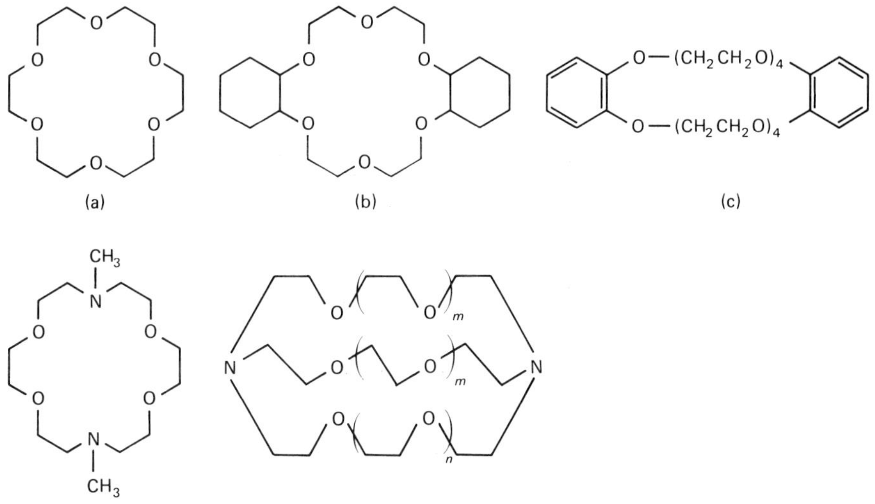

15.3 REFERENCE ELECTRODES

Figure 15.9 Cryptate molecules.

$$2H^+ + 2e = H_2(g)$$

and the standard potential is exactly 0.0000 V. Recall from Chapter 3 that the standard free energy change for the half-reaction is also 0.00 kJ/mole. The electrode is made by placing a "platinized" platinum foil electrode in a solution of hydronium ion

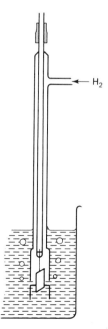

Figure 15.10 Standard hydrogen electrode.

(activity unity) and bubbling hydrogen gas over its surface at 1.00 atm pressure. Platinizing is a plating process: a platinum electrode is made the cathode of a full cell and platinum metal is plated onto its surface. The finely divided platinum plate is black, rather than the silvery metallic color one might expect.

The SHE is not very easy to use in routine laboratory work. The acid solution prepared for use in the SHE will not be exactly at unit activity and must be calibrated by titration. The pressure of hydrogen gas will not be exactly 1.00 atm, and corrections are needed. Furthermore, the electrode is rather delicate and the platinized surface is easily contaminated by solution impurities, which may change its potential.

If we were to design a new type of reference electrode after struggling with an SHE to make routine measurements in the laboratory or in the field, we would make several changes. First, we would certainly try to build an electrode that used an electrolyte which did not have to be standardized. We would build the electrode to have a very stable half-cell potential regardless of minor flaws in its construction or minor bumps and jolts in field work. We would want it to be able to withstand small current drains during potential measurements without changing its potential (more about this later). Finally, since so many solutions we work with in the laboratory contain chloride and/or nitrate, it would be a good idea to design an electrode whose electrolyte involved one of those ions.

An electrode with all of the properties just described is the *saturated calomel electrode,* or SCE. This electrode is based on the calomel (mercurous chloride) half-reaction,

$$Hg_2Cl_2(\text{sat'd}) + 2e = 2Hg + 2Cl^- \qquad E^f = +0.280 \text{ V } (1F \text{ KCl})$$

The SCE electrolyte is saturated (sat'd) with both mercurous chloride and potassium chloride, and the solids of both of these compounds must be present. We write the SCE half-reaction in the following way:

$$Hg_2Cl_2(\text{sat'd}) + 2K^+ + 2e = 2Hg + 2KCl(\text{sat'd}) \qquad E^0 = +0.2415 \text{ V}$$

As long as the solution is saturated in KCl and Hg_2Cl_2, the SCE potential is independent of chloride activity. However, since the degree of saturation depends on temperature, the SCE potential also depends on temperature:

$$E_{SCE} = +0.2415 - 7.6 \times 10^{-4} * (T - 25)$$

where T is the Celsius temperature.

Most commercially available SCEs have large reservoirs of KCl and calomel and will survive the passage of microampere-level currents with reliable and reproducible potentials. In most potentiometric measurements made with modern equipment, currents drawn are well below the microampere level (see below). A popular commercial

15.3 REFERENCE ELECTRODES

SCE design is shown in Fig. 15.11. The electrode connects to the test solution by an asbestos fiber junction.

Other electrodes based on the calomel half-reaction use less-than-saturated KCl solutions for an electrolyte. The decinormal calomel electrode (DNCE) contains $0.1000M$ KCl ($E^f = +0.334$ V vs. SHE), while the normal calomel electrode (NCE) contains $1.000M$ KCl ($E^f = +0.280$ V vs. SHE). These electrodes require electrolyte standardization and are less convenient than the SCE for routine use.

The silver chloride-silver metal electrode (AgCl/Ag) is another very useful electrode. Like the calomel electrode, the AgCl/Ag electrode is a class II electrode, with a half-reaction involving a metal and its insoluble salt,

$$AgCl + e = Ag + Cl^- \qquad E^0 = +0.2224 \text{ V}$$

The potential of the electrode depends on chloride activity,

$$E_{\text{electrode}} = E^0 - 0.05916 \log a_{Cl^-}$$

In saturated KCl the electrode has a potential of 0.197 V vs. SHE at 25°C. A silver wire can be coated with silver chloride by connecting it to the positive end of a flashlight

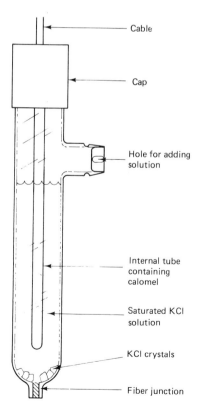

Figure 15.11 Saturated calomel electrode.

battery whose negative end is attached to another silver wire. Both wires are dipped into a dilute solution of KCl for a few minutes and the positive silver wire is oxidized. The resulting silver ions precipitate as AgCl on the wire surface. The AgCl-coated wire is then placed in a glass tube containing saturated KCl and equipped with a suitable junction (asbestos fiber or agar plug).

In experimental situations where chloride ion cannot be tolerated a mercurous sulfate/mercury electrode may be a good choice for a reference electrode. The half-cell reaction resembles that of calomel, but with sulfate ion taking the place of chloride. In a saturated solution of potassium sulfate the formal potential of the Hg_2SO_4/Hg electrode is +0.68 V vs. SHE.

Most conventional reference electrodes must be modified to work in nonaqueous solvents (Chapter 7). For example, although KCl is quite soluble in water, it is not very soluble in glacial acetic acid. The fiber flowing junction of an SCE plugs up with KCl when the SCE is placed in acetic acid solvent. Lithium chloride is much more soluble in acetic acid than is KCl, and an SCE filled with LiCl is quite effective in acetic acid. A silver nitrate/silver electrode is a good choice for studies in acetonitrile (CH_3CN) or in fused salts (e.g., molten potassium nitrate). All of these electrodes must be calibrated carefully; they lack the surety provided by the extent of saturation of KCl at 25°C and may also suffer from high liquid junction potentials.

15.4 EVALUATING ACTIVITIES AND CONCENTRATIONS BY POTENTIOMETRY

15.4.1 Analytical Methods

If the response of an indicator electrode to the activity of an ion has been well characterized by a series of experiments, the determination of an unknown amount of the ion can be rather easy. A process like that described in Chapter 6 for the calibration of a pH electrode with calibration standards can be applied to the analysis of many other ions. A series of solutions can be prepared which covers a wide range of analyte activities. The potential of an indicator electrode when placed in each solution can then be plotted in a "working curve" as a function of activity. The potential of the electrode in a test solution is then measured, and the activity corresponding to the potential is found on the working curve.

An alternative method is to use "standard addition." The analyst measures the potential of an indicator electrode in a sample solution both before and after adding a known amount of the ion to be analyzed. If the indicator response obeys the Nernst equation, the first measured potential is given by the expression:

$$E_1 = \text{constant} + (0.05916/n) \log (fC)$$

where f and C are the activity coefficient and molar concentration of the species. After the addition of the standard, the second potential, E_2, is measured:

15.4 EVALUATING ACTIVITIES AND CONCENTRATIONS BY POTENTIOMETRY

$$E_2 = \text{constant} + (0.05916/n) \log [f(C + \Delta C)]$$

where ΔC is the change in concentration resulting from adding the standard. When we combine the equations, we find

$$E = E_2 - E_1 = (0.05916/n) \log [(C + \Delta C)/C].$$

This last expression will be valid if the ionic strength of the first solution has not been changed by adding the standard. That is, the activity coefficient of the measured ion must remain the same.

If the working curve and standard addition methods are to be reliable, standards and samples must have the same ionic strength. Further complications may arise if the ions being analyzed are involved in competing equilibria (complexation of metal cations, for example). To overcome competing reactions it may be necessary to add a "releasing agent" to the sample. As a case in point, consider the analysis of fluoride ion in drinking water. Fluoride is known to form complexes with hard-water ions such as Fe^{3+}. Since a fluoride ion-responsive electrode may not respond to ferric fluoride complexes in the same way that it responds to free fluoride, water samples should be treated with a compound to release ("unbind") fluoride. The Leeds & Northrup Company sells a product to do just that. Their "total ionic strength adjustment buffer" (TISAB) contains disodium DCTA (*trans*-1,2-diaminocyclohexane-N,N,N′,N′-tetraacetic acid, a relative of EDTA) in an acetate and chloride solution at pH 5 to 6. DCTA will complex most metal ions better than will fluoride, and the fluoride is released.

15.4.2 Measuring Devices

Assuming that all solution activity problems have been solved, accurate potential measurements also require good potential measuring devices. One idea is extremely important: cell potentials cannot be measured accurately when electrical current flows through the cell. The flow of current can create two important problems. First, it can change the concentration of the measured ion immediately in front of the indicator electrode by causing an oxidation or reduction reaction to occur. Second, it can produce a counterpotential called an "iR drop," which will oppose the natural potential of the cell. All electrochemical cells have some electrical resistance, R_{cell}. When current, i, flows through the cell, a potential $E = i*R_{cell}$ results. Example 15.2 shows how the iR drop can affect measured potentials.

Reliable cell potentials may often be measured with devices called potentiometers. A prototype "student potentiometer" is shown schematically in Fig. 15.12. The operation of the potentiometer involves adjusting the voltage of a battery (AB in the figure) until it exactly equals the potential of the cell being measured. When the adjustment is exact, no current will flow in the circuit. The battery AB is connected across a large adjustable resistor, CD. The cell whose potential is to be measured is

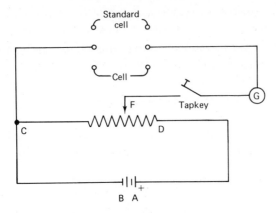

Figure 15.12 Student potentiometer.

placed between points C and F. A galvanometer is wired in series with the cell; whenever current flows in the circuit, the galvanometer needle will deflect. The circuit is interrupted by a tap key, which momentarily completes the circuit during a measurement. The adjustable resistor is moved gradually and the key is tapped momentarily until the deflection of the galvanometer ceases. At the exact balance point the following relationship holds:

$$E_{cell} = (R_{CF}/R_{CD})*E_{AB}$$

where R_{CF} is the resistance between points C and F and R_{CD} is the total resistance of the adjustable resistor.

If a standard cell (with potential E_{std}) is switched into the circuit and the cell to be measured is switched out, the adjustable resistor will have to be changed to find a new balance point. At the new balance point,

$$E_{std} = (R'_{CF}/R_{CD})*E_{AB}$$

where R'_{CF} is the new resistance between points C and F. Combining the two expressions gives the relationship between the potential of the standard cell and that of the test cell,

$$E_{cell} = (R_{CF}/R'_{CF})*E_{std}$$

When E_{std} is known accurately, E_{cell} may be determined accurately.†

---

† The most commonly used standard cell is the Weston cell, $(-)Cd(Hg)$, $CdSO_4 \cdot (8/3\ H_2O)$, $HgSO_4(sat'd)$, $Hg(+)$. The 8/3 is the hydration number of cadmium sulfate in this formulation. The cadmium amalgam, $Cd(Hg)$, is about 10 to 12% Cd metal dissolved in mercury. When saturated cadmium sulfate is used as an electrolyte the potential of the cell is 1.0186 V at 20°C. Most Weston cells used in routine work do not contain saturated cadmium sulfate and have potentials in the range 1.0180 to 1.0191 V.

15.4 EVALUATING ACTIVITIES AND CONCENTRATIONS BY POTENTIOMETRY

A less tedious and often more effective way to measure cell potentials is to use a voltmeter or multimeter with a high internal resistance. Most digital multimeters now available have internal resistances of at least 10^7 ohms, with a few models offering values as high as 10^{14} ohms. Glass electrodes may have resistances of 10^8 ohms or more. As the next example shows, the resistance of a measuring device must be much greater than the resistance of the cell being studied in order to avoid measurement errors.

EXAMPLE 15.2 Measuring Potentials

The true potential of a cell is 1.000 V. The cell, however, has an internal resistance of 10^6 ohms, most of which is resistance in the indicator electrode. If a voltmeter with an internal resistance of 10^7 ohms is used to measure the potential of the cell, what is the error in the measurement?

Solution

The current that flows through a resistance is given by Ohm's law, $i = E/R$, where i is current, E is potential, and R is resistance. In the measuring circuit drawn below there are two resistances, that of the cell and that of the meter.

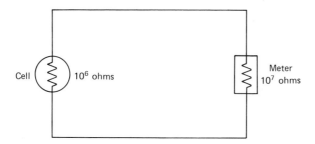

The sum of the resistances determines the current that can flow in the circuit,

$$i = E_{cell}/(R_{cell} + R_{meter}) = 1.000 \text{ V}/(10^6 + 10^7)$$
$$= 9.09 \times 10^{-8} \text{ ampere}$$

This small current flowing through the cell results in an iR drop in the cell, as mentioned earlier in the text,

$$E_{iRdrop} = i * R_{cell} = 9.09 \times 10^{-8} * 10^6 = 9.09 \times 10^{-2} \text{ V}$$

The iR drop opposes the real cell voltage, and the meter will read $(1.000 - 0.0909)$ V, or 0.909 V, about 10% less than the true value. Do you think an iR drop could ever make the measured value *greater* than the true value?

15.5 POTENTIOMETRIC TITRATION CURVES: DATA TREATMENT

Potentiometric titrations involve measurement of the potential of a cell during the delivery of titrant from a buret. Data are plotted as cell potential vs. volume of titrant and have the familiar sigmoid or S shape. The most important information for the analyst is the volume required to reach the steepest part of the sigmoid curve, what we call the "endpoint break." For reactions with 1:1 stoichiometry the endpoint of the titration may be the same as the equivalence point of the reaction being monitored. If the concentrations of analyte and titrant are sufficiently large and the equilibrium constant for the titration reaction is also large, there may be only a negligible titrant volume difference between the endpoint break and the equivalence point. Unfortunately, analysts are often forced to work with titrimetric reactions with small equilibrium constants and with potentiometric titration curves in which the endpoint is hard to locate. There are several numerical and graphical methods for treating data in such less-than-ideal situations, and we will consider some of them briefly in this section.

Figure 15.13 shows a plot of data obtained from the titration of a sample of potassium biphthalate (KHP) with standard base. The endpoint potential break is about 4 pH units, or 240 mV, and is quite distinct. The endpoint volume is easily read as 19.13 mL. Figure 15.14 shows data from a titration of the protonated form of tris(hydroxymethyl)amino methane, trisH$^+$, with standard sodium hydroxide.† The titration curve shows only a gradual break, and the point of maximum slope is not as obvious as in Fig. 15.13. In this situation it usually helps to determine the *derivative* of the titration curve, that is, the change in pH (or voltage) with titrant volume, ΔpH/ΔV. When the derivative is plotted against titrant volume a peak is observed at the volume where the original titration curve had its maximum slope. The first derivative of the TrisH$^+$/NaOH titration curve is also sketched in Fig. 15.14.

In the event that the first derivative peak does not show the endpoint clearly, a *second derivative* plot may help. The second derivative is the rate of change of the slope of the titration curve with added titrant. Recall from introductory calculus that the second derivative of a sigmoid-shaped function equals zero at the point of maximum slope. A second derivative plot of the TrisH$^+$/NaOH data is sketched at the top of Fig. 15.14. Table 15.4 gives data for all three plots in Fig. 15.14 and should help you

---

† Tris(hydroxymethyl)amino methane(tris) is a weak base and was discussed in Chapter 6. If dilute hydrochloric acid is added dropwise to a solution of tris, the weak base-strong acid (WB/SA) titration can be followed with a glass electrode. Figure 15.14 was prepared by following the *back titration* with standard NaOH and observing the weak acid-strong base titration. pK_a for trisH$^+$ is 8.10.

15.5 POTENTIOMETRIC TITRATION CURVES: DATA TREATMENT

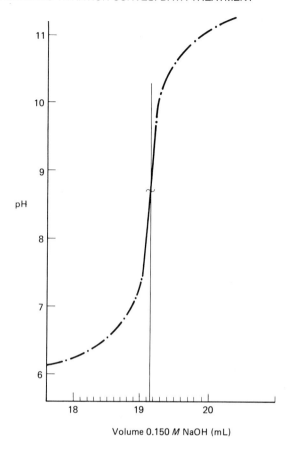

Figure 15.13 Titration of potassium biphthalate with 0.150*M* NaOH.

understand the simple numerical difference process used. Since the calculations are rather laborious, you should be sure that the results are worth the extra effort. Most titrations you will perform in undergraduate laboratories have sharp enough endpoints to be judged without numerical treatment.

Some interesting numerical methods called digital convolution techniques have been applied to titration curve analysis. They are particularly useful when working with large numbers of data points obtained with a laboratory computer. The simplest and most useful convolution methods for potentiometric data are the Savitsky-Golay (S-G) numerical "filters" (Refs. *11, 12*). To illustrate how they work, let us take the set of 11 potentiometric titration data points in Table 15.3. We will arrange the pH values in a linear array indexed according to the volume increment:

$$pH_1 pH_2 pH_3 pH_4 \cdots pH_{11}$$

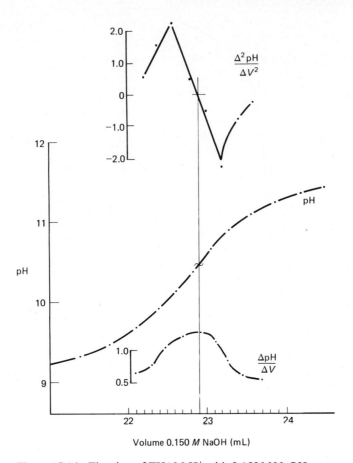

Figure 15.14 Titration of THAM-H$^+$ with 0.150M NaOH.

where point 1 is at 22.00 mL, point 2 at 22.20 mL, ..., and point 11 at 24.00 mL. Notice that the volume increments are all equal. We will multiply the pH elements in the array by integer coefficients of a special function we will call an S-G filter function. The simplest first derivative function we will use will operate on groups of five pH values in the array. Begin by multiplying the first five array terms by the coefficients 1, −8, 0, 8, −1, adding the products, and dividing the sum by the number 12:

$$(1*pH_1 - 8*pH_2 + 0*pH_3 + 8*pH_4 - 1*pH_5)/12$$

This quantity becomes the first element of a new array which will hold numbers proportional to the first derivative of the original data set. We give the first element in the new array a volume index which is the same as that of point 3 in the original data array (22.40 mL). We then calculate the next convoluted point by multiplying elements indexed at pH_2 through pH_6:

15.5 POTENTIOMETRIC TITRATION CURVES: DATA TREATMENT

TABLE 15.4 TITRATION CURVE DATA FROM FIG. 14.14 (DERIVATIVE CALCULATIONS)[a]

Volume (mL)	pH	$\dfrac{\Delta pH}{\Delta V}$	$\dfrac{\Delta^2 pH}{\Delta V}$
22.00	9.61		
.10		0.65	
.20	9.74		0.50
.30		0.75	
.40	9.89		1.50
.50		1.05	
.60	10.10		2.25
.70		1.20	
.80	10.34		0.50
.90		1.30	
23.00	10.60		−0.50
.10		1.20	
.20	10.84		−2.25
.30		0.75	
.40	10.99		−0.75
.50		0.60	
.60	11.11		−0.25
.70		0.55	
.80	11.22		−0.25
.90		0.40	
24.00	11.30		

[a] First and second derivative values are plotted at volumes to their left.

$$(1*pH_2 + 8*pH_3 + 0*pH_4 - 8*pH_5 - 1*pH_6)/12$$

We make this the second element of the new array and give it the pH index of point 4 (22.60 mL here). We move the coefficients on through the entire array, working with five points at a time, creating the new derivative array as we go. We lose two volume points at each end of the data set as a result of the filter technique, and so our first derivative array has only seven points.

Table 15.5 shows S-G filter coefficients for first and second derivatives and for smoothing data that have random error (noise). The last column contains the number which must be divided into each set of convoluted data. Several filters are shown for each process. In treating real data, there is a slight danger that important trends will be filtered out of the data. The danger becomes greater as the complexity of the experimental data curve increases. The larger the number of terms in the S-G function, the less likely it is that error will occur. The functions listed are generally safe for convoluting curves which are described by cubic or quartic equations. Notice that the

TABLE 15.5 SAVITSKY-GOLAY FILTER COEFFICIENTS[a]

Points				Coefficients					Norm
				First derivative					
				(use for up to cubic or quartic curve)					
5			1	−8	0	8	−1		12
7		22	−67	−58	0	58	67	−22	252
9	86	−142	−193	−126	0	126	193	142	−86	1188
				Second derivative					
				(use for up to quartic or quintic curve)					
5			−3	48	−90	48	−3		3
7		−117	603	−171	−630	−171	603	−117	99
				Data smoothing					
				(use for up to quartic or quintic curve)					
7		5	−30	75	131	75	−30	5	231
9	15	−55	30	135	179	135	30	−55	15	429

[a] More complete tables can be found in A. Savitsky and M. Golay, *Analytical Chemistry,* 36, 1627–1639 (1964).

price to be paid for the extra safety of using an S-G function with more terms is that more data points are discarded at the beginning and at the end of the data set. This can be a problem for students who collect only a dozen points in the vicinity of an endpoint. It is not a problem when a computer collects several dozen points over the same range.

There are two important requirements a data set must meet before a Savitsky-Golay function is applied. First, the data must be taken at regular intervals. In the case of a potentiometric titration curve, voltage values must be recorded at regular volume increments. Second, graphs of the plotted data must be continuous and rather smooth. The first requirement is easy to meet by planning the titration experiment. The second is usually met by titration curves.

If you have access to a microcomputer, the following BASIC routine should help you take a first derivative using the simplest S-G function. It will work as written on an IBM-PC. Supply your own input/output routines if you use a different computer.

```
100 'program convdrv.bas takes a derivative of a function
110 'using a Savitsky-Golay filter. LWPotts 6/85
120 DIM X(20),Y(20),DY(20)
130 'load array with x and y values
140 N=1:I=1
150 READ X(I),Y(I)
160 ON ERROR GOTO 190    'traps out of data error to quit loop
170 N=N+1:I=I+1
180 GOTO 150 'loop again
```

15.5 POTENTIOMETRIC TITRATION CURVES: DATA TREATMENT

```
190 'begin convolution using S-G 5 element deriv. filter
200 FOR I=3 TO N-3 'lose 2 data pts at start and at finish
210 DY(I-2)=(Y(I-2)-8*Y(I-1)+0*Y(I)+8*Y(I+1)-Y(I+2))/12
220 NEXT
230 'dump the derivative to a printer
240 LPRINT "X-VALUE", "DY/DX"
250 FOR I=3 TO N-3
260 LPRINT X(I),DY(I-2)
270 NEXT
280 DATA 22.00,9.61,22.20,9.74,22.40,9.89,22.60,10.10,22.80,10.34,23.00,
    10.60,23.20,10.84,23.40,10.99,23.60,11.11,23.80,11.22,24.00,11.30
999 END
```

A word of caution is in order: because this text is an introductory text the convolution method has been developed only qualitatively. If you wish to apply the S-G method to more complex experimental data than titration curves, you should read Refs. *11* and *12* very carefully.

15.5.1 If All Else Fails

In some titrations the endpoint may be so poorly defined that even derivative plots are vague. While this is usually a sign that the titrimetric method is inadequate for careful work and some alternative method should be found, there is one more graphical method that should be tried. Using the method of Gran (Ref. *13*), the endpoint is found by extrapolating data from before and after the endpoint, but not relying on data in its immediate vicinity. The functions to be plotted depend on the type of titration. In the case of the WA/SB titration the function plotted before the endpoint is $V*10^{k-pH}$ vs. V, while after the endpoint $(V_0 + V)*10^{pH-k}$ vs. V is plotted. In these expressions V_0 is the sample volume at the start of the titration, V is the volume at each point in the titration, and k is an arbitrary constant chosen to make the plotting convenient. A Gran plot of the titration data in Fig. 15.14 appears in Fig. 15.15. Gran functions for other types of titrations can be found in Ref. *14*. According to more recent work by Pehrsson (Ref. *14*), Gran plots for WA/SB titrations are most accurate ($\pm 0.1\%$) for pK_a values in the range 5 to 8.5, when the formal concentration of weak acid acid is 0.01 and the base is $0.1F$.

15.5.2 A Final Point about Titrimetric Errors

A subtle and often important source of error in analyses based on potentiometric titrations arises from the use of the inflection point of a curve to signal the equivalence point of a reaction. In many situations students encounter in the laboratory it is entirely correct to assume that the inflection point and equivalence point are the same,

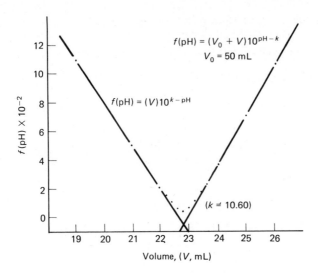

Figure 15.15 Gran plot for titration of THAM-H$^+$ with 0.150M NaOH.

particularly if the reactions are "isovalent" (i.e., reaction stoichiometry is 1:1 or 2:2 for titrant and analyte). An important exception is when a very weak or very dilute acid is titrated with a base. Meites and Goldman (Ref. *15*) have shown that in the titration of 0.1F weak acid with pK_a = 10 with 0.1F strong base, the inflection point will occur 1% before the equivalence point. Even if you use graphical or numerical methods to find the inflection point, it will not exactly coincide with the equivalence point. This represents a limit to the applicability of potentiometric titrations to analytical problems.

In precipitation titrations where stoichiometries are not 1:1 [for example, $Pb^{2+} + 2IO_3^- = Pb(IO_3)_2(s)$], inflection points are seldom identical with equivalence points. Physical effects such as the adsorption of ions on the precipitate near the equivalence point add more difficulties. Adsorption of halide ions in the titration of mixtures of halides was discussed in Chapter 11.

It should be apparent that a variety of factors must be weighed carefully in judging the accuracy of a particular titrimetric determination. With proper care and careful study, methods based on potentiometric titrations can be rapid, precise, and accurate.

REFERENCES

1. R. P. Buck, *Analytical Chemistry,* 45, 654 (1973).
2. R. P. Buck, *Analytical Chemistry,* 48, 23R (1976) (review article).
3. J. Vesely, D. Weiss, and K. Stulik, *Analysis with Ion-Selective Electrodes,* Ellis Horwood, Chichester, U.K., 1978.

4. G. Eisenman, *Glass Electrodes for Hydrogen and Other Cations,* Dekker, New York, 1967.
5. J. W. Severinghaus and A. F. Bradley, *Journal of Applied Physics,* 13, 515 (1958).
6. G. Guilbault and J. G. Montalvo, *Journal of the American Chemical Society,* 92, 2533 (1970).
7. C. Tran-Minh and G. Broun, *Analytical Chemistry,* 47, 1359 (1975).
8. G. Guilbault and W. Stokbro, *Analytica Chimica Acta,* 76, 237 (1975).
9. G. A. Rechnitz et al., *Analytical Chemistry,* 44, 1098 (1972).
10. E. Pungor, *Analytical Chemistry,* 39, 28A (1967).
11. A. Savitsky and M. Golay, *Analytical Chemistry,* 36, 1627–1639 (1964).
12. J. H. Noggle, *Physical Chemistry on a Microcomputer,* Little, Brown, Boston, 1985, chapter 7.
13. G. Gran, *Analyst,* 77, 661 (1952).
14. L. Pehrsson, F. Ingman, and A. Johnson, *Talanta,* 23, 769 (1976).
15. L. Meites and J. A. Goldman, *Analytical Chimica Acta,* 31, 297 (1964). This is the last in a series of articles on errors in titrations.

RECOMMENDED READING

General Books, Monographs

Bates, R. G. *Determination of pH,* 2nd ed., Wiley, New York, 1973.

Koryta, J. *Ions, Electrodes and Membranes,* Wiley, New York, 1982. A very good introductory level book on ISEs.

Koryta, J. *Ion Selective Electrodes,* Cambridge Univ. Press, London, 1975.

Ma, T. S., and S. S. M. Hassan. *Organic Analysis Using Ion-Selective Electrodes,* 2 vols., Academic Press, London, 1982.

Perrin, D. D., and B. Dempsey. *Buffers for pH and Metal Ion Control,* Chapman & Hall, London, 1979. Elementary treatment and practical information.

Vesely, J., D. Weiss, and K. Stulik. *Analysis with Ion-Selective Electrodes,* Ellis Horwood, Chichester, 1978.

Westcott, C. C. *pH Measurements,* Academic Press, Orlando, Fla., 1978. Excellent introduction to use of pH meters.

Review Articles

Arnold, M. A. and M. E. Meyerhoff. *Analytical Chemistry,* 56, 40R (1984). This is the 1984 edition of "Fundamental Reviews" published by *Analytical Chemistry* every other year.

Covington, A. K. *Critical Reviews in Analytical Chemistry*, 5, 355 (1974).

Koryta, J. *Analytica Chimica Acta*, 61, 329 (1972).

PROBLEMS

15.1. A Beckman E-2 glass membrane electrode gives the following response to pH in the presence of $1.0 M$ Li$^+$ at 25°C:

pH	E (mV) vs. SCE
8.0	−80
9.0	−138
10.0	−184
11.0	−230
12.0	−272
13.0	−296

(a) If a pH electrode gives a "Nernstian" response its potential will change by 0.0592 V per factor of 10 change in hydronium ion activity. Is the E-2 electrode giving a Nernstian response in solutions of lithium ion?

(b) Calculate the selectivity coefficient, K_{HLi}, in the pH range 9 to 11. Assume that the coefficient $m = 1$.

(c) Does the selectivity coefficient change with pH?

15.2. One of the glass formulations in Table 15.2 gives a selectivity coefficient of 2800 for potassium in the presence of sodium. Write expressions for the response of a potassium-sensitive electrode to two solutions: $0.01 M$ KCl and $0.01 M$ KCl containing $1 \times 10^{-4} M$ NaCl. What difference in potential can be ascribed to the presence of sodium ion in the second solution? (Note: You do not need to know the "constant" term in a Nernst expression or the potential of a reference electrode to answer this question. You are looking for a difference in potentials. Assume Nernstian behavior for the potassium electrode and a value of $m = 1$ in the selectivity coefficient equation.)

15.3. The standard potential of the silver chloride/silver metal reference electrode is 0.2224 V. Write a Nernst equation for the reference electrode. Explain how the following factors influence the potential of the electrode:

(a) A 1% error is made in standardizing the chloride solution in which the electrode is immersed.

(b) The electrode is used at 35°C instead of 25°C. (Hint: Look at the coefficient RT/nF in the Nernst equation.)

PROBLEMS

15.4. Knowing that the standard potential of the AgCl/Ag electrode is 0.2224 V and that the potential of the AgCl/Ag electrode is 0.197 V in saturated KCl solution, calculate the activity of chloride ion in saturated KCl solution.

15.5. The standard potential of the SCE is 0.2415 V. What is the potential of the SCE at 5 and at 35°C?

15.6. The standard potential of the ferricyanide/ferrocyanide couple is +0.36 V. What potential would you read from a voltmeter if you measured this standard potential against an SCE, AgCl/Ag, normal calomel, or mercurous sulfate reference electrode?

15.7. The following data were obtained in a potentiometric titration. Plot the data and locate the inflection point of the curve, first by inspection and then by using the first and second derivatives.

Volume (mL)	E (mV)
14.14	218
14.16	220
14.18	219
14.20	220
14.22	222
14.24	224
14.26	227
14.28	235
14.30	244
14.32	252
14.34	256
14.36	259
14.38	261
14.40	260
14.42	262

15.8. Savitsky-Golay filter coefficients for second derivatives and for data smoothing are presented in the chapter. Modify the BASIC program in the text to use these filters, and use the data in Problem 15.7 to test the results.

15.9. Take the data in Problem 15.7 and add "noise," for example, a random 2- or 3-mV addition to or subtraction from the potential readings. Apply the Savitsky-Golay smoothing filter to your noisy data and see if the added noise is filtered out. Experiment by adding more data points and by adding some nonrandom potential changes.

15.10. An experiment shows that the response of a particular fluoride-responsive electrode to fluoride ion activity is Nernstian over the concentration range 1×10^{-5} to $0.01 M$. When the electrode is dipped into 50 mL of a solution containing an unknown, the potential of the electrode is 0.124 V vs. SCE. Exactly 5 mL of a 1.00 mM solution of NaF is pipetted into the unknown sample. The electrode potential is found to be 0.169 V vs. SCE. What is the concentration of fluoride in the unknown, assuming constant ionic strength?

15.11. Two solutions of KCl are prepared. The first is made up to be 0.050M. The second is made by diluting 10 mL of the first solution to 100 mL with deionized water.

   (a) Calculate the concentrations and activities of chloride ion in the two solutions (see Chapter 3 for Debye-Hückel treatment).

   (b) If the response of a chloride-responsive electrode is a 59 mV potential change for a factor of 10 change in *activity*, by how many millivolts will the potential change when the *concentration* is decreased as specified in this problem?

Chapter 16

Redox Titrimetric Methods

16.1 INTRODUCTION

Now that we have discussed in detail the relationships between species activities and cell voltage in Chapter 14 and the design and operation of indicator and reference electrodes in Chapter 15, we are able to learn about some important analytical applications of oxidation-reduction reactions. The focus will be titrimetric methods involving either the use of indicator electrodes (potentiometric endpoint detection) or indicator color changes. While many other valuable methods are used to monitor redox titrations (such as the measurement of small currents, "amperometry"), they are left to advanced courses in analysis.

The chemistry discussed in this chapter is mainly inorganic chemistry because inorganic systems are somewhat simpler to understand and apply than are organic systems. While many inorganic redox systems react reversibly and rapidly, only a few organic reactions are so cooperative. More often organic systems are irreversible, sluggish except in an excess of titrant (a symptom of kinetic complexity), fail to react quantitatively, and are subject to side reactions which form several products. Several organic redox systems are nevertheless quite important and will be discussed later in the chapter.

16.2 FUNDAMENTALS OF REDOX TITRATIONS

From an analyst's perspective there are four important requirements that a redox titration procedure must meet. First, the reaction must have a known and reproducible stoichiometry. Second, the equivalence point of the titration must be easily and accurately monitored as an endpoint. The third requirement is that when the endpoint is reached, 99.9% of the analyte must have reacted. Finally, the reaction should be rapid at ordinary laboratory temperatures.

16.2.1 A Model System: Fe(II) + Ce(IV) = Fe(III) + Ce(III)

Potentiometric and indicator redox titrations can be illustrated by the oxidation of ferrous iron with standard cerium(IV) in $1M$ sulfuric acid. The formal potentials for the half-reactions are

$$Fe(III) + e = Fe(II) \qquad E^f = +0.68 \ (1M \ H_2SO_4)$$
$$Ce(IV) + e = Ce(III) \qquad E^f = +1.44 \ (1M \ H_2SO_4)$$

From the formal potentials we expect Ce(IV) to oxidize Fe(II) readily. In fact, at the $0.01F$ concentration level the reaction is quantitative. Both the oxidation and reduction half-reactions act reversibly, and equilibrium is achieved very rapidly at room temperature.

The following example shows that the Ce(IV)-Fe(II) reaction is quantitative at the $0.01F$ level.

EXAMPLE 16.1 Quantitative Reaction

Given the formal potentials in the text, is the titration of 50 mL of $0.01F$ Fe(II) with $0.01F$ Ce(IV) complete at the equivalence point?

Solution

The equilibrium constant can be calculated from the formal potentials, as shown in Chapter 14,

$$\log K = n_1 n_2 [E^f[Ce(IV)] - E^f(Fe(III))]/0.05916$$
$$K = 1*1(1.44 - 0.68)/0.05916 = 0.76/0.05916 = 10^{12.85}$$

The expression for K is

$$[Fe(III)][Ce(III)]/[Fe(II)][Ce(IV)] = K$$

At the equivalence point (almost) all the iron is in the +3 state, and its concentration is (almost) $0.01/2 = 0.005M$, as a result of dilution. The equilibrium concentration of Ce(III), the other product of the reaction, is also $0.005M$ at the equivalence point.

16.2 FUNDAMENTALS OF REDOX TITRATIONS

Fe(II) and Ce(IV) are also present in solution, but only to the extent permitted by the size of K. Since K is large in this system, [Fe(II)] and [Ce(IV)] are very small. They are also equal. Recognizing that [Fe(III)] must equal $0.005M -$ [Fe(II)] and that [Ce(III)] must equal [Fe(III)], we can write the following expression for K in terms of [Fe(II)] and the analytical concentration (accounting for dilution):

$$(0.005 - [Fe(II)])^2/[Fe(II)]^2 = 10^{12.85}$$

Solving for [Fe(II)] gives

$$[Fe(II)] = 1.9 \times 10^{-9}M$$

The equilibrium concentration of Fe(II) is more than six orders of magnitude smaller than its initial value, and we conclude that the reaction should be quantitative at the equivalence point.

The reaction of Ce(IV) with Fe(II) has simple 1:1 stoichiometry. Here is an example of a calculation of titration results.

EXAMPLE 16.2 Analytical Results

A sample containing iron is dissolved in dilute sulfuric acid and treated with stannous sulfate, Sn(II), to reduce the iron to the +2 state quantitatively. The resulting solution is diluted to 100 mL with dilute sulfuric acid. A 10-mL portion is transferred to a titration vessel and diluted with 40 mL of dilute sulfuric acid. The solution is titrated with standard ceric ammonium sulfate solution ($0.01034M$) and progress is monitored with a platinum electrode. The equivalence point is reached with the addition of 11.32 mL of titrant. Calculate the number of milligrams of Fe in the original sample.

Solution

Each millimole of Ce(IV) consumes a millimole of Fe(II). The 11.32 mL of Ce(IV) solution contains 11.32(0.01034) or 0.1171 mmole of Ce(IV). The weight of 0.1171 mmole of iron is (55.85 mg/mmole)(0.1171 mmole) = 6.540 mg. Since a 10-mL portion of the 100-mL solution containing the iron was taken for the titration, the original sample contains 6.540(10) = 65.40 mg of iron.

Calculations involving endpoint volumes and concentrations usually involve molar or millimolar concentration units. As you can see from the preceding example, the calculations are entirely parallel to those in acid-base titrations. An alternative concentration system, the normality system mentioned in Chapter 1, involves chemical equivalents. The normality of a solution is given by the number of gram equivalent weights of solute per liter of solution,

$$N = \text{equivalent weights/liter (also milliequivalent weights/mL)}$$

The equivalent weight of a solute which is used in a redox titration is found by dividing the gram formula weight of the solute by the number of electrons it donates or accepts in the balanced redox reaction. For example, in acidic solution dichromate ion takes part in a six-electron transfer, as shown by the half-reaction,

$$Cr_2O_7^{2-} + 14H^+ + 6e = 2Cr^{3+} + 7H_2O$$

The equivalent weight of potassium dichromate is thus equal to one-sixth its gram formula weight, $294.2/6 = 49.03$. A $0.10N$ solution of potassium dichromate will contain 4.90 g of $K_2Cr_2O_7$ per liter, and will be $0.0167M$.

Oxalic acid is oxidized in a two-electron process,

$$H_2C_2O_4 = 2CO_2 + 2H^+ + 2e$$

and its equivalent weight is one-half its gram formula weight, $90.03/2 = 45.02$. A $0.010N$ solution of oxalic acid will contain 0.450 g of oxalic acid per liter and will be $5.0 \times 10^{-3}M$.

Matters are complicated in systems such as permanganate, where several half-reactions are possible:

$$MnO_4^- + 8H^+ + 5e = Mn^{2+} + 4H_2O \qquad \text{eq. wt.} = \text{formula wt.}/5$$
$$MnO_4^- + 4H^+ + 3e = MnO_2(s) + 2H_2O \qquad \text{eq. wt.} = \text{formula wt.}/3$$
$$MnO_4^- + e = MnO_4^{2-} \qquad \text{eq. wt.} = \text{formula wt.}$$

The equivalent weight depends on which half-reaction is being used in the analysis (and the half-reaction depends on pH, as discussed below). A solution which is labeled "$0.10N$ $KMnO_4$" is therefore quite ambiguous. It might mean $0.020M$, $0.033M$, or $0.10M$, depending on the purpose intended by the person who prepared it.

The following example illustrates equivalent weight and normality calculations.

EXAMPLE 16.3 Normality and Molarity

A 632.2-mg sample of $KMnO_4$ is dissolved and diluted to 500 mL with deionized water and is subsequently used for the titration of Fe^{2+}. The balanced chemical reaction is

$$MnO_4^- + 8H^+ + 5Fe^{2+} = Mn^{2+} + 5Fe^{3+} + 4H_2O$$

16.2 FUNDAMENTALS OF REDOX TITRATIONS

(a) What are the molarity and normality of the permanganate solution?

(b) If 12.72 mL of the permanganate solution is used to reach the endpoint, how many millimoles, milliequivalents, and milligrams of iron are present in the sample?

Solution

(a) 632.2 mg/(158.04 mg/mmole) = 4.000 mmole. The concentration is 4.000 mmole/500 mL = $8.00 \times 10^{-3} M$. To find the normality, notice that five electrons are transferred to permanganate in the reaction. The equivalent weight of potassium permanganate is therefore 158.04/5 = 31.61. There are 20.00 meq of $KMnO_4$ in 632.2 mg. The normality is given by 20.00 meq/500 mL = 0.0400N.

(b) The number of milliequivalents of permanganate delivered by the endpoint is 12.72(0.0400) = 0.5088. This is the same as the milliequivalents of iron. Since each Fe^{2+} loses one electron, the milliequivalents of iron are the same as the millimoles of iron, that is, 0.5088. This corresponds to 0.5088(55.85) = 28.42 mg of iron in the sample.

If we had done this calculation only with millimoles, we would have recognized from the balanced equation that 12.72 mL(0.00800 mmole/mL) = 0.1018 mmole of permanganate would react with 5(0.1018) = 0.5088 mmole of iron.

16.2.2 Potentiometric Monitoring

There are three regions of interest in a simple potentiometric redox titration curve: the portion before the equivalence point, the equivalence point itself, and the portion after the equivalence point. These regions can be seen in Fig. 16.1, a simulated curve for the titration of 0.01F Fe(II) with 0.01F Ce(IV) in 1M H_2SO_4. In this figure the potential of a platinum indicator electrode is plotted as a function of the volume of titrant. The potential of the electrode is measured against a saturated calomel reference electrode. Potentials relative to the standard hydrogen electrode can be found by subtracting 0.244 V from the plotted values (i.e., E_{SCE} = 0.244 V vs. SHE). The platinum electrode is a "type 0" electrode referred to in Chapter 14. Platinum is not a direct part of either half-reaction; it responds only to the tendency of the solution to give or take electrons. The potential of the platinum indicator electrode will be controlled by the couple which is reversible, that is, the couple which has both of its components present in substantial concentrations. Initially the only iron species present at a significant concentration is Fe(II). What little Fe(III) is present might arise from air oxidation or incomplete prior reduction. There is, however, not enough Fe(III) present to allow the Fe(III)/Fe(II) couple to behave reversibly. Review the definitions of reversibility in Chapter 14 if you cannot remember why this is the case. When the

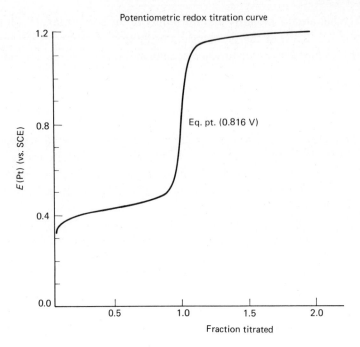

Figure 16.1 Potentiometric redox titration curve.

platinum electrode is in contact with only an irreversible couple, its potential wanders: the electrode is "poorly poised," in the jargon of electrochemists. Once a few drops of Ce(IV) have been added and some Fe(III) has been produced, the Fe(III)/Fe(II) couple becomes reversible and the electrode potential stabilizes. Since there is only a trifling amount of Ce(IV) before the equivalence point (see Example 16.1 for a way to calculate how much), the Ce(IV)/Ce(III) couple is irreversible and cannot control the potential of the platinum electrode.

The potential of the platinum electrode is described by a Nernst equation for the reversible Fe(III)/Fe(II) couple:

$$E_{Pt} = E^f_{Fe(III)/Fe(II)} - 0.05916 \log ([Fe(II)]/[Fe(III)])$$

Before moving on, notice that this equation predicts that the potential of the Pt electrode *does not change with dilution.* In real titrations, changes in ionic strength may change measured potentials, but such effects should be minor. An important consequence for calculations is that we might as well use X as the fraction titrated, and $1.0 - X$ as the fraction untitrated, as use formal concentrations. The simplified equation is therefore

$$E_{Pt} = E^f_{Fe(III)/Fe(II)} - 0.05916 \log [(1.0 - X)/X]$$

This equation applies from the first drop of titrant up to the equivalence point.

16.2 FUNDAMENTALS OF REDOX TITRATIONS

At the equivalence point the concentrations of both Fe(II) and Ce(IV) are too small to stabilize the potential of the Pt electrode. The electrode is poorly poised with respect to both half-reactions, and the potential drifts until a drop of Ce(IV) has been added in excess. The potential at the equivalence point is expressed in terms of both formal potentials:

$$E_{\text{Pt,eq.pt.}} = (n_1 E_1^f + n_2 E_2^f)/(n_1 + n_2)$$

When the two half-reactions of a titration involve the same number of electrons, the equivalence point potential is simply the average of the formal potentials. When one reaction involves a larger number of electrons, its formal potential weighs more heavily in the equivalence point potential. Under such circumstances the titration curve break is not symmetrical: rather than occurring at the inflection point, the equivalence point occurs closer to the formal potential of the couple with the greater number of electrons.

Past the equivalence point the potential of the Pt electrode is governed by the titrant couple, Ce(IV)/Ce(III) in the example being discussed. The Nernst equation for this portion of the curve is

$$E_{\text{Pt}} = E^f_{\text{Ce(IV)/Ce(III)}} - 0.05916 \log ([\text{Ce(III)}]/[\text{Ce(IV)}])$$

This equation predicts that the potential of the electrode will continue to rise as Ce(IV) is added (note: the ratio [Ce(III)]/[Ce(IV)] gets smaller). Since we are once again dealing with a dilution-independent case, we can express the Nernst equation in terms of fraction titrated, X,

$$E_{\text{Pt}} = E^f_{\text{Ce(IV)/Ce(III)}} - 0.05916 \log [1.0/(X - 1.0)]$$

where X will now have values greater than unity. Notice that at a fraction titrated of 2.0 (that is, 200% titrated), the ratio of the fractions is unity, and the electrode potential is equal to the formal potential of the reversible couple Ce(IV)/Ce(III). The Fe(III)/Fe(II) couple is irreversible at and after the equivalence point.

Irreversibility of either the titrant or analyte half-reaction can cause experimental difficulties in real systems. For example, the permanganate/manganese(II) couple, which involves a transfer of five electrons,

$$\text{MnO}_4^- + 8\text{H}^+ + 5e = \text{Mn}^{2+} + 7\text{H}_2\text{O}$$

is irreversible. In actual experiments a platinum electrode appears to be governed by the Mn(III)/Mn(II) couple ($E^f = 1.5$ V in $8M$ sulfuric acid), rather than the $\text{MnO}_4^-/\text{Mn(II)}$ couple.

Dichromate ion is another good case in point: the six-electron transfer,

$$\text{Cr}_2\text{O}_7^{2-} + 14\text{H}^+ + 6e = 2\text{Cr}^{3+} + 7\text{H}_2\text{O}$$

is quite irreversible. In titrations of Fe(II) with dichromate titrant, the potential of a platinum electrode after the equivalence point (when excess dichromate is added) wanders and is irreproducible. Laitinen's text (Ref. *1*) shows potentiometric titration curves obtained with irreversible half-reactions.

16.2.3 Redox Endpoint Indicators

Once the behavior of a redox titration system is understood it may save time and trouble in routine work to use redox indicators for endpoint detection rather than potential measurement. There are two major classes of redox indicators. First, there is a small group of indicators which react with one form of a redox couple to produce a characteristic color change. These reactions tend to be quite system-specific—for example, the reaction of starch with iodine to form a violet complex (see iodometric methods, Section 16.3.3). The second major class of redox indicators includes substances which themselves undergo oxidation or reduction and change color. This class is generally more useful and has been well investigated. Almost all of the popular indicators have positive formal potentials; their reduced forms are oxidized by strong oxidants, and their oxidized forms are rather easily reduced.

For our purposes redox indicators can be divided into three groups according to their formal potentials: high (1.0 to 1.2 V), moderate (0.7 to 0.85 V), and low (0.0 to 0.4 V). Several examples from each category are given in Table 16.1. Notice that

TABLE 16.1 SELECTED REDOX INDICATORS

Indicator	Color change (red → ox)	Formal potential (vs. SHE)
Ferroin[a]	Red to blue	+1.06 (1M H_2SO_4)
5-Nitroferroin	Violet red to blue	+1.25
4,7-Dimethylferroin	Red to yellow-green	+0.88
5,6-Dimethylferroin	Red to yellow-green	+0.97
5-Methylferroin	Red to yellow-green	+1.02
Diphenylamine	Colorless to violet	+0.76 (1M H_2SO_4)
Diphenylamine sulfonate	Colorless to violet	+0.85 (0.5M H_2SO_4)
4-Aminodiphenylamine	Colorless to dark blue	+0.76
4-Amino-3-methoxydiphenylamine	Colorless to blue	+0.64
Methylene blue	Blue/green to colorless	+0.101 (pH 5), +0.011 (pH 7), −0.050 (pH 9)
Phenol-*m*-sulfonate–indo-2,6-dibromophenol		+0.390 (pH 5), +0.270 (pH 7), +0.168 (pH 9)
Indigo disulfonate	Blue to colorless	−0.010 (pH 5), −0.125 (pH 7), −0.199 (pH 9)

[a] Ferroin is tris(1,10-phenanthroline)-Fe(II)

16.2 FUNDAMENTALS OF REDOX TITRATIONS

the compounds with higher formal potentials are related to the tris(o-phenanthroline)-Fe(II) complex described and discussed in Chapter 13. These compounds are called "ferroin indicators." Indicators with high formal potentials are needed when using strongly oxidizing titrants. For example, the formal potential of the Ce(IV)/Ce(III) couple in sulfuric acid is 1.44 V vs. SHE. The equivalence point in the titration of Fe(II) with Ce(IV) occurs at 1.06 V vs. SHE (+0.816 V vs. SCE), and an indicator which changes color at a potential close to 1.06 V is desirable. A high formal potential indicator could not be used in a titration involving a weaker oxidizer, such as dichromate in $1M$ sulfuric acid (formal potential 1.03 V). If a color change were observed, it would be quite late, and results would be high.

The ferroin compounds are among the most exhaustively studied compounds in analytical chemistry. Their effectiveness relies on the differences in the colors of Fe(II) and Fe(III) phenanthroline complexes. In the case of the simplest complex in the group, tris(o-phenanthroline)-Fe(II), the color of the reduced form is red-orange, while that of the oxidized form is pale blue (verging on colorless in most applications). The oxidation of the Fe(II) complex is chemically reversible, meaning that the Fe(II) complex can be oxidized to the Fe(III) complex, which can be reduced again to form the initial Fe(II) complex. As indicated in Table 16.1, the nature and position of substituents on the phenanthroline rings affect the formal potential of the complexes. 5-Nitroferroin has the highest formal potential of the ferroins, 1.25 V in $1M$ H_2SO_4. The 4,7-dimethyl derivative has the lowest formal potential of the ferroin indicators, 0.88 V. The formal potential of a redox indicator can be expected to differ somewhat from the potential at which the color is perceived to change. In the case of simple ferroin, the Fe(II) form is so much more intense than the Fe(III) form [molar absorptivity 11,700 liter/mole cm at 508 nm, as opposed to 600 liter/mole cm at 590 nm for the Fe(III) complex] that more than 90% of the indicator must be titrated with oxidant before the analyst judges the endpoint to have occurred. The potential for this point is about 0.050 V more positive than the formal potential, a value you might like to verify by examining a Nernst equation for the indicator. An excellent monograph on phenanthroline complexes, including redox indicators, is available in many libraries (Ref. 2).

The indicator compounds with moderate formal potentials are all related to diphenylamine. The diphenylamine compounds are quite useful as indicators in dichromate titrations. In a two-electron oxidation step, two diphenylamines are linked to form a colorless diphenylbenzidine compound,

$$2 \; \text{Ph-NH-Ph-SO}_3^- \rightarrow {}^-\text{O}_3\text{S-Ph-NH-Ph-Ph-NH-Ph-SO}_3^- + 2\text{H}^+ + 2e^-$$

The diphenylbenzidine then undergoes another two-electron oxidation to the violet end product,

$$\rightarrow {}^-\text{O}_3\text{S-Ph-N=Ph=Ph=N-Ph-SO}_3^- + 2\text{H}^+ + 2e^-$$

The color formation reaction is reversible with a formal potential of 0.76 V in $1M$ sulfuric acid. The closely related compound diphenylamine sulfonic acid is more soluble than diphenylamine. Its higher formal potential (0.85 V) makes it a better indicator for the determination of Fe(II) with dichromate. The oxidation of the indicator is known to be catalyzed by the presence of Fe(III). A practical consequence of this is that a dichromate solution used to determine iron should be standardized with standard Fe(II) solution rather than some other reducing agent. Diphenylamine sulfonate is used as an indicator in the titrimetric iron experiment at the end of the chapter.

The indicators with low formal potentials find many uses in biochemical redox titrations. Methylene blue is an excellent example. At pH 5 the formal potential of methylene blue is +0.101 V (30°C). At pH 7 the formal potential decreases to 0.011 V, and at pH 9 it is actually negative, −0.050 V.

Redox indicators are also quite useful when one must estimate the formal potential of a redox couple to which a metal indicator electrode will not respond (an enzyme-activated couple, for example). By observing which indicators change color in the presence of the redox system, it may be possible at least to estimate a range in which the formal potential lies. Colorimetric measurement of the concentrations of oxidized and reduced indicator forms in the presence of a redox couple may allow a very exact measurement of the formal potential of that couple.

16.3 ANALYTICAL METHODS AND REAGENTS

We will now begin a survey of important analytical methods and reagents. This major section of the chapter is divided into several parts. First, to gain some perspective, we will consider some popular methods involving the iodine/iodide couple. We can use these methods as models for direct and indirect titration procedures and make more sense of applications involving oxidizing and reducing reagents. We will then describe the use of several important oxidizing agents and reducing agents and learn about prior reduction and oxidation of samples. We will end the chapter with a second look at iodometric methods and a detailed discussion of the iodometric determination of copper.

16.3.1 Analytical Case Studies: Iodometric Methods

Before starting a discussion of oxidizing and reducing agents and introducing a great deal of descriptive chemistry, it makes sense to outline some typical titrimetric procedures. The discussion in this section will center on the reactions of iodide and iodine, two species related by a reversible electron exchange,

$$I_2(aq) + 2e = 2I^- \qquad E^0 = +0.6197 \text{ V}$$

The standard potential of the couple is rather modest: iodine is a good enough oxidizer to pull electrons from a variety of common reducing agents, among them sulfite, sulfide, and thiosulfate. At the same time, iodide is a good enough reducer to give electrons up to stronger oxidizers such as iodate and bromate. Although iodine is not very soluble in water, it forms a soluble brown-colored species with iodide ion called triiodide ion,

$$I_2(aq) + I^- = I_3^- \qquad K = 710 \qquad (25°C)$$

which is only slightly weaker as an oxidizing agent.

Iodine reacts with soluble starch and iodide to form a deep blue indicator complex. The absorptivity of the complex is so high that the color of a $10^{-5} M$ solution of iodine is plainly visible. The indicator is so sensitive and so easy to prepare that it is the indicator of choice in iodometric methods.

The reaction of iodine with thiosulfate to form iodide and tetrathionate occurs rapidly and quantitatively in slightly acidic solutions:

$$I_3^- + 2S_2O_3^{2-} = 3I^- + S_4O_6^{2-}$$

As one performs a titration of a sample of triiodide with thiosulfate, the brown color of the triiodide complex becomes fainter and fainter as the equivalence point is approached. When the triiodide color is a pale yellow, starch solution is added, and the deep blue color of the starch-iodine indicator complex is seen. Additional thiosulfate reduces iodine in the starch complex and when the endpoint is reached the solution is colorless.

There are some representative cases to consider, the direct and indirect redox determinations with iodine.

1. In the *direct* method standard iodine solution is added as a titrant to a solution containing a reducing agent. The first excess of iodine after the equivalence point can be seen with the help of starch indicator.

Titration: sought material + I_2 standard = sought oxidized form + $2I^-$

2. In the *indirect* method excess potassium iodide is added to a solution of oxidizing agent to be analyzed. Iodine formed in the reaction is then titrated (that is, back-titrated) with standard thiosulfate to the blue → colorless endpoint.

Sought material + excess KI = sought reduced form + I_2
Titration: $I_2 + 2S_2O_3^{2-} = 2I^- + S_4O_6^{2-}$

In a variation on the indirect method, a known amount of strong oxidizing agent is added in excess to a solution containing a reducing agent. Potassium iodide is then added to the solution to react with the unreacted oxidizing agent to form iodine. The iodine is then titrated with standard thiosulfate solution to the blue → colorless starch endpoint. Iodide/iodine is used as an intermediary in this case.

Other variations on the indirect method involve the use of iodine to determine hydrogen sulfide,

$$H_2S + I_3^- = S + 2H^+ + 3I^-$$

Standard triiodide solution may serve as a trap to collect hydrogen sulfide in a gas stream. When trapping is complete the iodine that has not been consumed by hydrogen sulfide is titrated with standard thiosulfate. Alternatively, standard potassium iodate might be used to trap the hydrogen sulfide. When trapping is complete potassium iodide is added. The iodide reacts with the excess iodate to form iodine, which is then back-titrated with standard thiosulfate. The fact that so many variations exist points to the great versatility of the iodine/iodide couple for redox methods.

We now have enough background information to discuss redox reagents and some interesting applications. Important details of iodometric methods will be covered near the end of the chapter.

16.3.2 Redox Reagents

In this section we will discuss several popular oxidizing and reducing reagents used in analytical chemistry.

Oxidizing Agents

Permanganate MnO_4^- is a very versatile oxidizing agent with very complicated chemistry. Manganese is formally in the +7 oxidation state in permanganate. In acidic solutions it is a very strong oxidizer,

$$MnO_4^- + 8H^+ + 5e = Mn^{2+} + 4H_2O \qquad E^0 = 1.51 \text{ V}$$

The five-electron reduction process consists of several steps and is irreversible; Mn^{2+} is reoxidized to products other than permanganate. The standard potential is calculated on the basis of free energy changes accompanying other manganese reactions.

In alkaline solutions permanganate is a weaker oxidizer,

$$MnO_4^- + 2H_2O + 3e = MnO_2(s) + 4OH^- \qquad E^0 = 0.588 \text{ V}$$

In strongly alkaline solutions permanganate ion is reduced in a one-electron process to manganate ion, Mn(VI),

$$MnO_4^- + e = MnO_4^{2-} \qquad E^0 = 0.504 \text{ V}$$

16.3 ANALYTICAL METHODS AND REAGENTS

While potassium permanganate is available in high purity, it is not a good primary standard material. Crystalline KMnO$_4$ will have surface contamination from manganese dioxide (MnO$_2$), which catalyzes the reduction of permanganate when in solution and gradually lowers the concentration of a standard solution. Reducing impurities on glassware surfaces and direct sunlight contribute to the formation of manganese oxides; in fact, it is these oxides that produce the dark stain on glassware used to store permanganate. It is standard practice to prepare a potassium permanganate solution, set it aside in the dark for a few days, then filter out manganese dioxide with a fritted glass filter (paper filters can be oxidized slightly in a process that makes more manganese dioxide). Neutral solutions of KMnO$_4$ are observed to be more stable than acidic or alkaline solutions.

Solutions of permanganate can be standardized conveniently with primary standard sodium oxalate (available from the National Bureau of Standards) according to the reaction:

$$2MnO_4^- + 5C_2O_4^{2-} + 16H^+ = 2Mn^{2+} + 10CO_2 + 8H_2O$$

This reaction has been studied in great detail and is known to be quite complicated. The first few drops of permanganate added to a solution of oxalate react very slowly; as manganous ion [Mn(II)] forms the reaction speeds up, suggesting that Mn(II) catalyzes the reaction between permanganate and oxalate. The Mn(II) is also able to react directly with permanganate in what is called the "Guyard reaction,"

$$2MnO_4^- + 3Mn^{2+} + 2H_2O = 5MnO_2 + 4H^+$$

This reaction occurs very slowly in acidic solutions but quite rapidly in neutral solutions.

Arsenious oxide, As(III), is another excellent primary standard for the standardization of permanganate solutions. The balanced reaction is

$$5As_2O_3 + 4MnO_4^- + 12H^+ = 5As_2O_5 + 4Mn^{2+} + 6H_2O$$

The addition of a small amount of iodide or iodate catalyzes the reaction. Samples of arsenious oxide must be dissolved in a few milliliters of $1M$ NaOH (free of reducing impurities), then made acidic with sulfuric acid. A drop or two of millimolar KIO$_3$ solution is sufficient to catalyze the reaction with permanganate. The arsenious oxide solution is titrated with permanganate until the first lingering rose color (excess permanganate) is observed.

EXAMPLE 16.3 Standardization of Permanganate

A 0.2256-g sample of primary standard arsenious oxide is dissolved in NaOH, made acidic by the addition of 10 mL of 1:1 sulfuric acid, and diluted to about 50 mL with deionized water. A drop of potassium iodate solution is added. Titration with permanganate solution requires 30.47 mL to reach the first lingering rose color. What is the concentration of the permanganate solution?

Solution

The gram formula weight of arsenious oxide is 197.84. The 0.2256 g sample corresponds to 1.140 mmole of arsenious oxide. The reaction stoichiometry tells us that 5 mmole of arsenious oxide require 4 mmole of permanganate. Therefore 1.140 mmole will require 0.9120 mmole of permanganate. The concentration of the permanganate solution is given by $0.9120/(30.47 \text{ mL}) = 0.02993 M$.

Cerium(IV) (Cerate) Cerium(IV) undergoes a one-electron reduction (without intermediates) to Ce(III). The oxidizing power of Ce(IV) depends on the electrolyte in which it is dissolved. In $1M$ perchloric acid the formal potential of the Ce(IV)/Ce(III) couple is 1.70 V, while in $1M$ nitric acid it is 1.61 V. In $1M$ sulfuric acid the formal potential is somewhat smaller, about 1.44 V. In $1M$ HCl the formal potential is only about 1.28 V. The interpretation of this behavior is that nitrate, sulfate, and chloride stabilize the Ce(IV) state by complexation and thus lower the formal potential for the couple (see arguments like this in Chapter 14). Chloride presents a special problem. Cerate ion is a strong enough oxidizer to oxidize chloride ion to chlorine, and Ce(IV) solutions in HCl are not very stable, particularly if they must be boiled. From a practical standpoint, however, Ce(IV) can be used to titrate compounds dissolved in HCl solutions without significant interference from chloride.

Cerate solutions can be prepared from either ammonium hexanitratocerate, $(NH_4)_2Ce(NO_3)_6$ (formula weight 548.26), or ceric ammonium sulfate, $Ce(SO_4)_2 \cdot 2(NH_4)_2SO_4 \cdot 2H_2O$ (formula weight 632.36). Solutions in sulfuric acid are quite stable. Nitric and perchloric acid solutions of cerate should be kept out of strong light, which induces a photochemical oxidation of water by Ce(IV) and a resulting decrease in concentration.

Both sodium oxalate and arsenious oxide can be used as primary standards for the standardization of cerate solutions. Unfortunately, the reactions are slow at room temperature. Iodine monochloride (ICl) and osmium tetroxide (OsO_4) are good catalysts for these titrations. Nitroferroin is a good indicator for the endpoints of cerate titrations.

Potassium Dichromate $K_2Cr_2O_7$ is available as a primary standard material from the National Bureau of Standards. Standard solutions can be prepared by drying the salt at 150°C and diluting with deionized water (free of reducing impurities). Potassium dichromate solutions are extremely stable and may be kept for years.

16.3 ANALYTICAL METHODS AND REAGENTS

The standard potential for the dichromate/chromium(III) couple is lower than that of permanganate or cerate,

$$Cr_2O_7^{2-} + 14H^+ + 6e = 2Cr^{3+} + 7H_2O \qquad E^0 = 1.36 \text{ V}$$

It should be noted that the formal potential in acidic solutions is seldom as large as the standard potential. In $0.1F$ HCl the formal potential is about 0.93 V, while in $1F$ sulfuric acid it is about 1.03 V. Reactions of dichromate involve complex intermediate compounds. As mentioned above, the dichromate/Cr(III) couple is irreversible, and monitoring a potentiometric titration curve with a platinum electrode can be quite frustrating. Dichromate reactions can also be slow to reach equilibrium at room temperature. In fact, most dichromate titrations could be carried out as well or better with Ce(IV) titrant. Potassium dichromate solutions are, however, easier and less expensive to prepare for routine work.

The most important application of dichromate in analysis is the oxidation of Fe(II). Directions for the method appear at the end of the chapter. Many other compounds can be determined indirectly by their reactions with Fe(II) or Fe(III). For example, nitrate can be determined by adding an excess of standard Fe(II) in $2M$ sulfuric acid with ammonium molybdate catalyst:

$$3Fe(II) + NO_3^- + 4H^+ = 3Fe(III) + NO + 2H_2O$$

The unreacted Fe(II) is then back-titrated with dichromate. See reference 3 for details.

Oxyhalogens Limitations of space make it impossible to give any more than a simple overview of the oxyhalogen compounds. This broad class of oxidizing agents has at least three important members: iodate, periodate, and bromate. The chemistry of these compounds has been studied extensively. Volume III of the series on volumetric methods by Kolthoff and Belcher (Ref. *3*) and the classic text of Laitinen (Ref. *1*) are excellent starting points for a deeper understanding and keys to the literature.

1. *Iodate.* Potassium iodate is a strong oxidizing agent in acidic solution and can be purchased as a primary standard from several chemical supply houses. Its reaction with iodide is rapid and quantitative in the presence of acid,

$$IO_3^- + 6H^+ + 5I^- = 3I_2 + 3H_2O$$

Notice that iodine is reduced from the +5 state in iodate to the 0 state in molecular iodine; it is easy to guess incorrectly that only five electrons are involved for each iodate because of the appearance of five iodides in the reaction. Actually, six equivalents of iodine are produced. KIO_3 is very important in iodimetric methods and will be discussed in a subsequent section. It is also used as a primary standard for the preparation of standard solutions of acids and is therefore one of the more versatile reagents in the laboratory. Each mole of iodate will react with 6 moles of hydronium ion.

In hydrochloric acid solutions iodate is reduced to iodine chloride:

$$IO_3^- + 6H^+ + 2Cl^- + 4e = ICl_2^- + 3H_2O \qquad E^0 = 1.23 \text{ V}$$

The product ICl_2^- is more stable in HCl solutions than is molecular ICl. Iodide and iodine are oxidized by iodate in HCl solutions,

$$2I^- + IO_3^- + 6H^+ + 6Cl^- = 3ICl_2^- + 3H_2O$$
$$2I_2 + IO_3^- + 6H^+ + 10Cl^- = 5ICl_2^- + 3H_2O$$

Several other reducing agents are determined with iodate, notably As(III) (see above), $Sb(III)Cl_5^-$, Hg_2Cl_2, hydrazine (N_2H_4), and thiocyanate (SCN^-).

2. *Periodate.* Periodate ion, containing iodine in the formal +7 oxidation state, is a stronger oxidizing agent than is iodate and has a higher standard potential:

$$H_5IO_6 + H^+ + 2e = IO_3^- + 3H_2O \qquad E^0 = \sim 1.6 \text{ V}$$

Periodate is particularly useful in the determination of alpha-diols and alpha-carbonyl alcohols. It is a strong enough oxidizing agent to split carbon-carbon bonds to form aldehyde or acid fragments,

$$\underset{\text{Propylene glycol}}{\begin{array}{c}CH_3\\|\\CHOH\\|\\CH_2OH\end{array}} + IO_4^- \rightarrow \underset{\text{Acetaldehyde}}{\begin{array}{c}CH_3\\|\\CHO\end{array}} + \underset{\text{Formic acid}}{HCHO} + IO_3^- + H_2O$$

$$\underset{\text{Glycerol}}{\begin{array}{c}CH_2OH\\|\\CHOH\\|\\CH_2OH\end{array}} + 2IO_4^- \rightarrow \underset{\text{Formaldehyde}}{2CH_2O} + \underset{\text{Formic acid}}{HCOOH} + 2IO_3^- + H_2O$$

Periodate is reduced to iodate rather than iodide in these reactions.

A good way to make the oxidation useful for analysis is to add a measured excess of standard periodate solution to a sample of a diol. After the reaction runs its course an excess of iodide and acid are added to the mixture. The unreacted periodate and the iodate formed by the reaction both oxidize iodide to iodine, which is then titrated with standard thiosulfate solution (see below). The periodate consumed can be determined because iodate produces three-fourths as much iodine as an equimolar amount of periodate. Alternatively, the aldehydes produced by the periodate reaction can be distilled into a reducing solution (bisulfite, for example), which can be assayed iodometrically.

3. *Bromate.* In acidic solution bromate ion is a very strong oxidizer,

$$BrO_3^- + 6H^+ + 5e = \tfrac{1}{2}Br_2 + 3H_2O \qquad E^0 = 1.52 \text{ V}$$

Molecular bromine (Br_2) is a good oxidizer in its own right,

$$Br_2 + 2e = 2Br^- \qquad E^0 = 1.09 \text{ V}$$

so in reactions with strong reducers bromate is reduced all the way to bromide ion.

When bromate is used as a titrant, the endpoint can be detected by the bleaching of the indicator methyl orange with bromine produced just after the equivalence point. If an excess of bromate is added, the reductant can be determined indirectly. Once the bromate reaction is complete, an excess of iodide and acid are added; unreacted bromate oxidizes the iodide to iodine, which is then titrated with standard thiosulfate.

The classic determination of phenol, now over 100 years old and still widely used, involves bromination,

$$C_6H_5OH + 3Br_2 = 3HBr + C_6H_2Br_3OH$$

Bromine is produced by the reaction of an excess of standard iodate with bromide. The bromine left after the reaction with phenol is determined by iodimetric titration, and the bromine consumed is determined by difference.

Reducing Agents There are many good reducing agents available to the analyst. By far the most useful reducing titrants are acidic solutions of titanous ion [Ti(III)] and chromous ion [Cr(II)]. Solutions of mercurous ion (Hg_2^{2+}), stannous ion [Sn(II)], vanadium(II), and ascorbic acid are all useful titrants for redox titrimetric methods, but are much less popular. All of these species react with atmospheric oxygen, and some are even strong enough reducers to react with water to produce hydrogen gas. As a result of this reactivity, their solutions are somewhat unstable and difficult to store. Storage under inert gas atmospheres is necessary for most of these compounds. Although suitable redox indicators are available, most titrations with these species are followed potentiometrically (platinum wire electrode). Let us consider some properties of three of the most important reducing agents. The references cited at the end of the chapter provide details about other systems.

Titanium(III) Ti(III) is a moderately strong reducer in acidic solutions,

$$E^f[\text{Ti(IV)}/\text{Ti(III)}] = 0.03 \text{ V} \qquad \text{in } 0.7M \text{ HCl}$$

Stock solutions of Ti(III) can be prepared from commercially available $TiCl_3$ (20% solution in HCl), and standardized with solutions of Fe(III) in HCl (potentiometric endpoint). Iron(III) can be prepared by dissolving primary standard grade iron wire in molar HCl, bubbling with chlorine gas, and boiling to volatilize any excess dissolved chlorine. While the Ti(III)-Fe(III) reaction is a little sluggish at room temperature, it is quite rapid and quantitative at 50 to 60°C.

Copper(II), nitro and nitroso compounds, organic peroxides, and sulfoxides can be determined by titration with Ti(III). The reduction of a nitroso compound is a simple example:

$$RNO + 4Ti(III) + 4H^+ = RNH_2 + 4Ti(IV) + H_2O$$

Chromium(II) Chromous ion is the strongest reducing agent used in routine titrimetric analysis. The Cr(III)/Cr(II) couple has a formal potential of -0.38 V in $1M$ sulfuric acid. In theory it is therefore a strong enough reducing agent to reduce hydronium ion. Cr(II) solutions are prepared by treating acidic solutions of Cr(III) with zinc metal ($E^0 = -0.76$ V) and can be standardized with Fe(III) (iron wire, as above). Cr(II) solutions must be stored under an inert atmosphere because they are so easily oxidized by oxygen.

Chromium(II) has been used to determine a variety of organic and inorganic oxidants. For example, Cr(II) reduces organic nitro compounds to amines and cleaves azo nitrogen bonds:

$$RN=NR' + 4Cr(II) + 4H^+ = RNH_2 + R'NH_2 + 4Cr(III)$$

Sodium Thiosulfate Thiosulfate, $S_2O_3^{2-}$, is a moderately strong reducing agent,

$$S_4O_6^{2-} + 2e = 2S_2O_3^{2-} \qquad E^0 = +0.08 \text{ V}$$

It is most often used as a reagent for the determination of iodine. The properties of thiosulfate and the preparation and storage of its solutions are discussed in section 16.3.3 on iodometric methods.

Ascorbic Acid Ascorbic acid (vitamin C) is a moderately strong reducing agent whose formal potential is sensitive to pH:

$$C_6H_6O_6 + 2H^+ + 2e = C_6H_8O_6 \qquad E^f \sim 0.19 \text{ V at pH 7}$$

Ascorbic acid solutions can be standardized with potassium iodate or iodine. Ascorbic acid has been used in the determination of silver, gold, and platinum salts. It is a sufficiently strong reducer to reduce these metal ions all the way to their metallic state:

$$2Ag^+ + C_6H_8O_6 = 2Ag + C_6H_6O_6 + 2H^+$$

The literature should be consulted for details.

Prior Oxidation or Reduction The success of all redox titrimetric procedures depends on the quantitative conversion of an analyte from one oxidation state to another. For example, if an analyst sets out to determine iron in a sample by oxidation of Fe(II) to Fe(III), it is of vital importance to start the titration with iron only in the +2 state; any Fe(III) at the beginning of the titration will not react with the titrant and will go undetected. To avoid this kind of error, an analyst can treat a sample just before titration with a compound which will bring the analyte to the desired oxidation state, a "prior oxidant" or "prior reductant." These compounds must have three important qualities if they are to be effective:

1. They must react quantitatively with the analyte. Standard or formal potentials should be consulted in choosing an appropriate prior oxidant or reductant.
2. It must be possible to remove them or mask them before the analytical titration is begun. Any species which is able, for example, to reduce Fe(III) to Fe(II) will also react with an oxidizing titrant. If the prior oxidant or reductant is in solution at the time of titration, analytical results will be high.
3. They must not react with other components of a sample to form interfering compounds. An analyst must know something about the likely contaminants in a sample before running an analysis and prepare to take proper precautions. Finding solutions to such problems makes the development of methods interesting and challenging.

In order to survey as quickly as possible the commonly used prior reductants and oxidants, we will rely on Table 16.2 and brief commentary.

Some of the most popular prior reactants are gases; they have the great virtue of being easy to remove after their work is done, often by boiling the analyte solution for a few minutes. Ozone, chlorine, and bromine are gaseous oxidants, and hydrogen sulfide and sulfur dioxide are gaseous reductants. Table 16.2 gives important information about them.

Many prior oxidants and reductants are dissolved in water and added dropwise to the analyte solutions. Solutions of the strong oxidizers discussed above (permanganate, periodate, and dichromate) are often used as prior oxidizers. In addition, hydrogen peroxide, sodium hypochlorite, and potassium chlorate are all useful oxidizers. Perchloric acid is a good oxidizer when hot and concentrated. Stannous chloride and chromous sulfate are popular reductant solutions. Hydrazine (as the hydrochloride salt) is a very strong reductant. It is particularly interesting because its oxidation product is nitrogen gas, which is quite unreactive in solution.

Metallic and nonmetallic solids have also been used as prior oxidants and re-

TABLE 16.2   PRIOR OXIDANTS AND REDUCTANTS

Gases[a]:

Reducers:
- H_2S: $S + 2H^+ + 2e = H_2S$ $E^0 = +0.14$ V
- SO_2: $SO_4^{2-} + 4H^+ + 2e = H_2SO_3 + H_2O$ $E^0 = +0.17$ V

Oxidizers:
- Br_2: $Br_2 + 2e = 2Br^-$ $E^0 = +1.09$ V
- Cl_2: $Cl_2 + 2e = 2Cl^-$ $E^0 = +1.36$ V
- O_3: $O_3 + 2H^+ + 2e = O_2 + H_2O$ $E^0 = +2.07$ V

Solutions[b]:

Reducers:
- Cr^{2+}: $Cr^{2+} + e = Cr^{3+}$ $E^0 = -0.41$ V
- $N_2H_5^+$: $N_2 + 5H^+ + 4e = N_2H_5^+$ $E^0 = -0.23$ V
- Sn^{2+}: $Sn^{4+} + 2e = Sn^{2+}$ $E^0 = +0.54$ V

Oxidizers:
- H_2O_2: $H_2O_2 + 2e = 2OH^-$ $E^0 = +0.88$ V
-        $H_2O_2 + 2H^+ + 2e = 2H_2O$ $E^0 = +1.77$ V
- MnO_4^-: $MnO_4^- + 8H^+ + 5e = Mn^{2+} + 4H_2O$ $E^0 = +1.51$ V
- $HOCl$: $HOCl + H^+ + 2e = Cl^- + H_2O$ $E^0 = +1.63$ V
- $S_2O_8^{2-}$: $S_2O_8^{2-} + 2e = 2SO_4^{2-}$ $E^0 = +2.01$ V

Solids[c]:

Reducers:
- Zn^0: $Zn^{2+} + 2e = Zn$ $E^0 = -.763$ V
- Cd^0: $Cd^{2+} + 2e = Cd$ $E^0 = -.403$ V
- Ni^0: $Ni^{2+} + 2e = Ni$ $E^0 = -.136$ V
- Cu^0: $Cu^{2+} + 2e = Cu$ $E^0 = +.337$ V

Oxidizers:
- PbO_2: $PbO_2 + 4H^+ + 2e = Pb^{2+} + 2H_2O$ $E^0 = +1.455$ V
- Bi_2O_4: $Bi_2O_4 + 4H^+ + 2e = 2BiO^+ + 2H_2O$ $E^0 = +1.59$ V

[a] Unreacted gases can be removed by boiling and/or bubbling with CO_2.
[b] Excess Cr^{2+} and hydrazine can be removed by air oxidation. Sn^{2+} can be reacted with $HgCl_2$. The oxidizers can be removed by boiling; permanganate is reduced by Mn^{2+} and will form MnO_2, which can be removed by filtering.
[c] Unreacted solid reducers and oxidizers can be removed by filtering.

ductants. Silver(II) oxide, lead dioxide, and sodium bismuthate are very strong oxidizers. Metals such as zinc, silver, and mercury are often the most convenient reducers available in the laboratory and deserve special attention. They are used in several forms: powder, shot, or sheet metal. Powder and shot are particularly important forms, because they can be packed into columns through which analyte solution can be run. Two important metal columns are the zinc (or Jones) reductor and the silver/silver chloride (or Walden) reductor.

Jones Reductor   The Jones reductor is prepared by packing a column with an aqueous slurry of zinc powder which has been coated with mercury. The standard potential of zinc is large and negative (−0.76 V) and indicates that zinc is a strong reducer. In fact, zinc reduces hydrogen ion in acidic solutions unless its surface is lightly amalgamated with mercury (1 or 2% by weight). Amalgamation slows the rate at which hydrogen ion and other species are reduced, but does not raise the standard potential. Hydrogen gas bubbles open channels and can even separate large sections of the reductor column bed, rendering the column useless.

Zinc is not a selective reducer. A glance at a table of standard potentials tells the story: Fe(III), Cr(III), Ti(IV), Sn(IV), and V(V) are all reduced quantitatively to lower oxidation states. A zinc reductor may therefore not help remove interferences. Some ions cannot be reduced before titration: metals such as Cu, Ag, Hg, Sb, and Bi are all plated out on a zinc amalgam surface and could not be titrated.

A real advantage of using a zinc reductor is that Zn(II) is the product of the oxidation of Zn metal and is not oxidized by ordinary oxidizing agents. Zinc(II) will not interfere in a titration of reduced analyte solution.

Walden Reductor The Walden reductor is prepared by packing an aqueous HCl solution slurry of silver powder into a column. Analyte is then reduced as it passes through the column. The Walden reductor relies on the silver chloride/silver metal couple,

$$AgCl + e^- = Ag + Cl^- \qquad E^0 = 0.222 \text{ V}$$

Many of the species which are reduced by a Jones reductor are not reduced by a Walden reductor. Iron(III) is reduced to Fe(II) quantitatively, but Ti(IV), Cr(III), and Sn(IV) are not reduced. Copper(II) is reduced only as far the soluble Cu(I) complex, $CuCl_2^-$.

It can be seen from the half-reaction that the reducing power of silver metal increases (that is, E^f decreases) as the concentration of chloride ion increases in solution. The shift in standard potential is accentuated by the tremendous increase in the activity coefficient of chloride in more concentrated HCl solutions. In $3M$ HCl the formal potential is about 0.184 V; in $9M$ HCl the formal potential is about 0.090 V.

16.3.3 Iodometric Methods—A Closer Look

As shown in an earlier section, the iodine/iodide couple is extremely versatile for analytical titrations. This versatility comes from two sources. First, the standard potential of the I_2/I^- couple is neither very large nor very small, meaning that many oxidizing and reducing agents will react with iodine and iodide:

$$\text{reducing analyte} + I_2 \text{ (std)} = \text{analyte oxidized form} + I^-$$
$$\text{oxidizing analyte} + I^- = \text{analyte reduced form} + I_2$$

Both direct and indirect methods are used. I_2 can be used as a standard solution (as triiodide ion I_3^-) or, when added in excess or produced in a reaction, can be titrated with a standard solution of thiosulfate,

$$I_2 + 2S_2O_3^{2-} = 2I^- + S_4O_6^{2-}$$

The second source of versatility is the excellence of visual endpoints for titrations involving iodine and iodide. Although many iodometric titrations can be monitored potentiometrically, the starch or extraction endpoints are usually much more convenient and at least as sensitive.

Standard Solutions

Standard Iodine Solutions Reagent grade iodine may be used as a primary standard. Because iodine solutions are subject to air oxidation and iodine is volatile, it is more common to prepare iodine solutions and then standardize them with primary standard As_2O_3. The reaction of iodine with As(III) is sensitive to solution pH,

$$H_3AsO_3 + I_2 + H_2O = HAsO_4^{2-} + 4H^+ + 2I^-$$

Quantitative results can be achieved between pH 4 and 9. In acidic solutions the reaction becomes inconveniently slow. If the pH is much greater than 9, iodine may disproportionate to form iodide and hypoiodous acid (HOI),

$$2I_2 + 2H_2O = 2HOI + 2H^+ + 2I^-$$

HOI may then decompose to form iodate. The best results are obtained at about pH 7 in solutions buffered to absorb hydronium ion produced by the reaction.

Iodine is not very soluble in water (~ 1.3 mM at room temperature), and titrant solutions at the $0.1M$ level must be prepared by complexing iodine with iodide to form the soluble triiodide ion, as explained earlier. These solutions are rather difficult to store. The vapor pressure of iodine is sufficiently large over triiodide solutions to require that they be kept in tightly sealed bottles. During short titrations no special precautions must be taken to avoid volatilization, however. Iodine oxidizes rubber, so storage bottles with ground glass stoppers are required.

Iodide is oxidized by atmospheric oxygen in a reaction catalyzed by species such as nitrogen oxides and accelerated by light and heat. Iodide solutions should therefore be stored in the dark and never heated. Nitrite ion is a particularly effective oxidizer, even in trace amounts. Iodide reduces nitrite to nitric oxide, which is oxidized by atmospheric oxygen to higher oxidation state compounds, and these react with more iodide to form nitric oxide, and so forth:

$$2NO_2^- + 2I^- + 4H^+ = 2NO + I_2 + 2H_2O$$

Iodometric titrations cannot be run in the presence of nitrogen oxides. They must be removed by boiling or by prior reduction to form harmless compounds.

Standard Thiosulfate Solution All indirect titrations of iodine use a standard solution of thiosulfate as titrant. In the pH range 5 to 9 the reaction of iodine and thiosulfate occurs rapidly and quantitatively. Above pH 9 the HOI formed by disproportionation of I_2 may react with thiosulfate to form sulfate and iodide,

$$S_2O_3^{2-} + 4IO^- + 2OH^- = 2SO_4^{2-} + 4I^- + H_2O$$

Below pH 5 iodide is more sensitive to oxidation by oxygen, but in the short times required for titrations the error is negligible.

Standard sodium thiosulfate solutions are quite stable when care is taken in their preparation. A major cause of decomposition is bacterial growth during storage. Boiling water prior to preparing the solution may help kill many bacteria. Adding some sodium bicarbonate or borax raises the pH and inhibits bacterial growth. Mercury salts such as HgI_2 are also quite effective.

Thiosulfate decomposes in acidic solutions,

$$S_2O_3^{2-} + 2H^+ = H_2O + SO_2 + S(ppt)$$

a reaction accelerated by light. Standard thiosulfate should never be acidified or stored in bright light.

While anhydrous sodium thiosulfate may be dried and used as a primary standard, it is more common to prepare solutions of the pentahydrate salt to about the desired molarity, let them stand overnight to allow any oxidizing impurities present in the water to react, and then standardize with a primary standard oxidizers. The most common oxidizers are KIO_3, $K_2Cr_2O_7$, and electrolytic copper [as Cu(II)]. Iodine, $KBrO_3$, and $K_3Fe(CN)_6$ [potassium ferricyanide, Fe(III)] are also used.

The iodate reaction,

$$IO_3^- + 5I^- + 6H^+ = 3I_2 + 3H_2O$$

goes to completion in acidic solution in the presence of an excess of potassium iodide. The iodine generated is then titrated with thiosulfate to the blue → colorless starch endpoint,

$$3I_2 + 6S_2O_3^{2-} = 6I^- + 3S_4O_6^{2-}$$

Notice that in the overall reaction each mole of iodate reacts with 6 moles of thiosulfate. Since five iodides appear in the first reaction and the oxidation state of iodine in iodate changes from 5 to 0, it is easy to mistake the stoichiometry for 1:5 rather than 1:6. A blank titration of the potassium iodide added in excess should be performed because potassium iodate may be an impurity in KI. Potassium iodide may be purchased in a purity grade certified to be "iodate free." Air oxidation of the acidic iodide solution may add to the blank.

EXAMPLE 16.4 Standardization of Thiosulfate Solution

A 1.6040-g sample of dried primary standard grade KIO_3 is dissolved in water and transferred to a 500-mL volumetric flask. A pipet is used to transfer exactly 25 mL of the standard solution to an Erlenmeyer flask. A few drops of $1M$ sulfuric acid are added, followed by 2 g of potassium iodide. The iodine generated is then titrated with 22.37 mL of sodium thiosulfate solution. What is the molarity of the thiosulfate solution?

Solution

The gram formula weight of KIO_3 is 214.00. The 25-mL portion of the stock solution contains

$$(25/500)*1604 \text{ mg}/214.00 = 0.3748 \text{ mmole}$$

This amount of iodate will react with 6(0.3748) or 2.249 mmole of thiosulfate. The volume of thiosulfate required is 22.37 mL, so the molar concentration is

$$2.249 \text{ mmole}/22.37 \text{ mL} = 0.1005M$$

Visualizing the Equivalence Point A colloidal suspension of starch is used as a sensitive endpoint indicator in iodine titrations. Iodine and iodide adsorb on the surfaces of starch particles to produce a blue color, which is visible at iodine concentrations as low as $10^{-7}M$ in the presence of $10^{-4}M$ iodide. Iodine by itself does not produce the blue species; iodide must be present. The blue color does not form as well in solutions of pH greater than about 8. Furthermore, conditions which coagulate colloids (heat, nonaqueous solvents, strong electrolytes) make the indicator less sensitive. The formation of the blue complex is quite reversible.

It is common to titrate iodine solutions with standard thiosulfate to the disappearance of the blue color at the endpoint. It is a good idea not to add the starch solution until the titration is nearly done and the iodine concentration is low (a faint yellow color in solution). If added too early, some starch may hydrolyze to form species that give a reddish color with iodine, which may mask the true endpoint.

Starch solutions are prepared by making a paste of potato starch or soluble starch with a little HgI_2 to kill molds which thrive on starch. About 2 g of the paste is stirred into 1 liter of boiling deionized water. After boiling for a while the solution will become clear. If the disposal of solutions containing mercuric salts makes their use objectionable, fresh starch solution may be prepared every few days.

An alternative to the use of starch is the *extraction endpoint*. Iodine is soluble in water-immiscible solvents such as carbon tetrachloride and chloroform and forms an intensely colored purple solution. Very low concentrations of iodine can be determined by adding a few milliliters of carbon tetrachloride to 50 mL of solution being titrated, shaking, and extracting the iodine into the CCl_4 layer. Thiosulfate is added slowly, shaking between additions, until the CCl_4 layer fails to turn purple. Special Erlenmeyer flasks called iodine flasks have ground glass stoppers and are designed to be inverted for endpoint detection. The CCl_4 layer, denser than water, settles into the narrow neck of the inverted iodine flask, and its color is easily observed. The extraction method requires more time than the starch indicator method but is about as sensitive. It is most useful in strongly acidic solutions, where starch is much less sensitive.

Determination of Copper When an excess of potassium iodide is added to a slightly acidic solution of Cu(II), iodine is produced according to the reaction

$$Cu^{2+} + 5I^- = 2CuI(s) + I_3^-$$

The iodine produced in the reaction is then titrated with standard thiosulfate to the starch endpoint. Notice that the iodine is complexed as triiodide and that insoluble cuprous iodide forms as a second product. Iodide ion serves two purposes in the reaction. First, it reduces cupric ion to the +1 state, and second, it complexes the cuprous ion produced to form insoluble CuI. It turns out that if it were not for the precipitation of CuI the redox reaction would not proceed spontaneously. The standard potential of the uncomplexed Cu(II)/Cu(I) couple

$$Cu^{2+} + e = Cu^+ \qquad E^0 = +0.15 \text{ V}$$

tells us that Cu^{2+} should be too weak an oxidizer to react with iodide. The stability of CuI, however, makes Cu^{2+} a much stronger oxidizer in the presence of iodide,

$$Cu^{2+} + I^- + e = CuI(s) \qquad E^0 = +0.86 \text{ V}$$

This value is much higher than that of the iodine/triiodide couple, +0.536 V.

The pH of the solution is an important consideration. While there is no explicit pH dependence in either half-reaction, in basic solutions hydroxide complexes Cu^{2+} and makes the reaction with iodide sluggish, causing a fading endpoint. If the pH is too low (<0.5) iodide is prone to oxidation by the oxygen in air. Impurities such as arsenic(V) and antimony(V) become much stronger oxidizers in acidic solution (see above) and may interfere if the solution pH is too low.

It is standard practice to keep the pH above 3.5 with either acetate or bifluoride buffers. Using ammonium bifluoride gives an extra advantage to the analyst: copper samples often contain significant quantities of iron, and Fe(III) is so strongly complexed by fluoride ion that it will not oxidize iodide.

The sharpness and accuracy of the starch endpoint may be hampered by adsorption of iodine on the surface of the precipitated cuprous iodide. The sluggishness of the desorption causes a fading endpoint. A very useful remedy is to add potassium thiocyanate, KSCN, just before the endpoint. CuSCN is less soluble than CuI and displaces iodide from the surface,

$$CuI(surface) + SCN^- \rightarrow Cu(SCN)(surface) + I^-$$

Cu(SCN) is far less able to adsorb iodine (as triiodide) than is CuI. The timing of the addition of thiosulfate is important, though: thiocyanate is capable of oxidizing iodine and producing low results if given enough time.

Oxides of nitrogen can be a source of serious error in the iodometric determination of copper. Many copper samples can be dissolved only with nitric acid, and all traces of nitrogen oxides must be removed before adding iodide to the sample. Boiling the solution after adding sulfuric acid will remove nitrogen oxides, but it takes time and extra patience. Urea and sulfamic acid are quite effective at removing HNO_2. For example,

$$2HNO_2 + CO(NH_2)_2 = 2N_2 + CO_2 + 3H_2O$$

Determining Water: The Karl Fischer Method In the 1930s the chemist Karl Fischer proposed that the reaction of iodine with sulfur dioxide be used to analyze samples for water:

$$I_2 + SO_2 + H_2O = SO_3 + 2HI$$

He observed that the reaction runs particularly well in a solvent composed of methanol and pyridine. The overall reaction shows that both methanol and pyridine (py) are involved in the reaction,

$$I_2 + SO_2 + H_2O + CH_3OH + 3py = 2pyH^+I^- + pyHSO_3OCH_3$$

It turns out that pyridine is absolutely essential for the reaction to proceed, but that solvents other than methanol work quite well. The details of the mechanism have been worked out but are not critical to understanding the reaction in the context of a quantitative analysis course. The analytical procedure involves adding "Karl Fischer reagent" (KF), a methanol solution containing iodine, sulfur dioxide, and pyridine, to a solution of a sample containing water, and titrating until the appearance of unreacted iodine. When using $0.01 F$ KF reagent, the color change from yellow to red-brown (iodine) is easy to see. In more dilute solutions colorimetric or potentiometric monitoring may be more satisfactory.

KF reagent is extremely hygroscopic and is therefore very hard to store. It helps to keep the components separate until the reaction is run. For example, a sample to be analyzed is dissolved in a solution containing SO_2 and pyridine, and then iodine dissolved in methanol (or another solvent) is added as titrant. Still, extraordinary precautions must be taken to avoid introducing atmospheric moisture. An alternative method is to mix SO_2, pyridine, potassium iodide, and the sample in methanol and then generate iodine electrochemically (a method called "coulometry," discussed in most instrumental methods courses). KF reagent solutions can be standardized by titrating samples of stable crystalline hydrates such as ammonium oxalate monohydrate. Daily standardization is necessary in routine applications. Recipes for preparing KF reagent can be found in the literature. Cheronis and Ma present a good experimental procedure (experiment 48 in Ref. 4).

The Karl Fischer method can be used to determine water in many solvents and many compounds. It is an extremely important tool for chemists who work with systems in nonaqueous solvents and must analyze solvents for water. There are some interferences which must be avoided. Any substance which reduces iodine will interfere with the KF method. Ascorbic acid, quinones, and mercaptans (thioalcohols) are the most important interferences. Most organic acids, alcohols, hydrocarbons (saturated and unsaturated), ethers, and halides do not interfere.

TABLE 16.3 MISCELLANEOUS IODOMETRIC METHODS

Direct methods (titrant: standard I_3^- solution)

$$SO_3^{2-} \rightarrow SO_4^{2-} + 2e^-$$
$$S^{2-} \rightarrow S + 2e^-$$
$$Sn^{2+} \rightarrow Sn_4^+ + 2e^- \quad \text{requires prereduction}$$
$$N_2H_4 \rightarrow N_2 + 4H^+ + 4e^-$$

Indirect methods (titrant: standard $S_2O_3^{2-}$ solution or other)

Ba^{2+}, Sr^{2+}	Add excess standard chromate, precipitate chromate salts, filter to remove. Acidify solution, add iodide, titrate iodine generated.
Pb^{2+}	Add excess chromate, precipitate $PbCrO_4$, collect, redissolve in acid, add iodide, and titrate iodine generated.
Zn^{2+}, Cd^{2+}, other metal ions	Precipitate sulfides, collect, dissolve in acid, react sulfide with excess iodine, and back-titrate with thiosulfate.
CN^-	$HCN + I_2 = ICN + I^- + H^+$; basic solution, extraction endpoint.
Cl^-	Add solid $AgIO_3$ to solution of Cl^-, iodate displaced by chloride ($AgCl$), reacts with added iodide. Titrate iodine generated with thiosulfate.
RCHO (aldehyde)	Add measured excess of HSO_3^-, titrate unreacted reagent with iodine solution. $RCHO + HSO_3^- = RCH(OH)-SO_3^-$

Miscellaneous Iodometric Methods There are far too many methods of analysis based on the reactions of iodine to discuss in detail in an introductory text. Many of the most important variations are listed in Table 16.3.

REFERENCES

1. H. Laitinen, *Chemical Analysis,* McGraw-Hill, New York, 1960.
2. A. F. Schilt, *Analytical Applications of 1,10-Phenanthroline and Related Compounds,* Pergamon, Oxford, 1969.
3. I. M. Kolthoff and R. Belcher, *Volumetric Analysis,* vol. III, Interscience, New York, 1957.
4. N. D. Cheronis and T. S. Ma, *Organic Functional Group Analysis,* Interscience, New York, 1964.

RECOMMENDED READING

Blaedel, W. J. and V. W. Meloche. *Elementary Quantitative Analysis,* 2nd ed., Harper & Row, New York, 1963. This old standard text has some excellent detailed experimental directions and a strong section on iodometric methods.

Cheronis, N. D., and T. S. Ma, *Organic Functional Group Analysis,* Interscience, New York, 1964. This book deals with micro and semimicro methods of analysis for organic compounds. It contains a wealth of information about redox reactions for specific groups.

Kolthoff, I. M., and R. Belcher, *Volumetric Analysis,* vol. III, Interscience, New York, 1957. This is the classic compendium of redox titrimetric methods. A book of this scope on this subject will probably never again be written.

Laitinen, H. *Chemical Analysis,* McGraw-Hill, New York, 1960. This is another classic in the field. Chapters 17–23 are relevant to the study of redox methods.

Schilt, A. F. *Analytical Applications of 1,10-Phenanthroline and Related Compounds,* Pergamon, Oxford, 1969. Chapter 4 is a discussion of ferroin indicators.

Wawzonek, S. *Potentiometry: Oxidation-Reduction Potentials,* in *Techniques of Chemistry,* Vol. 1, A. Weissberger and B. W. Rossiter, eds., Wiley, New York, 1971. Sections at the end of Chapter 1 dealing with organic redox systems and indicators are particularly good sources.

PROBLEMS

16.1. How would you prepare each of the following solutions, given the starting materials and the intended reactions?

PROBLEMS

(a) a 0.1050M solution of $FeCl_2 \cdot 4H_2O$

(b) a 0.0546N solution of $SnCl_2 \cdot 2H_2O$ for titration with Fe^{3+} ($Sn^{2+} = Sn^{4+} + 2e$)

(c) a 0.0957N solution of $KMnO_4$ for titrating in basic solutions [$MnO_4^- + 2H_2O + 3e = MnO_2(s) + 4OH^-$]

(d) a 0.1005M solution of $KMnO_4$

16.2. What is the relationship between the gram formula weight and equivalent weight of each of the following species:

(a) zinc metal (to be used as a reductant: $Zn = Zn^{2+} + 2e$)

(b) $Tl(NO_3)_3 \cdot 3H_2O$ ($Tl^{3+} + 2e = Tl^+$)

(c) potassium bromate ($BrO_3^- + 3H_2O + 6e = Br^- + 6OH^-$)

(d) sodium hypochlorite ($OCl^- + H_2O + 2e = Cl^- + 2OH^-$)

16.3. How many millimoles of Fe^{3+} will be reduced by 10.00 mL of 0.05052N stannous chloride solution?

16.4. Iodate is a strong oxidizer in acidic solutions. What is the molarity of a 0.05737N solution of potassium iodate used for the reaction $IO_3^- + 6H^+ + 5I^- = 3I_2 + 3H_2O$?

16.5. Arsenious oxide, As_2O_3, is available in primary standard grade. When dissolved in base it forms H_3AsO_3 and is used as a two-electron reducing agent,

$$H_3AsO_3 + H_2O = H_3AsO_4 + 2H^+ + 2e^-$$

What are the gram formula weight and equivalent weight of As_2O_3?

16.6. Copper metal dust weighing 483.6 mg is poured into a solution containing an excess of silver nitrate. What amount of silver metal will be formed by the complete reaction with Cu metal:

$$Cu + 2Ag^+ = Cu^{2+} + 2Ag$$

Express the amount in millimoles, milliequivalents, and milligrams.

16.7. Using the following standard potentials, answer the questions below.

$$E^0(Fe(III)/Fe(II)) = +0.77 \text{ V}$$
$$E^0(Fe(CN)_6^{3-}/Fe(CN)_6^{4-}) = +0.36 \text{ V}$$
$$E^0(I_2/2I^-) = +0.62 \text{ V}$$

(a) Calculate the equilibrium constant for the reaction of iodide with ferric iron. Is the reaction of 0.1F Fe(III) with 0.1F iodide complete?

(b) Will a reaction occur between 0.1F $Fe(CN)_6^{3-}$ and 0.1F iodide? Between 0.1F $Fe(CN)_6^{4-}$ and 0.1F triiodide?

16.8. The formal potential of the ferricyanide/ferrocyanide couple is surprisingly sensitive to acidity. The standard potential is +0.36 V, while the formal potential in 1F HCl is +0.71 V.

(a) One of the species, ferricyanide or ferrocyanide, is a weaker Brønsted base than the other. On the basis of the behavior of the formal potential, which species is the stronger Brønsted base?

(b) Predict the results of an experiment in which 0.1-g portions of KI are added to 50 mL of $0.1F$ $K_3Fe(CN)_6$ in water and to 50 mL of $0.1F$ $K_3Fe(CN)_6$ in $0.1F$ HCl. Explain the behavior.

16.9. A 50-mL sample of $0.05F$ Sn(II) in $1F$ HCl is to be determined by titration with $0.05F$ Fe(III), also in $1F$ HCl. The titration will be monitored with a platinum indicator electrode and a saturated calomel reference electrode (E_{SCE} = +0.244 V vs. standard hydrogen electrode). Calculate the potentials observed at 1, 20, 50, 90, 99, 100, 101, and 110% titrated. Is the reaction complete at the equivalence point?

$$E^f[Sn(IV)/Sn(II)] = +0.14 \text{ V } (1F \text{ HCl})$$
$$E^f[Fe(III)/Fe(II)] = +0.70 \text{ V } (1F \text{ HCl})$$

16.10. Assume that a redox indicator IN undergoes a one-electron reduction to IN$^-$ and that the oxidized and reduced forms have identical molar absorptivities. The formal potential of the indicator is +0.80 V vs. SCE.

(a) An experimentalist can see the color of the reduced form when the ratio of the concentrations of reduced to oxidized forms is 10:1 and the color of the oxidized form when the ratio of the concentrations of reduced to oxidized forms is 1:10. Calculate the range in potential over which the endpoint transition will be judged to occur.

(b) A second indicator has the same formal potential as IN, but the molar absorptivity of its oxidized form is about one-tenth that of its reduced form. Compare the potentials of the endpoints observed with these two indicators. Under what conditions might the use of the second indicator cause a determinate error in a redox titration analysis?

16.11. Plutonium is a poisonous radioactive element. Small amounts of plutonium can be concentrated from water using an ion exchange resin, then reduced in a Jones reductor [to form Pu(III)] prior to titration with Ce(IV) in $1M$ H_2SO_4.

(a) If ferroin can be used as an indicator and the oxidation of Pu(III) to Pu(IV) is complete at the endpoint, what is an approximate maximum value for $E^f[Pu(IV)/Pu(III)]$ in $1M$ sulfuric acid?

(b) 100 liters of water containing plutonium is passed through a bed of ion exchange resin. The plutonium (and other metals, which we assume will not interfere) is then washed from the bed with 50 mL of $1M$ sulfuric acid. The resulting solution is passed through a Jones reductor and titrated with standard Ce(IV) solution, 2.39 mL of $0.00500M$ solution being required to reach the ferroin endpoint. How many milligrams of Pu were present in the original 100 liters of water? (The gram atomic weight of Pu is 242.)

PROBLEMS

16.12. The uranium in a 1.0732-g sample is prereduced to U^{4+}, then titrated with $0.0563N$ permanganate according to the reaction:

$$5U^{4+} + 2MnO_4^- + 2H_2O = 5UO_2^{2+} + 2Mn^{2+} + 4H^+$$

If 8.97 mL of permanganate solution is required to reach the equivalence point, calculate percent uranium (as metal) in the original sample.

16.13. The As_2O_3 in a 672.0-mg sample of a mysterious powder is dissolved to form H_3AsO_3, which is subsequently titrated with $0.0205M$ triiodide solution,

$$H_3AsO_3 + I_3^- + H_2O = H_3AsO_4 + 2H^+ + 3I^-$$

If 27.32 mL of triiodide solution is required, what is the percent As_2O_3 in the powder sample?

16.14. Benzaldehyde reacts with sodium bisulfite to form an addition product,

$$C_6H_6CH{=}O + HSO_3^- = C_6H_5CH(OH)SO_3^-$$

120 mg of a sample containing benzaldehyde is placed in a 125-mL Erlenmeyer flask along with 10 mL of deionized water. Then 10 mL of a stock solution prepared by dissolving 1.556 g of $NaHSO_3$ in enough deionized water to make 100 mL of solution is delivered by pipet to the Erlenmeyer flask. After allowing 1 hour for the reaction to occur, starch indicator is added and the unreacted bisulfite is titrated with standard $0.0523M$ triiodide. To reach the blue endpoint, 21.37 mL is required. The reaction is

$$HSO_3^- + I_3^- + H_2O = 3I^- + HSO_4^- + 2H^+$$

How many milligrams of benzaldehyde were present in the original sample?

16.15. The lead in a 0.7352-g sample is precipitated as lead iodate, $Pb(IO_3)_2$, which is collected on filter paper and washed with deionized water. The solid is transferred to an acidic solution, and potassium iodide is added to form triiodide:

$$Pb(IO_3)_2 + 18I^- + 12H^+ = PbI_2 + 6I_3^- + 6H_2O$$

40.10 mL of standard $0.1030M$ sodium thiosulfate solution is required to titrate the triiodide generated by the reaction. Calculate the percentage of lead in the original sample.

16.16. Ascorbic acid can be determined iodometrically. A 50.0-mL solution containing ascorbic acid was acidified with 5 mL of $5M$ sulfuric acid. Then 25.00 mL of standard $0.01035M$ triiodide solution was added by pipet and the following reaction ensued:

$$C_6H_6O_6 + 2H^+ + I_3^- = C_6H_8O_6 + 3I^-$$

The unreacted triiodide ion was titrated with 5.18 mL of 0.0100M sodium thiosulfate,

$$2S_2O_3^{2-} + I_3^- = S_4O_6^{2-} + 3I^-$$

How many milligrams of ascorbic acid (GFW 176.1) were present in solution?

16.17. A 1.030-g sample containing a small amount of nitrate is dissolved in 50 mL of 2M sulfuric acid and reacted with 25.00 mL of a 0.1030M Fe(II) solution prepared from Mohr's salt (ferrous ammonium sulfate). The unbalanced reaction is

$$Fe^{2+} + NO_3^- + H^+ = Fe^{3+} + NO + H_2O$$

The unreacted ferrous iron is titrated with standard 0.01180M potassium dichromate solution. To reach the diphenylamine sulfonate endpoint 26.48 mL is required. Calculate the percentage of nitrate in the sample.

16.18. Selenite ion, SeO_3^{2-}, reacts with iodide in acidic solution to form iodine,

$$SeO_3^{2-} + 4I^- + 6H^+ = Se + 2I_2 + 3H_2O$$

Zirconium can be determined by precipitating $Zr(SeO_3)_2$, collecting the precipitate, and dissolving it in 6 mL of 1:1 sulfuric acid and water and 10 mL of 3% sodium fluoride. Several grams of solid potassium iodide are then added to the solution and the iodine generated is titrated with standard sodium thiosulfate solution.

A 0.5403-g sample containing zirconium is dissolved and precipitated as zirconium selenite. When the analytical procedure is complete 11.73 mL of 0.05135M sodium thiosulfate is required to react with the iodine generated. Calculate the percent Zr in the sample. (gram atomic weight of Zr = 91.22)

16.19. The water in a 50.0-g sample of acetonitrile solvent requires for titration 5.25 mL of Karl Fischer reagent containing 10.3 mg of iodine per milliliter. How many millimoles of water are there in 1 liter of the acetonitrile if its density is 0.7856 g/mL at 20°C?

16.20. Anhydrous barium chloride has a gram formula weight of 208.24. A stable hydrate form of $BaCl_2$ is to be analyzed for water of crystallization by the Karl Fischer method. A 0.2823-g sample of the pure hydrated salt is dissolved in anhydrous methanol and titrated with Karl Fischer (KF) reagent. Each milliliter of the reagent is known to react with 3.05 mg of water. If 13.62 mL of the KF titrant is required to reach the endpoint, what are the percentage of water in the sample and the formula of the hydrated salt?

PROBLEMS

16.21. Uranium in uranyl salts, U(VI), can be determined indirectly by titration with dichromate. A sample of uranyl salt is dissolved in molar sulfuric acid, then passed through a Jones reductor to form a mixture of U(IV) and U(III). The U(III) is then oxidized to U(IV) by bubbling the solution with air. An excess of ferric chloride is then added. The reaction is

$$U(IV) + 2Fe(III) = 2Fe(II) + U(VI)$$

The ferrous ion generated is titrated with standard potassium dichromate solution to the diphenylamine endpoint.

(a) What is an approximate value for the formal potential of the U(IV)/U(III) couple in molar sulfuric acid?

(b) A sample of salt containing U(VI) is treated according to the procedure outlined above, and 15.78 mL of 0.01537M potassium dichromate solution is required to reach the endpoint. Calculate the weight of U(VI) as the salt UO_2Cl_2 (GFW 340.93) present in the original sample.

(c) Describe a method for U(VI) using the Fe(III)/Fe(II) couple and Ti(III) as a reducing titrant.

16.22. A sample containing an unknown amount of phenol, C_6H_5OH (GFW 94.0), is treated with 25.00 mL of 0.0200 $KBrO_3$ solution, 0.5 g of KBr, and 5 mL of 2M sulfuric acid. The following reactions take place:

$$BrO_3^- + 5Br^- + 6H^+ = 3Br_2 + 3H_2O$$
$$C_6H_5OH + 3Br_2 = 3HBr + C_6H_2Br_3OH$$

No other bromophenol species form. After 10 minutes in a closed flask, 1 g of potassium iodide is added and the contents are shaken. The reaction of the unreacted bromate is

$$BrO_3^- + 6H^+ + 6I^- = 3I_2 + 3H_2O + Br^-$$

The iodine formed is then titrated to the starch endpoint with 11.25 mL of 0.04520M sodium thiosulfate. Calculate the number of milligrams of phenol in the original sample. (Note: Excess bromine formed by the bromate/bromide reaction reacts with iodide to form iodine: $3Br_2 + 6I^- = 3I_2 + 6Br^-$.)

16.23. The dimeric Ru(IV) anion, $Ru_2OCl_{10}^{4-}$, is reduced quantitatively but quite slowly by Ti(III) in 3M HCl. In order to avoid the complexities of the Ru(IV)-Ti(III) kinetics, the dimer can be reduced with an excess of Fe(II) (as Mohr's salt). The resulting Fe(III) is then titrated quickly and smoothly with Ti(III). The reactions are

$$Ru_2OCl_{10}^{4-} + 2H^+ + 2Fe(II) = 2RuCl_5^{2-} + 2Fe(III) + H_2O$$
$$Fe(III) + Ti(III) = Fe(II) + Ti(IV)$$

(a) A sample of Ru(IV) dimer in $3M$ HCl is reduced with Mohr's salt. The resulting Fe(III) requires 15.23 mL of $0.01723M$ Ti(III) in $3M$ HCl for reduction. Calculate the number of milligrams of $K_4Ru_2OCl_{10}$ present in the sample (GFW = 729.04).

(b) If the formal potential for the Ru(IV) dimer/Ru(III) couple is about 0.85 V in $3M$ HCl and that of the Fe(III)/Fe(II) couple is 0.68 V in $3M$ HCl, is the reaction

$$2Fe(II) + Ru_2OCl_{10}^{4-} = 2Fe(III)$$

quantitative if Fe(II) is initially 10 times more concentrated than Ru(IV) dimer? In terms of the titration with Ti(III), does it matter if the Fe(II) + Ru(IV) reaction is not quantitative?

(c) In the potentiometric curve for the titration of Fe(III) with Ti(III), the potentials on the upper plateau are determined by the Fe(III)/Fe(II) couple. An excess of Fe(II) must be added to drive the Ru(IV) reduction to Ru(III). The same excess lowers the plateau of the Fe(III)-Ti(III) titration curve and may make the endpoint less sharp. Explain the behavior using Nernst equations.

EXPERIMENT 16.1: Redox Titrimetric Determination of Iron (Volumetric or Gravimetric Procedure)

The Reaction

Iron can be determined directly by titration of Fe(II) with dichromate in acidic solutions. The equivalence point is signaled by a change in color of the redox indicator sodium diphenylamine sulfonate. Before titration with dichromate the iron sample may be treated with stannous chloride to reduce the iron to the +2 state. Excess Sn(II) is oxidized with mercuric chloride. The experiment illustrates the use of a redox indicator, the use of iron wire as a primary standard, and the use of stannous chloride as a prior reductant. The equivalence point is sufficiently sharp to justify the use of a gravimetric titration procedure, which is presented as an alternative.

The Titrant: Standard Potassium Dichromate

The preparation of dichromate solutions was discussed earlier in the chapter. In many applications we can treat potassium dichromate as a primary standard. Unfortunately, in this experiment we use an indicator which requires the presence of iron to catalyze the endpoint color change. We will therefore weigh out approximately what we need to dissolve to make a $0.017M$ solution of reagent grade potassium dichromate and standardize the solution with samples of iron(II) prepared from primary standard grade iron wire.

Weigh out 2.5 g of reagent grade $K_2Cr_2O_7$ on a triple-beam balance, transfer to a clean glass bottle, and dissolve in about 500 mL of deionized water.

Weigh out three individual 0.2- to 0.25-g samples of primary standard grade iron wire. Iron wire is wound on a spool. The most common gauge weighs about 0.32 g per meter. Your instructor will tell you how much 1 m of your iron wire will weigh. With clean dry hands, measure out the appropriate length of wire against a meter stick and clip with side cutters or scissors. Make a tight loop of each portion of wire, and weigh each loop as carefully as possible on an analytical balance. Place each loop in its own 250-mL Erlenmeyer flask, and add about 5 mL of deionized water and 10 mL of concentrated hydrochloric acid to each flask.

CAUTION: Work in a fume hood.

Heat each flask gently over a Bunsen burner flame to dissolve the wire samples more quickly. Do not boil the solutions: iron may be lost if the samples bump. When the wire samples are dissolved, cool the flasks with tap water until they are at room temperature. Rinse down the insides of the flasks with a few milliliters of deionized water, and add an additional 10 mL of concentrated hydrochloric acid to each sample. A few tiny flecks of carbon may be present after the wire dissolves. You may ignore them.

> NOTE: From this point work with only one sample at a time.

Heat one of the solutions of iron nearly to boiling and add $0.5M$ $SnCl_2$ in $6M$ HCl dropwise until the faint yellow color of Fe(III) disappears. Add one or two drops of stannous chloride solution in excess. Add deionized water until the volume is about 75 mL. Then, while swirling the solution, add 15 mL of 5% $HgCl_2$ solution. Stopper the flask and set it aside to react for about 5 minutes. The following reactions take place:

$$2Fe(III) + Sn(II) \rightarrow 2Fe(II) + Sn(IV) \text{ [excess Sn(II)]}$$

and

$$\text{excess Sn(II)} + 2Hg(II) + 2Cl^- \rightarrow Sn(IV) + Hg_2Cl_2(ppt)$$

If too much Sn(II) has been added some mercury metal will also form and give the solution a pearly lustre. Mercury metal will be oxidized gradually by dichromate and may cause determinate error if too much time is taken during the titration. Avoid adding too much Sn(II) in the first place.

> CAUTION: Dilution of phosphoric acid produces heat.

Add 15 mL of concentrated phosphoric acid. Stopper the flask and cool it to room temperature with tap water. Add three drops of diphenylamine sulfonate indicator and titrate with the dichromate stock solution. As the titration proceeds the solution turns green with Cr(III). Just before the endpoint the solution will take on a gray cast. The endpoint is the first deep violet-green color which lasts longer than 15 seconds. Dispose of samples in a special vessel provided by your instructor. Do not dispose of mercury and chromium solutions in a sink.

The phosphoric acid lowers the equivalence point potential just enough to bring it very close to the diphenylamine sulfonate indicator's formal potential.

The Unknown

Your instructor will issue you an unknown and give you directions for its preparation. Samples purchased from Thorn Smith, Inc. are quite popular and are prepared from Mohr's salt, which is 14.2% Fe. If your sample contains between 5 and 14% Fe, how much must you weigh out to require between 30 and 40 mL of dichromate titrant? Your instructor will tell you whether or not you should prereduce your unknown sample with stannous chloride and tell you the approximate range for unknown results.

EXPERIMENT 16.2: Iodometric Determination of Copper in Ore

Additional Notes

You may use either a volumetric technique or a gravimetric technique for this analysis. The volumetric method involves the use of a buret, which you have used in many other experiments. The molarity of the dichromate solution can be calculated in the following way:

$$M(\text{dichromate}) = (1/6)*(\text{mmole Fe})/(\text{mL to endpoint})$$

The millimoles of iron is found by dividing the weight of the iron wire sample by 55.85 mg/mmole. The millimoles of iron in the unknown sample can be calculated as follows:

$$\text{mmole Fe} = M(\text{dichromate})*(\text{mL to endpoint})*6$$

Milligrams of iron can be found by multiplying the number of millimoles by 55.85 mg/mmole, the gram atomic weight of iron.

Alternatively, you may transfer some of your dichromate stock solution to a 125-mL plastic dropping bottle (obtainable from most chemical supply houses). Weigh the dropping bottle and solution on an analytical balance or a top-loader. Squeeze the bottle to deliver dichromate to the sample solution. Be careful to transfer all the dichromate to the flask and none to the counter top. When the endpoint is reached, weigh the dropping bottle and its contents again. Calculate the concentration of dichromate as a "titer" in units of milligrams of iron per gram of dichromate solution. This simple ratio lets you ignore the 6:1 stoichiometry of the reaction and greatly simplifies calculations. You will find that the precision of the gravimetric process (± 1 mg in 140 g on a top-loader) is better than that of the buret (± 0.02 mL in 40 mL). Remember that a limiting factor is also the precision of your weighing of the iron wire samples, ± 0.1 mg in 250 mg.

EXPERIMENT 16.2: Iodometric Determination of Copper in Ore

The Reaction

Copper can be determined indirectly by measuring the amount of iodine generated by the reaction with excess iodide:

$$2Cu^{2+} + 5I^- = 2CuI(\text{ppt}) + I_3^-$$
$$(\text{unknown}) \qquad\qquad\qquad (\text{measured})$$

Iodine is titrated with standard sodium thiosulfate solution to the starch indicator endpoint (blue → colorless),

$$I_3^- + 2S_2O_3^{2-} = 3I^- + S_4O_6^{2-}$$

The reaction involves both an oxidation-reduction and a precipitation. As discussed in the text, the stabilization of Cu^+ with iodide makes Cu^{2+} a strong enough oxidizer to react quantitatively with iodide ion. If Cu^+ were not stabilized, Cu^{2+} would not react with iodide at all.

Solution pH must be controlled during the titration. If the pH gets much above 4.5, the formation of copper hydroxide complexes may cause an indistinct endpoint. If the pH is less than about 3.5, arsenic(V), if present, can oxidize iodide and cause high results. The pH is controlled by adding acetic acid and ammonium bifluoride to an ammoniacal solution prior to titration. Fluoride ion also complexes Fe^{3+}, which otherwise would be able to oxidize iodide.

The endpoint in the titration relies on the blue complex of starch, iodine, and iodide. Thiocyanate is added just before the endpoint to desorb iodine from the CuI precipitate and sharpen the endpoint, as explained in the text.

The Titrant: Standard Sodium Thiosulfate

The preparation and storage of thiosulfate solutions have been discussed earlier. Rather than treat sodium thiosulfate as a primary standard, we will weigh out a portion of sodium thiosulfate pentahydrate and dissolve it in freshly boiled water. It will be used as a titrant in the reaction of primary standard potassium iodate, as discussed in the text:

$$IO_3^- + 5I^- + 6H^+ = 3I_2 + 3H_2O$$

and

$$3I_2 + 6S_2O_3^{2-} = 6I^- + 3S_4O_6^{2-}$$

Notice that 1 mmole of iodate will consume 6 mmole of thiosulfate.

Preparation and Standardization of 0.05*M* Solution of Na$_2$S$_2$O$_3$

1. Dissolve in about 1 liter of freshly boiled and cooled deionized water 13 g of sodium thiosulfate pentahydrate. Add about 0.2 g of sodium carbonate to the solution.

2. Weigh out as carefully as possible 1.5 to 1.6 g of primary standard reagent grade potassium iodate into a clean beaker. Dissolve in a minimum of water and

EXPERIMENT 16.2: IODOMETRIC DETERMINATION OF COPPER IN ORE   **585**

transfer quantitatively to a 500-mL volumetric flask. Dilute to the mark with freshly boiled and cooled deionized water and mix thoroughly.

3. Pipet 25-mL portions of the potassium iodate solution to clean 250-mL Erlenmeyer flasks. This titration is quick and easy, and you may wish to do four or five replicates. From this point treat all samples individually.

4. Add about 2 g of solid reagent grade (iodate-free) potassium iodide and 10 mL of $1M$ sulfuric acid to each flask just before you are ready to titrate. Begin titrating with thiosulfate immediately, constantly swirling the solution as you titrate. When the color of the solution becomes a very pale yellow add 2 or 3 mL of starch indicator solution. The solution will turn deep blue-violet but will become colorless at the endpoint.

Preparing the Ore Sample

> EXERCISE EXTREME CAUTION: Protect your eyes. Work in hood. Copper ore samples are difficult to dissolve, requiring hot nitric acid. The nitric acid must be removed by boiling with concentrated sulfuric acid. Bromine water and ammonia are also used. YOU MUST PROTECT YOURSELF FROM THESE MATERIALS. Hot acids will burn the skin immediately. Resist the temptation to sniff vapors coming from flasks. Do not distract your neighbors in the hoods while they are working with these materials.

1. Dry your sample in a weighing bottle for 1 hour at 110°C.

2. Weigh out three portions of the ore sample in the range of 0.8 to 0.9 g (as carefully as possible) into clean 250-mL Erlenmeyer flasks. Add 10 mL of 1:1 nitric acid to each flask, and place the flasks on ring stands, over Bunsen burners, in a fume hood. Heat below boiling until the volumes are reduced to about 5 mL. (5 mL of water containing blue dye in a 250-mL Erlenmeyer flask makes a good model). Remove the flasks from the ring stands, and let them cool to room temperature.

3. Add 10 mL of concentrated sulfuric acid to each flask. Put the flasks back on the stands and heat them cautiously until bubbles appear. Nitric acid will volatilize, along with red-brown nitrogen dioxide. When dense white fumes of SO_3 appear nitrogen oxides will be gone. The solution will be extremely hot at this point, so be very careful.

4. Remove the flasks from the stands and let them cool to room temperature. Do not plunge them into ice water; they might break. Dilute their contents to about 50 mL with deionized water. Add 5 mL of bromine water (saturated) to each flask to oxidize As(III) to As(V). Boil gently for at least 5 minutes to remove residual bromine, which might later oxidize iodide to iodine. Now treat the samples individually.

5. Add ammonium hydroxide, diluted 1:1 with deionized water, down the inside wall of the flask until the solution turns light blue. You will begin to see some copper hydroxide precipitate when you add ammonium hydroxide, but it will redissolve as the desired copper-ammonia complex forms. Do not add too much ammonium hydroxide. Cautiously sniff the blue solution. If you smell ammonia, you have added too much ammonium hydroxide. Add dilute sulfuric acid dropwise until you no longer smell ammonia.

6. Cool the solution to room temperature. Add 10 mL of acetic acid, diluted 1:1 with water, and 3 g of ammonium bifluoride. Swirl the contents of the flask vigorously and proceed without hesitation to the next step.

7. Add 3 g of KI, swirl, and titrate immediately with standard thiosulfate solution. The KI will turn the solution a muddy brown, but as thiosulfate is added the solution color will become lighter and lighter. When the color is a light buff, add 3 mL of starch indicator solution and 3 g of KSCN. Swirl the solution to dissolve the salt and titrate drop by drop to the starch endpoint. The starch blue color will be changed by the presence of the precipitate, but the change should be quite clear when it occurs. The endpoint is the first absence of "blue" for 20 seconds. Iodide is oxidized by air and the solution will revert to the starch-iodine color in a matter of minutes if you have not added an excess of thiosulfate.

8. Repeat steps 5 to 7 for the two remaining samples. From the volumes of thiosulfate titrant, calculate percent Cu in the sample.

Chapter 17

Analytical Separations and an Introduction to Chromatography

17.1 INTRODUCTION

In this chapter we will examine some fundamental concepts of analytical separations. We will begin by defining some important terms with the aid of a very simple model for solvent extraction and then consider a separation of two solutes by solvent extraction. The next more complicated situation will involve multiple solvent extractions of two varieties, crosscurrent and countercurrent separations. Finally, we will consider some rapid and automatable methods of separation called chromatographic methods. The material presented is basic to the understanding of gas chromatography, liquid chromatography, and high-performance liquid chromatography and will serve as an introduction to more detailed discussion in more advanced courses.

17.2 SOLVENT EXTRACTION

Imagine a 150-mL separatory funnel which holds 50 mL of diethyl ether containing 10 mg of the carboxylic acid, maleic acid ($cis-$HOOCCH=CHCOOH). When 50 mL of aqueous $0.5F$ HCl is added to the separatory funnel and the contents are shaken vigorously, some of the maleic acid transfers from the ether phase to the aqueous phase.† Net transfer continues until an equilibrium state is reached. The separatory

---

† The ether and aqueous acid phases have some mutual solubility, and we will assume that they have been equilibrated with each other before the extraction process is begun. This is standard practice and will not be pointed out in every example in this text.

funnel is set aside to let the phases separate, and then the aqueous acid phase, containing an equilibrium amount of maleic acid, is drawn off. In the case of this particular system we find that about 9.1 mg of acid is transferred (or "partitions") to the aqueous phase, while about 0.9 mg stays behind in the ether. The ratio of the concentration of maleic acid in the aqueous acid to that in the ether at equilibrium is called the *concentration distribution coefficient, K*,

$$K = [\text{maleic}]_w/[\text{maleic}]_e \; (=9.1/0.9 \text{ or about } 10)$$

(Subscripts w and e stand for water and ether, respectively.)

The concentration distribution coefficient has a thermodynamic counterpart called simply the distribution coefficient. The free energy change for the distribution process, represented by

$$\text{maleic acid (ether)} = \text{maleic acid (aqueous)}$$

is expressed as

$$\Delta G = \Delta G^0 + RT \ln [a(\text{mal}, w)/a(\text{mal}, e)]$$

In this expression the letter a denotes activity. At equilibrium the free energy change is zero, and the ratio of the activities of maleic acid in aqueous acid and ether is determined by the standard free energy change for the distribution. The ratio of these activities is the thermodynamic distribution coefficient K^0. Like all equilibrium constants, K^0 is independent of concentration but does change with temperature. As we have done many times before, we recognize that the activity is a product of the concentration and a proportionality constant called an activity coefficient, f. Each species will have an activity coefficient for each phase. In the case of maleic acid,

$$a(\text{mal}, w) = f(\text{mal}, w) * [\text{maleic}]_w \quad \text{(in aqueous phase)}$$
$$a(\text{mal}, e) = f(\text{mal}, e) * [\text{maleic}]_e \quad \text{(in ether phase)}$$

Solutions in which $f = 1$ are called ideal solutions and are said to obey Raoult's law, a concept introduced in most general chemistry courses. Real solutions behave ideally only when they are very dilute (millimolar level). The relationship between K and K^0 can then be written as

$$K = K^0 * [f(\text{mal}, e)/f(\text{mal}, w)]$$

Since activity coefficients vary with concentration, K (but not K^0) should vary with concentration. We should not be surprised, then, to find one distribution coefficient when we extract maleic acid at the $0.1F$ level from ether and a different value when the extraction is done at the 0.005F level. If we can evaluate the activity coefficients, we can in principle calculate the thermodynamic constant from experimental data.

When the species which partition between two phases are also involved in secondary reactions with the solvents, it usually simplifies matters to define distribution coefficients in terms of analytical (total) concentrations. We call this type of coefficient a *distribution ratio* and designate it as D:

$$D = C_{t,w}/C_{t,e}$$

An example involves our model compound, maleic acid, which is a diprotic weak acid. The dissociation reactions

$$H_2Mal = H^+ + HMal^- \qquad K_{a1} = 1.2 \times 10^{-2}$$

and

$$HMal^- = H^+ + Mal^{2-} \qquad K_{a2} = 5.9 \times 10^{-7}$$

produce ionic species which are soluble in water but much less soluble in ether. When solution pH is high the conjugate base forms are favored, and there is greater partitioning to water than we might have expected by simply considering K. If we use a material balance expression for maleic acid in water,

$$C_{t,w} = [H_2Mal]_w + [HMal^-]_w + [Mal^{2-}]_w$$

and a second material balance in ether,

$$C_{t,e} = [H_2Mal]_e$$

then the distribution ratio, D, can be written as

$$D = ([H_2Mal]_w + [HMal^-]_w + [Mal^{2-}]_w)/[H_2Mal]_e$$

If we express $[HMal^-]$ and $[Mal^{2-}]$ in terms of K_{a1} and K_{a2} in the numerator, $[H_2Mal]_w/[H_2Mal]_e$ can be factored out, leaving

$$D = K[1 + (K_{a1}/[H^+]) + (K_{a1}K_{a2}/[H^+]^2)]$$

D functions as a conditional distribution coefficient, that is, one whose value applies at only one solution pH. In a weak acid system, D (ratio of concentration in water to that in ether) increases with increasing pH. Example 17.3 illustrates this effect.

Many methods for the analysis of metal ions at the trace level require complexation with some Lewis base, followed by extraction into an immiscible solvent. As pointed out in the chapter on complexation equilibria, Lewis bases are usually Brønsted bases as well. The determination of a metal ion is therefore complicated by solution pH, not only in the sense of protons competing for the Lewis base, but also in the effects of pH on the distribution equilibrium. You should derive an expression for the relationship between D and K_b for a weak base.

Some interesting relationships emerge when we express K in terms of mole fractions rather than in terms of concentrations. For the abstract case in which a species "A" partitions between water (subscript w) and an immiscible organic solvent (subscript o), we can write an expression for K in terms of moles (m) and volumes (V),

$$K = [A]_w/[A]_o = (m_{A,w}/V_w)*(V_o/m_{A,w})$$
$$= (m_{A,w}/m_{A,o})*V_o/V_w$$

We will then express the mole fraction of species A in water as

$$q_A = (m_{A,w})/(m_{A,w} + m_{A,o})$$

and the mole fraction of species A in the organic phase as

$$p_A = (m_{A,o})/(m_{A,w} + m_{A,o})$$

To express each fraction in terms of K, divide the numerator and denominator of each by $m_{A,o}$,

$$q_A = \frac{(m_{A,w}/m_{A,o})}{(m_{A,w}/m_{A,o}) + 1}$$

$$p_A = \frac{1}{(m_{A,w}/m_{A,o}) + 1}$$

Since separation depends ultimately on the amounts of solute in the system rather than their concentrations, we should define a new factor which combines K and the volume ratio,

$$k' = K(V_w/V_o) = (m_{A,w}/m_{A,o})$$

This factor is called the "capacity factor" and is quite important.

The fractions q_A and p_A can now be expressed even more simply,

$$q_A = k'/(k' + 1) \quad \text{and} \quad p_A = 1/(k' + 1)$$

The volume ratio will be used again in a later section on chromatographic separations. For the time being we have enough information to work on some example problems.

17.2 SOLVENT EXTRACTION

EXAMPLE 17.1 Extraction

A 10.0-mg sample of maleic acid is placed in a separatory funnel containing 50 mL of $0.5F$ HCl and 50 mL of diethyl ether. The separatory funnel is shaken for a minute to let the system equilibrate and then set aside to allow the phases to separate. Predict the mole fraction and the milligrams of maleic acid that will be in each phase after equilibration. The distribution coefficient of maleic acid is 9.7.

Solution

In $0.5F$ HCl the fraction of undissociated maleic acid is about 0.98. For simplicity we will ignore the 2% which is dissociated. Since the volumes of $0.5F$ HCl and ether are the same we can use the simple expression

$$q = K/(K + 1)$$

to find that the mole fraction in the aqueous phase is 0.91. The mole fraction in the ether phase must be $1.00 - 0.91 = 0.09$. Since we started with 10 mg of acid, $10*0.91 = 9.1$ mg will be in the aqueous phase, and $10*0.09 = 0.90$ mg will be in the ether phase. We are working with only one compound, so the weight fraction and mole fraction will be numerically equal.

EXAMPLE 17.2 Extraction with Unequal Volumes

Solve the problem in the last example using an aqueous acid volume of 50 mL and an ether volume of 10 mL.

Solution

Intuition suggests that the smaller the volume of a phase, the smaller will be the amount of solute it can hold. Using $k' = K(V_w/V_e) = 9.7(50/10) = 48.5$, we can calculate the fraction in the aqueous phase as

$$q = k'/(k' + 1) = 0.98$$

The fraction in the ether phase is therefore $1.00 - 0.98 = 0.02$. At equilibrium there will be 9.8 mg of maleic acid in the aqueous acid phase and 0.2 mg in the ether phase. As we had expected, less maleic acid can be held in this smaller ether phase.

EXAMPLE 17.3 Distribution Coefficient

Change the system in Example 17.1 to 50 mL of water buffered at pH 3 and 50 mL of ether. Starting with 10.0 mg of maleic acid, predict the amount of maleic acid in each phase after equilibration. $K = 9.7$ for maleic acid partitioning between $0.5F$ HCl and ether. For maleic acid $pK_{a1} = 1.95$, $pK_{a2} = 6.23$.

Solution

By substituting values for K_{a1}, K_{a2}, and $[H^+]$ into the equation for D in terms of K, we find

$$D = K(1 + 10^{1.05} + 10^{-2}) = 12.2 * K = 118$$

The distribution ratio is defined in terms of total concentrations,

$$D = 118 = (C_{t,w})/(C_{t,e}) = (\text{moles}_w/\text{moles}_e)$$

We can also write this as

$$D = 118 = (\text{mg}_w/\text{mg}_o)$$

Notice that D is much larger than K. Higher solution pH favors species which are soluble in the aqueous phase.

We let x be the number of milligrams of maleic acid (in all acid or base forms) in the aqueous layer and $10.0 - x$ be the number of milligrams of maleic acid in the ether phase. We can then solve the algebraic equation

$$118 = x/(10.0 - x)$$

to find $x = 9.92$ mg (aqueous) and $10.0 - x = 0.08$ mg (ether). Compare this result with that of the last example, in which we reduced the volume of ether. Raising the pH and decreasing the volume of ether both favor extraction into the aqueous phase.

17.3 SEPARATING TWO SOLUTES

The really interesting problems in analytical separations involve removing one compound from a mixture of several others. Whether or not two species can be separated by partitioning between phases depends on their distribution coefficients or distribution ratios. A convenient measure of the separability of two species A and B is the separation factor α (alpha),

17.3 SEPARATING TWO SOLUTES

$$\alpha = K_A/K_B \text{ or } D_A/D_B$$

When α is either much larger or much smaller than unity separations can be made quite efficiently. The simplest and best situation is when α is large as a result of a large K_A and a small K_B: quantitative separations may be possible with one or two equilibrations. When α is close to unity either a series of many batch extractions or a continuous extraction method may be needed to separate species. Remember that α is a ratio and it can be large even if both K_A and K_B are large. When this is the case both solutes partition strongly to the same phase, and separation can be quite difficult. Special techniques for such situations are beyond our requirements here, but can be found in several of the references listed at the end of the chapter. The best separations can be obtained when the product of the distribution coefficients is about unity, that is, when $K_A = 1/K_B$.

In the following sections we will discuss two important repetitive extraction techniques. Their names suggest the models of stream flow which they resemble: crosscurrent and countercurrent separations.

17.3.1 Crosscurrent Separations

In Example 17.1 we saw that about 90% of the maleic acid placed in a system of equal volumes of aqueous $0.5F$ HCl and diethyl ether partitioned to the aqueous phase and that the remainder partitioned to the ether. Suppose we then do a second extraction. We will replace the aqueous phase with 50 mL of fresh $0.5F$ HCl and equilibrate it with the ether phase remaining from the first extraction. Maleic acid will again partition between the solvent layers as required by the distribution coefficient, K. The mole fraction of maleic acid remaining in the ether phase after the second extraction is given by

$$p*p = p^2 \text{ (fraction left in ether)}$$

When we combine the second 50-mL portion of the aqueous phase with the 50 mL taken from the first separation, we find that the mole fraction extracted (in two separations) is

$$q(\text{total}) = 1 - p^2 \text{ (fraction in 100 mL of aqueous phase)}$$

If we were to repeat this process n times, the fractions of maleic acid in the two solvent phases would be given by

$$p = 1/(K + 1)^n \quad \text{(in 50 mL of ether)}$$
$$q = 1 - p^n \quad \text{[in } (n*50) \text{ mL of aqueous phase]}$$

EXAMPLE 17.4 Multiple Extractions

Calculate the fractions of maleic acid in the ether and combined aqueous acid phases under the initial conditions of example 17.1, after 3 separations. $K = 9.7$.

Solution

Begin by calculating the fraction left in 50 mL of ether:

$$p = 1/(K + 1)^n = 1/(10.7)^3 = 8.2 \times 10^{-4}$$

In the combined water phase (150 mL) the fraction will be $1.000 - 8.2 \times 10^{-4}$, or 0.9992. Since more than 99.9% has been removed we call the extraction quantitative.

If we had started with 10 mg of maleic acid, after three extractions we would have 8.2×10^{-3} mg in the ether phase and 9.992 mg in the aqueous acid phase. The concentration of maleic acid in ether is $8.2 \times 10^{-3}/50 = 1.64 \times 10^{-4}$ mg/mL, and that in the aqueous acid is $9.992/150 = 6.66 \times 10^{-2}$ mg/mL. The price we pay for a quantitative extraction is a solution which is more dilute by a factor of 3 (a factor of n in the general case).

Now let us make the situation more interesting by adding fumaric acid to the system. Fumaric acid has the same formula and molecular weight as maleic acid, but is the *trans*, rather than the *cis*, isomer.

$$\underset{\text{Maleic acid}}{\begin{array}{c}\text{HOOC}\quad\text{COOH}\\ \text{C=C}\\ \text{H}\quad\text{H}\end{array}}\qquad\underset{\text{Fumaric acid}}{\begin{array}{c}\text{HOOC}\quad\text{H}\\ \text{C=C}\\ \text{H}\quad\text{COOH}\end{array}}$$

The distribution coefficient for fumaric acid, $K(\text{fum})$, is 0.90, about one-tenth the value for maleic acid. When we perform multiple extractions on a mixture of maleic acid and fumaric acid, we will find that the ratio of the fractions in the ether phase after n equilibrations is

$$p(\text{mal})/p(\text{fum}) = [(K(\text{fum}) + 1)/(K(\text{mal}) + 1)]^n$$

The ratio of the fractions of maleic and fumaric acids in the combined aqueous phases after n separations is given by

$$q(\text{mal})/q(\text{fum}) = [1 - p(\text{mal})^n]/[1 - p(\text{fum})^n]$$

17.3 SEPARATING TWO SOLUTES

The ratio of the fractions in each phase will be an important consideration if we are planning to purify a solute by multiple batch extractions.

EXAMPLE 17.5 Multiple Extractions, Two Solutes

Calculate the number of milligrams of maleic acid and fumaric acid in the ether layer after four batch extractions. Assume that the process starts with 10.0 mg of each acid dissolved in 50 mL of diethyl ether and that 50 mL of 0.5F HCl is used in each extraction step. $K(\text{mal}) = 9.7$; $K(\text{fum}) = 0.90$.

Solution

After four extractions the ether phase has maleic and fumaric acids in the mole fraction ratio

$$p(\text{mal})/p(\text{fum}) = [(0.90 + 1)/(9.7 + 1)]^4 = 9.9 \times 10^{-4}$$

Since the original ratio was unity, it is clear that four extractions have dramatically increased the purity of fumaric acid in the ether phase. The actual fraction of fumaric acid in the ether phase is

$$p(\text{fum}) = 1/[K(\text{fum}) + 1]^n = 1/(1.90)^4 = 0.077$$

Starting with 10 mg, we have 10(0.077) or 0.77 mg left in the ether phase after four extractions. Notice that while the fumaric acid is quite pure, there is not much of it there! About 9.2 mg of fumaric acid has partitioned to the aqueous acid phase, along with virtually all of the maleic acid. The maleic acid is only slightly more pure than when we started.

Separatory funnel extractions are only one example of crosscurrent separations. The operations involved in using separatory funnels help illustrate the concept of a *crosscurrent* process: a series of aqueous phases are moved "across" a single ether phase, solutes distributing until an equilibrium state is reached at each "crossing." If you have taken an introductory course in organic chemistry you have probably performed other crosscurrent separations: recrystallizations, repeated distillations, and Soxhlet extractions are all crosscurrent methods. In each case a solution or a solid is repeatedly washed with successive portions of fresh solvent. Soxhlet extractions are also batch separations. Solid material in the Soxhlet sample holder is washed with solvent which is purified by continuous distillation from the extracted solution. A Soxhlet apparatus is shown in Fig. 17.1. The glass extractant return line is designed so that the contents of the Soxhlet chamber "flush" whenever the liquid level exceeds

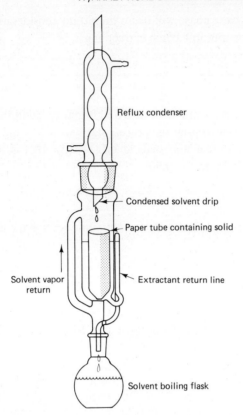

Figure 17.1 Soxhlet extractor.

a critical level. The chamber refills gradually with distilled solvent and extracts material from the sample holder until the critical level is again reached.

17.3.2 Countercurrent Separations

A countercurrent separation is one in which solutes partition between two immiscible solvents which flow by (or through) each other in opposite directions. This solvent stream flow can be understood in terms of a model in which batches of solvent are transferred through a series of separatory funnels. After each transfer step the contents of the immiscible solvent phases in each funnel are equilibrated. The separatory funnels act as discrete separation *stages,* that is, points in the process at which partitioning reaches equilibrium.

To describe a countercurrent separation graphically, we will represent a rack of separatory funnels by the boxes in Fig. 17.2. Each line of boxes shows the result of an operation (numbered 1 to 4). The upper half of each box is occupied by the less dense solvent of an immiscible solvent pair—for example, diethyl ether, if we are using water and ether as solvents. We will call this solvent 1. The lower half of each box represents

17.3 SEPARATING TWO SOLUTES

Process		Phase	Stage: 0	1	2	3
1a	Start	1	1			
		2	0			
1b	Equilibrate	1	p			
		2	q			
2a	Transfer (not equilibrated)	1	0	p		
		2	q	0		
2b	Equilibrate	1	pq	p^2		
		2	q^2	pq		
3a	Transfer (not equilibrated)	1	0	pq	p^2	
		2	q^2	pq	0	
3b	Equilibrate	1	pq^2	$2p^2q$	p^3	
		2	q^3	$2pq^2$	p^2q	
4a	Transfer (not equilibrated)	1	0	pq^2	$2p^2q$	p^3
		2	q^3	$2pq^2$	p^2q	0
4b	Equilibrate	1	pq^3	$3p^2q^2$	$3p^3q$	p^4
		2	q^4	$3q^3p$	$3p^2q^2$	p^3q

Figure 17.2 Countercurrent extraction process.

the second solvent, solvent 2. The separatory funnels are given stage numbers from left to right, starting with stage 0 in the very first equilibration. The separation procedure involves moving batches of solvent 1 (upper) to the right to contact fresh batches of solvent 2 (lower). Fresh solvent 1 is added to stage 0 at each transfer, where it contacts solvent 2 used in a prior extraction. The net effect is to move solvent 1 in batches toward the right and to move solvent 2 toward the left, creating (if you stretch your imagination) a flow of currents in opposite directions.

Operations 1a and 1b in Fig. 17.2 involve partitioning one solute between the two solvents in stage 0. After equilibration the fraction of solute in solvent 1 is p, while that in solvent 2 is q, the same fraction labels we used in the single batch extraction in the previous section. In the first transfer step (step 2a) solvent 1 is moved one stage to the right, where it contacts fresh solvent 2. Fresh solvent 1 is added to stage 0. Both stages 0 and 1 are shaken to reach equilibrium, and the resulting fractions are shown in step 2b. When we partition solute between two phases containing fractions 0 and q, we transfer the fraction p of q (or $p*q$) to solvent 1, and leave behind the fraction q of q (or $q*q$) in solvent 2. In stage 1 we move a fraction q of p ($p*q$) from solvent 1 to solvent 2, and leave behind a fraction p of p (or $p*p$) in solvent 1.

In the next transfer (step 3a) we move the solvent 1 phase of stage 1 to stage 2, where it contacts fresh solvent 2, and move solvent 1 of stage 0 to stage 1, where it contacts solvent 2 equilibrated in the previous operation. Fresh solvent 1 is added at stage 0, and all three separatory funnels are shaken to equilibrate. The fractions in both phases of all stages are shown in step 3b. In stage 0 we have moved a fraction p of $(q*q)$ from solvent 2 to solvent 1 and have left $(q*q*q)$ behind. In stage 2 we have moved the fraction $(q*p*p)$ from solvent 1 to solvent 2 and have left $(p*p*p)$ behind. Notice the symmetry in the fractions. Now the process becomes more difficult to follow. In stage 1 we started with $(p*q)$ in solvent 1 transferred from stage 0. Equilibration will transfer $q*(p*q)$ to solvent 2 and leave $p*(p*q)$ in solvent 1. At the same time stage 1 solvent 2 starts with $(p*q)$. Equilibration will transfer $p*(p*q)$ to solvent 1 and leave behind $q*(p*q)$ in solvent 2. When the fractions from both transfers are summed we find $(2p^2q)$ in solvent 1 and $(2pq^2)$ in solvent 2.

The next step (number 4a in Fig. 17.2) is to move the solvent 1 phases of all separatory funnels one stage to the right, add fresh solvent 1 to stage 0 and fresh solvent 2 to stage 3, and equilibrate all stages. The fractions that result are shown in step 4b. You may recognize the coefficients of the fractions as those of the binomial expansion, $(x + y)^n$ (1:3:3:1 in the last step). The binomial distribution was introduced in Chapter 2. Remember that for large values of n a binomial distribution becomes Gaussian. In a system with a great many separatory funnels solutes will distribute in bands, with widths that can be described in terms of standard deviations, with the units of stages. The rate at which a solute band moves to the right in the chain of stages depends directly on the fraction of solute in solvent 1 at each equilibration. Solutes which partition strongly to solvent 1 move to the right quickly. Those which partition strongly to solvent 2 move to the right much more slowly. As was the case in the crosscurrent method, differences in partitioning make it possible to separate two solutes. As you will see later, the countercurrent method is more efficient than the crosscurrent method; that is, better separations are possible in a small number of steps.

Some general relationships can be found which describe the behavior just explained. We will represent the stage number by r and the number of transfers by n. When successive transfers of solute in solvent 1 (the upper phase in Fig. 17.2) are performed, we describe the fraction of solute as

$$p(p + q)^n$$

where p and q are solute fractions in solvent 1 and solvent 2, respectively, that result from a single partitioning. Notice, for example, in the operation labeled 3b in Fig. 17.2 that the distribution in solvent 1 is $p(q^2 + 2pq + p^2)$, which is the same as $p(p + q)^2$.

17.3 SEPARATING TWO SOLUTES

When transfers of solute in solvent 2 are made, the fraction of solute in solvent 2 is described by

$$q(p+q)^n$$

In Figure 17.2, step 3b, the distribution in solvent 2 is $q(q^2 + 2pq + p^2)$. The coefficients of the terms in parentheses in any equilibration step can be calculated directly from the binomial theorem. In general, the expansion for the rth stage in a series of n total transfer steps is given by

$$(p+q)^n_r = (n!/(r!(n-r)!)) * p^r q^{n-r}$$

Recall that p and q can be expressed in terms of K (or D, see above),

$$q = K/(K+1) \quad \text{and} \quad p = 1/(K+1)$$

and write the fraction of solute in the rth stage after n transfers as

$$F_{n,r} = (n!/(r!(n-r)!)) * (K^{n-r})(1/(1+K))^n$$

It is easier to use this form of the expression for ordinary calculations. Notice that the fraction calculated is that of a solute in *both* solvents in the rth stage. In a particular stage the fraction of solute in solvent 1 is determined by multiplying $F_{n,r}$ by p [note $p = 1/(K+1)$ while the fraction in solvent 2 is determined by multiplying $F_{n,r}$ by q [i.e., $K/(K+1)$].

EXAMPLE 17.6 Countercurrent Separation
Consider again the separation of maleic acid and fumaric acid, but using countercurrent separations of 50-mL portions of 0.5F HCl and diethyl ether. K(maleic) = 9.7 and K(fumaric) = 0.90. Calculate the fractions of maleic acid and fumaric acid in stage 5 after 15 transfers.

Solution

For stage 5,

$$n! = 3.6288 \times 10^6, \quad r! = 120, \quad (n-r)! = 120$$
$$K(\text{mal})^{n-r} = 8.59 \times 10^4, \quad \{1/[1+K(\text{mal})]\}^{10} = 0.375$$

Using these values and the equation just given for the fraction of solute, we calculate the fraction of maleic acid to be 1.10×10^{-3} at stage 5.

The value of q is given by $K/(K + 1) = 0.91$, and $p = 0.09$. Therefore, in stage 5 the fraction of maleic acid in solvent 1 (ether) is $0.09 * 1.10 \times 10^{-3} = 9.9 \times 10^{-5}$, and in solvent 2 (0.5$F$ HCl) is $0.91 * 1.10 \times 10^{-3} = 1.00 \times 10^{-3}$.

By a similar calculation the fraction of fumaric acid in both phases at stage 5 is 0.243. In ether the fraction is $0.47 * 0.243 = 0.114$. In 0.5F HCl the fraction is $0.53 * 0.243 = 0.129$.

We conclude that most of the maleic acid has been removed from the fumaric acid. The fumaric acid is more dilute: it is spread out over several separatory funnels.

Figures 17.3a, b, and c show the distribution of maleic and fumaric acids between ether and 0.5F HCl for 10, 25, and 40 transfers. Notice that the distributions are not particularly symmetrical when a small number of transfers has been performed. This is especially true for maleic acid because it partitions so strongly to the left-moving solvent. When there are more transfers, the distributions begin to take on the appearance of symmetrical bands of solute. Also notice that when there are more transfers, the fractions in the stages are lower. The solutes spread out into more stages when there are many stages.

For really large numbers of stages the equation for fractions given above becomes unwieldy. An important computational limitation arises from the large factorials in the first part of the expression. Popular 8-bit microcomputers, for example, must work with real numbers smaller than about 10^{37}, a number exceeded by 34! Even worse, imagine calculating 34! by hand!! A better approach when a large number of stages is involved is to put the equation in the form of a Gaussian distribution,

$$F_{n,r} = (K + 1)/(\sqrt{2\pi nK}) * \exp[-(r_{max} - r)^2(K + 1)^2/2nK]$$

where r_{max} is the stage number at which the maximum fraction of solute with distribution coefficient K is to be found after n transfers. The value of r_{max} is given by the product $n * p$, where p is the fraction in solvent 1 after one equilibration (also $p = 1/K + 1$). The quantity $(r_{max} - r)$ is the number of stages that separate the stage whose fraction is being calculated from the stage at which the fraction is a maximum. Notice that when $r_{max} = r$, the fraction is a maximum and is equal to $(K + 1)/(\sqrt{2\pi nK})$.

In Chapter 2 the Gaussian distribution was written as

$$f(x_i) = (1/\sigma\sqrt{2\pi}) * \exp\{-\tfrac{1}{2}[(x_i - \mu)/\sigma]^2\}$$

where μ is the population mean, x_i is a member of the population, and σ is the standard deviation of the population, a measure of the breadth of the distribution. This equation is the same as the last equation for $F_{n,r}$ if we define σ in the following way:

$$\sigma = \sqrt{(nK)}/(1 + K)$$

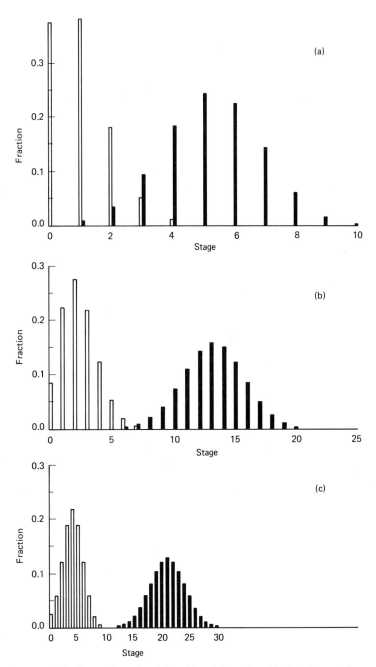

Figure 17.3 Separation of maleic acid and fumaric acid (a) in 10 transfers, (b) in 25 transfers, and (c) in 40 transfers. In (a) the maximum fraction of maleic acid is 0.386 at stage 1; the maximum fraction of fumaric acid is 0.242 at stage 5. In (b) the maximum fraction of maleic acid is 0.274 at stage 2; the maximum fraction of fumaric acid is 0.157 at stage 13. In (c) the maximum fraction of maleic acid is 0.216 at stage 4; the maximum fraction of fumaric acid is 0.126 at stage 21. In all cases it is assumed that K (maleic acid) = 9.7; K (fumaric acid) = 0.9.

It is important to see that the width of the Gaussian band increases as the square root of the number of transfers. σ is about one-fourth the width of a Gaussian band.

Figure 17.4 shows the effect of n on the height and breadth of countercurrent bands.

17.3.3 The Craig Countercurrent Apparatus

Much of the pioneering work in countercurrent separations was done by Lyman Craig and his associates in the 1940s and 1950s (Ref. *1*). Among their most important contributions was a glass apparatus which became the heart of a mechanized multiple extraction system. A model of the apparatus is shown in Fig. 17.5. As many as 1000 of these tubes were connected together to create a "Craig machine." The equilibration chamber of the Craig tube is labeled A in the figure and is about 12 inches long by 0.5 inch inner diameter. The two phases are loaded into the tube by the port labeled E. They are equilibrated by rocking the tube through an angle of about 35°. After equilibration the layers are allowed to separate, the tube is tilted clockwise to an angle of 90°, and the upper (less dense) phase pours off, through tube B, into the chamber labeled C. When the Craig tube is tilted back to the position in the figure, the upper phase (in chamber C) will flow through D to the next Craig tube connected in series, where it will meet fresh lower (denser) phase.

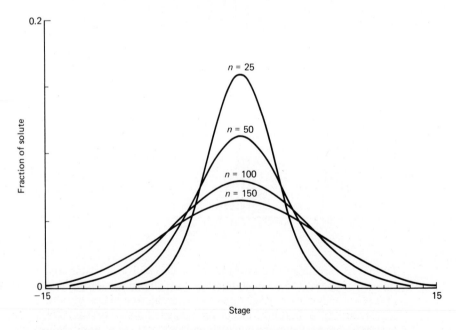

Figure 17.4 Effect of the number of stages on solute distribution. Stage numbers are normalized to the stage at which the fraction is a maximum for each experiment. See if you can measure the effect that doubling N will have on the standard deviation of a band.

17.4 SIMPLIFIED MODEL FOR COLUMN CHROMATOGRAPHY

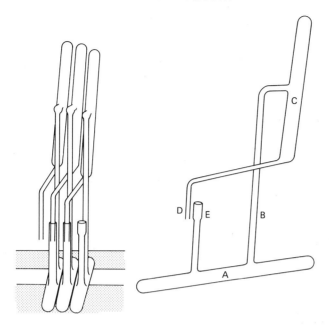

Figure 17.5 Craig countercurrent apparatus. (Reproduced from G. H. Morrison and H. Freiser, *Separations in Analytical Chemistry,* Interscience, New York, 1957 with the kind permission of G. H. Morrison.)

Some extremely difficult separations have been performed with Craig machines. For example, in a calibration test of a 220-tube device, Craig and co-workers separated a mixture of 10 amino acids (300-mg quantities of each), using water and *n*-butanol as solvents. A particularly challenging pair was leucine ($K = 0.223$) and isoleucine ($K = 0.190$), which required nearly 2800 transfers for resolution (Ref. *2*). It is important to realize that this work was done long before the invention of high-performance liquid chromatography and was really quite remarkable. A review written by Craig describes the flurry of activity in separation science that followed his invention (Ref. *3*). An article by Scheibel describes other mechanical separators devised in the same era (Ref. *4*).

17.4 SIMPLIFIED MODEL FOR COLUMN CHROMATOGRAPHY

It may help to understand chromatographic separations if we draw an analogy between chromatography and batch extractions. Visualize a chromatographic column as a long tube containing a series of separation stages. A stream of either gas or liquid (the "mobile phase") containing a narrow band of some solute is passed across the stages from left to right. At each stage equilibrium is achieved between the solute in the band

and solute in the stage. As the band traverses the series of stages it meets fresh material in new stages. Behind the band fresh mobile phase (without solute) is supplied. Solute will partition to the mobile phase from stages already passed by the solute band. The model corresponds to the line of separatory funnels we described in the discussion of countercurrent separations. We call the stages "theoretical plates," a name borrowed from distillation theory. The greater the number of theoretical plates, the easier it will be to separate two solutes with similar distribution coefficients.

This model is, of course, not very realistic. Real chromatography columns do not contain little plates or vanes or stages. They are packed with finely divided materials with active surfaces at which partitioning can occur. Since one phase moves past the other continuously and rapidly in a chromatographic column, it seems unreasonable to suppose that equilibrium states always govern the behavior of solutes. The idea of theoretical plates (TP) and the distance between them ("height equivalent to a theoretical plate" or HETP) is nevertheless still used in modern chromatography.

Let us consider a second model for chromatography, based on process rates, called the "random walk" model. We will pass a stream of mobile phase through a narrow tube packed with some material that will "sorb" solutes from the fluid stream.† We will add some solute to the mobile phase at the head of the column in a very narrow band. As the solute is swept along, it will be in either of two states, free or sorbed. Transfer from the free state to the sorbed state will have a particular rate, which need not be the same as the rate of the reverse transfer. In fact, when the rate of sorption is greater than the rate of desorption, the solute will move more slowly than the mobile phase. It should be possible to separate two solutes whose sorption rates are different.

An individual solute molecule will undergo many sorptions and desorptions as it moves through the column. Whenever it is free it moves at the speed of the mobile phase. Freedom corresponds to a forward step in a random walk. Whenever the molecule is sorbed, it cannot move, and effectively takes a step backward (relative to the bulk of material in the band) in the random walk. The result of steps forward and backward by molecules is that the narrow band spreads out. The variations in sorption and desorption rates produce spreading of the band: very slow desorptions, for example, cause the band to drag out (large steps backward), while very fast desorptions return solute to the mobile phase quickly (small steps backward). In the absence of some secondary factors which we will discuss later, the tendency is for a band to spread into a Gaussian shape. The width of the Gaussian band is expressed as a function of the standard deviation, σ (in units of length, e.g. cm). The standard deviation is proportional to the average distance between sorptions (l) and the square root of the number of sorptions (N),

$$\sigma = l * \sqrt{N}$$

† "Sorb" is a useful term here because it is so indefinite. It includes *adsorption*, a process which occurs at a surface of a phase, and *absorption*, a process which occurs in the body or the bulk of a phase.

The *variance* (σ^2) of the band is given by $l^2 * N$. The product lN is the distance over which the solute moves while in the column, and to a crude approximation is equal to the length of the column, L.† The average distance between sorptions can be thought of as the distance between theoretical plates (HETP, or simply "H"). We can then express σ in terms of H and L,

$$\sigma = \sqrt{H * L}$$

This result should look familiar; we saw its equivalent at the end of the last section. The width of a band increases as the square root of the distance the solute moves (the number of stages in the equilibrium extraction treatment).

17.5 CHROMATOGRAPHIC DATA

Now let us place a device at the end of the column to detect the band as it emerges from the column and plot the signal from the device versus time as in Fig. 17.6. What we see is a *peak* which appears Gaussian. The peak is characterized by a signal maximum which occurs at time t_r (measured from when the solute band was loaded on the column) and a width, W, now measured in seconds. The width is defined where tangents to the inflection points of the peak intersect the baseline. The inflection points of a Gaussian curve, incidentally, occur at a fraction 0.607 of the peak height. The width has a value of 4 standard deviations ($\pm 2\sigma$) at the base, and σ is sometimes called the "quarter peak width." The width at 60.7% of peak height is 2σ. At 50% of peak height the width is 2.354σ and is called the peak width at half height, W_h. The time elapsed between loading and the appearance of the peak maximum is called the *retention time* and is often expressed in seconds or minutes. If the mobile phase flow rate is known (in milliliters per second, for example), it is easy to convert the retention time to *retention volume*. This is the volume of mobile phase needed to elute the solute to the center of the band.

A second, much smaller peak has been drawn in Fig. 17.6 to represent the passage of a solute which is not sorbed by the column material. The time from loading to the peak time for this material corresponds to flow time of mobile phase through the column and is often called "dead time." The corresponding mobile phase volume is called "dead volume." As you will soon see, it is often to our advantage to subtract the dead time from retention times and study "corrected retention times," t'_r, or "corrected retention volumes," v'_r.

The number of theoretical plates "contained" in a column and the distance between them (H) are used as measures of efficiency. The plate height H in centimeters

---

† In a real column there will be motion in all directions relative to the stream flow. Cavities and channels in the column packing will divert some solute, as discussed later. The actual distance traveled by a molecule will be greater than the tube length.

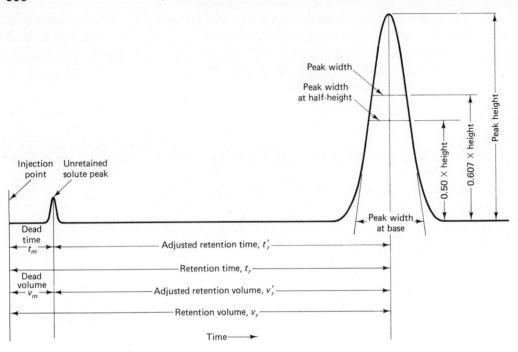

Figure 17.6 Chromatographic data.

is related to the variance of the band on the column (also in centimeters) and the length of the column, L:

$$H = \sigma^2/L$$

We can also express efficiency in terms of N, the number of theoretical plates,

$$N = L^2/\sigma^2$$

Most experimental chromatograms use time rather than distance as the x-axis variable. We will use the Greek letter τ (tau) to represent the standard deviation in units of time. The average rate at which molecules of solute travel is given by L/t_r, in centimeters per second. If we divide this average rate into the standard deviation in centimeters, we can then express τ in terms of σ:

$$\tau = \sigma/(L/t_r)$$

The width of a Gaussian curve in time units is approximately 4τ. Substituting this into the last equation and rearranging gives

$$\sigma = LW/4t_r$$

Since $H = \sigma^2/L$, we can write

$$H = LW^2/16 * t_r^2$$

Since $N = L/H$, we can also find N experimentally,

$$N = 16(t_r/W)^2$$

from two time measurements, t_r and W.

17.6 EFFICIENCY OF COLUMN SEPARATIONS

Given a mixture of two or more solutes, how good a separation can we achieve on a particular column? In section 17.3 we used the separation factor, α (alpha), the ratio of the distribution coefficients of the compounds, as a measure of how easy it is to separate two compounds. Remember that the best separations are obtained when one distribution coefficient is about equal to the inverse of the second distribution coefficient. In column chromatography the experimental separation factor for two compounds A and B is often defined as the ratio of their adjusted retention volumes,

$$\alpha = v'_{rB}/v'_{rA}$$

B is taken to be the compound more strongly retained by the column, forcing α always to be greater than unity. The corrected retention volume is calculated from the corrected retention time (t'_r) and the mobile phase flow rate, u,

$$v'_r = t'_r * u \quad (t'_r \text{ in sec, } u \text{ in mL/sec})$$

17.6.1 Resolution

Resolution (R) is another widely used experimental measure of separation efficiency in chromatography. It is the difference in retention times (or volumes) of two solute peaks, divided by the average of their baseline widths

$$R = (t_A - t_B)/[(W_A + W_B)/2] = 2*(t_A - t_B)/(W_A + W_B)$$

See Fig. 17.7. A resolution $R = 1$ provides a reasonably good separation of peaks. If the two peaks are the same width at baseline, then

$$R = (t_A - t_B)/4\sigma \quad \text{and} \quad t_A - t_B = 4\sigma$$

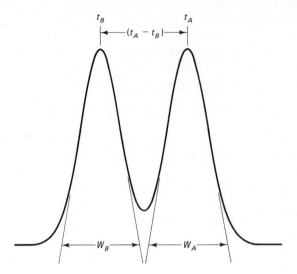

Figure 17.7 Graphical determination of peak resolution.

Only the tails of the peaks beyond 2σ overlap, a mixing of about 2% of one compound with the other (remember that $\pm 2\sigma$ is approximately 96% of a population). A column separation with R of 1.5 is essentially complete and is called a *baseline separation*. How much contamination is there when $R = 1.5$?

17.6.2 Effects of Column Capacity

In the discussion of crosscurrent separations we accounted for unequal phase volumes by modifying the distribution coefficient,

$$K' = K * V_w/V_o$$

where V_w and V_o are the volumes of the aqueous and organic phases, respectively. In column chromatography there is an analogous quantity, the *capacity factor*, which we define as

$$K' = K * V_s/V_m$$

where V_s is the volume of the stationary phase, and V_m is the volume of the mobile phase in the column.

While we will not derive it here, there is a simple experimental relationship between the retention times of the solute and an unretained solute, and K',

17.6 EFFICIENCY OF COLUMN SEPARATIONS

$$t_r = t_m * (1 + K') \quad \text{or} \quad K' = (t_r - t_m)/t_m$$

When the volume of the stationary phase is made larger, K' is made larger, and the retention time (t_r) increases relative to the dead time (t_m). We can increase the column capacity by using a more finely divided stationary phase. As particle size decreases, surface area increases dramatically (recall Chapter 9 on precipitation), and the amount of sorption increases. In situations where the stationary phase is coated with a material which absorbs solute, we can increase column capacity by making thicker layers of coating. Specific systems will be discussed in the next chapter.

The effect of changing the capacity factor of a column on the resolution of peaks can be examined by modifying the equation for resolution,

$$R = (t_B - t_A)/W_{avg}$$

We know that for a Gaussian curve $W_{avg} = 4*t_r/\sqrt{N}$. Therefore we can write

$$R = (t_B - t_A)\sqrt{N}/4*t_r$$

We also know that $t_r = t_m(1 + K')$ for each solute. We can incorporate this into the equation for R:

$$R = (K'_B - K'_A)\sqrt{N}/4*(1 + K'_B)$$

If we hold K'_A and N constant, some trial calculations show that R increases as K'_B increases. The price to be paid, though, is that the retention time also increases [remember that $t_r = t_m(1 + K')$]. If we want to achieve better resolution by increasing the capacity factor, we must wait longer for the analysis to be done.

17.6.3 Band Broadening in Real Chromatograms

From the ideas just presented it follows that any factor which causes a band to broaden will lower the efficiency of a separation and that we should find ways to control such factors. Chromatographic bands are broadened by the phenomena called *longitudinal diffusion* and *resistance to mass transfer.* The variables which affect the importance of these phenomena are mobile phase flow rate, temperature, size of the stationary phase particles, and thickness of the stationary phase liquid coating (if such a coating is present). The most general mathematical relationship between the physical phenomena and HETP is called the Van Deemter equation,†

---

† The original equation of van Deemter dates from the mid-1950s and does not look much like the equation given here. For an excellent discussion of the evolution of our understanding of the equation and the factors behind it, see the article by Hawkes (Ref. 5).

$$H = B/u + C_s u + C_m u$$

The B/u term is the contribution of longitudinal diffusion, $C_s u$ is the contribution of sluggish mass transfer to and from the stationary phase, and $C_m u$ is the contribution of complex mass transport in the mobile phase. The flow rate of mobile phase is represented by u. The equation was developed originally for gas chromatography, but it can also serve to characterize liquid chromatographic efficiency. An idealized plot of H vs. flow rate for one solute is shown in Fig. 17.8. The "check mark" shape is generally observed in gas chromatography and shows that there is a flow rate at which H is a minimum. At this point the number of theoretical plates (N) is a maximum and peak width a minimum. Two peaks are most easily resolved when the separation is performed at a flow rate near the plot minimum. At any flow rate the value of H can be broken down into contributions from the component terms. This is shown graphically in Fig. 17.8 by the dashed lines representing the B, C_s, and C_m terms. At low flow rates the B term dominates, while at high flow rates the C terms are most important. Let us examine the terms and phenomena they represent individually.

Longitudinal Diffusion (the B/u Term) This term takes into account the tendency of solute molecules to migrate away from the center of a solute band (where their concentration is high) to the edges of a band (where their concentration is low). In gas chromatography longitudinal diffusion is important when long columns are used because diffusion rates are very high in the gas phase. Solutes which are in a column for a long time are apt to diffuse away from the band and cause it to disperse. Short

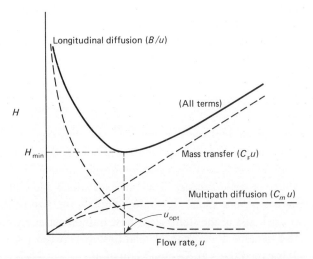

Figure 17.8 Van Deemter plot (applied to gas chromatography). Note: in liquid chromatography longitudinal diffusion is unimportant. As a result, a minimum in an LC van Deemter plot is seldom seen.

columns and high flow rates minimize the effect. Since the rate of diffusion of a molecule increases with temperature, decreasing a column's temperature may help keep peaks narrow. Diffusion rates in liquids are about 10,000 times slower than in gases. As a result, longitudinal diffusion is almost never an important factor in liquid chromatography.

Mass Transport Involving the Stationary Phase (the $C_s u$ Term) Bands are also broadened because solute molecules cannot be transferred between stationary and mobile phases instantaneously. At the leading edge of a quickly moving band, solute will be carried farther than it would be if sorption could occur instantaneously. At the tailing edge of a band, desorption may not be able to keep pace with the stream flow, and the band will stretch out ("tailing"). The $C_s u$ term applies to the stationary phase; if solute has difficulty moving in and out of the stationary phase, $C_s u$ will be large. Thin stationary phase coatings help speed mass transfer, but also cause low column capacity. Unless small samples are used with thinly coated columns, broad peaks will result from "overloading." Figure 17.9 shows peaks that are distorted by slow desorption and by overloading. Raising column temperature speeds mass transfer, but the gain in efficiency may be offset by increased peak broadening due to diffusion (in gas chromatography).

Mass Transfer in the Mobile Phase ($C_m u$ Term) This is the most complex term in the van Deemter equation and is still an object of debate among chromatographers. There are several contributing factors. One of the most important is called "multipath peak broadening." It is caused by the existence of a variety of flow paths available to solute molecules as they move through a column. Figure 17.10 shows how two dramatically different paths might exist in a column. Path 1 is much shorter than path 2, and at constant mobile phase flow rate a molecule on path 1 will move through the column more quickly. A molecule on path 2 will be caught in an indirect route and move through the column slowly. The result is broadening of the solute band. A second factor is the existence of pockets of solvent in columns packed with large particles; solute can be trapped in such pockets and leave only very slowly. Pockets can be minimized by using packing materials of small uniform size. At low flow rates solutes can migrate by diffusion between channels and pockets, and multipath broadening becomes less important, as indicated by the slope of the C_m line in Fig. 17.8. Columns in which the stationary phase coats the walls of a column and in which there is no packing material display negligible multipath broadening and are capable of extremely efficient separations.

17.7 QUALITATIVE AND QUANTITATIVE ANALYSIS

Chromatography has become the most important method for the separation of small amounts of very similar chemical compounds. When coupled with sensitive methods

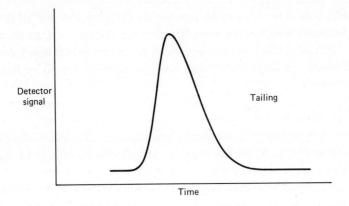
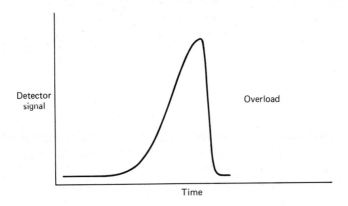

Figure 17.9 Distorted chromatographic peaks.

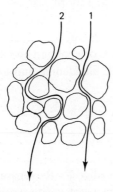

Figure 17.10 Mechanism for multipath transport.

of detection, chromatographic techniques are among the most powerful available to analytical chemists.

The qualitative information about a solute given by a chromatogram is its retention time in a column. It is possible to identify at least some compounds in simple mixtures by comparing measured retention times of unknowns and standards (knowns) under the same mobile phase/stationary phase conditions. Additional evidence can be obtained by comparing the behavior of knowns and unknowns with different mobile and stationary phases; if retention times of an unknown and known match under two sets of conditions, the chances are quite good that they are the same compound. The best strategy, however, is to collect the unknown compound as it elutes from the column, then confirm its identity by a spectroscopic technique. Several "hyphenated methods" have been developed which use chromatographic separation as a first step. Perhaps the most important is gas chromatography-mass spectrometry (GC-MS), in which separated solutes are passed on to a mass spectrometer for molecular weight and structural analysis. Gas chromatography-infrared spectrometry (GC-IR) is another important technique. In GC-IR infrared spectra are taken of solutes as they emerge from the column, and the spectra are used to analyze structure. Among the most powerful techniques are the "doubly hyphenated" GC-MS-MS (two mass spectrometers for very high resolution) and GC-IR-MS. The hyphenated methods rely on the tremendous technical advances made in computers and in Fourier transform IR spectrometry in the past decade. While the equipment is quite costly, analyses can now be performed routinely which would have been impossible only a few years ago (Ref. 6).

Quantitative chromatography is based on a comparison of peak heights or areas from an unknown sample with those of a known or standard sample. If the conditions which affect peak width are carefully controlled (i.e., temperature and flow rate), peak height should be directly proportional to concentration over a limited concentration range. The most important limitation, particularly in the analysis of compounds which elute quickly, is the rate at which samples are loaded at the top of the column. Samples are ordinarily injected by small-volume syringes (5 to 20 microliters), and variations in the rate of injecting samples can change peak widths. Slower injections produce broader peaks and may change the proportionality between peak height and concentration.

Peak areas are less influenced by temperature and flow rate than are peak heights and are more reliable quantitative measures. Unfortunately, it is much harder to measure areas than heights. Modern computerized equipment will integrate peaks electronically: voltage signals from a detector are measured at regular intervals (say 10 times per second) and are summed over the peak elution time. The computer must be programmed to read a baseline and begin to sum whenever the signal departs sufficiently from the baseline voltage. The summation ends when the signal returns to within a certain range of the baseline. If two peaks are so poorly resolved that the signal does not return to baseline, the computer must be programmed to decide how

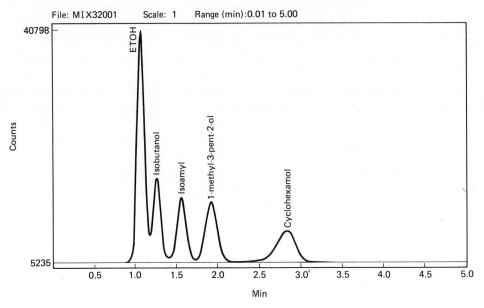

Figure 17.11 Computerized data treatment chromatogram of a mixture of alcohols. (Gas chromatogram from a Carle model 8500 gas chromatograph; data recorded and plotted with an IBM CS9000 laboratory computer.)

to separate peaks. An example is shown in the chromatogram in Fig. 17.11, which was obtained with the help of a laboratory computer.

Before microprocessors were available a variety of mechanical integrating devices were used in quantitative analysis. They were all less accurate and reproducible than electronic integrators. A surprisingly good and extremely inexpensive method is to use scissors to cut apart a recorded chromatogram and weigh the peaks on an analytical balance. The weight of a paper peak from an unknown can be compared with that from a standard, and composition can be estimated within a few percent. If peaks are symmetrical a "quick and dirty" method is to measure the area graphically by multiplying peak height by width at half-height. All methods of integration run into trouble when peaks are seriously malformed or when several peaks are unresolved. The problem then becomes one of finding the conditions under which the peaks are well enough formed to have their areas measured.

A logical approach to take in quantifying simple unknowns is to try to prepare a sample from pure compounds which mimics the chromatographic behavior of an unknown. One might begin by guessing the composition, preparing a composite known, and comparing peak areas of the known and unknown. The next step would be to prepare a second sample using a refined guess based on the first composite sample.

The limiting factor in this approach is the irreproducibility of small-volume syringes. It helps to fill a sample loop which can be switched into the mobile phase stream, rather than to rely on a 10-microliter syringe.

It appears that the best quantitative analyses are performed using internal standards and careful injection techniques. A carefully measured amount of a standard compound, different from anything in the original sample, is added to both the standard and the unknown. In the chromatogram the standard peak should lie close to the unknown peak but should be fully resolved. Since a detector may have a different sensitivity to the internal standard than to the compound to be analyzed, the ratio of peak areas in the standard solution is absolutely vital to the analysis.

REFERENCES

1. L. C. Craig and D. Craig, in *Techniques of Organic Chemistry* (A. Weissberger, ed.), vol. III, part I, 2nd ed., Interscience, New York, 1956.
2. L. C. Craig, W. Hausmann, E. H. Ahrens, and E. J. Harfest, *Analytical Chemistry*, 23, 1236 (1951).
3. L. C. Craig, *Analytical Chemistry*, 23, 29A (1951).
4. E. G. Scheibel, in *Techniques of Chemistry*, vol. XII, 3rd ed., (E. S. Perry and A. Weissberger, eds.), Wiley, New York, 1978, chapter III, section 4.
5. S. J. Hawkes, "Modernization of the van Deemter Equation for Chromatographic Zone Dispersion," *Journal of Chemical Education*, 60, 393–398 (1983).
6. T. Hirschfeld, "The Hy-phen-ated Methods," *Analytical Chemistry*, 52, 297A (1980).

RECOMMENDED READING

Borman, Stuart A., ed. *Instrumentation in Analytical Chemistry,* American Chemical Society, Washington, D.C., 1982. The section called "Hyphenated Methods" (pp. 271–293) is a good introduction to those methods. The book is a compilation of articles published in *Analytical Chemistry* which were aimed at the general analytical chemistry audience.

Hawkes, S. J. "Modernization of the van Deemter Equation for Chromatographic Zone Dispersion," *Journal of Chemical Education,* 60, 393–398 (1983). Although this is a challenging article for students, it is an excellent treatment of the phenomena leading to zone broadening and the history of van Deemter's ideas.

Karger, B. L., L. R. Snyder, and C. Horvath. *An Introduction to Separation Science,* Wiley, New York, 1973. This text combines fundamentals and applications for advanced undergraduates who have studied physical chemistry and calculus.

Miller, J. M. *Separation Methods in Chemical Analysis,* Wiley, New York, 1975. This text is aimed at advanced undergraduates and graduate students.

Morrison, G. H., and H. Freiser. *Solvent Extraction in Analytical Chemistry,* Wiley, New York, 1957. This was the first comprehensive book on solvent extraction and is suitable for undergraduates. It has become a classic in the field of solvent extractions and is also a good source of early literature citations.

Scott, R. P. W. *Contemporary Liquid Chromatography,* Techniques of Chemistry series, vol. XI, Wiley, New York, 1976. Chapter 2 contains an excellent rigorous discussion of the van Deemter equation and column efficiency.

PROBLEMS

Many of the following problems require the calculation of a factorial. The following BASIC code for calculating x! (called "XFACT") may be useful:

```
10 IF X=0 THEN XFACT=1: GOTO 13
11 M=1:FOR N=1 TO X:M=M*N:NEXT N
12 XFACT=M
13 REM on to procedure
```

17.1. Two solutes, A and B, partition between water and diethyl ether with distribution coefficients 5 and 0.5, respectively. A countercurrent separation procedure is carried out in 25 transfers. In each stage equal volumes of water and ether are used.

(a) Calculate the fractions of A and B in the water phase after one equilibration (i.e., the fractions in stage zero).

(b) In which stages will A and B have maximum values (r_{max} values)?

(c) Calculate the fractions of A and B in their respective r_{max} stages after 25 transfers.

(d) Calculate the fractions of A and B in the water phase after one equilibration if the initial volume of water is 10 mL and the initial volume of ether is 50 mL.

17.2. A monoprotic weak acid with $pK_a = 4.8$ has a distribution coefficient of 10 between 0.1F HCl (in water) and chloroform. Calculate the distribution ratio of the weak acid between an aqueous solution buffered at pH 6 and chloroform.

17.3. 20 mmole of a solute is equilibrated between 50 mL of deionized water and 50 mL of toluene; 2.45 mmole of solute is found in the water phase.

(a) Calculate the capacity factor for the solute between the two solvents.

(b) If 10 mL of water were used instead of 50 mL, what would be the distribution coefficient for the solute between the solvents?

(c) If 100 mL of water were used instead of 50 mL, what would be the distribution coefficient?

17.4. A neutral monobasic weak base with $pK_b = 3.60$ has a distribution coefficient of 20 between $0.1F$ NaOH and toluene. Calculate the distribution coefficients for this base between aqueous buffers at pH 8, 9, 10, 11, and 12, and tolune.

17.5. An anionic weak base, B^-, has a pK_b of 5.40. When 50 mL of pH 8.60 buffer containing a total of 20 mmole of the weak base and its neutral conjugate acid are equilibrated with an equal volume of chloroform, it is discovered that 0.24 mmole of HB are left in the aqueous buffer solution. What is the distribution coefficient of HB between the two solvents?

17.6. Two weak acids, HA and HB, have distribution coefficients of 10 and 5, respectively, when partitioned between aqueous $0.1F$ HCl and chloroform. HA has a pK_a of 3, while HB has a pK_a of 8.
(a) A 50-mL sample of $0.1F$ HCl contains 0.01 mole of HA and 0.01 mole of HB. Three extractions are performed into 50-mL portions of chloroform. How many moles of HA and HB are left in the aqueous acid phase after three equilibrations?
(b) Describe a simple modification you could make to accomplish the separation of the two acids in fewer steps.

17.7. Two solutes, A and B, have distribution coefficients of 8 and 0.1, respectively, between water and chloroform. If one starts with an equimolar mixture of A and B (say 0.001 mole of each), how many crosscurrent separations are needed to isolate a sample of A which is 99.9% pure? Assume 50 mL portions of both solvents are used in each equilibration.
(a) What is the concentration of A in the aqueous phase which contains the 99.9% pure A?
(b) What are the concentrations of A and B in the combined chloroform phases?

17.8. A gaussian-shaped chromatographic peak elutes from a 2.00-m chromatography column in 135 s (center of peak) and has a baseline width of 12 s.
(a) Calculate the widths of the peak at half height and at the inflection points.
(b) Calculate the height equivalent to a theoretical plate (H).
(c) Calculate the number of theoretical plates in the column.
(d) Calculate the number of theoretical plates required to produce a peak width of 6 s with the same retention time.

17.9. Two gaussian chromatographic bands elute in 193 s (solute A) and 217 s (solute B) from a 1.00-m column when the mobile phase flow rate is 0.23 mL/s. A nonretained solute elutes in 18 s.
(a) Calculate the corrected retention times, corrected retention volumes, and the experimental separation factor for the two solutes.
(b) If the baseline width of the first peak is 38 s and that of the second is 40 s,

calculate the resolution (R) of the peaks. Does the column provide a baseline separation?
- (c) Calculate H and N for each solute.
- (d) Calculate the capacity factor for each solute. If N values for the solutes stay the same, would increasing both capacity factors by, say, 10% improve the resolution?
- (e) Baseline separation of the peaks could be achieved if their baseline widths could be reduced to 24 s. How many theoretical plates would be needed to obtain such narrow peak widths? What changes would you make to increase N?

Chapter 18

Chromatographic Equipment

18.1 INTRODUCTION

In Chapter 17 the theory of chromatographic separations was developed in considerable detail. In this chapter we will use that theory as a basis for a brief survey of the equipment needed to perform column chromatographic separations. The discussion will focus on the two major varieties of chromatography, gas chromatography (GC) and high-performance liquid chromatography (HPLC), and the hardware needed to do them. Both techniques have become very popular for the separation and analysis of mixtures of compounds. As you will see, HPLC and GC equipment components are designed to achieve the same ends (namely separating mixtures), but are physically rather different. In GC a sample containing several solutes is injected into a high-temperature injection port and vaporized. The sample is carried along into a column by a stream of gas called the carrier gas. Once in the column, the solutes are separated by a mechanism such as adsorption on active column material or partition in a liquid coating on inert packing material. The gas stream carrying the separated components then passes through a detector, which is specially designed to respond to solute molecules in the gas phase but not to the carrier gas. The signal from the detector is recorded over time as a *chromatogram.*

From the point of view of an instrument operator, the separation procedure in HPLC is quite similar to that in GC: samples are injected into an instrument, pass through a column for separation, are detected by a device designed to respond to some physical property of the solutes, and chromatographic peaks are recorded. The major physical differences between HPLC and GC are consequences of the use of the liquid phase in HPLC and the gas phase in GC. High-pressure pumps supply solvent in

HPLC, while tanks of compressed gas supply carrier gas in GC. Furthermore, separations on columns in HPLC can make use of mechanisms other than adsorption and partitioning; ion exchange and the exclusion of solutes by size are two mechanisms which are widely used. Methods of detection are also quite different: HPLC detectors are designed to make use of physical properties characteristic of condensed phases, such as electrical conductivity, refractive index, and optical absorbance.

Let us begin with a more detailed look at GC equipment and then consider HPLC separately.

18.2 GAS CHROMATOGRAPHY

18.2.1 Equipment

As mentioned above, gas chromatography is often a useful method for the separation and analysis of mixtures of volatile compounds. A simple schematic diagram of a gas chromatograph appears in Fig. 18.1. The basic components are a supply of carrier gas, pressure regulators, an injection port, a column, and a detector. The injection port, column, and detector are usually housed in a thermostated oven whose temperature can be held quite steady or varied (see temperature regulation, below). The adjustable experimental variables in GC are the carrier gas flow rate, choice of column packing material, and temperature of the column.

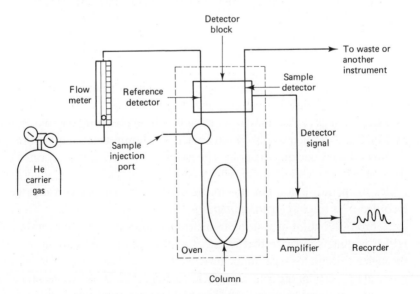

Figure 18.1 Schematic diagram of a basic gas chromatograph.

18.2 GAS CHROMATOGRAPHY

Mobile Phases (Carrier Gases) The gas used to carry the sample into and through the column must be chemically inert and must produce only a small background signal with a particular detector. Helium is the most common carrier gas, although nitrogen, hydrogen, and argon are also used. These gases can be purchased in pressurized cylinders and are available in high purity. Pressures at the injection port usually range from 10 to 50 psig (pounds per square inch, gauge pressure, that is, above atmospheric pressure). Such pressures provide a flow rate between 5 and 50 mL per minute, depending on the size of the particles which pack the column.

Flow rates can be measured with commercially available flow meters or (less expensively) with a soap bubble flow meter made from a spare side-arm buret, as shown in Fig. 18.2.

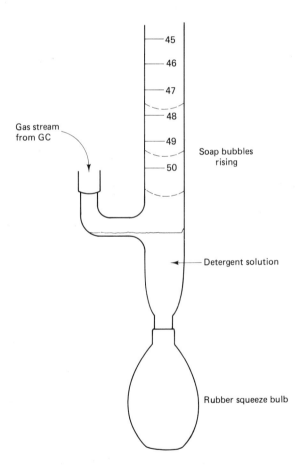

Figure 18.2 Soap bubble flow meter. A 50-mL side-arm buret is used for this flow meter. The rubber bulb is squeezed to raise the level of the detergent solution. The gas stream will pick up a bubble and raise it in the buret. Flow rate is measured by measuring the time it takes to traverse, say, 10 mL.

Injector Design Samples are most often introduced with a microsyringe through a self-sealing rubber disk called a septum. The injection port is designed so that the sample is swept quickly into the column. It is usually heated to help volatilize liquids or solids that have been injected. Figure 18.3 shows a simplified injection port. Some injectors use switching valves and sample loops; sample is injected into a sample loop, which is switched into the gas stream when the sample is to be loaded onto the column.

It is important that the sample be neither too large for the capacity of the column (see Chapter 17) nor too small to be detected. For most routine laboratory work with analytical columns, samples of 1 to 10 μL will suffice. Capillary columns (see below) require even smaller samples, and special devices called sample splitters are often needed to load perhaps 1% of a 1-μL sample onto the column. The remaining 99% is lost as waste.

Samples should be injected and vaporized quickly so that they are introduced as a "slug" of vapor at the top of the column. Slow injection spreads a sample out in the column and leads to poor separations.

Column Designs Columns are long tubes of metal (copper, stainless steel, or aluminum) or glass. The two major column types are packed columns and open tubes (OT) or capillary columns. Packed columns are commonly available in lengths of 2 to 5 m (wrapped in coils 15 to 20 cm in diameter), although some difficult separations

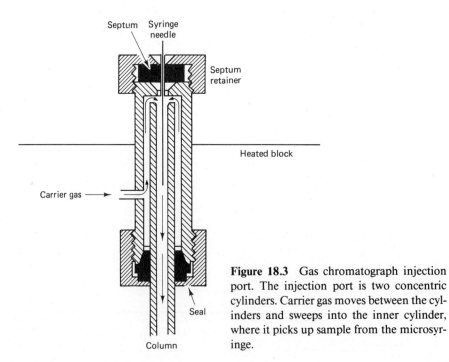

Figure 18.3 Gas chromatograph injection port. The injection port is two concentric cylinders. Carrier gas moves between the cylinders and sweeps into the inner cylinder, where it picks up sample from the microsyringe.

may require lengths up to 20 m. Common inside diameters are 2 to 4 mm. Columns used for gas-liquid chromatography (GLC) are packed with an inert solid material (called a "support") which is coated with a liquid stationary phase. For gas-solid chromatography (GSC) a solid stationary phase is used without a liquid coating. The support material should be in the form of very small particles in order to expose a very large surface area to the mobile phase. The best supports are small uniform spheres which are mechanically strong and chemically inert at high temperatures. The most useful particle sizes are in the ranges of 60 to 80 mesh (0.25 to 0.17 mm) and 80 to 100 mesh (0.17 to 0.15 mm). Such small particles may provide active areas of 1 to 4 m^2 per gram of support. Columns packed with particles much smaller than 100 mesh, however, may be less efficient. To achieve reasonable flow rates with very small particles it is necessary to use very high carrier gas pressures, and so small-particle packings are seldom used. A uniform particle size produces very orderly packing and minimizes multipath effects. This improves column efficiency (see the van Deemter treatment in Chapter 17). A well-packed GC column should provide 500 to 1000 theoretical plates per meter of column length.

The earliest but still most popular supports for GC are made from diatomaceous earth, a natural material consisting of the skeletons of tiny marine organisms called diatoms. There are several ways to grind, screen, and pretreat diatomaceous earths, and three grades (with different surface activities) are sold under the trade names of Chromosorb P, W, and G. Other solids used as supports include Teflon microbeads, glass microbeads, molecular sieves, and graphitized carbon. The latter materials have extremely high active surface areas, 100 to 1000 m^2/g. You should consult the literature if you are interested in the differences in their properties.

Open tube columns are not packed with a support. Rather, their inside surface is coated with a thin layer of stationary phase. They are therefore often called wall-coated open tubule (WCOT) columns. The WCOT columns are commonly quite long (10 to 150 m, wrapped in coils) and have very narrow inner diameters (0.2 to 0.5 mm). The stationary phase coating is usually only a fraction of a micrometer thick and therefore has a very small volume. This limited phase volume means that the capacities of WCOTs are low and so only very small quantities of materials can be separated. The absence of packing and the use of very thin stationary phases leads to very great column efficiency by eliminating multipaths and speeding stationary phase transport (again, see Chapter 17). Capillary columns of 0.25 mm inner diameter provide 1000 to 4000 theoretical plates per meter, roughly a factor of 4 more than packed columns. Columns 100 m long with 300,000 or more theoretical plates are commercially available.

Stationary Phase Liquids Several hundred different liquids have been used as the stationary phase in gas-liquid chromatography. The choice of the right stationary phase liquid is critical to many separations and presents a real challenge. Remember from the last chapter that the quality of a separation of two solutes depends on both the size of and the difference between their partition coefficients in the stationary phase. Large partition coefficients lead to long retention times; even if two solutes are separated, the time required to wait for them to emerge may be prohibitively long. On the other hand, even rather large differences between two small partition coefficients may make the separation of two solutes impossible. How then can we predict retention? The best predictor of retention of a solute on a column is how similar its polarity is to that of the stationary phase. For example, nonpolar molecules like simple hydrocarbons will be retained on low-polarity stationary phase materials like squalene (a large alkane, $C_{30}H_{62}$) and poly(methyl siloxane)($\cdots-O-Si(CH_3)_2-O-\cdots$). Slightly polar molecules (e.g., esters, ethers) will not be retained as well by squalene, but will be retained by a slightly polar material such as the polyethylene glycol compound called Carbowax. More polar halogenated compounds and alcohols will be retained by polar stationary phases such as poly(cyanopropylsiloxane).

The elution order of a series of solutes of similar polarity usually follows the order of their normal boiling points. Therefore we would expect n-pentane (bp 36.1°C) to elute before n-hexane (bp 68.7°C) and n-hexane to elute before n-heptane (bp 98.4°C). If two solutes with the same boiling point but different polarities are to be separated, the stationary phase material must have a stronger affinity for one than the other, either because of dipole interaction or another mechanism such as hydrogen bonding.

A second factor to consider in selecting a stationary phase is its volatility at the temperatures required to do the separation. For example, say that a mixture of hydrocarbons must be separated. Both squalene and OV-1 [a poly(methylsiloxane)] are available. Squalene volatilizes ("bleeds off") support material at temperatures above about 150°C and will be picked up by the detector as a large background signal. On the other hand, OV-1 can be used up to 350°C without volatilizing. Squalene would be a fine stationary phase for separating low-boiling hydrocarbons, while OV-1 would be the necessary choice for high-boiling hydrocarbons.

The thickness of stationary phase coating on the support can also have an effect on the usefulness of a column. Thick coatings produce columns with high capacities to hold solute. But remember from the last chapter that thick coatings may cause sluggish mass transfer and lower column efficiency (the C_s term in the van Deemter equation is increased by inefficient mass transfer). Therefore, it may be harder to separate solutes if the column can hold larger samples. As mentioned above, capillary columns use very thin coatings to maximize efficiency. However, they have a very low capacity for solutes.

Hundreds of stationary phases are available commercially. Table 18.1 lists a few of the more popular ones, along with their relative polarities, their maximum operating temperatures, and some classes of compounds they might be used to separate.

Temperature Regulation As a rule, the temperature of a column should be a few degrees warmer than the average boiling point of the solutes in a mixture which is to

TABLE 18.1 SOME STATIONARY PHASES FOR GAS CHROMATOGRAPHY

Trade name	Composition	Polarity	Separations	Maximum temperature (°C)
Squalene	$C_{30}H_{62}$	Low	Hydrocarbons	150
OV-1	Poly(methylsiloxane)	Low	Low-polarity solutes (general)	350
OV-17	Methylphenylsiloxane	Medium	General	325
Poly-A	Polyamide	Medium	Nitrogen compounds	250
DEG-adipate	Diethylene glycol adipate	Medium	Esters, fatty acid esters	200
Carbowax 20M	Polyethylene glycol	High	Alcohols	250
OV-210, QF-1	Trifluoropropylsilicone	High	Halogenated compounds, ketones	275

be separated. Low temperatures result in longer retention times and longer analyses. High temperatures shorten retention times, but may also prevent the separation of low-boiling solutes. When a sample is known to have solutes with a wide range of boiling points it may be necessary to change the temperature during the elution. This is done by starting at a low temperature (allow the separation of low-boiling solutes) and raising the temperature at a rate of several degrees per minute to speed the elution of the high-boiling solutes. Figure 18.4 shows how dramatic an effect programming temperature can have on the resolution of a mixture of solutes.

Detectors Devices for detecting solutes as they emerge from a GC column must be quite sensitive and quick to respond. Bear in mind that the quantities of solute in the carrier gas are quite small, perhaps only a few parts per million, and that they pass through a detector in only a few seconds. Even though the detector response must be quick, it must also be stable and reproducible. It should also respond linearly to solute concentration over a wide range of concentrations. While no single device meets all of these criteria well, several designs are quite useful. The most important are the thermal conductivity, flame ionization, and electron capture detectors.

Thermal Conductivity Detector (TCD) This detector consists of an electrically heated element in the stream of carrier gas at the end of the column. The element may be a platinum or tungsten wire or a thermistor (a semiconducting device). The current to the element is controlled carefully, and the temperature of the element is determined by how well the carrier gas can carry heat away from it. Hydrogen and helium are very good conductors of heat and make very good carrier gases for TCD detectors. They are far better thermal conductors than organic compounds being separated on the column. When even small amounts of these solutes pass by the detector, its temperature rises and its electrical resistance changes. Thermal conductivity detectors are normally used in pairs, one in the gas stream before the injector and one in the stream after the column. Pairing allows one detector to serve as a reference for the

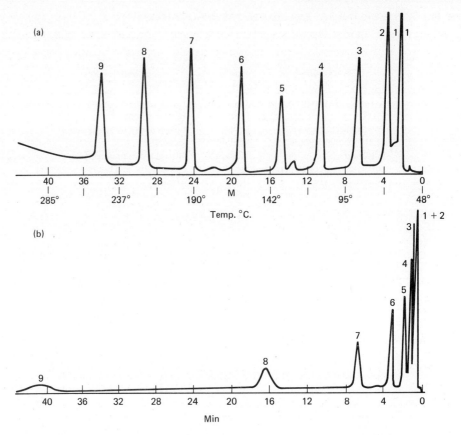

Figure 18.4 Programmed temperature gas chromatography. (a) Programmed temperature chromatogram of nine alcohols: (1) methanol, (2) ethanol, (3) 1-propanol, (4) 1-butanol, (5) 1-pentanol, (6) cyclohexanol, (7) 1-octanol, (8) 1-decanol, and (9) 1-dodecanol. Temperature 48 to 245°C at 6°C per minute. (b) Constant-temperature chromatogram at 165°C of the same alcohol mixture. From [S. Dal Nogare and C. E. Bennett, *Analytical Chemistry*, 30, 1157–1158 (1958); reproduced with the permission of the American Chemical Society.]

other, and this minimizes the effects of changes in gas flow rate, temperature, and electric current. The two TCDs are wired into a simple resistance bridge circuit.

The TCD has a number of advantages over other detectors. It has a simple design, is inexpensive, and provides a very wide range of linear response to solute concentration (five orders of magnitude). One of its chief advantages is that it does not destroy solutes as it detects them. This means that after separation the solutes can be collected for other studies. The major disadvantage of the TCD is that it is not very sensitive. The detection limit of the TCD is about 10^{-8} g/mL of carrier gas, about 10^3 times less sensitive than other designs.

Flame Ionization Detector (FID)   The FID uses a small hydrogen-oxygen flame to burn sample solutes as they emerge from a GC column. Most carbon-containing

18.2 GAS CHROMATOGRAPHY

compounds produce charged molecular fragments when burned, and these fragments allow flames to conduct electricity. A high voltage (about 300 V) is applied across electrodes close to the flame, and the ions formed in the flame cause a current to flow between the electrodes. The current is very small (10^{-9} ampere or less) and must be amplified before it can be recorded.

The main advantage of an FID is its great sensitivity. The main disadvantage is that solutes are destroyed during detection, unless the gas stream is split just before the detector. A number of compounds cannot be detected with an FID, notably those containing large numbers of oxygen, sulfur, and halogen atoms. These elements tend to react with carbon radicals in a flame and lower the signal. This is not always a disadvantage, however. Water cannot be detected and thus would not interfere when aqueous solutions are injected into the instrument.

Figure 18.5 shows a cutaway view of a Varian Instruments FID.

Electron Capture Detector (ECD) The ECD is a rather specialized detector which has become important in the analysis of chlorinated compounds such as pesticides and polychlorinated biphenyls (PCBs). The carrier gas stream from the GC column flows through the chamber of an ECD, where it is subjected to a flux of electrons from a radioactive source (a beta particle source). Nickel-63 foil and titanium foil containing adsorbed tritium (^{3}H) make excellent beta sources. The beta electrons ionize the carrier gas molecules, which release electrons in a 100-V electric field. The resulting current is amplified as a background level current. A solute species which can capture electrons from the gas stream in the chamber will lower the background

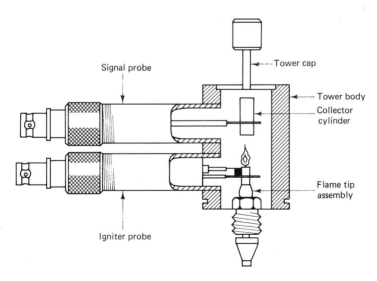

Figure 18.5 Cutaway view of a flame ionization detector. (Courtesy of Varian Associates, Inc., Instrument Group, Sunnyvale, California.)

current, and the current decrease will be recorded. The ECD is very sensitive to solutes which have halogens, peroxides, and nitro groups, because these groups absorb electrons efficiently. The ECD is thus a very selective detector. This is not a serious limitation, however, because compounds which do not contain such groups can be reacted with molecules containing them and thereby determined with an ECD detector. For example, trifluoroacetic acid reacts with amino acids, alcohols, and phenols to form derivatives which can be detected with the ECD.

While the ECD is much more sensitive than the FID (the detection limit of the ECD is about 10^{-12} g/mL of carrier gas), it has the distinct disadvantage that its response is linear over a very narrow concentration range. This makes quantitative analysis with the ECD more challenging than with the FID or TCD.

18.2.2 Applications

Gas chromatography has become a very important tool for qualitative and semi-quantitative analysis of a vast number of complex mixtures. The applications literature is growing very rapidly as more and more problems are discovered in environmental contamination, occupational safety, drug and alcohol abuse, and industrial process monitoring.

Figure 18.6 shows five gas chromatograms. Chromatograms (a) to (c) show the separations of chlorinated compounds which are often found in water supplies contaminated with hazardous waste. Many of the haloethylene compounds in chromatogram (a) have been used as solvents for grease and inks for several decades and have been disposed of improperly in sanitary landfills. Notice that the column is a 2-mm inner diameter (ID) glass column, 10 feet long, and packed with Chromosorb W coated with OV-351, a low-polarity siloxane. Chromatogram (b) shows the separation of a mixture of chlorinated phenols on an uncoated polymeric adsorbant packing called Tenax. A number of the chlorinated phenols in chromatogram (b) are in the waste streams of pesticide producers and have also ended up in ground water. A similar mixture is separated on a WCOT column coated with OV-351 in chromatogram (c). Even though the time scales on chromatograms (b) and (c) are different, it should be apparent that the peaks in chromatogram (c) have no tails and are much sharper. Notice also that peaks 3 and 4 in both chromatograms correspond to the same two compounds. Which column affords a better separation?

Chromatogram (d) shows the separation of six nitrosoamines. Some nitrosoamines are known to be cancer-causing agents, and there is evidence that they are associated with nitrate additives in foods. A WCOT column coated with Carbowax 20M is used in this separation. Notice also that the column temperature is programmed.

The analysis of drugs and mixtures of drugs is important in the criminal justice system, and GC has played an increasingly important role in drug identification. Chromatogram (e) shows the separation of five barbiturates on a WCOT coated with

GB-1, a compound analogous to the low-polarity siloxane OV-1. Without the high efficiency of the capillary column, amobarbital and pentobarbital (compounds 2 and 3) would probably be inseparable.

18.3 HIGH-PERFORMANCE LIQUID CHROMATOGRAPHY

18.3.1 Equipment

Until the late 1960s liquid chromatography relied on the gravity feed of solvent through columns which were 1 cm or more wide and 10 to 500 cm long. In order to achieve solvent flow rates of more than a few milliliters per hour, particle sizes were kept rather large, 0.2 mm or larger. Smaller particle sizes were known to produce more efficient columns, but the flow of solvent became intolerably slow in such columns. The development of small-particle technology along with improved high-pressure solvent pumps made it possible in the 1970s to achieve separations of the same efficiency as those in GC while staying in the liquid phase. With particles only a few micrometers in diameter it is necessary to apply pump pressures of several thousand pounds per square inch (psi) to achieve flow rates of a few milliliters per minute. Clearly, to operate under pressures like this, pumps, columns, and their associated plumbing must be quite well engineered and manufactured. As a consequence, HPLC components also tend to be rather expensive. Figure 18.7 shows in schematic form the important components of a high-performance liquid chromatograph. We will discuss the components in detail.

Solvent Reservoirs, Purification, and Delivery An HPLC system must have a reservoir of solvent. Solvents must be free of dust and other solids and must contain only low concentrations of dissolved gases. Dust and solids cause moving parts in the high-pressure pump to wear out prematurely. Solvents should be filtered through micrometer filter pads before being drawn into an HPLC system. Dust filters called "metal stones" are also used in solvent uptake lines to filter out dust. Dissolved gases may produce bubbles wherever there is a sudden pressure drop in the system. The worst places for pressure drops are where the solvent stream expands, that is, at the top of the column and in the detector. A gas bubble in the column can partially block solvent flow and drastically reduce column efficiency. Bubbles in the detector can cause large pulses to appear in the output and distort or obscure chromatographic peaks. Bubbling the solvent for 15 minutes with helium (a process called "sparging") will displace most of the dissolved air. Helium is much less soluble than air in most solvents and is less apt to create bubbles. Alternatively, heating a solvent to boiling and then cooling in a sealed vessel is often adequate protection.

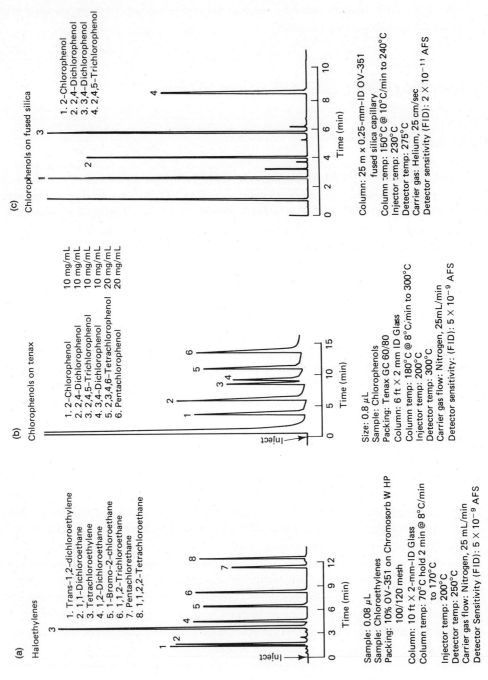

Figure 18.6 Gas chromatography applications. (Chromatograms courtesy of the Foxboro Company, North Haven, Connecticut.)

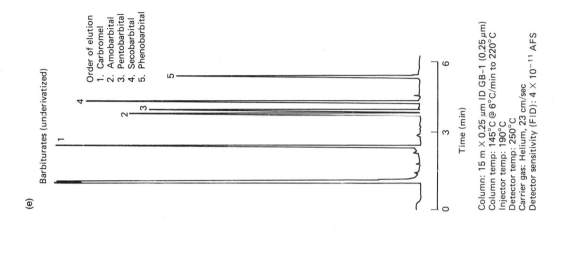

Figure 18.6 (*continued*)

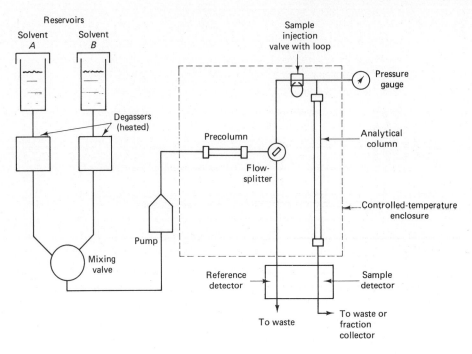

Figure 18.7 Schematic diagram of an HPLC system with gradient elution capabilities.

For many HPLC separations, elution with a single solvent will give adequate separation. Single-solvent separations are called *isocratic elutions*. In some situations two or more solvents with different polarities are blended by a technique called *gradient elution*. For example, at the beginning of a separation, a pure solvent (solvent 1) with low polarity is pumped into a system. A second solvent (solvent 2), with higher polarity, is added in increasingly large quantities until after perhaps 10 or 15 minutes the solvent stream is pure solvent 2. In the early part of the separation low-polarity solutes are swept along quickly, while late in the separation higher-polarity solutes are swept along more quickly. The process is the HPLC counterpart of programmed temperature GC, described in an earlier section.

Pumps Pumps for HPLC must be able to generate pressures of 6000 psi at flow rates of 0.1 to 10 mL per minute, reproducible to within a few tenths of 1%. Furthermore, they must be essentially pulse-free. Both reciprocating piston and screw-driven piston designs are popular, but detailed descriptions are beyond the scope of this text.

Injection Systems Injecting samples with a microsyringe directly into a solvent stream at a pressure of several thousand psi can be quite challenging. In "stop-flow" methods, solvent pressure is relieved momentarily while sample is injected into a port at the head of a column. Pressure is then reapplied to move the sample through the column. A simpler and more reproducible method is to have a short length of tubing called a sample loop attached to a high-pressure valve, so that the loop can be switched in and out of the solvent stream. When the loop is out of the stream it can be filled at low pressure with a syringe. When the valve is turned the loop is switched into the solvent stream (high pressure), and the sample is swept in a "slug" onto the top of the column. Sample loops are available from 0.2 µL to 2.0 mL in volume. Figure 18.8 shows a schematic representation of a sample valve and loop.

Detectors The most popular detectors for HPLC are based on the interactions of solutes with light. One design makes use of changes in the refractive index of the solvent caused by the presence of solute. It comes close to being a universal detector in the sense that any solute with a refractive index different from that of the solvent will be detectable. Unfortunately, it is one of the least sensitive detectors, responding to about 10^{-7} g of solute per milliliter. This kind of detector cannot be used in gradient elution studies because the refractive index of the solvent stream changes constantly during elution.

A more widely used detector is based on the absorption of ultraviolet or visible light. Simple photometers making use of fixed wavelengths (for example, the 254 or 280 nm spectral lines of a mercury lamp) are inexpensive and popular. Many organic molecules, especially those containing aromatic rings, absorb light at these wavelengths. More expensive and versatile spectrophotometric detectors are available which allow the analyst to change the detecting wavelength. Photometric detectors have detection limits of about 10^{-10} g/mL, and can be used in gradient elution studies if none of the solvents absorbs light at the detector wavelength.

Electrochemical detectors are becoming more and more popular. A voltage is

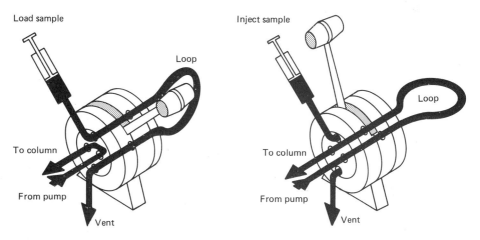

Figure 18.8 HPLC sample valve and loop. (Courtesy of Beckman Instruments, Inc., Fullerton, California.)

applied between a pair of electrodes in the solvent stream, and solutes which can be reduced or oxidized at the applied voltage will do so, causing a small current to flow. The current can be amplified and recorded with time in the form of a chromatogram. Detection limits are about 10^{-12} g/mL.

Columns Columns used in HPLC are usually made of stainless steel and are 4 mm to 1 cm in inner diameter and 5 to about 30 cm long. Columns are packed with much smaller particles than are used in GC, typically 3 to 10 μm in diameter. It is this small particle size that requires the use of such high pressures to achieve flow rates of milliliters per minute. Well-packed columns may have 50,000 plates per meter, a better ratio than capillary GC columns. Remember, though, that HPLC columns are never longer than 30 cm, while GC capillary columns may be 100 m or longer. Pumps cannot provide enough pressure to move solvent through longer packed columns.

Common packing materials are silica particles, alumina, porous polymer microbeads, and ion exchange resins. Silica and alumina are often coated with thin films of organic materials which may be bonded either physically or chemically to the surface.

18.3.2 Separation Mechanisms and Applications

Several important separation mechanisms are used in HPLC: partition, adsorption, ion exchange, and size exclusion. Partition is the mechanism most widely used and is discussed here in some detail.

High-Performance Partition Chromatography (HPPC) Separations based on this mechanism are usually divided into two categories, liquid-liquid and bonded-phase chromatography. In liquid-liquid chromatography, an organic phase is simply adsorbed (low-energy interactions) to the packing material. In bonded-phase chromatography, there are actual chemical bonds holding the liquid phase to the packing surface. Developments during the past decade in the chemistry of bonding phases to supports has led to their becoming extremely popular.

Bonded-phase packings can be prepared by reacting organochlorosilane compounds with hydroxide groups on the surface of silica particles. For example, in hot, slightly acidic solution a reaction such as the following can be performed:

$$-\underset{|}{\text{Si}}-\text{OH} + \text{Cl}-\underset{\underset{\text{CH}_3}{|}}{\overset{\overset{\text{CH}_3}{|}}{\text{Si}}}-(\text{C}_8\text{H}_{17}) \rightarrow -\underset{|}{\text{Si}}-\text{O}-\underset{\underset{\text{CH}_3}{|}}{\overset{\overset{\text{CH}_3}{|}}{\text{Si}}}-(\text{C}_8\text{H}_{17}) + \text{HCl}$$

The long-chain hydrocarbon (an octyl group) is therefore bonded to the silica particle by the Si—O—Si bond. This provides a very stable nonpolar coating for the silica particle. Other organic groups which have been bonded to silica are octadecyl ($-C_{18}H_{37}$), propylamine, aromatic hydrocarbons, ethers, and nitriles (cyanopropyl). These materials all have different polarities, making bonded-phase packings quite versatile.

Depending on the relative polarities of the stationary phase and the solvent stream, it is possible to classify partition chromatography as being either *normal phase* or *reversed phase* chromatography. In the early days of chromatography it was the usual practice to work with rather polar stationary phases (e.g., triethylene glycol) and low-polarity solvents such as ethers or hexane. Traditionally this has been called normal phase partition chromatography. In more recent times low-polarity stationary phases (such as the octadecylsilyl bonded phase) have been used with rather polar solvents such as methanol, acetonitrile, and water. This is called reversed phase partition chromatography. In normal phase separations, low-polarity solutes elute first and high-polarity solutes elute last. In reversed phase separations the order is just the opposite. Figure 18.9 shows the separation of four aflatoxins on a normal phase (silica) and on a reversed phase (octadecyldimethylsilyl bonded phase) column. What can you conclude about the relative polarities of these aflatoxin species? Aflatoxins are extremely poisonous natural products, and their determination in foods is of great importance.

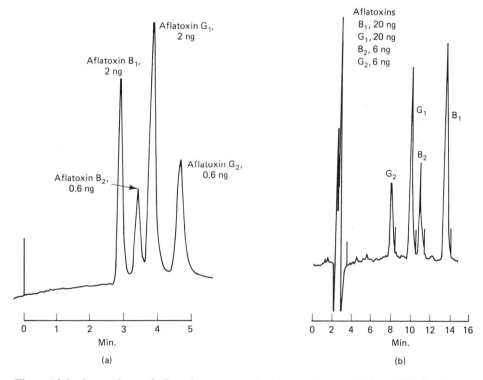

Figure 18.9 Separations of aflatoxins on normal phase and reversed phase HPLC columns. (Courtesy of Supelco, Inc., Bellefonte, Pennsylvania.)

Some generalizations are possible when predicting elution order from normal and reversed phase columns. The following order of increasing polarity should be a guide:

aliphatic hydrocarbons < olefins < aromatics < halides < ethers
    < nitro compounds < esters = aldehydes = ketones < alcohols
            = amines < sulfoxides < amides < carboxylic acids

Remember that polarities are determined not only by the nature of a substituent but also by its position in a molecule.

Choosing a combination of stationary phase and solvent (or mixture of solvents) to separate a group of solutes is seldom very easy. Complex solute-stationary phase and solute-mobile phase interactions must be considered. One strategy for selecting columns and solvents is called the GKS strategy.† With this strategy, most solute mixtures can be separated by using either of two columns (one normal phase, one reversed phase) and eight solvents. For the reversed phase the solvents are (in order of increasing polarity): tetrahydrofuran, acetonitrile, methanol, and water. For the normal phase the solvents are (in order of increasing polarity): n-hexane, chloroform, dichloromethane, and diethyl ether.

Using some simple mixtures and mixture adjustments, it is possible to arrive at an optimum combination after only a few experiments. The details of the selection process and the theory of solvent strength are better left to texts in instrumental methods of analysis. DuPont Analytical Instruments uses the GKS strategy in promoting a sophisticated multisolvent gradient elution HPLC system.

Some of the most difficult separations are those of optical isomers. It is possible to make an chirality-selective bonded-phase column by bonding an optically active molecule to a silica support. Figure 18.10 shows the separation of three D/L amino acid enantiomers on a Supelcosil LC-(R)-Urea column.

High-Performance Adsorption Chromatography The earliest work in chromatography involved studies of the adsorption of solutes on finely divided polar material such as alumina and silica. Polar solutes adsorb strongly on these surfaces and are removed by polar solvent molecules, which compete with the solute for surface sites. Since the stationary phase composition is fixed, variations in solvent polarity and column temperature are the only experimental variables which affect retention time. The most important applications of adsorption chromatography are in the separation of rather nonpolar solutes which differ in the number and position of their functional groups. A typical example is shown in Fig. 18.11, in which a mixture of the three isomers of nitroaniline is separated on a silica column.

---

† Named for J. L. Glajch, J. J. Kirkland, and L. R. Snyder, who developed the method in a series of papers appearing in the *Journal of Chromatography* and the *Journal of Chromatographic Science* between 1978 and 1982.

18.3 HIGH-PERFORMANCE LIQUID CHROMATOGRAPHY

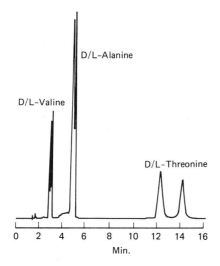

Figure 18.10 (a) Chiral bonded phase and (b) resolution of several pairs of D/L-PTH-amino acid enantiomers in one analysis. (Courtesy of Supelco, Inc., Bellefonte, Pennsylvania.)

High-Performance Ion Exchange Chromatography Ion exchange separations are based on equilibria between ions in solution and ions of similar sign attached to insoluble solids in contact with the solution. A number of natural ion exchange materials, such as zeolites and clays, have been used for decades to purify and soften water. A number of synthetic ion exchange materials ("ion exchange resins") have been created from synthetic polymers. Both anion and cation exchange resins are available in large-bead form or packed as microbeads into HPLC columns. The so-called strong acid exchangers are polystyrene and divinylbenzene copolymers containing sulfonic acid groups ("RSO_3H"). When the cation exchange resin is in contact with metal ions in solution, an equilibrium can be established between free metal ion and metal ion bound to the resin:

$$(RSO_3H)(solid) + M^+(solution) = (RSO_3M)(solid) + H^+(solution)$$

The position of this equilibrium depends on the relative concentrations of metal ions and hydronium ions in the mobile phase, as well as the inherent affinity of metal ions

NITROANILLINE ISOMERS

Operating Conditions
Column: IBM's normal phase guard 3.2 × 50 mm
Mobile phase: 50/25/25 heaxane/methylene chloride/THF
Flow rate: 8 mL/min
Detection: 254 nm

Peak identification

1. *o*-Nitroaniline
2. *m*-Nitroaniline
3. *p*-Nitroaniline

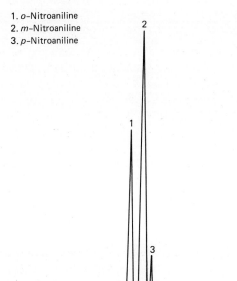

Figure 18.11 Adsorption chromatography (silica column). (Courtesy of IBM Instruments, Danbury, Connecticut.)

for the sulfonic acid material. As you might predict, the affinity of cations for the resin increases with cation charge, $M^+ < M^{2+} < M^{3+}$. For any particular positive change, it is much more difficult to predict the order of affinity, however. Among the dipositive cations the general order of affinity is

$$Mg^{2+} < Zn^{2+} < Cu^{2+} < Cd^{2+} < Ni^{2+} < Ca^{2+} < Sr^{2+} < Pb^{2+}$$

while for singly positive cations it is

$$Li^+ < H^+ < Na^+ < NH_4^+ < K^+ < Rb^+ < Cs^+ < Ag^+ < Tl^+$$

Weak acid resins are also available. Their exchange sites are carboxylic acid groups or phosphonic acid groups [$-PO(OH)_2$] rather than sulfonic acid groups.

18.3 HIGH-PERFORMANCE LIQUID CHROMATOGRAPHY

Anion exchange resins are also available. One of the so-called strong base anion exchange resins has trimethylamine groups bonded to polystyrene-divinylbenzene copolymers. The exchange equilibrium can be represented by

$$R-N(CH_3)_3^+OH^- + X^- = R-N(CH_3)_3^+X^- + OH^-$$

where the resin is understood to be in the solid phase and the anion and hydroxide are in the liquid phase. Strong anion exchangers show the following affinity order:

$$F^- < OH^- < CH_3COO^- < Cl^- < Br^- < NO_3^- < I^- < C_2O_4^{2-} < SO_4^{2-}$$

An example of a high-performance ion exchange chromatographic separation is shown in Fig. 18.12, in which a mixture of seven dipositive transition metals is separated. The signal recorded is an electrical conductance signal (units of microsiemens, or micro reciprocal ohms). The mobile phase stream passes through a detector, which measures the electrical conductance of the solution between two small platinum electrodes. An alternative method of detection is to react ions after they elute from the column with a complexing agent which forms colored complexes and then measure their optical absorbance in the visible region of the spectrum. Details can be found through the bibliography at the end of this chapter.

High-Performance Size Exclusion Chromatography (HPSEC) HPSEC is a new and potentially very powerful liquid chromatographic technique, particularly as applied to the separation of large molecules such as proteins. Column packings are porous silica or polymer particles which have networks of channels of various sizes. Solvent molecules and solutes of only certain sizes can migrate into these channels and are thus removed temporarily from the mobile phase. Molecules which are much smaller than the pore channels can move farthest into the packing material and will migrate out only slowly. They are thus retained on the column for a long time and will be the last eluted. Larger solute molecules will not be able to migrate into the packing as far and will move through the column as rapidly as the mobile phase. As a result, separation occurs according to the size and (to a lesser extent) shape of molecules. Chemical interactions are neither important nor very desirable in size exclusion chromatography.

A number of size exclusion packings are available. Some are hydrophilic polymers (polyacrylamides and polyvinyl alcohol gels) and are suitable for determining water-soluble molecules such as carbohydrates, starches, and polysaccharides. Many people refer to chromatography based on aqueous solvents as *gel filtration*. Other packings are hydrophobic (polystyrene gels) and are used in conjunction with nonaqueous solvents in what have been called *gel permeation* experiments. Polymer molecular weight standards are available for the calibration of retention times. Figure 18.13a shows an HPSEC separation of a mixture of proteins (and one dipeptide, no. 5).

HPSEC can also be used for a quick determination of molecular weights of polymers. Elution times can be calibrated with a series of standard compounds that

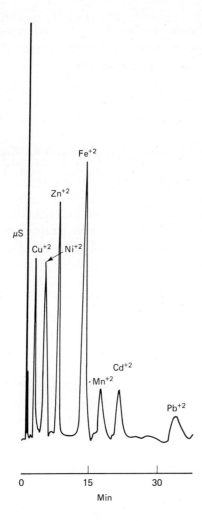

Figure 18.12 High-performance ion exchange chromatogram. (Courtesy of Interaction Chemicals, Mountain View, California.)

have similar chemical behavior. Figure 18.13b shows the chromatogram of a polystyrene molecular weight standard. If the chemical behavior of a sample differs from that of a standard (e.g., it might adsorb on the packing), serious misjudgment of the molecular weight can result.

18.4 COMPARISON OF THE CHROMATOGRAPHIC TECHNIQUES

As a summary to this chapter, let us compare GC and HPLC. Both methods can be made extremely efficient and highly selective. Both are applicable to the analysis of a wide variety of materials. When materials are thermally stable and volatile, GC offers the advantages of low cost and speed in preparing to do the work. Remember that

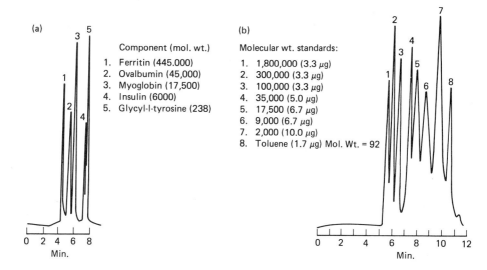

Figure 18.13 High-performance size exclusion chromatogram. (Courtesy of Supelco, Inc., Bellefonte, Pennsylvania.)

HPLC equipment is rather expensive and that solvent preparation (mixing, filtering, and degassing expensive solvents) can take a great deal of time. Both methods use very small samples; microliter and nanogram quantities are often sufficient. Both methods can be nondestructive, meaning that samples can be passed on to other analytical methods (for example, mass spectrometry).

HPLC has some advantage over GC in terms of versatility. Ion exchange and size exclusion methods open doors to trace-level inorganic analysis and the analysis of macromolecules. On the other hand, the resolving power of capillary columns in GC cannot be matched by HPLC.

RECOMMENDED READING

General Treatments

Karger, B. L., L. R. Snyder, and C. Horvath. *An Introduction to Separation Science.* Wiley, New York, 1973. More detailed and quantitative than the following book. A classic in the field.

Miller, J. N. *Separation Methods in Chemical Analysis.* Wiley, New York, 1975. A good, readable introduction to the subject.

Gas Chromatography

Lee, M. L., F. J. Yang, and K. D. Bartie, *Open Tubular Column Gas Chromatography.* Wiley, New York, 1984.

Perry, J. A. *Introduction to Analytical Gas Chromatography.* Dekker, New York, 1981.

High-Performance Liquid Chromatography

Krstulovic, A. M., and P. R. Brown. *Reversed Phase High Performance Liquid Chromatography.* Wiley, New York, 1982.

Snyder, L. R., and J. J. Kirkland. *Introduction to High Performance Liquid Chromatography,* 2nd ed. Chapman & Hall, New York, 1982.

Ion Exchange Chromatography

Smith, F. C., and R. C. Chang. *The Practice of Ion Chromatography.* Wiley, New York, 1983.

Size Exclusion Chromatography

Yao, W. W., J. J. Kirkland, and D. D. Bly. *Modern Size Exclusion Liquid Chromatography.* Wiley, New York, 1979.

PROBLEMS

18.1. Define the following terms used in chromatography:
- (a) partition
- (b) adsorption
- (c) gas-liquid chromatography
- (d) gas-solid chromatography
- (e) septum
- (f) WCOT
- (g) TCD
- (h) FID
- (i) ECD
- (j) isocratic elution
- (k) gradient elution
- (l) stop-flow
- (m) bonded phase
- (n) reversed phase
- (o) strong acid exchangers
- (p) HPSEC
- (q) gel filtration
- (r) gel permeation

18.2. Explain as carefully as possible the differences in the design and operation of a TCD and an ECD. Why is helium used as a carrier gas in TCD instruments, while nitrogen is used in ECD instruments?

PROBLEMS

18.3. Capillaries present some advantages over packed columns for gas chromatography. At the same time there are some disadvantages. Explain.

18.4. A sample of petroleum contains both very low boiling and very high boiling components. None of the components is a very polar molecule.
   (a) If the minimum boiling point is about 35°C and the maximum boiling point is about 200°C, at about what temperature would you set the thermostated oven for the column? Explain.
   (b) If you were to set the column temperature to 200°C, speculate on the effect on the separation of low boiling compounds.
   (c) If you could program the oven temperature to change with time, how might you program it for the petroleum sample?
   (d) Using the stationary phase materials in Table 18.1, select an appropriate material for the petroleum sample.

18.5. An analyst is concerned with the routine analysis of mixtures of chlorinated and unchlorinated organic compounds at very low concentrations. Both qualitative and quantitative analyses are needed. Gas chromatography with ECD and/or FID detectors is being considered.
   (a) In what ways would the ECD be superior to the FID?
   (b) In what ways would the FID be superior to the ECD?
   (c) Instruments are available which use both detectors in the same gas stream and send their signals to different recorder pens. How would you design a system with both detectors in one stream? What benefits would there be?

18.6. A silica column is used for adsorption chromatography. Compare the retention times of a moderately polar molecule such as nitrophenol with dioxane as a solvent as opposed to methyl ethyl ketone as a solvent (see Chapter 7 for more information on these solvents).

18.7. In reversed phase HPLC a compound is found to have too long a retention time when pure methanol is used as a solvent. Which solvents would you consider adding to methanol to decrease the retention time?

18.8. In normal phase HPLC a compound is found to have too long a retention time when a 1:1 mixture of chloroform and dichloromethane is used as a solvent. How might you alter the mixture to decrease the retention time? Explain.

18.9. In HPLC it is possible to lower retention times by changing solvent polarity. How might you change a solvent used in ion exchange chromatography to lower the retention times of metal ions?

18.10. In what order will the following species elute from a normal phase HPLC column?
   (a) toluene, benzaldehyde, benzoic acid
   (b) chlorobenzene, benzoic acid, phenol
   (c) dibromomethane, dimethyl sulfoxide, dimethyl ether

18.11. In what order should the following species elute from a reversed phase HPLC column?
  (a) 1-pentene, methyl ethyl ketone, 1-pentanone
  (b) 1,1,1-trichloro propane, 1,2,3-trichloropropane

18.12. In the text it is said that gradient elution in HPLC is analogous to programmed temperature in GC. Explore this analogy using some of the concepts developed quantitatively in Chapter 17.

18.13. How might changing the *anion* in the mobile phase in ion exchange chromatography influence the retention time and elution order of metal cations?

18.14. Think of the surface of a reversed phase octadecylsilyl packing in a low-polarity solvent: long carbon chains wave about rather freely in the low-polarity environment. What do you think will happen to this environment and the appearance of the octadecylsilyl coating when a more waterlike solvent is used? How might the polarity of the solvent influence the partition coefficient of a solute in this surface?

Appendix I

Weak Acid Dissociation Constants

Weak acid[c]	Structure	GFW	pK_{a1}	pK_{a2}	pK_{a3}
Acetic	CH$_3$COOH	60.05	4.76		
Adipic	HOOC(CH$_2$)$_4$—COOH	146.14	4.409	5.296	
Alanine (prot. cation)	CH$_3$C(NH$_3^+$)(H)—COOH	90.09	2.34	9.69	
Aminobenzoic (3-)	(benzene ring with COOH and NH$_2$)	137.13	3.124	4.744	
Aminobenzoic (2-)		137.13	2.108	4.946	
Aminobenzoic (4-)		137.13	2.413	4.853	
Aminopyridinium (2-)	(pyridinium ring with NH$_2$)	95.11	6.71[a]		
Aminopyridinium (3-)		95.11	6.03[a]		
Aminopyridinium (4-)		95.11	9.114		
Ammonium	NH$_4^+$	18.04	9.245		
Anilinium (prot. cation)	C$_6$H$_5$—NH$_3^+$	94.12	4.596		

Weak acid[c]	Structure	GFW	pK_{a1}	pK_{a2}	pK_{a3}
Anisic (3-)	(3-methoxybenzoic acid structure)	152.14	4.088		
Anisic (2-)		152.14	4.094		
Anisic (4-)		152.14	4.492		
Arginine (prot. cation)	(structure)	175.20	2.17	9.04	12.48
Arsenic	$(HO)_3As=O$	141.93	2.30	7.08	11.5[a]
Arsenious	$(HO)_3As$	125.93	9.22		
Ascorbic	(structure)	176.12	4.30	11.82	
Aspartic (prot. cation)	$HOOC(CH_2)\overset{NH_3^+}{C}H-COOH$	134.10	1.99	3.90	10.00
Barbituric	(structure)	128.09	4.00[a]		
Benzilic	(structure)	228.24	3.24		
Benzoic	(C$_6$H$_5$)COOH	122.12	4.121		
Boric	$B(OH)_3$	61.84	9.234		
Bromic	$HOBrO_2$	128.92	str.[b]		
Bromoacetic	$BrCH_2COOH$	138.96	2.903		
Bromobenzoic (3-)	(structure)	201.03	3.81		
Bromobenzoic (2-)		201.03	2.854		
Bromobenzoic (4-)		201.03	4.002		

WEAK ACID DISSOCIATION CONSTANTS

Weak acid[c]	Structure	GFW	pK_{a1}	pK_{a2}	pK_{a3}
Bromophenol (2-)	(2-bromophenol structure)	173.02	8.425		
Butylammonium (n-)	$CH_3(CH_2)_2CH_2-NH_3^+$	74.14	10.60		
Butyric (n-)	$CH_3(CH_2)_2COOH$	88.10	4.817		
Carbonic	H_2CO_3	62.03	6.35	10.33	
Catechol	(catechol structure)	110.11	9.449		
Chloroacetic	$ClCH_2COOH$	94.50	2.866		
Chlorobenzoic (3-)	(3-chlorobenzoic acid structure)	156.57	3.824		
Chlorobenzoic (2-)		156.57	2.943		
Chlorobenzoic (4-)		156.57	3.986		
Chlorophenol (3-)	(3-chlorophenol structure)	128.57	9.023		
Chlorophenol (2-)		128.57	8.48		
Chlorophenol (4-)		128.57	9.38		
Chlorous	$HOClO$	68.46	2.0[a]		
Chromic	$(HO)_2CrO_2$	117.97	-1[a]	6.49	
Cinnamic (cis)	(cis-cinnamic acid structure)	148.15	3.88		
Cinnamic (trans)		148.15	4.44		
Citric	$HOOC-CH_2-\underset{\underset{COOH}{\vert}}{\overset{\overset{OH}{\vert}}{C}}-CH_2-COOH$	192.12	3.128	4.761	6.396
Cresol (3-)	(3-cresol structure)	108.13	10.09		
Cresol (2-)		108.13	10.29		
Cresol (4-)		108.13	10.26		
Crotonic (α)	(crotonic acid structure)	86.09	4.698		
Cyanic (HOCN)	$HOC\equiv N$	43.03	3.92		

Weak acid[c]	Structure	GFW	pK_{a1}	pK_{a2}	pK_{a3}
Cyanoacetic	N≡C—CH$_2$—COOH	85.06	2.469		
Cyanobenzoic (3-)	(3-cyanobenzoic acid structure)	147.13	3.598		
Cyanobenzoic (4-)		147.13	3.551		
Cysteine (prot. cation)	HS—CH$_2$CH(NH$_3^+$)—COOH	122.16	1.96	8.18	10.28
Cystine (prot. cation)	HOOC—C(NH$_3^+$)—CH$_2$—S—S—CH$_2$C(NH$_2$)—COOH	241.30	1.65	2.26	7.85
Dichloroacetic	Cl$_2$CHCOOH	128.95	1.30[a]		
Dichlorobenzoic (2, 4)	(2,4-dichlorobenzoic acid structure)	191.01	2.76		
Dichlorobenzoic (2, 6)		191.01	1.82		
Dichlorobenzoic (3, 4)					
Dichlorophenol (2, 4)	(2,4-dichlorophenol structure)	191.01	3.68		
		163.00	7.850		
Diethanolammonium	(HOCH$_2$CH$_2$)$_2$—NH$_2^+$	106.14	8.883		
Diethylacetic	(CH$_3$CH$_2$)$_2$CHCOOH	116.16	4.736		
Diethylammonium	(CH$_3$CH$_2$)$_2$NH$_2^+$	74.14	10.93		
Diphenylacetic	(C$_6$H$_5$)$_2$CHCOOH	212.24	3.939		
Ephedrine (prot. cation)	(ephedrine structure)	166.23	9.54		
Ethylammonium	CH$_3$CH$_2$—NH$_3^+$	46.08	10.63		
Ethylenediammonium	H$_3^+$N—CH$_2$—CH$_2$—NH$_3^+$	62.10	6.85	9.93	
Ethylmalonic	CH$_3$CH$_2$CH(COOH)$_2$	132.12	2.96	5.90	
Fluoroacetic	FCH$_2$COOH	78.04	2.586		

WEAK ACID DISSOCIATION CONSTANTS

Weak acid[c]	Structure	GFW	pK_{a1}	pK_{a2}	pK_{a3}
Formic	HCOOH	46.02	3.752		
Fumaric	HOOC−CH=CH−COOH (trans)	116.07	3.019	4.384	
Glutamic (prot. cation)	HOOC−(CH$_2$)$_2$−CH(NH$_3^+$)−COOH	147.13	2.30	4.28	9.67
Glutaric	HOOC−(CH$_2$)$_3$−COOH	132.11	4.343	5.272	
Glycine (prot. cation)	$^+$H$_3$NCH$_2$COOH	76.07	2.350	9.78	
Glycylglycine (prot. cation)	H$_3^+$NCH$_2$C(=O)−N(H)−CH$_2$COOH	133.12	3.148	8.252	
Hydrazine	H$_2$N−NH$_2$	32.05	8.07[a]		
Hydrobromic	HBr	80.91	str.[b]		
Hydrochloric	HCl	36.45	str.[b]		
Hydrocyanic (HCN)	HCN	27.03	9.40		
Hydrofluoric	HF	20.01	3.14[a]		
Hydrogen peroxide	HOOH	34.02	11.60		
Hydrogen sulfide	H$_2$S	34.08	7.2[a]	14.0	
Hydroiodic	HI	127.91	str.[b]		
Hydroquinone	HO−C$_6$H$_4$−OH	110.11	10.0[a]		
Hydroxylammonium	NH$_2$OH$_2^+$	34.03	5.96		
Hypobromous	HOBr	96.92	8.68		
Hypochlorous	HOCl	52.47	7.25		
Hypoiodous	HOI	143.91	10.6		
Iodic	HOIO$_2$	175.93	0.78[a]		
Iodoacetic	ICH$_2$COOH	185.96	3.175		
Iso-leucine (prot. cation)	CH$_3$CH$_2$CH(CH$_3$)−CH(NH$_3^+$)−COOH	132.17	2.318	9.758	
Lactic	CH$_3$−CH(OH)−COOH	90.08	3.862		

Weak acid[c]	Structure	GFW	pK_{a1}	pK_{a2}	pK_{a3}
Leucine (prot. cation)	(CH$_3$)$_2$CHCH$_2$C(NH$_3^+$)(H)COOH	132.17	2.328	9.744	
Maleic	HOOCC(H)=C(H)COOH (cis)	116.07	1.921	6.225	
Malonic	HOOCCH$_2$COOH	104.06	2.855	5.696	
Mandelic	C$_6$H$_5$—C(OH)(H)—COOH	152.14	3.411		
Methoxyacetic	CH$_3$OCH$_2$COOH	90.08	3.571		
Nicotinic (prot. ion)	pyridinium-COOH	124.11	2.07	4.81	
Nitric	HONO$_2$	63.02	str.[b]		
Nitrobenzoic (3-)	3-NO$_2$-C$_6$H$_4$-COOH	167.12	3.450		
Nitrobenzoic (2-)		167.12	2.170		
Nitrobenzoic (4-)		167.12	3.442		
Nitrophenol (3-)	3-NO$_2$-C$_6$H$_4$-OH	139.11	8.399		
Nitrophenol (2-)		139.11	7.234		
Nitrophenol (4-)		139.11	7.149		
Nitrous	HONO	47.02	3.3[a]		
Oxalacetic	HOOC—C(=O)—CH$_2$COOH	132.07	2.555	4.370	
Oxalic	HOOC—COOH	90.04	1.271	4.266	
Periodic	(HO)$_5$I=O	227.96	1.64[a]		
Phenol	C$_6$H$_5$OH	94.11	9.998		
Phenoxyacetic	C$_6$H$_5$—O—CH$_2$COOH	152.14	3.171		
Phenylacetic	C$_6$H$_5$—CH$_2$COOH	136.14	4.312		
Phosphoric	(HO)$_3$P=O	98.00	2.148	7.198	12.38
Phosphorous	(HO)$_2$P(H)=O	82.00	1.80	6.16	

WEAK ACID DISSOCIATION CONSTANTS

Weak acid[c]	Structure	GFW	pK_{a1}	pK_{a2}	pK_{a3}
Phthalic (3-)	(benzene-1,2-dicarboxylic acid structure)	166.13	3.70	4.60	
Phthalic (2-)		166.13	2.950	5.408	
Phthalic (4-)		166.13	3.54	4.46	
Picolinic (prot. cation)	(pyridine-NH$^+$ with COOH)	124.11	1.01	5.32	
Picric	O$_2$N-(C$_6$H$_2$)(NO$_2$)$_2$-OH	229.11	0.29		
Pimelic	HOOC—(CH$_2$)$_5$—COOH	160.17	4.509	5.312	
Piperidinium	(piperidine-NH$_2^+$)	86.15	11.12		
Proline (1-) (prot. cation)	(pyrrolidine-COOH, NH$_2^+$)	116.13	1.952	10.64	
Propionic	CH$_3$CH$_2$COOH	74.08	4.874		
Pyridinium	(pyridine-NH$^+$)	80.10	5.22		
Pyrophosphoric	(HO)$_2$P—O—P(OH)$_2$ (with =O on each P)	177.99	0.85	1.96	6.68
Pyruvic	CH$_3$C(=O)—COOH	88.06	2.49		
Quinolinium	(quinoline-NH$^+$)	130.15	4.882		
Salicylic	(phenol-COOH, ortho-OH)	138.12	2.98	13.4[a]	
Serine (prot. cation)	HOCH$_2$—CH(NH$_3^+$)—COOH	106.09	2.186	9.208	
Silicic	(HO)$_2$Si=O (polymeric)		9.7[a]		
Suberic	HOOC—(CH$_2$)$_6$—COOH	174.19	2.524	5.327	
Succinic	HOOC—(CH$_2$)$_2$—COOH	118.09	4.207	5.635	
Sulfamic	H$_2$N—SO$_2$—OH	97.10	0.99		

Weak acid[c]	Structure	GFW	pK_{a1}	pK_{a2}	pK_{a3}
Sulfanilic	H$_2$N—C$_6$H$_4$—SO$_3$H	173.84	3.227		
Sulfuric	(HO)$_2$SO$_2$	98.08	str.[b]	1.921	
Sulfurous	(HO)$_2$SO	82.08	1.764	7.192	
Tartaric	HOOC—CH(OH)—CH(OH)—COOH	150.09	3.036	4.366	
Taurine (prot. cation)	$^+$H$_3$N—(CH$_2$)$_2$—SO$_3$H	126.14	9.061		
THAM (prot. cation)	(HOCH$_2$)$_3$CNH$_3^+$	125.17	8.075		
Thiocyanic	HSCN	59.09	str.[b]		
Threonine (dl-) (prot.)	CH$_3$CH(OH)—CH(NH$_3^+$)—COOH	120.12	2.088	9.100	
Toluic (3-)	C$_6$H$_4$(CH$_3$)COOH	136.14	4.272		
Toluic (2-)		136.14	3.908		
Toluic (4-)		136.14	4.373		
Toluidinium (3-)	C$_6$H$_4$(CH$_3$)NH$_3^+$	108.15	4.683		
Toluidinium (2-)		108.15	4.394		
Toluidinium (4-)		108.15	5.091		
Trichloroacetic	Cl$_3$CCOOH	163.40	0.70[a]		
Triethanolammonium	H$^+$N(CH$_2$CH$_2$OH)$_3$	150.19	7.762		
Triethylammonium	H$^+$N(CH$_2$CH$_3$)$_3$	102.19	10.72		
Trifluoroacetic	F$_3$CCOOH	114.03	0.23[a]		
Trimethylacetic	(CH$_3$)$_3$C—COOH	102.13	5.03		
Trimethylammonium	H$^+$N(CH$_3$)$_3$	60.11	9.80		
Tyrosine (prot. cation)	HO—C$_6$H$_4$—CH$_2$CH(NH$_3^+$)COOH	182.19	2.20	9.11	10.07
Valeric (n-)	CH$_3$(CH$_2$)$_3$COOH	102.13	4.84		

WEAK ACID DISSOCIATION CONSTANTS

Weak acid[c]	Structure	GFW	pK_{a1}	pK_{a2}	pK_{a3}
Valine (prot. cation)	$(CH_3)_2CH-\underset{H}{\overset{NH_3^+}{C}}COOH$	118.15	2.286	9.719	
Vanillin (3-)	HO—C₆H₃(OCH₃)—CHO	152.14	8.889		

[a] Means formal pKa, not thermodynamic.
[b] "str." means strong acid, negative pKa.
[c] Numbers in parentheses are positions on an aromatic ring

In disubstituted aromatic compounds, 1,2 (or 1,6) substituents are *ortho* to each other, 1,3 substituents are *meta* to each other, and 1,4 substituents are *para* to each other.

Appendix II

Solubility Product Constants Listed by Anions

Thermodynamic Values, 25°C

Arsenates
- Ag_3AsO_4 6.2×10^{-21}
- $Ba_3(AsO_4)_2$ 7.7×10^{-51}
- $Cd_3(AsO_4)_2$ 2.2×10^{-33}
- $Cu_3(AsO_4)_2$ 7.6×10^{-36}
- $Pb_3(AsO_4)_2$ 4.1×10^{-36}

Bromates
- $AgBrO_3$ 5.4×10^{-5}
- $Pb(BrO_3)_2$ 2.9×10^{-4}

Bromides
- $AgBr$ 5.0×10^{-13}
- $CuBr$ 5.9×10^{-9}
- Hg_2Br_2 6.24×10^{-23}
- $HgBr_2$ 8×10^{-20}
- $PbBr_2$ 6.2×10^{-6}
- $TlBr$ 3.65×10^{-6}

Carbonates
- Ag_2CO_3 7.7×10^{-12}
- $BaCO_3$ 2.6×10^{-9}
- $CaCO_3$ 6.9×10^{-9}
- Li_2CO_3 1.1×10^{-3}
- $MgCO_3$ 8×10^{-6}
- $SrCO_3$ 5.7×10^{-10}

Chlorides
- $AgCl$ 1.77×10^{-10}
- $CuCl$ 1.8×10^{-7}
- Hg_2Cl_2 1.43×10^{-18}
- $HgCl_2$ 6.0×10^{-14}
- $PbCl_2$ 1.6×10^{-5}
- $TlCl$ 1.87×10^{-4}

Chromates
- Ag_2CrO_4 1.6×10^{-12}
- $BaCrO_4$ 1.1×10^{-10}
- Hg_2CrO_4 2.0×10^{-9}
- $PbCrO_4$ 3.3×10^{-18}

Cyanides
- $AgCN$ 1×10^{-15}
- $Hg_2(CN)_2$ 5.0×10^{-40}

Fluorides
- BaF_2 1.0×10^{-6}
- CaF_2 5×10^{-11}
- MgF_2 6.5×10^{-9}
- PbF_2 3.7×10^{-8}
- SrF_2 2.5×10^{-9}

Hydroxides

AgOH	2×10^{-8}
$Al(OH)_3$	2.20×10^{-32}
$Cd(OH)_2$	5.9×10^{-15}
$Cr(OH)_3$	7×10^{-31}
$Cu(OH)_2$	4.2×10^{-20}
$Fe(OH)_2$	5.4×10^{-15}
$Fe(OH)_3$	3.5×10^{-37}
$Mg(OH)_2$	8.9×10^{-12}
$Pb(OH)_2$	4.4×10^{-15}
$Zn(OH)_2$	4.5×10^{-17}

Iodates

$AgIO_3$	5.0×10^{-7}
$Ba(IO_3)_2$	3.9×10^{-9}
$Ca(IO_3)_2$	6.1×10^{-6}
$Cu(IO_3)_2$	5.6×10^{-7}
$Pb(IO_3)_2$	3.2×10^{-13}
$Hg_2(IO_3)_2$	2.7×10^{-14}
$Hg(IO_3)_2$	5×10^{-14}
$Sr(IO_3)_2$	1.1×10^{-7}
$TlIO_3$	2.6×10^{-6}

Iodides

AgI	8.7×10^{-17}
BiI_3	8.1×10^{-19}
CuI	1.4×10^{-12}
Hg_2I_2	5.2×10^{-29}
PbI_2	7.9×10^{-9}
TlI	9.4×10^{-8}

Oxalates

$Ag_2C_2O_4$	1.1×10^{-11}
BaC_2O_4[a]	1.5×10^{-8}
CaC_2O_4[a]	2.3×10^{-9}
MgC_2O_4[b]	5.5×10^{-6}
ZnC_2O_4[b]	1.5×10^{-9}

Phosphates

$Ag_3(PO_4)$	2.3×10^{-18}
$AlPO_4$	3.7×10^{-41}
$Ca_3(PO_4)_2$	1.2×10^{-29}
$Pb_3(PO_4)_2$	8.1×10^{-47}
$Zn_3(PO_4)_2$	5.1×10^{-36}

Sulfates

Ag_2SO_4	1.2×10^{-5}
$BaSO_4$	1.9×10^{-10}
$CaSO_4$	2.4×10^{-6}
$Hg_2(SO_4)$	6.5×10^{-7}
$PbSO_4$	1.3×10^{-8}
$SrSO_4$	3.4×10^{-7}

Sulfides

Ag_2S	1×10^{-50}
Bi_2S_3	1×10^{-102}
CdS	1.1×10^{-29}
CoS	5×10^{-22}
CuS	1.1×10^{-36}
FeS	1×10^{-17}
PbS	6.7×10^{-29}
HgS	1.8×10^{-52}
NiS	1×10^{-20}
SnS	4.9×10^{-30}
SnS_2	6×10^{-63}
Tl_2S	6.6×10^{-21}
ZnS	9.1×10^{-22}

Thiocyanates

AgSCN	1.0×10^{-12}
CuSCN	9.3×10^{-15}
$Hg_2(SCN)_2$	1.9×10^{-20}

Most of these values have been calculated from standard free energies cited in *Standard Potentials in Aqueous Solution*, A. Bard, R. Parsons, J. Jordan, Eds. Marcel Dekker, New York, 1985. A few were found in *The Handbook of Analytical Chemistry*, L. Meites, Ed., McGraw-Hill, New York, 1963. (NOTE: Superscript *a*: monohydrate salt, $MC_2O_4 \cdot H_2O$; *b*: dihydrate salt, $MC_2O_4 \cdot 2H_2O$.)

Appendix III

Formation Constants for Metal Ion Complexes

Log K_f Values Listed by Ligand

Acetate
Ag	0.4, −0.2 (var)
Al	
Ba	0.4 (0.2)
Ca	0.5 (0.2)
Cd	1.0, 0.9, −0.1, −0.5 (1)
Co(2)	1.1, 0.4 (0.1)
Co(3)	
Cu(2)	1.7, 1.0, 0.4, −0.2 (1)
Fe(2)	
Fe(3)	3.4, 2.7, 2.6 (0.1)
Hg(2)	
Mg	0.5 (0.2)
Mn(2)	0.5, 0.9 (0.1)
Ni	0.7, 0.55 (1)
Pb	1.9, 1.4 (0.5)
Sr	0.4 (0.2)
Zn	1.3, 0.8 (0.1)

Acetylacetonate[a]
Ag	
Al	8.1, 7.6, 5.5
Ba	
Ca	
Cd	3.4, 2.6

Acetylacetonate[a] (*Continued*)
Co(2)	5.0, 3.9
Co(3)	
Cu(2)	7.8, 6.5
Fe(2)	4.7, 3.3
Fe(3)	9.3, 8.6, 7.2
Hg(2)	
Mg	3.2, 2.3
Mn(2)	3.8, 2.8
Ni	5.5, 4.3, 2.1
Pb	4.2, 4.2
Sr	
Zn	4.6, 3.6

Alanine[a]
Ag	3.4, 3.5
Al	
Ba	0.4
Ca	0.8
Cd	4.5, 3.5, 1.5
Co(2)	4.4, 3.7
Co(3)	
Cu(2)	8.1, 6.6
Fe(2)	1–2:7.0
Fe(3)	11.0
Hg(2)	1–2:18.2

Alanine[a] (*Continued*)

Mg	
Mn(2)	3.0, 2.7
Ni	5.6, 4.4, 3.0
Pb	4.6
Sr	0.3
Zn	4.8, 4.1

Ammonia

Ag	3.40, 4.00 (0.1)
Al	
Ba	
Ca	
Cd	2.6, 2.1, 1.4, 0.9, −0.3, −1.7 (0.1)
Co(2)	2.1, 1.6, 1.0, 0.7, 0.1, −0.7 (0.1)
Co(3)	7.3, 6.7, 6.1, 5.6, 5.1, 4.4 (2)
Cu(2)	4.1, 3.5, 2.9, 2.1 (0.1)
Fe(2)	1.4, 0.8 (0)
Fe(3)	
Hg(2)	8.8, 8.7, 1.0, 0.9 (2)
Mg	
Mn(2)	0.8, 0.5 (var)
Ni	2.8, 2.2, 1.7, 1.2, 0.7, 0 (0.1)
Pb	
Sr	
Zn	2.3, 2.3, 2.4, 2.1 (0.1)

2,2′-Bipyridine

Ag	3.6, 3.6 (0.1)
Al	
Ba	
Ca	
Cd	4.3, 3.6, 2.7 (0)
Co(2)	6.1, 5.4, 4.6 (0.1)
Co(3)	
Cu(2)	8.0, 5.6, 3.5 (0.1)
Fe(2)	4.4, 1–3:17.6 (0.33)
Fe(3)	
Hg(2)	9.6, 7.1, 2.8 (0.1)
Mg	0.5 (0.5)

2,2′-Bipyridine (*Continued*)

Mn(2)	4.1, 3.8, 3.6 (1)
Ni	7.1, 6.9, 6.5 (0.1)
Pb	2.9 (0.1)
Sr	
Zn	5.4, 4.4, 3.5 (0.1)

Bromide

Ag	4.15, 2.95, 0.85, 0.95 (0.1)
Al	
Ba	
Ca	
Cd	1.56, 0.54, 0.06, 0.37 (0.75)
Co(2)	−0.13, −0.42 (0.69)
Co(3)	
Cu(2)	
Fe(2)	
Fe(3)	−0.3 (1)
Hg(2)	9.05, 8.25, 2.40, 1.3 (0.5)
Mg	
Mn(2)	0.27, −0.26 (0.69)
Ni	
Pb	1.1, 0.3, 1.2
Sr	
Zn	0.06, −0.88, 1.1 (4)

Chloride

Ag	2.9, 1.8, 0.3, 0.9 (0.2)
Al	
Ba	
Ca	
Cd	1.6, 0.5, −0.6, −0.6 (0.1)
Co(2)	
Co(3)	
Cu(2)	0.1, −0.6 (1)
Fe(2)	0.4 (2)
Fe(3)	0.6, 0.1, −1.4 (1)
Hg(2)	6.7, 6.5, 0.9, 1.0 (0.5)
Mg	
Mn(2)	0.6, 0.2, −0.4 (0.7)
Ni	
Pb	1.2, −0.6, 0.6 (0.1)
Sr	
Zn	−0.2, −0.4, 0.6 (3)

FORMATION CONSTANTS FOR METAL ION COMPLEXES

Citrate[b]

Ag	
Al	20
Ba	
Ca	
Cd	11.3
Co(2)	12.5
Co(3)	
Cu(2)	18
Fe(2)	15.5
Fe(3)	25.0
Hg(2)	
Mg	
Mn(2)	
Ni	14.3
Pb	12.3
Sr	
Zn	11.4

Cyanide

Ag	1–2:21.1, 0.7, −1.1 (0.3)
Al	
Ba	
Ca	
Cd	5.5, 5.1, 4.7, 3.6 (3)
Co(2)	1–6:19.1 (5)
Co(3)	
Cu(2)	
Fe(2)	1–6:24 (0)
Fe(3)	1–6:31 (0)
Hg(2)	18.0, 16.7, 3.8, 3.0 (0.1)
Mg	
Mn(2)	
Ni	1–4:31.3 (0.1)
Pb	1–4:10 (1)
Sr	
Zn	1–4:16.7 (0.1)

DCTA[a]

Ag	8.15
Al	17.6
Ba	8.0
Ca	12.5
Cd	19.2

DCTA[a] (*Continued*)

Co(2)	18.9
Co(3)	
Cu(2)	21.3
Fe(2)	18.2
Fe(3)	29.3
Hg(2)	24.3
Mg	10.3
Mn(2)	16.8
Ni	19.4
Pb	19.7
Sr	10.0
Zn	18.7

Dien[a]

Ag	6.1
Al	
Ba	
Ca	
Cd	8.45, 5.40
Co(2)	8.1, 6.0
Co(3)	
Cu(2)	16.0, 5.3
Fe(2)	6.23, 4.13
Fe(3)	
Hg(2)	21.8, 3.25
Mg	
Mn(2)	4.0, 2.8
Ni	10.7, 8.2
Pb	
Sr	
Zn	8.9, 5.4

DTPA[a]

Ag	8.70
Al	18.51
Ba	8.8
Ca	10.6
Cd	19.0
Co(2)	19.0
Co(3)	
Cu(2)	20.5
Fe(2)	16.0
Fe(3)	27.5
Hg(2)	27.0

DTPA[a] (*Continued*)
- Mg 9.3
- Mn(2) 15.5
- Ni 20.0
- Pb 18.9
- Sr 9.7
- Zn 18.0

EDTA[a]
- Ag 7.3
- Al 16.1
- Ba 7.8
- Ca 10.7
- Cd 16.5
- Co(2) 16.3
- Co(3)
- Cu(2) 18.8
- Fe(2) 14.3
- Fe(3) 25.1
- Hg(2) 21.8
- Mg 8.7
- Mn(2) 14.0
- Ni 18.6
- Pb 18.0
- Sr 8.6
- Zn 16.5

EGTA
- Ag 6.9
- Al 13.9
- Ba 8.4
- Ca 11.0
- Cd 15.6
- Co(2) 12.3
- Co(3)
- Cu(2) 17
- Fe(2) 11.9
- Fe(3) 20.5
- Hg(2) 23.2
- Mg 5.2
- Mn(2) 11.5
- Ni 12.0
- Pb 13.0
- Sr 8.5
- Zn 12.8

Ethylenediamine (EN)
- Ag 4.7, 3.0 (0.1)
- Al
- Ba
- Ca
- Cd 5.5, 4.6, 2.1 (0.1)
- Co(2) 5.9, 4.8, 3.1 (0.1)
- Co(3) 1–3:46.9 (0.1)
- Cu(2) 10.6, 9.1 (0.1)
- Fe(2) 4.3, 3.3 (0.1)
- Fe(3)
- Hg(2) 1–2:23.4 (0.1)
- Mg
- Mn(2) 2.7, 2.1, 0.9 (0.1)
- Ni 7.7, 6.4, 4.5 (0.1)
- Pb
- Sr
- Zn 5.7, 4.7, 1.7 (0.1)

Fluoride
- Ag
- Al 6.1, 5.1, 3.8, 2.7, 1.7, 0.3 (0.53)
- Ba
- Ca
- Cd 0.5, 0.1
- Co(2)
- Co(3)
- Cu(2) 0.7 (0.5)
- Fe(2) <1.5 (var)
- Fe(3) 5.3, 4.0, 2.7 (0.5)
- Hg(2) 1.0 (0.5)
- Mg 1.3 (0.5)
- Mn(2)
- Ni 0.7 (1)
- Pb 1.5 (1)
- Sr
- Zn 0.7 (0.5)

Formate
- Ag
- Al 1.8 (1)
- Ba 1.4 (0)
- Ca 1.43 (0)

Formate (*Continued*)

Cd	0.65, −0.25, 0.92 (1)
Co(2)	
Co(3)	
Cu(2)	2.0 (0.5)
Fe(2)	
Fe(3)	1.9, 1.8, 0.3, 1.5 (1)
Hg(2)	
Mg	1.4 (0)
Mn(2)	0.8 (1)
Ni	
Pb	0.85, 0.13, 0.17 (1)
Sr	1.39 (0)
Zn	1.97 (0.1)

Glycine[a]

Al	
Ba	0.4
Ca	1.0
Cd	4.4, 3.8, 1.6
Co(2)	4.7, 3.8, 2.5
Co(3)	
Cu(2)	8.1, 7.0, 0.6
Fe(2)	3.9, 3.3
Fe(3)	
Hg(2)	10.5, 9.0
Mg	3.1, 3.0
Mn(2)	3.0, 2.1
Ni	5.8, 4.8, 3.8
Pb	5.1, 3.1
Sr	0.5
Zn	5.0, 4.1

HEDTA[a]

Ag	6.7
Al	14.4
Ba	6.2
Ca	8.0
Cd	13.0
Co(2)	14.4
Co(3)	
Cu(2)	17.4
Fe(2)	12.2
Fe(3)	19.8
Hg(2)	20.1

HEDTA[a] (*Continued*)

Mg	5.2
Mn(2)	10.7
Ni	17.0
Pb	15.5
Sr	6.8
Zn	14.5

Hydroxide

Ag	2.3, 1.3, 1.2 (0)
Al	1–4:33 (2)
Ba	0.7 (0)
Ca	1.3 (0)
Cd	4.3, 3.4, 2.6, 1.7 (3)
Co(2)	5.1, 1–3:10.2 (0.1)
Co(3)	
Cu(2)	6.0 (0)
Fe(2)	4.5 (1)
Fe(3)	11.0, 10.7 (3)
Hg(2)	10.3, 11.4 (0.5)
Mg	2.6 (0)
Mn(2)	3.4 (0.1)
Ni	4.6 (0.1)
Pb	6.2, 4.1, 3.0 (0.3)
Sr	0.8 (0)
Zn	4.4, 1–4:15.5 (0)

8-Hydroxyquinolate

Ag	
Al	
Ba	2.1 (0)
Ca	3.3 (0)
Cd	7.8 (0)
Co(2)	9.1, 8.1 (0.01)
Co(3)	
Cu(2)	12.2, 11.2 (0.01)
Fe(2)	8.0, 7.0 (0.01)
Fe(3)	12.3, 11.3, 10.6 (0.01)
Hg(2)	
Mg	4.5 (0.01)
Mn(2)	6.8, 5.8 (0.01)
Ni	9.9, 8.8 (0.01)
Pb	9.0 (0)
Sr	2.6 (0.1)
Zn	8.6 (0)

IDA[a]

Ag	
Al	
Ba	1.7
Ca	2.6
Cd	5.3, 4.2
Co(2)	6.9, 5.4
Co(3)	
Cu(2)	10.5, 5.7
Fe(2)	5.8, 4.3
Fe(3)	
Hg(2)	10.8
Mg	2.9
Mn(2)	
Ni	8.3, 6.3
Pb	
Sr	2.2
Zn	7.3, 5.3

Iodide

Ag	6.6, 5.1, 2.0, 0.6 (0)
Al	
Ba	
Ca	
Cd	2.3, 1.6, 1.1, 1.1 (0)
Co(2)	
Co(3)	
Cu(2)	
Fe(2)	
Fe(3)	1.9 (0)
Hg(2)	12.9, 11.0, 3.8, 2.2 (0.5)
Mg	
Mn(2)	
Ni	
Pb	2.0, 1.2, 0.7, 0.5 (0)
Sr	
Zn	

NTA[a]

Ag	5.16
Al	11.4
Ba	4.8
Ca	6.4
Cd	10.1, 4.4

NTA[a] (*Continued*)

Co(2)	10.6, 3.7
Co(3)	
Cu(2)	12.7, 3.6
Fe(2)	8.8
Fe(3)	15.9, 24.3
Hg(2)	14.6
Mg	5.4
Mn(2)	7.4
Ni	11.3, 4.5
Pb	11.8
Sr	5.0
Zn	10.5, 3.0

***o*-Phenanthroline**

Ag	
Al	
Ba	
Ca	
Cd	6.4, 5.2, 4.2 (0.1)
Co(2)	7.0, 6.7, 6.4 (0.1)
Co(3)	
Cu(2)	9.1, 6.7, 5.2 (0.1)
Fe(2)	5.9, 5.2, 10.2 (0.1)
Fe(3)	1–3:14.1 (0.1)
Hg(2)	
Mg	
Mn(2)	4.1, 3.1, 3.2 (0.1)
Ni	8.8, 8.3, 7.7 (0.1)
Pb	5.1, 2.4, 1.5 (0.1)
Sr	
Zn	

Oxalate

Ag	
Al	1–2:11.0, 3.6 (0.5)
Ba	0.6 (1)
Ca	
Cd	2.9, 1.8 (0.5)
Co(2)	3.5, 2.3 (0.5)
Co(3)	
Cu(2)	4.5, 4.4 (0.5)
Fe(2)	3.1, 2.1 (1)
Fe(3)	8.0, 6.3, 4.2 (0.5)
Hg(2)	<4.0 (0.1)

FORMATION CONSTANTS FOR METAL ION COMPLEXES

Oxalate (*Continued*)
Mg	2.4 (0.5)
Mn(2)	2.7, 1.4 (0.5)
Ni	4.1, 3.1 (1)
Pb	3.3, 1.7 (1.5)
Sr	1.3, 0.6 (1)
Zn	3.7, 2.5 (0.5)

Pyridine
Ag	2.0, 2.4 (0)
Al	
Ba	
Ca	
Cd	1.4, 0.5, 0.3 (0.1)
Co(2)	1.1, 0.4 (0.5)
Co(3)	
Cu(2)	2.5, 1.9, 1.3, 0.9 (0.5)
Fe(2)	
Fe(3)	
Hg(2)	5.1, 4.9, 0.4 (0.5)
Mg	
Mn(2)	
Ni	1.9, 0.9, 0.6, 0.1
Pb	
Sr	
Zn	1.4, −0.3, 0.5, 0.3 (0.1)

Salicylate[a]
Ag	
Al	14 (var)
Ba	
Ca	
Cd	5.6
Co(2)	6.8, 4.7
Co(3)	
Cu(2)	10.6, 7.9
Fe(2)	6.6, 4.7
Fe(3)	15.8, 11.7, 7.8 (3)
Hg(2)	
Mg	
Mn(2)	5.9, 3.9
Ni	7.0, 4.8
Pb	
Sr	
Zn	6.9

Sulfate
Ag	0.2
Al	
Ba	
Ca	2.3 (0)
Cd	0.85 (3)
Co(2)	2.5 (0)
Co(3)	
Cu(2)	1.0, 0.1, 1.2 (1)
Fe(2)	
Fe(3)	4.0, 1.4 (0)
Hg(2)	1.3, 1.1
Mg	2.4 (0)
Mn(2)	2.3 (0)
Ni	2.3 (0)
Pb	3.5
Sr	
Zn	2.3 (0)

Sulfosalicylate[a]
Ag	
Al	12.9, 10.0, 6.1
Ba	
Ca	
Cd	4.7
Co(2)	6.0, 3.8
Co(3)	
Cu(2)	9.5, 7.0
Fe(2)	5.9, 4.1
Fe(3)	14.4, 10.8, 7.0 (3)
Hg(2)	
Mg	
Mn(2)	5.2, 3.0
Ni	6.4, 3.8
Pb	
Sr	
Zn	6.1, 4.5

Tetren[a]
Ag	7.4
Al	
Ba	
Ca	
Cd	14.0

Tetren[a] (*Continued*)

Co(2)	15.1
Co(3)	
Cu(2)	24.3
Fe(2)	11.4
Fe(3)	
Hg(2)	27.7
Mg	
Mn(2)	7.62
Ni	17.6
Pb	10.5
Sr	
Zn	15.4

Thiocyanate

Ag	7.6, 2.5, 1.0 (2.2)
Al	
Ba	
Ca	
Cd	1.4, 0.6, 0.6 (3)
Co(2)	1.0 (1)
Co(3)	
Cu(2)	1.7, 0.8, 0.2, 0.3 (0.5)
Fe(2)	1.0 (var)
Fe(3)	2.3, 1.9, 1.4, 0.8 (var)
Hg(2)	1–2:16.1, 2.9, 1.9 (1)

Thiocyanate (*Continued*)

Mg	
Mn(2)	1.2 (0)
Ni	1.2, 0.4, 0.2 (1.5)
Pb	0.5, 0.4, −1.0, 1.9 (2)
Sr	
Zn	0.5, 0.3, 0, 1.3 (2)

Trien[a]

Ag	7.7
Al	
Ba	
Ca	
Cd	10.8, 3.1
Co(2)	11.0
Co(3)	
Cu(2)	20.4
Fe(2)	7.8
Fe(3)	21.9
Hg(2)	25.3
Mg	
Mn(2)	4.9
Ni	14.0
Pb	10.4
Sr	
Zn	12.1

Values are $\log(K_{f_1})$, $\log(K_{f_2})$, ..., except when notation like "1–N:" appears. This means the log of the product of the first N formation constants, $\log(N)$. Ionic strengths appear in parentheses.

[a] All values for this ligand at $0.1 M$ ionic strength

[b] All values for this ligand at $0.5 M$ ionic strength

Most of these values have been taken from A. Ringbom and E. Wainninen, "Complexation Reactions," in *Treatise on Analytical Chemistry*, 2nd. Ed., Part 1, Volume 2, Ch. 20, I. M. Kolthoff and P. J. Elving, Eds, New York, Wiley, 1979, and references they cite. Many values are taken from L. G. Sillen and A. E. Martell, *Stability Constants of Metal–Ion Complexes*, London, The Chemical Society, 1964.

Appendix IV

Standard and Formal Potentials (25°C)

Formal Potentials Are in Parentheses

Half reaction	Standard potential
Aluminum	
$Al^{3+} + 3e = Al(c)$	-1.676
$Al(OH)_4^- + 3e = Al(c) + 4OH^-$	-2.35
Antimony	
$Sb_2O_5 + 6H^+ + 4e = 2SbO^+ + 3H_2O$	$+0.605$
$SbO^+ + 2H^+ + 3e = Sb(c) + H_2O$	$+0.204$
$SbO_2^- + 2H_2O + 3e = Sb(c) + 4OH^-$	-0.6389
$SbCl_4^- + 3e = Sb(c) + 4Cl^-$	$+0.17$
$Sb(c) + 3H^+ + 3e = SbH_3$	-0.510
Arsenic	
$H_3AsO_4 + 2H^+ + 2e = HAsO_2 + 2H_2O$	$+0.560$
$AsO_4^{3-} + 2H_2O + 2e = AsO_2^- + 4OH^-$	-0.67
$HAsO_2 + 3H^+ + 3e = As(c) + 2H_2O$	$+0.240$
$As(c) + 3H^+ + 3e = AsH_3(g)$	-0.225
Barium	
$Ba^{2+} + 2e = Ba(c)$	-2.92
Beryllium	
$Be^{2+} + 2e = Be(c)$	-1.95
Bismuth	
$Bi^{+5} + 2e = Bi^{3+}$	$+2.0 \pm 0.2$
$Bi^{3+} + 3e = Bi(c)$	$+0.3172$

Half reaction	Standard potential

Bismuth (*Continued*)

$BiOCl^+ + 2H^+ + 3e = Bi(c) + H_2O + Cl^-$	+0.1697
$BiOH^{2+} + 3e = Bi(c) + OH^-$	+0.072
$BiCl_4^- + 3e = Bi(c) + 4Cl^-$	+0.199
$Bi(c) + 3H^+ + 3e = BiH_3(g)$	−0.97

Boron

$2H_3BO_3(aq) + 12H^+ + 12e = B_2H_6(g) + 6H_2O$	−0.519
$2B(s) + 6H^+ + 6e = B_2H_6(g)$	−0.150
$H_3BO_3(aq) + 3H^+ + 3e = B(s) + 3H_2O$	−0.890
$B(OH)_4^- + 3e = B(s) + 4OH^-$	−1.811

Bromine

$BrO_4^- + 2H^+ + 2e = BrO_3^- + H_2O$	+1.853
$2BrO_3^- + 12H^+ + 10e = Br_2(l) + 6H_2O$	+1.478
$BrO_3^- + 5H^+ + 4e = HOBr + 2H_2O$	+1.447
$HOBr + H^+ + 2e = Br^- + 2H_2O$	+1.341
$OBr^- + H_2O + 2e = Br^- + 2OH^-$	+0.766
$Br_2(aq) + 2e = 2Br^-$	+1.0874
$Br_2(l) + 2e = 2Br^-$	+1.0652
$Br_3^- + 2e = 3Br^-$	+1.0503

Cadmium

$Cd^{2+} + 2e = Cd(c)$	−0.4025
$Cd(CN)_4^{2-} + 2e = Cd(c) + 4CN^-$	−0.943
$Cd(NH_3)_4^{2+} + 2e = Cd(c) + 4NH_3$	−0.622
$Cd(OH)_2(s) + 2e = Cd(c) + 2OH^-$	−0.824
$CdCl_4^{2-} + 2e = Cd(c) + 4Cl^-$	−0.453
$CdBr_4^{2-} + 2e = Cd(c) + 4Br^-$	−0.488
$CdI_4^{2-} + 2e = Cd(c) + 4I^-$	−0.580

Calcium

$Ca^{2+} + 2e = Ca(c)$	−2.84

Cerium

$Ce(IV) + e = Ce(III)$ (1F HCl)	(+1.28)
(1F HClO$_4$)	(+1.70)
(1F HNO$_3$)	(+1.61)
(0.5F H$_2$SO$_4$)	(+1.44)

Cesium

$Cs^+ + e = Cs(c)$	−2.923

Chlorine

$ClO_4^- + 8H^+ + 8e = Cl^- + 4H_2O$	+1.388
$ClO_4^- + H_2O + 2e = ClO_3^- + 2OH^-$	+0.374

STANDARD AND FORMAL POTENTIALS (25°C)

Half reaction	Standard potential
$ClO_3^- + 2H^+ + e = ClO_2(g) + H_2O$	+1.175
$ClO_3^- + H_2O + 2e = ClO_2^- + 2OH^-$	+0.295
$2ClO_3^- + 12H^+ + 10e = Cl_2(g) + 6H_2O$	+1.47
$HOCl + H^+ + 2e = Cl^- + H_2O$	+1.49
$2HOCl + 2H^+ + 2e = Cl_2(g) + 2H_2O$	+1.630
$OCl^- + H_2O + 2e = Cl^- + 2OH^-$	+0.890
$Cl_2(g) + 2e = 2Cl^-$	+1.35828
$Cl_2(aq) + 2e = 2Cl^-$	+1.396
$Cl_3^- + 2e = 3Cl^-$	+1.4152

Chromium

Half reaction	Standard potential
$CrO_4^{2-} + 4H_2O + 3e = Cr(OH)_3(s) + 5OH^-$	−0.11
$Cr_2O_7^{2-} + 14H^+ + 6e = 2Cr^{3+} + 7H_2O$	+1.36
(0.1F HCl)	(+0.93)
(1.0F HCl)	(+1.00)
(0.1F H$_2$SO$_4$)	(+0.92)
(4F H$_2$SO$_4$)	(+1.15)
$Cr^{3+} + e = Cr^{2+}$	−0.42
$Cr^{2+} + 2e = Cr(c)$	−0.90

Cobalt

Half reaction	Standard potential
$Co^{3+} + e = Co^{2+}$	+1.842
$Co(OH)_3(s) + e = Co(OH)_2(s) + OH^-$	+0.17
$Co(NH_3)_6^{3+} + e = Co(NH_3)_6^{2+}$	+0.1
$Co(CN)_6^{3-} + e = Co(CN)_6^{4-}$	−0.84
$Co(en)_3^{3+} + e = Co(en)_3^{2+}$ (0.1F ethylenediamine(en), 0.1F KCl)	(−0.2)
$Co^{2+} + 2e = Co(c)$	−0.277

Copper

Half reaction	Standard potential
$Cu^{2+} + 2e = Cu(c)$	+0.340
$Cu^{2+} + e = Cu^+$	+0.159
$Cu^+ + e = Cu(c)$	+0.520
$Cu^{2+} + I^- + e = CuI(s)$	+0.861
$Cu^{2+} + 2CN^- + e = Cu(CN)_2^-$	+1.12
$Cu^{2+} + Cl^- + e = CuCl(s)$	+0.559
$Cu^{2+} + 2Cl^- + e = CuCl_2^-$	+0.463
$CuCl(s) + e = Cu(c) + Cl^-$	+0.121
$Cu^{2+} + Br^- + e = CuBr(s)$	+0.654
$Cu(CN)_2^- + e = Cu(c) + 2CN^-$	−0.44
$CuCl_2^- + e = Cu(c) + 2Cl^-$	+0.177
$Cu(EDTA)^{2-} + 2e = Cu(c) + EDTA^{4-}$ (0.1F EDTA pH 4–5)	(+0.13)

Half reaction	Standard potential
Fluorine	
$F_2 + 2e = 2F^-$	+2.87
Gold	
$Au^{3+} + 2e = Au^+$	+1.36
$Au^{3+} + 3e = Au(c)$	+1.52
$Au^+ + e = Au(c)$	+1.83
$AuCl_2^- + e = Au(c) + 2Cl^-$	+1.154
$AuBr_2^- + e = Au(c) + 2Br^-$	+0.959
$Au(CN)_2^- + e = Au(c) + 2CN^-$	−0.60
Hydrogen	
$2H^+ + 2e = H_2$	0.0000
$2H_2O + 2e = H_2 + 2OH^-$	−0.828
Iodine	
$H_5IO_6 + H^+ + 2e = IO_3^- + 3H_2O$	+1.603
$2IO_3^- + 12H^+ + 10e = I_2(aq) + 6H_2O$	+1.195
$IO_3^- + 3H_2O + 6e = I^- + 6OH^-$	+0.257
$2ICl^{2-} + 2e = I_2(s) + 4Cl^-$	+1.07
$2IBr_2^- + 2e = I_2(s) + 4Br^-$	+0.874
$I_2(aq) + 2e = 2I^-$	+0.6197
$I_2(s) + 2e = 2I^-$	+0.5345
$I_3^- + 2e = 2I^-$	+0.5355
Iron	
$Fe(III) + e = Fe(II)$	+0.771
(1F HCl)	(+0.70)
(1F H$_2$SO$_4$)	(+0.68)
(1F HClO$_4$)	(+0.735)
(2F H$_3$PO$_4$)	(+0.46)
$Fe(CN)_6^{3-} + e = Fe(CN)_6^{4-}$	+0.361
(1F HCl)	(+0.71)
$Fe(EDTA)^- + e = Fe(EDTA)^{2-}$ (0.1F EDTA pH 4–6)	(+0.12)
$Fe^{2+} + 2e = Fe(c)$	−0.440
Lead	
$PbO_2(s) + SO_4^{2-} + 4H^+ + 2e = PbSO_4(s) + 2H_2O$	+1.690
$PbO_2(s) + 4H^+ + 2e = Pb^{2+} + 2H_2O$	+1.460
$Pb^{2+} + 2e = Pb(c)$	−0.1263
(1F sodium acetate)	(−0.32)
$PbSO_4(s) + 2e = Pb(c) + SO_4^{2-}$	−0.3563
$PbCl_2(s) + 2e = Pb(c) + 2Cl^-$	−0.268

STANDARD AND FORMAL POTENTIALS (25°C)

Half reaction	Standard potential
$PbBr_2(s) + 2e = Pb(c) + 2Br^-$	−0.280
$PbI_2(s) + 2e = Pb(c) + 2I^-$	−0.365
Lithium	
$Li^+ + e = Li(c)$	−3.040
Magnesium	
$Mg^{2+} + 2e = Mg(c)$	−2.36
$Mg(OH)_2(s) + 2e = Mg(c) + 2OH^-$	−2.69
Manganese	
$MnO_4^- + 4H^+ + 3e = MnO_2(s) + 2H_2O$	+1.70
$MnO_4^- + 8H^+ + 5e = Mn^{2+} + 4H_2O$	+1.51
$MnO_4^- + e = MnO_4^{2-}$	+0.564
$MnO_2(s) + 4H^+ + 2e = Mn^{2+} + 2H_2O$	+1.23
$Mn(III) + e = Mn(II)$ (8F H_2SO_4)	(+1.51)
$Mn^{2+} + 2e = Mn(c)$	−1.18
$Mn(CN)_6^{4-} + e = Mn(CN)_6^{5-}$ (1.5F NaCN)	(−1.06)
Mercury	
$Hg^{2+} + 2e = Hg(l)$	+0.8535
$Hg_2^{2+} + 2e = 2Hg(l)$	+0.7960
(1F HClO$_4$)	(+0.776)
$2Hg^{2+} + 2e = Hg_2^{2+}$	+0.9110
$Hg_2Cl_2(s) + 2e = 2Hg(l) + 2Cl^-$	+0.26816
(0.1F KCl)	(+0.334)
(1F KCl)	(+0.280)
$Hg_2Cl_2(s) + 2K^+ + 2e = 2Hg(l) + 2KCl(s)$ (saturated calomel electrode)	(+0.2415)
$Hg_2SO_4(s) + 2e = 2Hg(l) + SO_4^{2-}$	+0.6151
$Hg_2(CH_3COO)_2(s) + 2e = Hg(l) + 2CH_3COO^-$ (acetate)	+0.5113
Nickel	
$NiO_2(s) + 4H^+ + 2e = Ni^{2+} + 2H_2O$	+1.68
$Ni^{2+} + 2e = Ni(c)$	−0.228
$Ni(CN)_4^{2-} + e = Ni(CN)_4^{3-}$ (1F KCN)	(−0.82)
$Ni(OH)_2(s) + 2e = Ni(c) + 2OH^-$	−0.720
Nitrogen	
$2NO_3^- + 4H^+ + 2e = N_2O_4(g) + 2H_2O$	+0.803
$NO_3^- + 3H^+ + 2e = HNO_2 + H_2O$	+0.94
$NO_3^- + 4H^+ + 3e = NO(g) + 2H_2O$	+0.96
$N_2O_4(g) + 2H^+ + 2e = 2HNO_2$	+1.07
$N_2O_4(g) + 4H^+ + 4e = 2NO(g) + 2H_2O$	+1.039

Half reaction	Standard potential
Nitrogen (*Continued*)	
$N_2O_4(g) + 8H^+ + 8e = N_2(g) + 4H_2O$	+1.357
$HNO_2 + H^+ + e = NO(g) + H_2O$	+0.996
$2NO_2^- + 3H_2O + 4e = N_2O(g) + 6OH^-$	+0.15
$2HNO_2 + 6H^+ + 6e = N_2(g) + 4H_2O$	+1.454
$N_2O(g) + 5H_2O + 4e = 2NH_2OH + 4OH^-$	−1.05
$N_2(g) + 5H^+ + 4e = N_2H_5^+$	−0.23
$2NH_3OH^+ + H^+ + 2e = N_2H_5^+ + 2H_2O$	+1.41
Osmium	
$OsO_4(s) + 4H^+ + 4e = OsO_2(s) + 2H_2O$	+1.005
$OsCl_6^{2-} + e = OsCl_6^{3-}$	+0.85
Oxygen	
$O_3(g) + 2H^+ + 2e = O_2(g) + H_2O$	+2.075
$O_3(g) + H_2O + 2e = O_2(g) + 2OH^-$	+1.246
$O_2(g) + 4H^+ + 4e = 2H_2O$	+1.229
$O_2(g) + 2H^+ + 2e = H_2O_2$	+0.695
$O_2(g) + e = O_2^-$	−0.284
$H_2O_2 + 2H^+ + 2e = 2H_2O$	+1.763
$HO_2^- + H_2O + 2e = 3OH^-$	+0.867
$HO_2 + e = HO_2^-$	−0.744
Palladium	
$PdCl_6^{2-} + 2e = PdCl_4^{2-} + 2Cl^-$ (1F HCl)	(+1.47)
$PdCl_4^{2-} + 2e = Pd + 4Cl^-$ (1F HCl)	(+0.62)
$Pd^{2+} + 2e = Pd(c)$	+0.915
Phosphorus	
$H_3PO_4 + 2H^+ + 2e = H_3PO_3 + H_2O$	−0.276
$H_3PO_3 + 2H^+ + 2e = H_3PO_2 + H_2O$	−0.499
$H_3PO_2 + H^+ + e = P + 2H_2O$	−0.508
$P(\text{red, c}) + 3H^+ + 3e = PH_3(g)$	−0.111
$4P(\text{red, c}) + 2H^+ + 2e = P_4H_2(c)$	−0.633
Platinum	
$PtCl_6^{2-} + 2e = PtCl_4^{2-} + 2Cl^-$	+0.726
$PtBr_6^{2-} + 2e = PtBr_4^{2-} + 2Br^-$	+0.613
$Pt^{2+} + 2e = Pt(c)$	+1.188
$Pt(OH)_2(s) + 2H^+ + 2e = Pt + 2H_2O$	+0.98
$PtCl_4^{2-} + 2e = Pt(c) + 4Cl^-$	+0.758
$PtBr_4^{2-} + 2e = Pt(c) + 4Br^-$	+0.698
Potassium	
$K^+ + e = K(c)$	−2.924

STANDARD AND FORMAL POTENTIALS (25°C)

Half reaction	Standard potential
Radium	
$Ra^{2+} + 2e = Ra(c)$	−2.92
$RaO + 2H^+ + 2e = Ra(c) + H_2O$	−1.32
Rubidium	
$Rb^+ + e = Rb$	−2.924
Selenium	
$SeO_4^{2-} + 4H^+ + 2e = H_2SeO_3 + H_2O$	+1.151
$SeO_4^{2-} + H_2O + 2e = SeO_3^{2-} + 2OH^-$	+0.031
$H_2SeO_3 + 4H^+ + 4e = Se(c) + 3H_2O$	+0.739
$Se(c) + 2H^+ + 2e = H_2Se$	−0.082
$Se(c) + 2e = Se^{2-}$	−0.670
Silver	
$Ag^{2+} + e = Ag^+$	+1.980
(4F HNO_3)	(+1.927)
$Ag^+ + e = Ag(c)$	+0.7991
(1F $HClO_4$)	(+0.792)
$AgCl(s) + e = Ag(c) + Cl^-$	+0.2223
$AgBr(s) + e = Ag(c) + Br^-$	+0.0711
$AgI(s) + e = Ag(c) + I^-$	−0.1522
$Ag_2O(s) + H_2O + 2e = 2Ag(c) + 2OH^-$	+0.342
$Ag_2S(s) + 2e = 2Ag(c) + S^{2-}$	−0.71
Sodium	
$Na^+ + e = Na(c)$	−2.713
Strontium	
$Sr^{2+} + 2e = Sr(c)$	−2.89
Sulfur	
$S_2O_8^{2-} + 2e = 2SO_4^{2-}$	+1.96
$S_4O_6^{2-} + 2e = 2S_2O_3^{2-}$	+0.080
$SO_4^{2-} + 4H^+ + 2e = H_2SO_3 + H_2O$	+0.158
$SO_4^{2-} + H_2O + 2e = SO_3^{2-} + 2OH^-$	−0.936
$2H_2SO_3 + 2H^+ + 4e = S_2O_3^{2-} + 3H_2O$	+0.400
$2SO_3^{2-} + 3H_2O + 4e = S_2O_3^{2-} + 6OH^-$	−0.576
$SO_3^{2-} + 3H_2O + 4e = S(s) + 6OH^-$	−0.659
$SO_2(g) + 4H^+ + 4e = S(s) + 2H_2O$	+0.451
$S(s) + 2e = S^{2-}$	−0.447
$S(s) + 2H^+ + 2e = H_2S(aq)$	+0.144
Thallium	
$Tl^{3+} + 2e = Tl^+$	+1.25
$TlCl_3(s) + 2e = TlCl(s) + 2Cl^-$	+0.890

Half reaction	Standard potential
Thallium (*Continued*)	
$Tl_2O_3(s) + 3H_2O + 4e = 2Tl^+ + 6OH^-$	+0.02
$Tl^+ + e = Tl(c)$	−0.3363
$TlCl(s) + e = Tl(c) + Cl^-$	−0.5568
$TlBr(s) + e = Tl(c) + Br^-$	−0.658
$TlI(s) + e = Tl(c) + I^-$	−0.752
Tin	
$Sn^{4+} + 2e = Sn^{2+}$	+0.15
$SnCl_6^{2-} + 2e = SnCl_4^{2-} + 2Cl^-$ (1F HCl)	(+0.14)
$SnO_3^{2-} + 3H^+ + 2e = HSnO_2^- + H_2O$	+0.375
$Sn^{2+} + 2e = Sn(c)$	−0.1375
$HSnO_2^- + H_2O + 2e = Sn + 3OH^-$	−0.91
Titanium	
$Ti(IV) + e = Ti(III)$ (5F H$_3$PO$_4$)	(−0.15)
$TiO^{2+} + 2H^+ + e = Ti^{3+} + H_2O$	+0.100
$TiO^{2+} + 2H^+ + 2e = Ti^{2+} + H_2O$	−0.135
$Ti^{3+} + e = Ti^{2+}$	−2.3
$Ti^{2+} + 2e = Ti(c)$	−1.63
Tungsten	
$W(VI) + e = W(V)$ (12F HCl)	(+0.26)
$2WO_3(s) + 2H^+ + 2e = W_2O_5(s) + H_2O$	−0.029
$WO_4^{2-} + 4H_2O + 6e = W + 8OH^-$	−1.074
$WO_3(s) + 6H^+ + 6e = W + 3H_2O$	−0.090
$W(V) + e = W(IV)$ (12F HCl)	(−0.3)
$W_2O_5(s) + 2H^+ + 2e = 2WO_2(s) + H_2O$	−0.031
$W(CN)_8^{3-} + e = W(CN)_8^{4-}$	+0.457
Uranium	
$UO_2^{2+} + 4H^+ + 2e = U^{4+} + 2H_2O$	+0.27
$UO_2^{2+} + e = UO_2^+$	−0.16
$UO_2^+ + 4H^+ + e = U^{4+}$	+0.38
$U^{4+} + e = U^{3+}$	−0.52
(1F HCl)	(−0.64)
$U^{4+} + 4e = U(c)$	−1.38
$U^{3+} + 3e = U(c)$	−1.66
Vanadium	
$VO_2^+ + 2H^+ + e = VO^{2+} + H_2O$	+1.000
$VO^{2+} + 2H^+ + e = V^{3+} + H_2O$	+0.337
$V^{3+} + e = V^{2+}$	−0.255
$V^{2+} + 2e = V(c)$	−1.13

STANDARD AND FORMAL POTENTIALS (25°C)

Half reaction	Standard potential
Zinc	
$Zn^{2+} + 2e = Zn(c)$	-0.7626
$Zn(NH_3)_4^{2+} + 2e = Zn(c) + 4NH_3$	-1.04
$Zn(CN)_4^{2-} + 2e = Zn(c) + 4CN^-$	-1.34
$Zn(OH)_4^{2-} + 2e = Zn(c) + 4OH^-$	-1.285
$Zn(OH)_2(s) + 2e = Zn(c) + 2OH^-$	-1.246

All ions should be taken as aquated [e.g., $Ag^+(aq)$]. Most values are taken from *Standard Potentials in Aqueous Solution*, Ed. A. Bard, R. Parsons, and J. Jordan, Marcel Dekker, New York, 1985, in cooperation with the International Union of Pure and Applied Chemistry (IUPAC).

Appendix V

Experimental Evaluation of Close pK_a Values for Diprotic Acids

When successive pK_a values for a diprotic weak acid differ by less than ~2.7 units, a curve for the titration of the weak acid with strong base will not show two distinct breaks. In such cases, in the sequential titration process,

$$H_2A + OH^- \rightarrow HA^- + H_2O$$

$$HA^- + OH^- \rightarrow A^{2-} + H_2O$$

HA^- is so similar in strength to H_2A that the second step is well underway before the first step is finished.

Even though the reaction sequence is not cleanly separated, we can evaluate (as we did in Chapter 6) the pK_a values from measured pH at the 25% and 75% points in the titration. This time, however, we will need to develop and solve a pair of equations which involve *both* K_1 and K_2. These equations can be derived from proton balances at each point, as follows:

(a) Let us say that the total concentration of weak acid (in all forms) is C_t. At the 25% titrated point, enough NaOH has been added so that the concentrations of H_2A and HA^- are *nominally* equal:

$$C_{H_2A} = C_{HA^-} = \tfrac{1}{2} C_t$$

The proton balance at 25% titrated will be†

---

† This is a proton balance for the dissociation of a small amount of H_2A. The nominal concentration of A^{2-} is *zero* at this point, but in reality it is finite and must be supplied by the dissociation of some H_2A.

$$[H^+] = [OH^-] + ([HA^-] - \tfrac{1}{2}C_t) + 2[A^{2-}]$$

Rearrange this expression to get

$$[HA^-] + 2[A^{2-}] = (\tfrac{1}{2}C_t + [H^+] - [OH^-])$$

We will call the right-hand side $M_{0.25}$ to simplify the expressions that follow. Next, express $[HA^-]$ and $[A^{2-}]$ as the products of their fractions and C_t,

$$[HA^-] = \alpha_{HA^-} C_t = \frac{K_1[H^+]}{D} C_t$$

$$[A^{2-}] = \alpha_{A^{2-}} C_t = \frac{K_1 K_2}{D} C_t$$

where $D = [H^+]^2 + K_1[H^+] + K_1 K_2$.

Substitute these last expressions into the equation for $M_{0.25}$ and get

$$M_{0.25} = (\alpha_{HA^-} + 2\alpha_{A^{2-}}) C_t = C_t \left[\frac{K_1[H^+] + 2K_1 K_2}{[H^+]^2 + [H^+]K_1 + K_1 K_2} \right]$$

Next, in three steps, multiply both sides of the last equation by the denominator of the right-hand side, collect terms containing $K_1[H^+]$ and $K_1 K_2$, and divide through by $[H^+]$. The result is

$$\frac{M_{0.25}[H^+]}{C_t - M_{0.25}} = K_1 + K_1 K_2 \left[\frac{2C_t - M_{0.25}}{(C_t - M_{0.25})[H^+]} \right] \tag{A5.1}$$

Remember that K_1 and K_2 are not thermodynamic constants at this point. They are related to thermodynamic constants by activity coefficients, however,

$$K_1^\circ = K_1 \frac{f_{HA^-} f_{H^+}}{f_{H_2A}}, \qquad K_2^\circ = K_2 \frac{f_{A^{2-}} f_{H^+}}{f_{HA^-}}, \qquad \text{and} \qquad K_1 K_2 = \frac{K_1^\circ K_2^\circ f_{H_2A}}{f_{H^+}^2 f_{A^{2-}}}$$

Putting this information into equation (A5.1) gives

$$\frac{M_{0.25}[H^+]}{C_t - M_{0.25}} = K_1^\circ \left(\frac{f_{H_2A}}{f_{HA^-} f_{H^+}} \right) + K_1^\circ K_2^\circ \left(\frac{f_{H_2A}}{f_{H^+}^2 f_{A^{2-}}} \right) \left[\frac{2C_t - M_{0.25}}{(C_t - M_{0.25})[H^+]} \right]$$

If we multiply both sides by f_{H^+}, we arrive at the relationship between K_1°, K_2°, and the experimental pH (i.e., $-\log[H^+]f_{H^+}$):

EXPERIMENTAL EVALUATION OF CLOSE pK_a VALUES

$$\frac{M_{0.25}[H^+]f_{H^+}}{C_t - M_{0.25}} = K_1^o \frac{1}{f_{HA^-}} + K_1^o K_2^o \left(\frac{1}{f_{A^{2-}}}\right)\left[\frac{2C_t - M_{0.25}}{(C_t - M_{0.25})[H^+]f_{H^+}}\right] \quad (A5.2)$$

(b) The equation for the 75% titrated point can be derived analogously, but starting from the proton balance†:

$$[H^+] = [OH^-] + ([A^{2-}] - \tfrac{1}{2}C_t) - [H_2A]$$

The result of the derivation is

$$\frac{M_{0.75}[H^+]f_{H^+}}{C_t - M_{0.75}} = K_1^o \left(\frac{1}{f_{HA^-}}\right) + K_1^o K_2^o \left(\frac{1}{f_{A^{2-}}}\right)\left[\frac{2C_t - M_{0.75}}{(C_t - M_{0.75})[H^+]f_{H^+}}\right] \quad (A5.3)$$

where $M_{0.75} = \tfrac{3}{2}C_t + [H^+] - [OH^-]$.

Equations (A5.2) and (A5.3) are solved simultaneously for K_1^o and K_2^o when activity coefficient estimates, experimental pH values, and values for C_t at both titration points are substituted.

Example

Use the titration data in Fig. 6.11 to determine K_1^o and K_2^o for succinic acid. Initially, 2.26 mmole of succinic acid were dissolved in 50.0 mL of deionized water, then titrated with $0.1427M$ NaOH. To reach the equivalence point, 31.50 mL of base were required.

Solution

Attack this problem in a series of steps. At the 25% point:

$$C_t = \frac{2.26 \text{ mmole}}{(50.0 + 7.9) \text{ mL}} = 0.0390M$$

$$\text{experimental pH} = 4.00$$

Ionic strength contributions:

$$[Na^+] \sim 1.13 \text{ mmole}/58 \text{ mL} = 0.0195M$$
$$[HA^-] \sim 0.0195M$$

---

† We have C_t mol of H_2A initially and have added $\tfrac{3}{2}C_t$ mol of NaOH. This is equivalent to mixing $\tfrac{1}{2}C_t$ mol of HA^- and $\tfrac{1}{2}$ mol of A^{2-} and equilibrating.

$$[H^+] \sim 0.0001 M$$
$$\mu \approx 0.02 M$$

Using the extended Debye–Hückel equation gives the approximate values $f_{HA^-} = 0.87$ and $f_{A^{2-}} = 0.58$. We evaluate $M_{0.25}$ as follows:

$$M_{0.25} = (\tfrac{1}{2}C_t + [H^+] - [OH^-]) = 0.0195 + 0.0001 = 0.0196$$

Substituting into equation (A5.2) gives

$$\frac{0.0196(10^{-4.00})}{(0.0390 - 0.0196)} = K_1^\circ\left(\frac{1}{0.87}\right) + K_1^\circ K_2^\circ\left(\frac{1}{0.58}\right)\left[\frac{0.0780 - 0.0196}{(0.0390 - 0.0196)10^{-4.00}}\right]$$

$$\frac{1.96 \times 10^{-6}}{0.0194} = 1.15 K_1^\circ + 1.72 K_1^\circ K_2^\circ \left(\frac{0.0584}{1.94 \times 10^{-6}}\right)$$

or

$$\boxed{1.01 \times 10^{-4} = 1.15 K_1^\circ + 5.18 \times 10^4 K_1^\circ K_2^\circ} \qquad \text{[Equation 1]}$$

At the 75% point:

$$C_t = \frac{2.26 \text{ mmole}}{50.0 + 23.6 \text{ mL}} = 0.0307 M$$

experimental pH = 5.42

ionic strength contributions:

$$[Na^+] = \frac{3.37 \text{ mmole}}{73.6 \text{ mL}} = 0.0458 M$$

$$[HA^-] \approx \frac{1.7 \text{ mmole}}{73.6 \text{ mL}} = 0.020 M$$

$$[A^{2-}] \simeq \frac{1.7 \text{ mmole}}{73.6 \text{ mL}} = 0.020 M$$

$$[H^+] \sim 10^{-5} M$$

$$\mu = \tfrac{1}{2}(0.0458 + 0.020 + 0.080) = 0.073 M$$

The extended Debye–Hückel equation gives

EXPERIMENTAL EVALUATION OF CLOSE pK_a VALUES

$$f_{HA^-} = 0.80 \qquad f_{A^{2-}} = 0.42$$

We then evaluate $M_{0.75}$

$$M_{0.75} = \tfrac{3}{2}C_t + [H^+] - [OH^-] = 0.0461 + 10^{-5} + 10^{-9}$$
$$= 0.0461$$

Substituting into equation (A5.3) gives

$$\frac{0.0461(10^{-5.42})}{(0.0307 - 0.0461)} = \frac{K_1^\circ}{0.80} + \frac{K_1^\circ K_2^\circ}{0.42} \frac{0.0614 - 0.0461}{(0.0307 - 0.0461)10^{-5.42}}$$

$$\frac{1.75 \times 10^{-7}}{-(1.54 \times 10^{-2})} = 1.25 K_1^\circ + 2.38 K_1^\circ K_2^\circ (-2.61 \times 10^5)$$

$$\boxed{-1.14 \times 10^{-5} = 1.25 K_1^\circ - 6.21 \times 10^5 K_1^\circ K_2^\circ} \qquad \text{[Equation 2]}$$

Solving the two equations for K_1° and K_2° gives

$$K_1^\circ = 8.00 \times 10^{-5}; \qquad pK_{a1} = 4.10 \quad (\text{lit. } 4.21)\dagger$$
$$K_2^\circ = 2.23 \times 10^{-6}; \qquad pK_{a2} = 5.65 \quad (\text{lit. } 5.64)$$

† L. Meites, Ed., *Handbook of Analytical Chemistry*, McGraw-Hill, New York, 1963, sect. 1.

Appendix VI

Solutions to Higher-Order Equations

Many of the simplest equilibrium calculations shown in this text require the solution of only linear or quadratic (second-order) equations. In equilibrium calculations which involve more species, there are often higher-order terms. Fortunately, as you have seen in Chapters 5 and 8, in many cases it is possible to solve approximate expressions in which higher-order terms can be ignored. In fact, one of the main reasons we use graphical methods in acid-base equilibria is to see which higher-order terms are small enough to be dropped from a proton balance expression.

If the higher-order terms are too large to be eliminated, what can we do? There are explicit expressions for finding the roots (that is, the solutions) of equations up to order four (x^4), but these equations are cumbersome and difficult to use. A popular and generally useful method for equations derived from chemical systems, called the *Newton method* (or the *Newton–Raphson method*), is the subject of this appendix. The Newton method is an iterative method. The idea is that we use the equation we are trying to solve to create a formula, and then guess an approximate root and plug it into the formula to obtain the next approximation that we hope is closer to the root. This second approximation is then used to get a third approximation, and so on, until the approximations converge on the root or we can tell that convergence will not occur. Sometimes convergence will occur so slowly that we simply give up and accept a crude approximation. Ideally, the iterative scheme will converge quickly and accurately to the root, regardless of how good our first guess was for the root.

The Newton formula is quite simple. For a function $f(x)$ whose derivative is $f'(x)$, the formula is

$$x_{i+1} = x_i - \left[\frac{f(x_i)}{f'(x_i)}\right]$$

Using the iterative process, we make an initial guess x_i for the root and calculate $f(x_i)$ and $f'(x_i)$ from the function and its derivative. We then subtract the quotient of these values from our first guess (x_i) to obtain a second approximation, which we call x_{i+1}. We then make our new x_{i+1} value the x_i for the next iteration and repeat the process until the ith and (ith + 1) values of x are very close, say a 1% relative difference.

Note that there will be some computational trouble if the derivative is zero or even if it is small. The derivative is the slope of the function, and if it is very small in the vicinity of the root, the iteration may take a long time to converge. If the slope is zero, convergence cannot occur at all. A computer program that is based on the Newton algorithm should check the value of the derivative at each step and, if the slope is too small, stop the iteration process and print out the current values for the function and its derivative.

A second source of trouble lies in the accuracy of the initial guess: if the first guess is absurdly large or small, convergence will either be very slow or the algorithm may accidentally converge to an incorrect root. You should use some of your knowledge about chemistry to make the first guess.

As an example, consider the calculation of the titration curve for the titrimetric determination of cadmium with cyanide illustrated in Figure 12.10. The problem is this: How does one calculate the concentration of uncomplexed (free) cadmium in a solution buffered at pH = 10 and to which quantities of sodium cyanide are added? It is known that cadmium forms four cyanide complexes,

$$Cd^{2+} + CN^- = CdCN^+ \qquad K_1 = 10^{5.5}$$
$$CdCN^+ + CN^- = [Cd(CN)^2] \qquad K_2 = 10^{5.1}$$
$$[Cd(CN)^2] + CN^- = [Cd(CN)^{3-}] \qquad K_3 = 10^{4.6}$$
$$[Cd(CN)^{3-}] + CN^- = Cd(CN)_4^{2-} \qquad K_4 = 10^{3.6}$$

and a rather unstable hydroxide complex,

$$Cd^{2+} + OH^- = CdOH^+ \qquad K_{OH} = 10^{2.3}$$

In addition to these equilibria, cyanide is the conjugate base of the weak acid, HCN:

$$HCN = H^+ + CN^- \qquad K_a = 6.3 \times 10^{-10}$$

To describe the titration curve we will simulate a series of solutions of cadmium ion and cyanide ion in which the molar ratio of CN^- to Cd^{2+} is varied from 1:1 to about 7:1. We will assume that each solution reaches equilibrium, and then we will calculate the concentration of uncomplexed (free) Cd^{2+} in each solution. We begin by writing a cadmium ion material balance:

$$C_{Cd} = [Cd^{2+}] + [CdOH^+] + [CdCN^+] + [Cd(CN)_2] + [Cd(CN)_3^-] + [Cd(CN)_4^{2-}]$$

SOLUTIONS TO HIGHER-ORDER EQUATIONS

Next we write a cyanide material balance:

$$C_{CN} = [CN^-] + [HCN] + [CdCN^+] + 2[Cd(CN)_2] + 3[Cd(CN)_3^-] + 4[Cd(CN)_4^{2-}]$$

We will rearrange the cadmium material balance and express the concentrations of all the complex species in terms of the free cadmium (which will be plotted as $-\log[Cd^{2+}]$) and free cyanide concentrations:

$$[Cd^{2+}] = \frac{C_{Cd}}{1 + [OH^-]*K_{OH} + K_1[CN^-] + \beta_2[CN^-]^2 + \beta_3[CN^-]^3 + \beta_4[CN^-]^4}$$

For each solution we will use this expression to calculate the free cadmium concentration using an approximate value for $[CN^-]$. To find $[CN^-]$ we take the cyanide material balance, express the concentrations of all the complex forms in terms of free cadmium and free cyanide concentrations, and the formation constants in the form of a function $f(CN)$:

$$f(CN) = [CN^-] + ([H^+][CN^-]/K_a) + K_1[Cd^{2+}][CN^-] + 2\beta_2[Cd^{2+}][CN^-]^2 \\ + 3\beta_3[Cd^{2+}][CN^-]^3 + 4\beta_4[Cd^{2+}][CN^-]^4$$

The last expression is a quartic equation which we can use to solve for $[CN^-]$ with the help of the Newton method. Its derivative with respect to cyanide ion concentration is

$$f'(CN) = 1 + ([H^+]/K_a) + K_1[Cd^{2+}] + 4\beta_2[Cd^{2+}][CN^-] \\ + 9\beta_3[Cd^{2+}][CN^-]^2 + 16\beta_4[Cd^{2+}][CN^-]^3$$

The cyanide concentration will then be used in the cadmium material balance to calculate $[Cd^{2+}]$. This value will then be used in a new iterative calculation of $[CN^-]$, and the process will be repeated until successive values of $[Cd^{2+}]$ agree within about 1%.

The solution to the problem is presented in the program listing which follows. Comments are attached to help explain the flow of the program. Note that there are inner and outer iteration loops. The outer loop is used to calculate $[Cd^{2+}]$ using a value for $[CN^-]$ which is obtained from the inner loop. The inner loop uses the Newton method to calculate a value for $[CN^-]$ using the result of the outer loop value for $[Cd^{2+}]$. Control of the program is passed between the inner and outer loops by the statement in line 1200:

```
IF ABS((CD(2)-CD(1))/CD(1)) < 0.01 THEN 5000
```

This statement compares the values of successive estimations of $[Cd^{2+}]$; if the relative difference is less than 1%, the calculation is complete. If it is not, a new excursion into the inner (Newton) loop is begun. Each time the Newton iteration is done the result is checked against the previous result in line 2500,

$$\text{IF ABS ((CN2(1)-CN2(2))/CN2(1)) < 0.01 THEN 2700}$$

If the absolute value of the relative difference is greater than 1%, the Newton iteration is repeated (line 2200). The ith value becomes the ith + 1 value in line 2400, and a new ith value is calculated on the next iteration (line 2100). The extra coding in the program is necessary to designate and initialize variables, and, unfortunately, makes the program quite hard to read.

Although the details are rather difficult to understand at first, you should see that this method can be used as a general approach to solving difficult equilibrium equations. For example, go back to the treatment of soluble silver chloride complexes in Chapter 8 and find the material balance equations for silver ion and for chloride. Using the method described in this appendix, try to find the free silver ion concentration when 1.0 mmole of silver nitrate and 3.0 mmole of sodium chloride are added to 50 mL of water.

```
   5  'Program CD-CN-TI is used to calculate curve for
         titration of Cd2+
  10  'with cyanide ion. Enter data at line 10000. See lines
         9998, 9999 for
  20  'details. For/next loop starting at 800 changes mole
         ratio.
  30  ' LWPOTTS 9/86
 500  READ LK1, LK2, LK3, LK4, PH
 600  K1=10∧(LK1)
 610  K2=10∧(LK2)
 620  K3=10∧(LK3)
 630  K4=10∧(LK4)
 699  'CALCULATE PRODUCTS OF FORMATION CONSTANTS
 700  B1=K1
 710  B2=K2*K1
 720  B3=K3*K2*K1
 730  B4=K4*K3*K2*K1
 790  PRINT ''RATIO CL/CM'', ''CD CONC'', ''pCD''
 800  FOR CM=.005 TO .0015 STEP -.00015      ' (TITRATION STEPS)
 810  CL=.01-CM
1000  ' outer iteration begins
1050  H=10∧(-PH)
1060  OH=10∧(PH-14)
1065  N=1
1070  CN(N)=CL
1080  CD(N)=8.999999E-03
1100  CD(3-N)=CM/(1+OH*10∧2.3+B1*CN(N)+B2*CN(N)∧2+B3*CN(N)
         ∧3+B4*CN(N)∧4)
```

SOLUTIONS TO HIGHER-ORDER EQUATIONS

```
1200  IF ABS((CD(2)-CD(1))/CD(1))<.01 THEN 5000
2000  'inner iteration begins
2050  M=1
2100  CN2(M)=CN(N):G=3-N
2140  DERIV=((H/6.3E-10)+1+B1*CD(G)+4*B2*CD(G)*CN2(M)
         +9*B3*CD(G)*CN2(M)^2+16*B4*CD(G)*CN2(M)^3)
2160  IF DERIV<.00001 THEN 5000
2200  T=CN2(M)+(CL-CN2(M)*((H/6.3E-10)+1+B1*CD(G)
         +2*B2*CD(G)*CN2(M)+3*B3*CD(G)*CN2(M)^2
         +4*B4*CD(G)*CN2(M)^3))/(DERIV)
2400  CN2(M)=T
2500  IF ABS((CN2(1)-CN2(2))/CN2(1))<.01 THEN 2700
2570  M=3-M
2600  GOTO 2140
2700  CN(3-N)=CN2(M)
2750  N=3-N
2770  GOTO 1100
5000  PRINT
5010  PRINT CL/CM,CD(3-N),-.4342945*LOG(CD(3-N))
5500  NEXT
6000  END
9998  'ENTER DATA LOGK1,LOGK2,LOGK3,LOGK4,pH
9999  'SET FOR CADMIUM CYANIDES NOW
10000 DATA 5.5,5.1,4.6,3.6,10
```

Appendix VII

Some Compounds of Analytical Importance

Compound	Formula	Weight
Acetic acid	CH_3COOH	60.05
Aluminum chloride	$AlCl_3 \cdot 6H_2O$	241.43
Aluminum oxide	Al_2O_3	101.96
Ammonia	NH_3	17.03
Ammonium chloride	NH_4Cl	53.49
Ammonium nitrate	NH_4NO_3	80.04
Antimony potassium tartrate	$K(SbO)C_4H_4O_6 \cdot \frac{1}{2}H_2O$	333.93
Antimony trioxide	Sb_2O_3	291.50
Arsenic trioxide	As_2O_3	197.84
Ascorbic acid	$C_6H_8O_6$	176.13
Barium chloride	$BaCl_2 \cdot 2H_2O$	244.28
Barium sulfate	$BaSO_4$	233.40
Bismuth trioxide	Bi_2O_3	465.96
Cadmium oxide	CdO	128.40
Calcium carbonate	$CaCO_3$	100.09
Calcium chloride	$CaCl_2 \cdot 2H_2O$	147.02
Calcium oxalate	$CaC_2O_4 \cdot H_2O$	146.12
Calcium oxide	CaO	56.08
Calcium sulfate	$CaSO_4 \cdot 2H_2O$	172.17
Ceric ammonium nitrate	$(NH_4)_2Ce(NO_3)_6$	548.23
Chromic nitrate	$Cr(NO_3)_3 \cdot 9H_2O$	400.15

Compound	Formula	Weight
Cobalt chloride	$CoCl_2 \cdot 6H_2O$	237.93
Copper(I) chloride	$CuCl$	99.00
Copper(II) oxide	CuO	79.55
Copper(II) sulfate	$CuSO_4 \cdot 5H_2O$	249.68
Dimethylglyoxime	$CH_3C(:NOH)C(:NOH)CH_3$	116.12
EDTA (ethylenediaminetetraacetic acid), disodium salt	$Na_2H_2Y \cdot 2H_2O$	372.24
Ferric ammonium sulfate	$FeNH_4(SO_4)_2 \cdot 12H_2O$	482.19
Ferric oxide	Fe_2O_3	159.69
Ferrous ammonium sulfate	$Fe(NH_4)_2(SO_4)_2 \cdot 6H_2O$	392.14
Hydrazine sulfate	$N_2H_4 \cdot H_2SO_4$	130.12
Hydroxylamine hydrochloride	$NH_2OH \cdot HCl$	69.49
Hydroxylamine sulfate	$(NH_2OH)_2 \cdot H_2SO_4$	164.14
Lanthanum oxide	La_2O_3	325.84
Lead chromate	$PbCrO_4$	323.18
Lead nitrate	$Pb(NO_3)_2$	331.20
Lead sulfate	$PbSO_4$	303.25
Lithium carbonate	Li_2CO_3	73.89
Magnesium chloride	$MgCl_2 \cdot 6H_2O$	203.30
Magnesium oxide	MgO	40.30
Manganous sulfate	$MnSO_4 \cdot H_2O$	169.01
Mercuric oxide	HgO	216.59
Mercuric chloride	$HgCl_2$	271.50
Mercurous chloride	Hg_2Cl_2	472.08
Nickel chloride	$NiCl_2 \cdot 6H_2O$	237.71
Oxalic acid	$H_2C_2O_4 \cdot 2H_2O$	126.07
Potassium biphthalate	$KOCOC_6H_4COOH$	204.23
Potassium bitartrate	$KOCO(CHOH)_2COOH$	188.18
Potassium bromate	$KBrO_3$	167.00
Potassium bromide	KBr	119.01
Potassium chloride	KCl	74.56
Potassium chromate	K_2CrO_4	194.20
Potassium dichromate	$K_2Cr_2O_7$	294.19
Potassium ferricyanide	$K_3Fe(CN)_6$	329.26
Potassium ferrocyanide	$K_4Fe(CN)_6 \cdot 3H_2O$	422.41
Potassium hydroxide	KOH	56.11
Potassium iodate	KIO_3	214.00
Potassium iodide	KI	166.01
Potassium nitrate	KNO_3	101.11
Potassium periodate	KIO_4	230.00
Potassium permanganate	$KMnO_4$	158.04

SOME COMPOUNDS OF ANALYTICAL IMPORTANCE

Compound	Formula	Weight
Potassium persulfate	$K_2S_2O_8$	270.32
Potassium dihydrogen phosphate	KH_2PO_4	136.09
Potassium monohydrogen phosphate	K_2HPO_4	174.18
Potassium thiocyanate	KSCN	97.18
Pyridine	C_5H_5N	79.10
Silver acetate	$AgOCOCH_3$	166.91
Silver bromate	$AgBrO_3$	235.77
Silver bromide	AgBr	187.77
Silver chloride	AgCl	143.32
Silver iodate	$AgIO_3$	282.77
Silver iodide	AgI	234.77
Silver nitrate	$AgNO_3$	169.87
Silver oxide	Ag_2O	231.74
Sodium acetate	$NaOCOCH_3 \cdot 3H_2O$	136.08
Sodium arsenate	$Na_2HAsO_4 \cdot 7H_2O$	312.01
Sodium bicarbonate	$NaHCO_3$	84.01
Sodium bisulfate	$NaHSO_4 \cdot H_2O$	138.07
Sodium bisulfite	$NaHSO_3$	104.06
Sodium borate [borax]	$Na_2B_4O_7 \cdot 10H_2O$	381.37
Sodium bromide	NaBr	102.89
Sodium carbonate	Na_2CO_3	105.99
Sodium chloride	NaCl	58.44
Sodium dihydrogen phosphate	$NaH_2PO_4 \cdot 2H_2O$	156.01
Sodium fluoride	NaF	41.99
Sodium hydroxide	NaOH	40.00
Sodium oxalate	$Na_2C_2O_4$	134.00
Sodium perchlorate	$NaClO_4$	122.46
Sodium sulfate	Na_2SO_4	142.04
Sodium thiosulfate	$Na_2S_2O_3 \cdot 5H_2O$	248.18
Stannous chloride	$SnCl_2 \cdot 2H_2O$	225.63
Strontium carbonate	$SrCO_3$	147.63
Sulfamic acid	NH_2SO_3H	97.09
Sulfuric acid	H_2SO_4	98.08
Uranyl acetate	$UO_2(CH_3COO)_2 \cdot 2H_2O$	424.15
Urea	NH_2CONH_2	60.06
Vanadium pentoxide	V_2O_5	181.88
Zinc chloride	$ZnCl_2$	136.28
Zinc nitrate	$Zn(NO_3)_2 \cdot 6H_2O$	297.47
Zinc oxide	ZnO	81.37

Answers to Selected Problems

Chapter 1

1.1 (a) 2.10; (c) 0.35; (e) 2.0009; (g) 0.2493
1.3 (a) 3.47; (b) 0.194
1.5 (a) Weigh by difference
(c) Weigh by addition
1.7 (a) Sample may absorb moisture
(c) Bottle may be warmer than chamber, and cooling
1.9 Fingerprints, buoyancy more serious at 0.01 mg level; warmth a problem at both levels.
1.11 1002.3 mL at 3 pm; (a) 0.1017M at 3 pm; (b) 999.8 mL
1.13 Buoyancy corrected volume at 24°C: 50.153 mL
1.15 (a) 0.056 gram KI
(b) 5.60×10^{-5} g/mL; 5.33×10^5 g/g
(c) $3.37 \times 10^{-4}M$.

Chapter 2

2.1 (a) Range: 4.8 g/mL
(b) Mean: 120.6 g/mL
(c) Median: 120.2 g/mL
(d) Std. dev.: 1.9 g/mL
Absolute error 1.5 g/mL; relative error 1.2%.
2.3 Absolute error = 1.48 g/eq
2.5 Positive proportional error
2.7 Absolute error = 0.05%; relative error = 1.7 ppt; std. dev. of mean = 0.173%; rel. std. dev. = 9.8 ppt

2.9 (a) Sum of squares: 0.0873%; range: 0.089%
(b) $Q = 0.556$, reject
(c) $c.i. = 0.857 \pm 0.081$; two results fall outside

2.11 Geometric mean = 10.43; arithmetic mean = 10.80
s (upper) = 1.313, s (lower) = 0.762
$c.i.$ (upper) = 10.43 + 0.78; $c.i.$ (lower) = 10.43 − 0.46

2.13 The means are different at the 95% level (5% significance)

2.15 (a) There is a difference at the 95% level (5% significance)
(b) Yes, in method B

2.17 (a) Mean: 104.3; median: 104.1; range: 1.80; std. dev.: 0.612; s (rel): 5.9 ppt;
$c.i. = 104.3 \pm 0.56$ (range coefficient method)

2.19 ±12 ppb interval: 6 determinations
±6 ppb interval: 13 determinations

2.21 0.00335 g.

Chapter 3

3.1 −34.35 kJ/mole
3.3 33.3 joule/K
3.5 Yes. When the activities of all species are unity the free energy change is negative.
3.7 1512 K
3.9 $10^{-3.7}$
3.11 (a) $K = 710$; (b) $1.41 \times 10^{-3} M$
3.13 (a) $4.03 \times 10^{-5} M$ (activity correction made)
(b) No; 3.2% unreacted
3.15 1.0×10^6 at 200°C
3.19 DHLL: 0.77; EDHL: 0.81
3.21 (a) 0.166; (b) overestimate K

Chapter 4

4.4 Chloroacetic acid $K_a = 1.64 \times 10^{-3}$; stronger than acetic acid
4.9 Ammonia is a much stronger base than is water and levels the strengths of acetic acid and HCl to the strength of ammonium ion as an acid.
4.11 Protonated forms (lyonium ions) of the solvents are the strongest acids.
4.12 Arrange the bases in order of increasing K_f for the complexes. Stronger bases make stronger complexes with the metal ion.
4.14 (a) To the right; (b) to the left.

Chapter 5

5.1 (a) $[H^+] = 0.05 M$; $[OH^-] = 2.0 \times 10^{-13} M$; pH = 1.41; pOH = 12.59
(c) $[H^+] = 1.0 \times 10^{-13} M$; $[OH^-] = 0.10 M$; pH = 12.9; pOH = 1.08
5.3 $K_a = 8.9 \times 10^{-6}$
5.5 $K_a = 1.69 \times 10^{-5}$

5.7 (b) Material balance: $C_T = [NH_4^+] + [NH_3] = 0.080M$
Charge balance: $[NH_4^+] + [H^+] = [OH^-] + [Cl^-]$
(d) Material balance: $C_T = [H_3PO_4] + [H_2PO_4^-] + [HPO_4^{2-}] + [PO_4^{3-}]$
Charge balance: $[H^+] + [Na^+] = [H_2PO_4^-] + 2[HPO_4^{2-}] + 3[PO_4^{3-}] + [OH^-]$

5.9 $0.08708F$ acetic acid

5.11 $K_a = 2.3 \times 10^{-11}$

5.13 $pK_a = 6.20$

5.15 (a) 10.62; (c) 2.12; (e) 11.3; (f) 4.88

5.17 Formic acid $= 1.66 \times 10^{-6}M$; formate $= 0.05M$

5.19 Nicotinic acid $= 0.0452M$; nicotinate $= 0.0360M$

5.21 (a) $[H^+] = [NO_2^-]$; pH $= 2.65$
(b) $[OH^-] = [HNO_2]$; pH $= 7.7$
(c) $[HNO_2] = [NO_2^-]$; pH $= 3.35$

5.25 (a) pH $= 2.25$
(b) pH $= 4.7$
(c) pH $= 11.9$
(d) 10 mmole Na_2HPO_4 + 10 mmole NaH_2PO_4 in 100 mL water
(e) $pK_{a2} = 6.97$ when ionic strength taken into account

5.27 (a) Phthalic acid: 1.26×10^{-3}; biphthalate: 0.447; phthalate: 0.552
(b) Add 0.550 mmole strong base
(c) pH $= 3.30$

5.29 (a) pH $= 7.20$
(b) pH $= 6.83$
(c) Water pH $= 7.0$; HCl solution pH $= 1.3$

5.31 Ratio pyridine to pyridinium $= 0.01$

5.33 pH $= 4.55$ (solve the quadratic equation)

5.35 (a) pH $= 8.08$; (b) pH $= 8.07$; (c) pH $= 0.70$

5.37 4.6×10^{-7}

5.39 0.15

5.43 pH $= 7.25$

Chapter 6

6.1 (a) 5.000 mL of 1.5073 M HCl, dilute to 100 mL with water
(c) 5.25 mL of 50% NaOH, dilute to 1 liter with water
(e) 226 mg of $Ba(OH)_2$, dissolve, dilute to 250 mL with water

6.3 (a) 820 mg; (c) 1.41 g

6.5 (a) 100.46 (GFW); (c) 45.02 (GFW/2)

6.7 (a) 5.00 mL of $1.20M$ H_2SO_4, dilute to 100 mL
(c) 5.540 g citric acid (0.0865 eq), dissolve, dilute to 0.500 liter
(e) 42.8 mg, dissolve, dilute to 50 mL

6.9 (a) $0.01797N$; (b) 2.835 g in 1 liter

6.11 (a) 2.5, 7.0; (c) 2.87, 2.6; (e) 10.3, 8.3

6.13 pK_1 ca. 2.0, pK_2 ca. 4.5

6.15 (a) Weak base with strong acid: glycine with HCl
(c) Pyridine with HCl

6.17 Solution is a mixture of anilium ion and unreacted HCl. Equivalence point (indistinct) at pH ca. 3, after 12.44 mL. Equivalence point sharp at pH 8.4 after 39.57 mL.

- **6.19** (a) Fraction adipate ion is $10^{-3.67}$; 99.98% complete
  - (b) Fractions of diprotic and monoprotic forms are both large at first eq. pt.; species not well differentiated.
  - (c) pH ca. 3.0
  - (d) Not complete. Fraction of diprotic form only 0.96
- **6.21** (a) Break is too gradual; not a good method
  - (b) Fraction = 0.96 at eq. pt.
  - (c) Sharp break at pH 5.55, suitable for analysis
- **6.23** 2.21 mmole piperidine, 6.55 mmole pyridine
- **6.25** Hydroxide is not a strong enough base. In $0.1F$ OH^- the fraction of bicarbonate is about 0.3%. The eq. pt. break is too gradual to be useful.
- **6.27** The reaction produces 4.58 mmole HPO_4^{2-} and 3.20 mmole PO_4^{3-}. 16.00 mL is required for the original phosphate, 22.90 mL + 16.00 mL to titrate original HPO_4^{2-} and that produced from phosphate. 38.90 mL needed to titrate $H_2PO_4^-$ produced in first two steps.
- **6.29** 28.06 mmole Na_2CO_3; 16.20 mmole NaOH
- **6.31** 4.639 mmole Na_3PO_4; 1.681 mmole Na_2HPO_4
- **6.33** 3.00% crude protein
- **6.35** 130 g/mole
- **6.39** $0.0248M$
- **6.41** 1.95% wt/vol

Chapter 7

- **7.1** Acidic solvents
- **7.3** Ion pairs form in low dielectric constant solvents. Homoconjugation may also occur in such solvents.
- **7.5** Dielectric constant, weaker solvating power
- **7.7** Phenolate: $pK_b = pK_s - pK_a = 3.4$

Chapter 8

- **8.1** Ag_2CrO_4: 4.82×10^{-12}
- **8.3** (a) 60 mg/100 mL; (c) 1.26 g/L
- **8.5** 3.13×10^{-8}
- **8.7** 1.64×10^{-10}
- **8.9** (a) $5.1 \times 10^{-4}M$; (c) $2.8 \times 10^{-12}M$
- **8.11** (a) $5.9 \times 10^{-7}M$; (b) $4.9 \times 10^{-11}M$; (c) AgBr first; (d) $8.3 \times 10^{-7}M$
- **8.13** (a) 0.10 M
  - (b) Chloride $5.96 \times 10^{-4}M$, iodate $0.0993M$
  - (c) $6.0 \times 10^{-4}M$
- **8.15** 1.1985 g
- **8.17** $0.0148M$
- **8.19** $1.4 \times 10^{-3}M$
- **8.21** 0.2924 g

ANSWERS TO SELECTED PROBLEMS 695

Chapter 9

9.3 (a) Primary sulfate, secondary sodium
(c) Primary silver, secondary nitrate
9.7 53.4 m^2
9.9 (a) A barium precipitate, e.g., $BaSO_4$
(b) Hg_2Cl_2 should adsorb $CN^- > SCN^-$
9.11 Yes. $[Ag^+]$ is $6 \times 10^{-5} M$; K_{sp} is not exceeded when barium precipitated
9.13 pH > 7.2

Chapter 10

10.1 1.1244 g AgBr, 1.4142 g AgI, 0.7574 g As
10.3 (a) 0.6573 g; (b) 0.3552 g; (c) 0.4260 g; (d) 0.3284 g; (e) 0.4762 g
10.5 19.49% Cu
10.7 Step 1: 0.0768 g; step 2: 0.0597 g; step 3: 0.0939 g
10.9 29.38% SiO_2
10.11 1.86% Sn
10.13 9 waters of hydration
10.15 33.26% Fe_2O_3, 66.74% Al_2O_3
10.17 4.2 chlorines
10.19 25.00% AgI, 22.24% AgBr, 52.76% AgCl

Chapter 11

11.1 (a) 9.98 mL; (b) pAg = 8.04; (c) Yes, it must be <8.77; (d) pAg = 14.6; (e) pAg = 8.28; (f) pAg = 4.87
11.3 Color is visible about 0.03 mL past equivalence point
11.5 (a) 12.3 mmole
(c) 1.2 g KSCN per liter
(e) Cu^{2+} might compete with Fe^{3+} and delay or obscure color change. Run a blank with Cu^{2+} present.
11.7 Indirect method avoids coprecipitation of indicator which makes endpoint indistinct.
11.9 29.1% fluoride
11.11 58.3 mg CHI_3
11.13 12.2 mg veronal
11.15 (a) 1 meq F^- reacts with 1 meq Th^{4+}; 1 meq Th^{4+} = 0.25 mmole Th^{4+}.
11.17 (a) 29.06 mL; (b) 24.74 mL; (c) 51.49 mL
11.19 (a) 0.0387 M Pb^{2+}; (c) 0.1072 M Cl^-; (e) 0.0234 M PO_4^{3-}

Chapter 12

12.1 0.02785M
12.3 (a) $1.809 \times 10^{-3} M$; (b) Raise the pH

12.5 (a) 0.520; (b) 3.28×10^8; (c) 3.28×10^8; (d) Yes
12.7 DCTA complex
12.9 (a) An entropy effect; (b) 0.970
12.11 (a) pH ca. 2.7; (b) pH ca. 2.5
12.13 Concentrations of $Zn(en)_2$ and $Zn(en)$ are too similar for endpoint to be sharp.
12.15 43.04% Zn, 24.80% Cu, 32.17% Sn

Chapter 13

13.1 (a) 5100 Å, 0.510 μm
(b) 5.88×10^{14} sec^{-1}
(c) 2.254×10^{10} cm/sec in water
13.3 219.5 kJ/mole
13.5 Yellow materials absorb in the blue-violet.
13.7 500 nm corresponds to 20000 cm^{-1}. A 2000 cm^{-1} vibration will have a wavelength of 5000 nm. A 200 cm^{-1} rotation will be at 50,000 nm, or 50 μm.
13.9 (a) 10% T; (c) 97.7% T; (e) 1% T
13.11 1.91
13.13 4.182×10^{-3} dL/mg cm; 129.7 L/mole cm
13.15 $2.5 \times 10^{-7} M$
13.17 (a) $1.67 \times 10^{-5} M$
(b) Intercept of Beer's law plot
(c) Larger variance increases data scatter
13.19 $K_f = 1.15 \times 10^6$
13.21 Fraction dissociated = 0.214, pK_a = 4.92
13.23 Curve is flat for 1.25 mmole EDTA, then rises for next 2.05 mmole, then flattens again with excess EDTA added.
13.25 0.05 mL: A = 0.498; 0.075 mL: A = 0.744; 0.10 mL: A = 0.990; 0.125 mL: A = 0.988; 0.150 mL: A = 0.985.
13.27 1.71×10^{-7} m
13.29 (a) 0.065 degree; (b) 1.23 nm/mm

Chapter 14

14.3 (a) Pt/H$_2$, H$^+$//Ag$^+$/Ag; $E°$ = +0.7995 V
(b) Zn/Zn^{2+}//Tl$^+$/Tl; $E°$ = +0.427 V
(c) Pt/H$_2$, H$^+$//AgCl/Ag; $E°$ = +0.222 V
(d) Pt/I$_2$(aq), I$^-$//S$_2$O$_8^{2-}$, SO$_4^{2-}$/Pt $E°$ = +1.39 V
14.5 +0.72 V
14.7 +1.562 V
14.9 +1.050 V
14.11 1.64×10^{-8}
14.13 +0.665 V
14.15 Ratio 0.01, E = 0.08 V; ratio 10.0, E = 0.17 V
14.17 (a) Platinum electrode; (b) 1.46×10^8; (c) $2.97 M$
14.19 (a) 22 cells

(b) No chloride concentration dependence
(c) Depletion of mercurous chloride or loss of silver electrode.

Chapter 15

15.1 (a) Between pH 8 and pH 9 it is close to Nernstian
(b) $K = 1.8 \times 10^{-11}$
15.3 (a) At the unit activity level a 1% error causes a 0.2 mV change in potential.
(b) The Nernst coefficient is 0.0611 at this temperature. If all activities are unity this will have no effect, however.
15.5 (a) 5°C: 0.2567 V; 35°C: 0.2339 V
15.11 (a) $[Cl^-] = 0.050M$, $f_{Cl} = 0.81$, $a_{Cl^-} = 0.040$
$[Cl^-] = 0.0050M$, $f_{Cl} = 0.92$, $a_{Cl^-} = 0.0046$
(b) 51.4 mV change

Chapter 16

16.1 (a) 15.201 g $FeCl_2 * 4H_2O$ per liter of solution
(c) 5.041 g $KMnO_4$ per liter of solution
16.3 0.5052 mmole Fe^{3+}
16.5 GFW = 197.84, GEW = 197.84/4 = 49.46
16.7 (a) $10^{5.08}$; not complete
(b) Iodine as oxidizer, ferrocyanide as reducer
16.9 (c) 50% titrated, $E = +0.14$ V
(e) 99% titrated, $E = +0.199$ V
(h) 110% titrated, $E = +0.623$ V
Yes, complete at equivalence point
16.11 (a) +0.88 V; (b) 2.90 mg
16.13 8.24%
16.15 9.70%
16.17 1.40%
16.19 3.35 mmole/L
16.21 (a) −0.76 V (Jones reductor); (b) 248 mg
16.23 (a) 95.65 mg; (b) Yes; no

Chapter 17

17.1 (a) Fraction of $A = 0.83$, fraction of $B = 0.33$
(b) A: stage 4, B: stage 17
17.3 (a) $K = 0.14$; (b) $K' = 0.028$; (c) $K' = 0.28$
17.5 0.0246
17.7 3 separations
(a) $[A] = 0.014M$
(b) $[A] = 0.0199M$; $[B] = 0.0667$
17.9 (a) For A: 175 s, 40.3 mL; for B: 199 s, 49.9 mL
Separation factor = 1.24

- (b) $R = 0.62$, poor separation
- (c) A: 413 plates; B: 471 plates
- (d) Capacity factors: $A = 9.72$, $B = 11.06$
  10% increase in both factors does not improve resolution
- (e) 1035 plates; use a longer column

Chapter 18

18.7 Add solvents of lower polarity; for example, dioxane
18.9 Make the solvent more acidic
18.11 Polar solutes elute before nonpolar solutes
 (a) MEK, 1-pentanol, 1-pentene
 (b) 1,1,1 isomer elutes before 1,2,3 isomer
18.13 Choosing an anion that complexes (or more strongly associates with) metals ions should speed elution.

Index

Absolute error of a result, 56
Absorbance, 434
Absorptivity, 435
  molar, 222, 436
Accuracy vs. precision, 40
Acetic acid
  as solvent for titrations, 263, 276
  dissociation constants of acids in, 276 (Table 7.3)
  titration with sodium hydroxide as model system (Fig. 6.8), 234
Acetonitrile as a solvent for acid-base titrations, 279
Acetylacetone (2,4 pentanedione) as ligand, 387
Acid-base competition in solubility equilibria, 307
Acid-base indicators: *see* Indicators, acid-base
Acids and bases, conjugate pairs, 145, 146 (Table 4.2)
Activities and concentrations in solutions, 115, 127–136
Activity coefficient, 158, 204
  mean, 129
  of uncharged species, 134
  at very high ionic strengths, 133
Activity corrections, 256
  in solubility equilibria, 287
Activity quotient, 115
Adsorbed ion layer, 320
Adsorption, 319
  coprecipitation, 322
  factors that promote, 320
  indicator (precipitation titrations), 369
Alcohols as solvents, 272
  dissociation constants of weak acids in, 273 (Table 7.2)
Alcohols, determination of
  acetylation and acid-base titration, 246
  oxidation with periodate, 562
Alkaline error (glass electrodes), 522
Aluminum hydroxide (gel), 318
  homogeneous precipitation of, 328
Amines, titration with strong acid, 247
Aminocarboxylate complexes, formation constants, 391 (Table 12.1)
Aminocarboxylate ligands, 390 (Figure 12.5)
  weak acid dissociation constants of, 390 (Figure 12.5)
4-Aminopyridine as a primary standard base, 217
Ammonia as a solvent, 148, 150, 277
Ammonium hexanitratocerate, 560
Amphiprotic (amphoteric) compounds, 145
  as solvents, 264
Amplitude (light), 427
Anode, 487
Aprotic solvents, 278
Arrhenius, S., 142
  theory of acids and bases, 142–144
  limitations of theory, 144
Arsenic trichloride as a solvent, 149
Arsenious oxide as primary standard, 559, 560

Ascorbic acid
  model system for pH calculations, 186
  pH of a solution of (algebraic), 188
  pH of a solution of (graphical), 185
  reducing titrant, 564
  reduction of silver ion, 565
Aspirator, 24
Asymmetry potential (glass electrode), 520
Autoprotolysis, 146, 159, 168
  constants, 146, 147, 268, 269
  effect on pH of weak acid solution, 172
  of water, temperature dependence, 160 (Table 5.1)
Azo compounds, Kjeldahl determination of, 247

Back titrations, 244
  complexometric, 411
  Volhard (precipitation), 371
Balance
  capacities, 15, 16 (Table 1.3)
  sensitivities, 15, 16 (Table 1.3)
Band broadening, chromatography, 609
Bandwidth, 449
  effective, 449
  relation to resolution, 450
Barium iodate
  solubility calculation, 289
  solubility model for common ion effect, 294
Base dissociation constant, 147, 163
Baseline separation, chromatographic peaks, 608
Battery, 506
Beer's law, 435
  applied to analysis of mixtures, 438
  chemical effects, 446
  deviations due to instrumentation, 449
  deviation due to large bandwidth, 451
  deviation used to calculate pKa, 448
  effect of changing refractive index, 445
  physical limitations in studies of, 445
Benedetti-Pichler, A. A., (propagation of errors), 86
Benzoic acid (model system for acid-base equilibria), 163
  as primary standard acid, 214
Beta notation for formation constants, 299
  in complexation equilibria, 397
Bias, 40
2,2'-Bipyridine as ligand, 387
Bismuth sulfide, solubility calculation, 290
Blank, 46, 47
  titration indicators, 44, 221
Blaze angle and wavelength (grating), 458
Bonded phase materials (HPLC), 634
Boric acid as trap for ammonia, 248

Boyle, R., theory of acids and bases, 141
Brass sample, dissolving and treating, 355
Bromate as oxidizing reagent, 563
  reaction with Cr(III), 354
Bromcresol green, 236, 242, 248, 258
  determination of pKa, colorimetric method, 478 (Experiment 13.3)
  structure of, 222, 481
Bromine determination of phenols, 563
Bromphenol blue, 476
Bromthymol blue, 229
Brønsted, J., 144
Brønsted–Lowry definition of acids and bases, 144–148
  limitations of, 148
Buffer capacity, 198
Buffer index (Van Slyke), 199
  function of pH, 201 (Figure 5.7)
Buffers, 193
  applications of, 194
  criteria for effectiveness, 198
  effect of dilution on pH of, 202
  how to prepare, 196
  National Bureau of Standards, 520
  universal, 202
Burets, 19

Cady and Elsey definition of acids and bases (solvent system), 149
Calcium, determination of
  gravimetric, as calcium oxalate monohydrate, 349
  titrimetric, with EDTA, 421 (Experiment 12.1)
  titrimetric, with EDTA, presence of magnesium, 422 (Experiment 12.2)
Capacity factor, 590
  column chromatography, 608
  effect on peak resolution, 609
Carbonate nuisance in acid-base titrations, 212
Carbowax, 624
Carrier gas (gas chromatography), 621
Carriers, postprecipitation of radioactive lead, 324
Catalysts and reaction rates, 126
Cathode, 487
Cation release precipitation method, 328
Cell, electrochemical
  alkaline manganese, 508
  concentration (zinc amalgam), 504
  conventions (IUPAC), 488
  Daniell, 506
  diagrams, 487
  electrolysis, 491
  fuel, 506

galvanic (voltaic), 491
junction, 487
lead–acid (Plante), 508
Leclanche, 507
lithium, 510
lithium–iodine, 510
lithium–silver chromate, 510
mercury, 508
nickel–cadmium (NiCad), 509
primary, 506
rechargeable power, 506
secondary, 506, 508
silver–zinc, 509
spontaneity, 491
used for titrations, 360
Cell emf
activity corrections, 497
effect of complexation on, 499
and formation constants, 501
and free energy change, 488
internal resistance, 535
measurement of, 535
presence of two soluble complexes, 501
Centroidal point (in linear regression), 90
Cerate as oxidizing reagent, 560
Ceric ammonium sulfate, 560
Charge balance (electroneutrality), 165
Chelate effect, 395
Chelates and chelating agents, 386
Chemical wastes, disposal of, 29
  incompatible, 30 (Table 1.8)
Chloride, gravimetric determination of, as silver chloride, 351
Chromasorb, 623
Chromate, homogeneous generation, 354
Chromatogram, 619
Chromatography, qualitative information, 613
Chromatography, quantitative information, 614
Chromium and manganese in steel, colorimetric method, 439
Chromium(II) as reducing titrant, 564
Clinical chemistry, 3
Colloids, 318
  stabilizing for adsorption indicators, 370
  surface areas of, 319
Colorimeter, visual, 464
Column chromatography
  countercurrent extraction model, 603
  random walk model, 604
Columns, chromatographic
  efficiency, 607
  gas chromatography, 622
  HPLC, 634
  open tube (OT), 623
  packed, 623

Common ion effect, 293
Comparison of a mean with a true value, 69
Comparison of precision of two methods (F-test), 75
Comparison of two sample means, 71
Completeness of a reaction, 118
Complexation, effect in solubility equilibria, 296–300
Complexometric titration curves
  algorithm for, 403
  calculation of, 402
Complexometric titrations
  auxiliary complexing agents, 411
  back titrations, 411
  displacement, 413
  indirect, 410
  indirect for anions with EDTA, 413
  masking in, 413
Concentration units, 7–9
  formality, 8
  molarity, 8
  normality, 9
Conditional formation constant, 398
Conditional solubility product constant, 397
Confidence interval, 62–67, 83
  about the regression line, 91
  calculated from range, 66
Confidence level, 63
Coning and quartering, 97
Conjugate pairs (acids and bases), 145
Constituent, categories of, 12
Control chart, 83
Control limits, 83
Controls, experimental, 47
Convolution techniques, potentiometric titration curves, 537
Coordination number, 385
Copper ore, sample preparation, 585
Copper, iodometric determination 571, 583 (Experiment 16.2)
  interferences, 571
Copper, photometric titration of, with EDTA, 441, 481 (Experiment 13.4)
Coprecipitation, 322
Countercurrent separations, 596
  large number of stages, 600
  stage model, 597
Covariance, 90
Craig countercurrent apparatus, 602
Crosscurrent separations, 593
Crown ether, 18-crown-6 as ligand, 388
Crucibles
  filter, 24
  Gooch, 25

Crucibles (*Continued*)
   sintered glass, 25 (Table 1.6)
   unglazed, 26
Cuprous chloride complexes, log concentration diagram, 305 (Figure 8.3)
Cuprous iodide, Standard Potential, 571
Cuvette (spectrometer cell), 459
Cyanide ion as complexometric titrant, 415
Cylinders, graduated, 19
Czerny–Turner monochromator, 454

Daniell cell, 506
Dark current, 461, 466
Davies equation (activities), 133, 162, 227, 480
Davy, H., composition of acids, 142
DCTA, 533
Dead volume, chromatography, 605
Debye–Hückel equation, 128, 227, 233, 258, 288
   extended form, 130–133
   model, 128
Degrees of freedom, computational, 57
Demasking, 413
Denticity, 386
Derivative spectra, 463
Derivatives of some common mathematical functions, 88 (Table 2.8)
Detectors, UV-visible, 459
Devarda's alloy (Kjeldahl), 247
Deviation
   average, of a sample, 56
   of a result, 56
Dextrin (hydrolyzed starch), 370, 379
1,2-Diaminocyclohexanetetraacetic acid (DCTA), 390
Diatomaceous earth, 623
Dichlorofluorescein, indicator for Fajan's method for halides, 369, 378
Dichromate as titrant, 550, 553
   irreversibility, effects of, 561
   preparation and standardization, 581
Dielectric constant, 127, 128, 263, 269
Diethylenetriamine (dien), 387
Diethylenetriaminepentaacetic acid (DTPA), 390
Differentiating ranges of solvents, 266
Dilution value of buffer, 203
Dimethyl sulfoxide as a solvent for acid-base titrations, 278
Dimethylformamide as a solvent for acid-base titrations, 278
Dimethylglyoxime, homogeneous generation to precipitation Ni(II), 329
Direction of a reaction and initial activities, 117
Dispersion (of data), measures, 54

Dispersion (optical), 428
   linear, 455
Dissociation constant, 161
   overall (nonaqueous solvents), 270
Distribution coefficient, 588
   conditional (distribution ratio), 589
   and free energy changes, 588
   mole ratio form, 590
Distribution ratio, 589
   weak acid species, 592
Distribution (statistical population)
   binomial, 52
   compound, 61
   equation for normal, 53
   inversely normal, 61
   lognormal, 60
   lognormal vs. normal, 62 (Figure 2.10)
   normal (Gaussian), 42, 53
   standard normal curve, 54
   Student-t, 62
   Student-t vs. normal, 63 (Figure 2.11)
DTPA diethylenetriaminepentaacetic acid, 390
Dynode, 461

ECD: *see* Electron capture detector in gas chromatography
EDTA ethylenediaminetetraacetic acid, 481, 533
   as colorimetric reagent, 481
EGTA ethyleneglycol bis(2-aminoethylether) tetracetic acid, 390
Electrical double layer (colloids), 321
Electrochemical cell, 486
Electrochemical detector for HPLC, 634
Electrodes
   classes, 494, 495
   CO_2-responsive (Severinghaus), 523
   combination glass, 523
   crown ether ion-selective, 528
   enzyme, 523
   heterogeneous membrane, 524
   irreversible, 494
   liquid ion exchange membrane, 527
   neutral carrier, 527
   reversible, 494
   rubber disk (Pungor), 526
   single crystal ion-selective, 524
   soluble redox couple (class O), 495
Electrolytes for washing precipitates, 325
Electromagnetic radiation
   as particles, 429
   as waves, 426
Electromagnetic spectrum, 430

INDEX

703

Electron capture detector (ECD) in gas chromatography, 627
Electron donation (Lewis), 151
Elution, effect of temperature in gas chromatography, 624
emf (electromotive force), 486
Endothermic reactions, 107
Endpoint, titration, 211
Energy density, 506
Enthalpy, 107
Enthalpy of formation, standard molar, 107
Enthalpy of reaction, 107
Entropy, 109
Entropy change with temperature change, 110
Eosin, adsorption indicator, 370
Equilibrium constant, 115
    effect of temperature, 120
    related to Standard Potential, 548
    thermodynamic, 161
Equilibrium, dynamic, 126
Equivalence point, in titrations, 211, 262
    vs. endpoint, potentiometric titration curves, 542
    in precipitation titrations, 361
Equivalent weight
    of an acid or base, 218
    in precipitation reactions, 372
    in redox reaction, 9
    in redox titration, 550
    of a weak acid, 256 (Experiment 6.1)
Eriochrome Black T ("Erio T") metallochromic indicator, 405
Eriochrome Blue Black R (calcon) indicator, 408
Errors
    of a calculated result, 84, 86
    absolute, 42
    analyst, 45
    constant, 46
    constant vs. proportional, 45
    constant, effect on working curve, 48
    drug effects, 44
    equipment and reagents, result of, 45
    method, result of, 43
    "personal," 45
    propagation of, 84, 86
    proportional, 46
    proportional, effect on working curve, 48
    relative, 42
    storage, result of, 45
    systematic (determinate), 43
Esters, saponification and acid-base titration, 246
Ethanol as a solvent for acid-base titrations, 274
Ethylenediamine as a solvent, 267, 277
    as a ligand, 387

Ethylenediaminetetraacetic acid (EDTA)
    effect on solubility of calcium oxalate, 309
    fraction of tetraanion vs. pH, 391
    as ligand, 388
    preparation of standard solution, 422
    structure, 390
Ethyleneglycol bis(2-aminoethylether) tetracetic acid (EGTA), 390
Exothermic reactions, 107
Eye protection, 28

F-test of precision, 75
F-values, 76 (Table 2.4)
Fajans method for halides, 369
    for chloride, 378 (Experiment 11.1)
    interferences, 379
Ferric hydroxide (gel), 318
Ferrocene, 394
Ferrous ammonium sulfate (Mohr's salt), 476
FID: see Flame ionization detector
Filter paper, 26, 337
    ashless, 26
    how to fold, 27
    ignition, 27–28
    porosities, 26
    pulp, 26
Filter photometer, 454, 464
Filters (optical)
    absorbance, 455
    cutoff, 455
    interference, 455
    order sorters, 458
Filtration, 24, 337
Flame ionization detector (FID) in gas chromatography, 626
Flasks, volumetric, 20
Flocculating ions and charges, 321
Flocculation, 318, 364
Flow rates, gas, in gas chromatography, 621
Fluoride ion as masking agent, 415
    as masking agent for Fe(III), 572
Foreign ligand competition in solubility equilibria, 304
Forensic chemistry, 3
Formal dissociation constant, 162
Formal potential, 497
    relationship to Standard Potential, 498
Formation constant, 297
    activity corrections, 398
    completeness of reaction, 399
    effects of charge density on, 393
    effects of ring formation, 395
    effects of solvent strength on, 396

Formation constant (*Continued*)
  standard free energy changes, 392
  steric effects on, 393
Fraction titrated, 227
Fractions of acid-base species
  pH dependence, 175, 176
  polyprotic systems, 187
  polyprotic systems, general equations, 188
Free energy, 111
  change, standard, 115
  changes and the equilibrium constant, 116
  concentration changes, 113
Frequency (light), 427
Fumaric acid system, 192
  as model for crosscurrent separation, 594

Gas chromatography (GC), 619
  applications, 628
  detectors, 625
  equipment, 620
  injector, 622
  mobile phases (carrier gases), 621
  programmed temperature, 625
  temperature regulation in, 624
Gaussian distribution, 604, 605
  shape of chromatographic band, 600, 604
Gel filtration chromatography, 639
Gel permeation chromatography, 639
Gels, 318
Geometric isomers of metal ion complexes, 386
Glacial acetic acid: *see* Acetic acid
Glass electrode, 158, 196, 225, 256, 266, 518
  calibration of, 158, 226, 257
  glass composition and ion response, 522
Glass filter crucible, 337
  cleaning, 352
Glass, thermal expansion, 24
Glycerol, redox determination with periodate, 562
Gosset, W. S. ("Student"), 62
Gradient elution, 632
Gram formula weight (GFW), 8
Gran plot, potentiometric titration curves, 541
Gratings, 456
  equation, 457
Gravimetric factor, 340
Gravimetric methods
  advantages and limitations, 344
  criteria for accurate and precise, 335
Gravimetry as process, 336
Guyard reaction (permanaganate), 559

Half-reactions, 486
Halides, gravimetric determination, 342

Halides, titration of a mixture of, 365
Hard and soft acid-base theory (HSAB), 152–153 (Table 4.3)
Henderson–Hasselbalch equation, 195, 232
HETP (height equivalent to a theoretical plate), 604, 605
  minimizing, 610
High-performance adsorption chromatography, 636
High-performance ion exchange chromatography, 637
High-performance liquid chromatography (HPLC), 619, 629
  choice of solvents, 636
  columns, 634
  detectors, 633
  injection systems, 632
  pumps, 632
  solvent delivery, 629
High-performance partition chromatography, 634
Homoconjugation, 272
Homogeneous precipitation, 325, 327
  advantages and limitations, 330
HPLC: *see* High-performance liquid chromatography
HPPC: *see* High-performance partition chromatography
Hydrazines, Kjeldahl determination of, 247
Hydrogen lamp (UV source), 453
Hydrogen–ascorbate ion, pH of a solution of (algebraic), 190
  pH of a solution of (graphical), 186
Hydrolysis, 384
Hydroxide ion balance, 164
Hydroxylamine, 476
8-Hydroxyquinoline as ligand, 387
"Hyphenated methods," chromatography, 613

Iminodiacetic acid (IDA), 390
Index of refraction, 428
Indicator blank, 44, 221
Indicator blocking, 412
Indicator electrode, 517
Indicators, acid-base, 221
  adsorption (precipitation titrations), 369
  color range, acid-base, 221–224 (Figure 6.2; p. 223)
  dissociation constant, acid-base, 221
  metallochromic, 404
  mixed and screened acid-base, 224 (Table 6.2)
  single acid-base, 221
Indigo carmine indicator in calcium titration, 408

INDEX

Inert solvents, 264
Intercept of working curve, 90
Internal energy, 106
Intrinsic solubility, 283–287
   inorganic salts, 286
   organic ligand–metal complexes, 286
Iodate
   oxidation of chloride, 562
   oxidation of iodide, 561, 569
Iodide, reaction of nitrite with, 569
Iodine
   disproportionation to iodide and HOI, 568
   intermediate in indirect methods, 557
   as oxidizer, 556
   as primary standard, 568
   reaction with arsenious acid, 568
   reaction with thiosulfate, 569
   solubility, 568
   as titrant: direct reaction, 557
Iodine–triiodide equilibrium, 568
Iodometric methods, 556, 567
   direct vs. indirect, 568
   miscellaneous, 573 (Table 16.3)
Ion exchange resins
   anion, 639
   order of preference for metal ions, 638
   strong acid, 637
   weak acid, 639
Ionic strength, 128, 226
   in buffers, 197
   calculation, 129
   effects in solubility equilibria, 287
Ion pair, 149
Ion product, 285
Ion selective electrode, 518
Iron citrate complexes, cell emf, and formation constants, 501
Iron(II) oxidation by cerium(IV) as model system, 548
Iron, determination of
   colorimetric method, 475 (Experiment 13.2)
   colorimetric method, interferences, 478
   photometric titration with permanganate, 442
   redox determination of, 581 (Experiment 16.1)
Isosbestic point, 447
Isocratic elution, 632
Isoelectric point, 364
Isomorphous replacement (coprecipitation), 322, 367
Isopropanol as a solvent for acid-base titrations, 274
Isopropanol, effect on titration curves of succinic acid, 271

Jones reductor (amalgamated zinc), 566

Karl Fischer reagent, 572
Kielland radii for Debye–Hückel equation, 132 (Table 3.1)
Kinetics, 123–127
Kjeldahl method for nitrogen (acid-base), 246
   catalysts for, 247

Labile complexes (vs. inert complexes), 392
Laboratory operations, basic, 14
Laboratory safety, 28
Lambert's law, 435
Lavoisier, A., theory of composition of acids, 142
Lead storage battery, 508
Lead, gravimetric determination in brass, 354
Least-squares line fitting, 89
LeChatelier's principle, 119
Leclanche cell, 506, 507
Leveling effect, 148, 262, 265, 267
Level of significance, 68
Lewis, G. N., definition of acids and bases, 151
   limitations, 151
Liebig, J., theory of composition of acids, 142
Ligand:metal combining ratios
   continuous variations method, 442, 443
   mole-ratio method (Yoe–Jones), 442
   photometric determination, 442
   slope-ratio method, 444
Ligands, bidentate, 386
Ligands, monodentate, 386
Light, monochromatic, 449
Linear regression, 89, 90
Linear regression (applied to Beer's law), 438
Liquid junction potential, 158, 503
Lithium–chromate cell, 510
Lithium–iodine cell, 510
Littrow monochromator, 454
Logarithmic concentration diagram
   ammonium acetate, 182 (Figure 5.4)
   ascorbic acid, 185 (Figure 5.5)
   autoprotolysis, 177
   benzoic acid, 177 (Figure 5.3)
   fumaric acid, 193 (Figure 5.6)
   how to draw, 179
   system point, 177
Logarithmic concentration diagram for silver chloride complex system, 303
Longitudinal diffusion, 609
Lowry, T., theory of acids and bases, 144
Lyate ion, 269
Lyonium ion, 266, 268

Magnesium–EDTA system as model, 399
Magnesium, gravimetric determination, 341

Magnesium oxalate
 postprecipitation on calcium oxalate, 323
 separation from barium oxalate, 331
Maleic acid as model for crosscurrent separation, 594
 distribution between ether and water, 587
Malonic acid titration with strong base as model system, 237–239
Manganese cell, alkaline, 508
Manganese in steel, colorimetric method, 473 (Experiment 13.1)
Masking and masking agents, 5, 413, 414 (Table 12.4), 584

Mass transfer, mobile phase (chromatography), 611
Mass transfer, stationary phase (chromatography), 611
Material balance (acid-base equilibria), 164
Mean, 41, 42
 geometric, 60, 129
 precision of the, 59
 standard deviation of the (standard error of), 59
Median, 41, 42
Mercuric chloride as oxidizer, 582
Mercurous sulfate reference electrode, 532
Mercury cell, 508
Metal ion complexes, geometry of, 385
Metal sulfides
 postprecipitation, 324
 solubilities, 290
Metallochromic indicators, 404
 color range of, 407
 effect of pH on, 407
 sharpness of color change, 407
Metastannic acid, 356
Methanol as a solvent, 273
Methods of analysis
 classical, 6
 colorimetric, 6
 definitive, 42, 44
 field, 44
 gasometric, 6
 gravimetric, 6
 physicochemical, 6
 reference, 44
 volumetric, 6
Methyl isobutyl ketone as a solvent for acid-base titrations, 267, 279
Methyl orange as redox indicator, 563
Methyl orange–xylene cyanole mixed indicator, 242
Methyl red, 229
 structure of, 222
Methyl yellow, 229

Methylene blue as screening dye, 224
Mineral acid concentrations, 216 (Table 6.1)
Mixed bases titrated with strong acid, 242
Mixed-crystal precipitates, 323
Mohr method for chloride, 367, 381
 effect of pH, 368
 pH control in, 369
Molar absorptivity, 222, 436
 and detection limit of a method, 436
Monobasic base equilibria, 167
Monochromators, UV-visible, 453
Monodentate ligands, 386
Monoprotic acid equilibria, 167

Nernst equation, 360, 496, 517, 518, 552
 glass electrode response, 226
Neutralization, 142, 150
Neutralization (solvent system), 149
Nickel–cadmium cell, 509
Nitrate
 determination with Fe(II) and dichromate, 561
 Kjeldahl determination, 247
Nitrilotriacetic acid (NTA), 387
Nitrite, Kjeldahl determination of, 247
Nitro compounds, redox determination with Cr(II), 564
Nitrobenzene, coating for precipitate in Volhard method, 372, 380
Nitroso compounds, redox determination with Ti(III), 564
Normal calomel electrode (NCE), 530
Normal phase chromatography, 635
Normality
 of acid and base solutions, 218
 limitations of use, 220
 vs. molarity in redox reactions, 551
 of permanganate as redox titrant, 550
 in redox titration, 550
Nucleation, 315
Null hypothesis, 68
 acceptance or rejection, 75
Number of replicates, 80–82
 and precision of the mean, 59

Occlusion (coprecipitation), 322
Octahedral coordination, 385
Operations, scale of analytical, 11–12 (Figure 1.1)
Optical isomers of metal ion complexes, 386
Order number (grating), 458
Ostwald ripening of precipitates, 325
Outlying results (Q-test), 78
Oxalic acid, oxidation of, 550
Oxidation, 485
Oxidation prior to titration, 565

INDEX

Paneth–Fajans–Hahn rule (adsorbed layer), 320
Particle size, in GC columns, 623
  of colloids, 319
Peak integration, chromatography, 614
Peak width, chromatography, 605
Peaks, distorted (chromatography), 612
Pearson, R. G., hard–soft acid-base theory, 152, 392
2,4-Pentanedione (acetylacetone) as ligand, 387
Peptization, 322, 325, 338
Perchloric acid as acidic titrant in acetic acid solvent, 277
Periodate, 473
  oxidation of alcohols, 562
  as oxidizing reagent, 562
Permanaganate
  half-reactions of, 558
  as a redox reagent, 558
  as redox titrant, 553
  standardization with arsenious oxide, 559
  standardization of solutions of, with oxalate, 559
Peroxodisulfate, 473
pH buffers: *see* Buffers
pH meter, 225, 256
pH of weak acid solution, approximate calculations, 172
pH
  and activities, 158
  Sorensen definition of, 157
1,10-Phenanthroline
  –iron(II) complex model system for absorption spectrometry, 425
  –iron complexes, Standard Potentials, 502
  as ligand, 387, 475
Phenol
  determination with bromine, 563
  titration in ethylenediamine, 263
  titration of (model system), 262
Phenolphthalein, 229, 257
  structure of, 222
Phosphate
  complexation of Fe(III), 582
  gravimetric determination, 342
  as masking agent, 473
Photodiode (light detectors), 461
  arrays, 462
Photometric detector for HPLC, 633
Photometric titrations, 441
Photomultiplier tubes (light detectors), 461
Photons, 429
Phototubes (light detectors), 460
Photovoltaic cells (light detectors), 460
Pi-electron acceptor ligands, 395
Pipets
  mechanical micro-, 19
  TD and TC, 19

Planck's constant, 429
Platinization of electrodes, 530
Polybasic weak base-strong acid titrations, 241
Polyprotic acids, 185
Polyprotic weak acid–strong base titrations, 236
Population (statistical), 50
Porphine molecule as ligand, 388
Postprecipitation, 322, 323
Potassium biiodate as primary standard acid, 215
Potassium biphthalate, 257
  as primary standard acid, 214
Potassium dichromate as a primary standard, 560
Potassium iodate as a primary standard, 561
Potassium sulfosalicylate as primary standard, 215
Potassium thiocyanate
  as indicator for Volhard method, 371
  as primary standard, 380
Potentiometer, 533
Potentiometric titration curves
  data treatment, 536
  equivalence point potential, 553
  redox, 551
  redox, effects of irreversibility, 553
Power density (electrochemical cell), 506
Precipitates
  digesting, 325
  drying, 338
  structural changes during digestion, 326
  washing, 325
Precipitation, rate of, and supersaturation, 317
Precipitation titrations, adsorption indicators, 369
Precipitation titration curves
  effect of adsorption on accuracy of, 364, 366
  effects of dilution on, 363
  formation of mixed crystals, 367
Primary standards, qualities of, 212
Prior oxidants and reductants, 566 (Table 16.2)
  criteria for usefulness, 565
Prisms, 459
Probability (and random error), 51
Propagation of error, 84, 86
Protogenic solvents, 264
Proton balance, 164
Protophilic solvents, 264
Protophobic solvents, 264

Q-test for outliers, 78–80
Quadratic formula, 170
Quality control, 82
Quantitative vs. qualitative, 1
Quartz–halogen lamp (visible region source), 453

Random error, 49
Range, 55, 78

Rate constants, 124
    relationship to equilibrium constant, 126
Recovery factor, 330
Redox indicators
    diphenylamine compounds, 555
    ferroin compounds, 554
    methylene blue, 556
Redox titrations
    completeness of, 549
    criteria for, 547
    effects of dilution, 552
    indicators, 554
Reducing agents, 563
Reduction, 485
    prior to titration, 565
Reference electrode, 517
Reference level (acid-base), 166
Reference materials
    secondary, 44
    standard, 44, 46
Reference states, 107
Refractive index, 428
    detector for HPLC, based on, 633
Releasing agent, 533
Reprecipitation to purify precipitates, 326
Residual (least squares), 90
Resistance to mass transfer (chromatography), 609
Resolution, chromatographic peaks, 607
Retention time, 605
Retention volume, 605, 607
Reversed phase chromatography, 635
Reversibility, 123, 551
    chemical vs. electrochemical 493
Riffle, 97
Rings, formation of ligand, 395
Risk, 63
Rounding numbers, rules for, 13
Round-robin testing (interlab testing), 83, 84

Salt bridge (agar recipes), 504
Salting coefficient, 134, 161, 234
Salts, pH of solutions of, 169
Sample, for analysis
    dissolving, 5
    macro, 12
    meso, 12
    micro, 12
    obtaining, 4
    purifying, 5
    storage (preservation), 97–98
    submicro, 12
Sample of a population, 50
Sample thief, 96
Samples
    composite, 95 (Figure 2.15)
    grab, 95

Samples vs. populations, 50
Sampling gases, liquids, and solids, 95–96
Sampling, goals and requirements, 94
Saturated calomel electrode (SCE), 494, 530
    in nonaqueous solvents, 532
    temperature dependence of emf, 530
Saturation, 316
Savitsky–Golay derivative function, 258
    numerical filters, 537
SCE: see Saturated calomel electrode
Schwartzenbach classification of acids and bases (complexes), 392
Selectivity coefficient (glass electrode), 520
Separation by precipitation, 292–293, 330
Separation factor, 330, 592, 607
Separations, effects of phase volumes, 590
Separatory funnel extractions, 595
Septum, 622
Significant digits, 13–14
    rules for logarithms, 14
    rules for rounding, 13
    rules for using, 13
    rules when adding, 14
    rules when multiplying or dividing, 14
Silicic acid (gel), 318
Siloxane stationary phases, 624
Silver ammonia complexes, 304
Silver chloride
    light sensitivity of, 354
    as reducing agent, 567
    reference electrode, 531
    separation from lead chloride, 293
Silver chloride complexes, 297–300
    log concentration diagram, 303 (Figure 8.2)
Silver chromate
    conditional solubility product for, 368
    indicator for Mohr method, 367
    precipitation of, 291
    solubility calculation, 290
Silver halides, Standard Potentials, 499
Silver nitrate as primary standard, 380
Size-exclusion chromatography, 639
Slits, spectrometer, 449
Soda ash (sodium carbonate), 258
Sodium ascorbate
    pH of a solution of (algebraic), 191
    pH of a solution of (graphical), 186
Sodium carbonate
    as primary standard base, 216
    and sodium bicarbonate mixtures, back titration, 244
    and sodium bicarbonate mixtures, titrations of, 243
    titration with strong acid as model system, 241
Sodium hydroxide solutions as titrant
    preparation, 213
    storage, 214

INDEX

Sodium methoxide as basic titrant, 275
Sodium oxalate as a primary standard
  base, 218
  reductant, 559, 560
Sodium tetraborate decahydrate (borax) as a primary standard base, 217
Sodium thiosulfate
  decomposition of solutions, 569
  preparation and standardization of solutions of, 584
  as primary standard, 569
  as reducing titrant, 564
Sols, 318
Solubility and crystal size, 318
Solubility equilibria, graphical methods, 300
Solubility product constant, 284
  determination from potentiometric titration curve, 361
  standard free energies and, 284
Solution environment, 127
Solution interactions, ion–dipole vs. ion–ion, 127
Solvating ability of a solvent, 272
Solvation of ions, 384
Solvent extraction, 587
Solvent system definition of acids and bases, 149–150
  limitations, 150
Solvents for titrations, classification of, 263
Solvents, inherent acidity of, 265
Sources, UV-visible, discrete, and continuous, 453
Soxhlet extractor, 595
Spectrometer, UV-visible
  block diagram, 452
  double beam, 466
  single beam, 465
Spectronic 20 (Bausch and Lomb), 465
Spectrum
  absorbance, 432, 434
  absorbance, of tris(1,10-phenanthroline)Fe(II), 434
  emission, 432
  origin of atomic, 432
  origin of molecular, 431
Speed of light, 427
Spontaneity, 109
  electrochemical cell reaction, 491
  and entropy changes, 110
  and free energy changes, 112
Squalene, 624
Stability constant: see Formation constant
Standard addition, method of, 532
Standard deviation
  calculated from range, 58
  of a point, 91
  pooled, 71, 72
  of a population, 54
  about a regression line, 91
  relative, 58
  of a sample, 57
Standard free energy, 161
Standard hydrogen electrode (SHE), 489, 530
Standard oxidation potential and Standard Potential, 489
Standard Potential, 489
  combining, 490
  and equilibrium constant, 491
  and formation of metal ion complexes, 393
  relationship to standard free energy, 490
Standard reference material (SRM), 44
Standard states, 107
Stannous chloride as reducing agent, 582
Starch indicator, 558, 570
  extraction endpoint, 571
  –iodine reaction, 570
  solution, preparation, 570
State functions, 109
Stationary phase liquids, gas chromatography, 623
  effects of thickness, 624
Statistical inference, 39
Statistical testing, 67–82
Statistical tests, two-tailed vs. one-tailed, 68
Statistics, descriptive, 39, 41
Stop flow, 632
Strength of acids and bases, 143
Strength of an acid (Brønsted–Lowry), 148
Strength and dissociation of electrolytes, 143
Strong acid titrant preparation, 215
Strong acid–strong base titration, 227
  effects of dilution on sharpness, 230
  equivalence point, 228
Strong acids or bases and their salts, 167
Succinic acid titration with strong base as model system, 240
Sulfate, homogeneous generation, 328
Sulfation, 508
Sulfide, homogeneous generation from thioacetamide, 328
Supersaturation, 316
  effect on precipitate purity, 325

t-Butanol as a solvent for acid-base titrations, 274
t-Values, 65 (Table 2.2)
TCD: see Thermal conductivity detector in gas chromatography
Tetrabutylammonium hydroxide as a basic titrant, 271, 275
Tetrahedral coordination, 385
Thallous chloride, solubility of, 295
Tris(hydroxymethyl)amino-methane (THAM), 258

Tris(hydroxymethyl)amino- (*Continued*)
  buffer made from, 197
  primary standard base, 217
  titration with strong acid as model system, 237
Theoretical plates, 604
Thermal conductivity detector in gas chromatography (TCD), 625
Thermodynamics, 106
Thermodynamics, first law of, 106
Thermodynamics, second law of, 109
Thermodynamic states, 106
Thermodynamic universe, 106
Thiocyanate, 379
  endpoint enhancement in iodometric copper, 572
  See also Potassium thiocyanate
Thiourea as masking agent, 415
Titanium(III) as reducing titrant, 563
Titrant, 211
Titration, analytical criteria, 212
Total ionic strength adjustment buffer, 533
Transmittance, 434
Triethylenetetramine (trien), 388
Triiodide ion
  formation from iodine, 557
  reaction with sulfide, 558
True value (statistics), 40
Tungsten lamp (visible region source), 453

Urea
  hydrolysis of, 327, 349
  reaction with nitrogen oxides, 572

Valinomycin, 527
Vanadium V/IV couple, pH dependence, 502
Van Deemter equation (chromatography), 609
van't Hoff equation (temperature and equilibrium constant), 120
Variance
  of a chromatography band, 605
  of a population, 57
  of a sample, 57, 90
Viscosity, 128
Visible spectrum and colors, 430
Volhard method
  for chloride, 379
  for halides, 371
Voltaic pile, 506
Volumetric glassware
  calibration of 21–24
  cleaning of, 20–21

grades, 22 (Table 1.4)
use, 19–20
von Weimarn equation (rate of precipitation), 317

Walden reductor (silver chloride), 567
Wall-coated open tubule columns (WCOT), 623
Water as a solvent, 261
Water, density of, 23 (Table 1.5)
Water hardness, 410
Water, Karl Fischer determination of, 572
Wavelength (light), 427
Wave number, 429
Weak acid (monoprotic), 169
Weak acid dissociation constant, evaluation from titration data, 232, 239
Weak acid, pH of solution of extremely, 172
Weak acid–strong acid mixture, pH of, 180
Weak acid–strong base titration, 231
  effect of dilution on sharpness, 234
Weak acid–weak base mixture, 182
Weak base, pH of solution of monobasic, 174
Weak base–strong acid titration, 235
  relationship to weak acid–strong base, 236
Weighing, 15–19
  by addition, 16
  buoyancy correction, 18, 23
  by difference, 15
  effect of error on results, 212
  error from fingerprints, 18
  error from pan oscillations, 17
  error from static electricity, 17
  error from temperature differences, 17
  error from water adsorption, 17
  liquids, 16
  techniques of, 15
Weight percentage
  wt/vol, 10
  wt/wt, 10
  ppt, ppm, ppb, 10 (see also Figure 1.1)
Welch's t^* test, 73–75
Working curve, 47, 89, 437, 532
  slope of, 90
  slope of, confidence interval, 92
  slope of, standard deviation of, 92

z-factor, 54, 63, 65 (Table 2.2)
Zero-level of protons, 166
Zinc metal as reducing agent, 566
Zwitterions, 389

Physical and Chemical Constants

Quantity	Symbol	Value
Unified atomic mass unit	u (or amu)	1.661×10^{-27} kg
Avogadro constant	N or N_A	6.022×10^{23} mol^{-1}
Electronic charge	e	1.602×10^{-19} C
Faraday constant	F	9.648×10^4 C mol^{-1}
Gas constant	R	8.314 J mol^{-1} K^{-1} 0.0821 L atm mol^{-1} K^{-1}
Mass of electron	m_e	9.110×10^{-31} kg
Mass of neutron	m_n	1.675×10^{-27} kg
Mass of proton	m_p	1.673×10^{-27} kg
Planck constant	h	6.626×10^{-34} J s
Rydberg constant	R_H	1.098×10^7 m^{-1}
Speed of light in a vacuum	c	2.998×10^8 m s^{-1}

Units and Conversion Factors

Base Units

Quantity	SI Unit	Symbol	Conversion Factors
Length	meter	m	1 cm = 10^{-2} m 1 nm = 10^{-9} m 1 Å = 10^{-10} m 1 inch = 2.54×10^{-2} m
Mass	kilogram	kg	1 g = 10^{-3} kg 1 mg = 10^{-6} kg 1 lb = 0.454 kg
Time	second	s	1 day = 8.6×10^4 s
Temperature	kelvin	K	0°C = 273.15 K
Amount	mole	mol	
Electric current	ampere	A	

Derived Units

Quantity	SI Unit	Symbol	Conversion Factors
Volume	cubic meter	m^3	1 L = 10^{-3} m^3 = 1000 cm^3 1 mL = 1 cm^3
Energy	joule	J	1 cal = 4.184 J 1 L atm = 101.3 J 1 eV = 1.602×10^{-19} J
Pressure	pascal	Pa	1 atm = 1.01×10^5 Pa = 760 Torr or mmHg
Force	newton	N	
Frequency	hertz	Hz	1 s^{-1} = 1 Hz
Electric charge	coulomb	C	
Electric potential difference	volt	V	